*linear control systems*

## *McGraw-Hill Series in Electronic Systems*

John G. Truxal and Ronald A. Rohrer, Consulting Editors

*Huelsman:* Theory and Design of Active RC Circuits
*Meditch:* Stochastic Optimal Linear Estimation and Control
*Melsa and Schultz:* Linear Control Systems
*Schultz and Melsa:* State Functions and Linear Control Systems
*Stagg and El-Abiad:* Computer Methods in Power System Analysis
*Timothy and Bona:* State Space Analysis: An Introduction

# linear control systems

## James L. Melsa
Information and Control Sciences Center
Southern Methodist University

## Donald G. Schultz
Department of Electrical Engineering
University of Arizona

## McGraw-Hill Book Company
New York, St. Louis, San Francisco, London,
Sydney, Toronto, Mexico, Panama

## Linear Control Systems

*Library of Congress Catalog Card Number* 68-8664

ISBN 07-041481-5

10 11 12 13   KPKP   798

*Dedicated to Kathy*

# *preface*

This book is an outgrowth of the conviction of the authors that many of the concepts of modern linear control theory can be presented in a meaningful way to undergraduate students. "Linear Control Systems," therefore, is an introduction to the representation, analysis, and design of linear control systems, making use of both modern and classical methods. One of the prime objectives in writing this book is to interrelate so closely the classical, transfer-function methods with the modern, state-variable methods that the reader will think of them not as different methods but, as is more proper, as differing only in depth of treatment.

State-variable methods, introduced at the beginning of the book, are emphasized throughout. On the other hand, the classical root-locus, Nyquist, Bode, and Routh-Hurwitz methods are also treated in detail. When linear state-variable feedback is used as the basic system configuration, many of the classical methods must be interpreted in a new

light.   The result of this reexamination is an emphasis on the fundamental concepts of the methods rather than on the rote mechanics that
have become so standard in the classical treatment.   In addition, it is
demonstrated that this configuration may result in significant improve-
ments in system performance.

In writing this book, the authors have consistently attempted to
remember the reader.   It is intended that the book be easy to read and,
to a great extent, be self-teaching.   To assist in this process, numerous
simple exercises with answers are included at the end of each technical
section.   To ensure that a deeper understanding is also achieved, at the
end of each chapter are more comprehensive problems which require the
interplay of many of the previously developed topics.   The solutions
for these problems are presented in a solutions manual.   To provide
computational assistance in the study of this book, a set of computer
programs has been compiled and is available upon request.   Every effort
has been made to present the material in as simple a form as possible,
consistent with a sound technical treatment.   The book has been written
for the student, not the teacher.

This book is intended for a one-semester course in control theory
offered to seniors or possibly to advanced juniors.   The only prerequisite
is a working knowledge of the Laplace transform and some appreciation
for frequency-domain concepts.   The material in the book has been
taught at Southern Methodist University and the University of Arizona
for the past $2\frac{1}{2}$ years.

The authors wish to express their appreciation for the suggestions
obtained from the many undergraduate students who have used the
rough drafts of the manuscript.   They also acknowledge the assistance
provided by their colleagues and graduate students at both Southern
Methodist University and the University of Arizona, especially Dr.
Charles R. Hausenbaver for his aid in the development of Chap. 8.   A
special acknowledgment is due to Katherine S. Melsa for her dedication
to the typing and retyping of the manuscript.   Finally each author
wishes to state that any errors that remain are the complete responsi-
bility of the other.

*James L. Melsa*
*Donald G. Schultz*

# contents

# *one*    *introduction to automatic control systems*

## *1.1  Introduction*

The term *automatic control system* is intended to be somewhat self-explanatory.  The word *system* implies not just one component but a number of components that work together in a systematic fashion to achieve a particular goal.  This goal is the control of some physical quantity, and the control is to be achieved in an automatic fashion, often without the aid of human supervision.

The topic of automatic control is certainly a romantic one.  In this age of missiles, nuclear submarines, and spacecraft, the need for the control of such dynamic systems is not only evident; it is dramatized repeatedly in the mass-communication media of newspaper, radio, and television.    As a spaceship settles on the moon within the designated target area, the country and the world marvel.    Similarly, as a satellite tumbles out of control, millions are instantly aware of the fact and wait in fear that control may not be regained and that the passengers may perish.

In the more prosaic area of civilian and earthbound endeavors, the subject of automatic control is romanticized by the word *automation*.    By automation is meant the automatic production of processed material. The end product of an automated plant may vary between such wide extremes as high-octane gasoline and transistor radio sets.    However, the unifying feature is that the process is under control every step of the way and often by automatic means.

Unfortunately, the study of automatic control systems is not as romantic as the uses to which control theory has been applied.    Because the systems and processes that must be controlled are dynamic rather than static, their behavior is described by differential rather than alge- braic equations.    In wrestling with the complexities inherent in such con- trolled dynamic systems, students have been known to sweat and to swear, far from being romantic.    Yet the rewards to be gained from such a study are immense.    Satellites and missiles and automatic factories do not just happen; they evolve from men's ideas and understanding.    And a surprisingly short time ago the men presently in charge of exotic control applications were in the same situation as the present readers of this book. They were students, and they were struggling; their victory was under- standing.    One object of this book is to make the reader's own victory in the understanding of automatic control systems as painless as possible.

The purpose of the introductory material in this first chapter is to give the reader an intuitive feeling for what is meant by an automatic control system and to outline the approach used in this book for the sys- tematic study of this subject.

## 1.2    *Closed-loop vs. open-loop control*

This book deals almost exclusively with *closed-loop automatic control* sys- tems, and it is therefore necessary at the outset to distinguish between open- and closed-loop systems.    This is most easily done by means of an example with which we are all familiar.    Consider the control of the tem- perature of a fluid within a tank in which incoming fluids are at different temperatures.    In terms of everyday experience, one might think of adjusting the water in the bathtub to a desired temperature.

One method of realizing the desired temperature is to open the hot-water tap a specified amount, open the cold-water tap a specified amount, and let the water run for a fixed time or until the water has reached a given level. This is the way in which a person in a rush would fill the tub, and if he had been rushed a sufficient number of times in the past, he might know rather well the necessary settings of the hot- and cold-water faucets to realize the desired water temperature. This is an open-loop control system.

An open-loop control system is one in which neither the output nor any of the other system variables has any effect on the control of the output. In this example the output is the temperature of the water in the tub, and control is exercised by setting the valves on the hot- and cold-water lines. If for any reason the water in the tub is not at the desired temperature, this is not known, and no control is exercised to force the actual water temperature toward the desired water temperature.

In this familiar example there are a number of factors that might affect the final water temperature, the most obvious of which is the amount of hot water available. Suppose that the normal hot-water supply were depleted, because someone else had just finished a bath or because the washing machine had just completed its cycle. Then, of course, when the water reached its final level, it would be too cold.

Someone in less of a rush might feel the water in the tub at several intervals while the tub is filling. If the water were not at the right temperature, adjustments could then be made to either the hot- or the cold-water faucet, as required. This is an approach toward closed-loop control. In a closed-loop control system, the system output and other system variables effect the control of the system. In this case the loop would be closed intermittently by the person making measurements and taking corrective control action.

A person with lots of time might proceed in yet a different fashion. Instead of feeling the water in the tub only at intervals, he might continuously stir the contents of the tub and continuously measure the average water temperature, at the same time making the necessary valve adjustments. Since it is assumed that the desired water temperature is higher than that of the tub, the incoming water must initially be at a higher temperature than the desired temperature in order to overcome the thermal capacity of the tub itself. As the tub heats up to the desired water temperature, the temperature of the incoming water must be decreased, and thus the person exercising the control would find that a continuous adjustment is necessary to maintain a constant water temperature as the tub is filling. This system is a closed-loop system, where the loop is actually closed by the human operator. Because a human operator is needed to make the system function properly, such a system is not classified as automatic.

Even in the latter situation, where the person is devoting his full time to controlling the water temperature, he might end up with a less than ideal situation. Assume, for example, that the supply of hot water is nearly depleted at the start of the operation. As the temperature of the hot water being supplied begins to diminish, the person stirring eventually notices that the temperature of the water in the tub is decreasing, and he might well increase the demand for hot water. If indeed the "hot" water were by this time cold, his action would produce the opposite of the desired result. A more astute tub filler might measure not just the temperature of the water in the tub but the temperature of the water coming from the tap as well. In other words, he might measure more than one variable and might therefore be able to perform his task of control more effectively. As a further refinement, he might even profit by measuring the rate at which the temperature of the hot-water supply is decreasing as the hot water is used up.

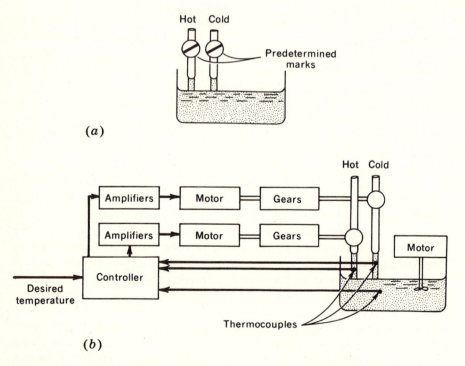

**Fig. 1.2-1** Liquid-temperature-control system. (*a*) Open loop; (*b*) closed loop.

Consider briefly now the automatic control of the water temperature in the tub. In view of the discussion above, the hardware requirements of such a control operation are evident. First we need to measure the temperature of the water in the tub, and, if we wish to do a good job of control, it might be well to measure the temperature of the incoming water and the rate of change of the temperature of the hot water. Next, a power element of some sort is necessary so that the control valves, the hot- and cold-water faucets, are positioned automatically. Assume that an electric motor is used as the power element. Since an electric motor is inherently a high-speed device and the valves move at relatively slow speeds, a gear train is necessary to make the velocities compatible. To ensure that the temperature transducer in the tub measures the average water temperature, a stirring device, such as a small outboard motor, might be clamped over the side of the tub. To accomplish the actual control, the measured water temperature in the tub and the other measured state variables must be combined and compared with the desired tub temperature. This is accomplished in the controller, and the resulting control signal is then amplified to a sufficient level to enable it to drive the system power element, which in turn positions the water faucets.

The open- and the automatic closed-loop control systems for controlling the water temperature in the tub are illustrated in Fig. 1.2-1. Recall that the open-loop control required only the initial positioning of the two control valves, the water faucets. This example is typical of open-loop-control situations, where success depends upon two things:

1. The calibration of the control element
2. The repeatability of related events over an extended period of time, that is, the lack of external disturbances

In the example being considered, for each desired water temperature, there is a corresponding setting of the control valves. If successful open-loop control is to be achieved, this calibration must remain fixed. But even if the calibration remains perfect, open-loop control is not successful unless all other parameters associated with the system remain fixed over an extended period of time. If either the hot or the cold water or the tub itself is at a different temperature than it was on the day of calibration, the end result is not as predicted.

It is obvious from Fig. 1.2-1 that the closed-loop control system is considerably more complicated than the open-loop system. The "trade-off" here is between complexity and performance. This is most easily seen if one considers the simplest case in which only the output, the tub water temperature, is fed into the controller. In this case assume that the controller is just a subtracting circuit, so that the control signal is proportional to the error between the required water temperature and the actual water temperature. If the required and the actual temperatures are the same, no error is present to drive the valve settings to new positions. Calibration is still necessary on the input and output transducers but not on the control elements, and the ultimate behavior of the system is not dependent upon the hot- or cold-water temperature or the tub temperature.

The distinguishing feature between the open- and closed-loop systems of Fig. 1.2-1 is the utilization of the state of the output to effect the input in the closed-loop-control case. Such utilization is termed *feedback*, and it is not clear that the final error is not affected if other system variables are fed back in addition to the output. This is, however, the case, and the feedback of the other system variables in addition to the output is of considerable aid in ensuring that the system under consideration actually can be controlled.

This last remark brings up the important question of stability and instability. In the open-loop case the settings of the control elements are programmed according to the desired output. In the example discussed above, the control elements were set once and for all, and there is no question as to whether the system is stable or not. In a closed-loop system, this need not be the case. In fact, one of the major difficulties in the design and synthesis of high-performance closed-loop control systems is the necessary compromise between desired performance and required stability. On the basis of the simple single example that has been discussed, it may be difficult for the reader to appreciate that a stability problem might exist.

To illustrate the possible occurrence of such instability, let us suppose that we control only the hot-water tap, with the cold-water rate set at some predetermined position. Let us assume, in addition, that we are slow to make a correction but when we do decide to make the correction, we make a very large one by turning the tap either full on or full off. If the water in the tub is initially warmer than desired, we turn the hot water completely off. As the water begins to cool, because of our slow response time, we might let it become too cool before we turn the hot water full on again, and so forth, so that we oscillate between water that is too hot and water that is too cold.

This example illustrates the two most common causes of instability in automatic control systems: delay and high gain. Delay in this case was caused by our slow response speed, and high gain is the result of our plan to turn the faucet full on or full off. Each of these items will be examined in detail in later chapters.

This section has attempted to give the reader an intuitive feeling for just what a closed-loop control system is by comparing it with an open-loop case.  On the surface, it may appear that the open-loop system is indeed the better.  Not only is it less complex, but no stability problem exists in open-loop systems.  The only negative feature is that performance may be degraded if any of the system parameters deviate from their expected values.  Actually, the comparison made on the basis of complexity, performance, and stability is somewhat of an academic question and is useful only to demonstrate the nature of a closed-loop system.  In many instances, open-loop control cannot even be considered.  The only cases in which open-loop control is possible are those in which the desired performance is known in advance.  In a great many instances the desired output is not known in advance.  Suppose that in our tub example the temperature of the water is required to be a function of some other variable, as the weather or the pH of a liquid in some adjacent tank.  Since either of these variables is unpredictable and beyond our control, it would be impossible to program the desired output in advance so that the valve settings could be programmed.  A closed-loop system is necessary to produce the desired response.

A fire-control system is another obvious situation in which open-loop control simply does not work.  In a fire-control system, a gun is made to follow a moving target such as an aircraft.  Since the flight path of the target aircraft is unknown and may even be intentionally evasive, it is evident that no open-loop control system can be programmed in advance to follow the plane's trajectory.  If a human operator were used to close the loop, his response time would usually be much too slow.  Thus, in spite of the complexity and the stability problems associated with automatic closed-loop control, there is often no alternative.  The design portions of this book are concerned with means of ensuring adequate performance and at the same time minimizing complexity and maintaining stability.

## *1.3  Historical and mathematical background*

The preceding section served to distinguish between open- and closed-loop control systems on the basis of feedback.  Feedback is a concept or principle that seems to be fundamental in nature and not necessarily peculiar to engineering.  In the social and political organizations of man, for example, a leader is the leader only as long as he is successful in realizing the desires of the group.  If he fails, another is elected or by other means obtains the effective support of the group.  The system output in this case is the success of the group in realizing its desires. The actual success is measured against the desired success, and if the two are not closely aligned, that is, if the error is not small, steps are taken to ensure that the error becomes small. In this case control is accomplished by deposing the leader. Individuals act in much the same way. Studying for this course involves feedback. If your study habits do not produce the desired understanding and grades, you change your study habits so that actual results become the desired results.

Because feedback is so evident in both nature and man, it is impossible to determine when feedback was first *intentionally* used. Newton, Gould, and Kaiser[1] cite the use of feedback in water clocks built by the Arabs as early as the beginning of the Christian era, but their next reference is not dated until 1750. In that year Meikle invented a device for automatically steering windmills into the wind, and this was followed in 1788 by Watt's invention of the flyball governor for regulation of the steam engine.

However, these isolated inventions cannot be construed as reflecting the application of any automatic control *theory*. There simply was no theory although, at roughly the same time as Watt was perfecting the flyball governor, both Laplace and Fourier were developing the two transform methods that are now so important in electrical engineering and in control in particular. The final mathematical background was laid by Cauchy, with his theory of the complex variable. It is unfortunate that the readers of this text cannot be expected to have completed a course in complex variables, although some may be taking this course at present. It is expected, however, that the reader is versed in the use of the Laplace transform. Note the word *use*. Present practice is to begin the use of Laplace transform methods early in the engineering curriculum so that, by the senior year, the student is able to use the Laplace transform in solving linear, ordinary differential equations with constant coefficients. But not until complex variables are mastered does a student actually appreciate how and why the Laplace transform is so effective in solving the problems that it does. In this text we assume that the reader has no knowledge of complex variables but that he does have a working knowledge of Laplace transform methods. A short summary of the more commonly used Laplace transform theorems is included in Appendix A, and Appendix B is a table of direct and inverse Laplace transforms. Although the Laplace transform is the mathematical language of the control engineer, in using this book the reader will not find it necessary to use more transform theory than appears in these two appendixes.

[1] G. C. Newton, L. A. Gould, and J. F. Kaiser, "Analytical Design of Linear Feedback Controls," John Wiley & Sons, Inc., New York, 1957.

Although the mathematical background for control engineering was laid by Cauchy (1789–1857), it was not until about 75 years after his death that an actual control theory began to evolve. Important early papers were "Regeneration Theory," by Nyquist, 1932, and "Theory of Servomechanisms," by Hazen, 1934. World War II produced an ever-increasing need for working automatic control systems and thus did much to stimulate the development of a cohesive control theory. Following the war a large number of linear-control-theory books began to appear, although the theory was not yet complete. As recently as 1958 the author of a widely used control text stated in his preface that "Feedback control systems are designed by trial and error."

With the advent of the new or modern control theory about 1960, advances have been rapid and of far-reaching consequence. The basis of much of this modern theory is highly mathematical in nature and almost completely oriented to the time domain. A key idea is the use of state-variable-system representation and feedback, with matrix methods used extensively to shorten the notation. In this text we retain the more conventional use of the frequency domain, at the same time embracing the ideas of state-variable representation and feedback and the use of the convenient matrix notation. In view of these innovations, it is now possible to state that feedback control systems no longer need be designed by trial and error.

## 1.4  Outline of the book

Broadly speaking, this book is divided into three sections: modeling, analysis, and synthesis. The section on modeling is made up of the two chapters immediately following this introductory chapter. These chapters are entitled Plant Representation and System Representation. By the *plant* is meant the physical hardware that is associated with the quantity being controlled. Usually the plant is considered unalterable. The elements that are added to effect control are designated as the *controller*, and the plant and the controller taken together constitute the entire closed-loop control system, or simply the *system*.

The discussion of plant representation emphasizes the different means by which the same plant can be described in terms of differential equations or transfer functions and stresses the idea of the state variable. Matrix manipulations are introduced and related both to the block diagram and to linear changes of variable. The separate discussion of system representation in Chap. 3 serves to emphasize the effects of feedback. There a general control configuration is postulated and then interpreted in three ways, involving both block diagrams and matrix elements.

Chapters 4 to 7 constitute the analysis portion of the book. Chapter 4 may be considered as something of a brute-force approach, as there we attempt to determine the total time response for a given input. Although the language of the control engineer may be the Laplace transform and the associated frequency domain, the actual control system exists in the time domain, and this chapter serves to emphasize the close relationship between the two. Emphasis is placed on the relationship between poles and zeros on the $s$ plane and the elements of the resolvent and state transition matrices.

It is often unnecessary to obtain the total system time response; in fact, it may even be impossible, if the plant is unknown. Chapter 5 returns to the frequency domain as a means of plant identification. The Bode methods explained there serve as the basis for one method of system synthesis discussed in Chap. 10.

The important question of stability is answered in Chap. 6, both in terms of frequency-domain methods and in terms of the roots of the characteristic equation. The root-locus methods of Chap. 7 serve as a transition between the analysis and synthesis sections of the book. In Chap. 7 the root locus is discussed from the point of view of analysis. However, in view of the information previously discussed in Chap. 4 concerning the time response, the root locus definitely serves as a lead-in to means of specifying closed-loop response in terms of desired closed-loop poles and zeros.

The synthesis portion of the book is introduced by a chapter on specifications (Chap. 8). Although both time- and frequency-domain specifications are discussed earlier in the text as appropriate, they are reviewed and placed in one section for emphasis and continuity. Steady-state errors, disturbances, and sensitivity receive particular attention and prepare the way for the following chapter on system design using state-variable feedback. This chapter (Chap. 9) assumes that all the state variables are available and uses as a design criterion the desired closed-loop response. In cases where all the state variables are not available, minor-loop equalizers result naturally, and if only the output is available, an equivalent series equalizer is specified.

The use of series equalizers is discussed in Chap. 10 and related to the results of the preceding chapter. As mentioned earlier, the design technique used is the Bode diagram, although at that point the reader should be fully equipped to evaluate the resulting design by a variety of methods.

# *two*   *plant representation*

## *2.1  Introduction*

In this chapter we consider the most fundamental part of any control problem, namely, the representation of the plant by appropriate *mathematical models*. The concept of a mathematical model is not new to the reader.  Before any engineering analysis or design may be undertaken, it is necessary to abstract from the physical object in question a description in terms of mathematical formulas.  For example, the mathematical model for an electrical circuit is the set of loop or node equations that describe the circuit.

Although we deal almost exclusively with the mathematical model for a system, one must not forget that the ultimate interest must be in the control of real, physical systems. One does not control a set of equations; rather, one controls a radar antenna or a rocket engine. Throughout this book various practical features of control-system design, which are not obvious from the mathematical description, are illustrated by means of examples or exercises for the reader.

Often more than one mathematical model may be found for the same physical object. Consider, for example, the loop and node representations of electrical circuits. In such cases, the various models tend to complement each other, with one model being better for one use whereas another model is better for another use.

In terms of control theory, two different means of representation are of interest. These two means of representation differ more in degree than in nature. To be more precise, the first method, known as the *input-output-relation method,* provides less detail and therefore is less complete than the second method, known as the *state-variable method.* As with electrical circuits, these two methods complement rather than conflict with each other.

This last statement may seem strange since it may appear that the more complete state-variable method would be superior in all situations. That this is not true may be traced to the fact that one simply does not need and often cannot profitably use the more complete representation. Therefore there is considerable need for both methods. This chapter contains not only a development of each of these two means of representation but also a considerable discussion of how the two methods are related.

The input-output-relation means of plant representation is discussed first, since it is easier and probably more familiar to the reader. Following this development, a convenient graphical representation of the input-output-relation information is achieved by the introduction of the concept of a block diagram.

Before the state-variable method of representation is introduced, matrices and matrix algebra will be discussed briefly, since much of the state-variable material can be presented most effectively and compactly in terms of these tools. Then, following an introduction to the concept of state, the state-variable method of representation will be discussed. Throughout this development the relation of the state-variable approach to the input-output-relation approach is emphasized.

This chapter is restricted to plant representation but *not* because the methods discussed apply only to that case. On the contrary, the methods of this chapter form the entire basis of the subject of closed-loop-system representation presented in the next chapter. The restriction has been made here for the sake of simplicity and to emphasize the important and natural distinction between the uncontrolled plant and the controlled plant, i.e., the closed-loop system.    By restricting our attention in this fashion we hope that the reader will be able to grasp the fundamental concepts of representation with the least chance of confusion.

## 2.2  *Transfer functions and block diagrams*

One of the principal tools of the input-output representation is the transfer function.    The idea of using transfer functions to represent physical systems is a natural outgrowth of the use of Laplace transform, operational methods to solve linear differential equations.    These operational methods have been so successful in simplifying and systematizing the problem of obtaining the time response of a system that it appears reasonable that they should also be valuable in system representation.

The general configuration of a closed-loop control system is shown in Fig. 2.2-1.    The system consists of two basic elements, the plant and the controller.    The plant encompasses the unalterable portion of the system being controlled.    The controller is added by the designer to achieve proper performance of the overall system.    In this chapter we restrict our attention to the description of the plant, for simplicity; the entire system will be considered in the next chapter.

In order to understand how transfer functions are used in plant representation, let us suppose that we have a general $n$th-order plant with an input $u(t)$ and an output $y(t)$, as shown in Fig. 2.2-1.    The plant may be described by an $n$th-order linear, time-invariant differential equation, relating the input and the output, of the form

$$\frac{d^n y(t)}{dt^n} + a_n \frac{d^{n-1} y(t)}{dt^{n-1}} + \cdots + a_1 y(t)$$

$$= c_{m+1} \frac{d^m u(t)}{dt^m} + c_m \frac{d^{m-1} u(t)}{dt^{m-1}} + \cdots + c_1 u(t) \quad (2.2\text{-}1)$$

Here all the $a_i$'s and $c_i$'s are assumed to be constant.

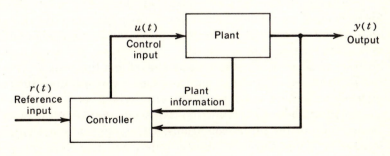

*Fig. 2.2-1*  General closed-loop system.

This differential equation completely describes the plant since, for any given input and initial conditions, the output may be determined. In actuality, the differential equation (2.2-1) is already a mathematical model of the plant. This model, however, is rather unwieldy and therefore is seldom used in this form.

A more basic starting point for our development of transfer functions is the physical plant itself. Such an approach tends to obscure the basic concept of transfer functions with an extensive development of plant components and their describing equations. It is assumed that the reader is somewhat familiar with the modeling of electromechanical systems, and therefore a separate development on components is not included here. On the other hand, several of the examples and exercises begin with real physical systems in order to acquaint the reader with the formulation of some of the more common control-system elements.

If we take the Laplace transform of Eq. (2.2-1), assuming that all initial conditions are zero, we obtain

$$s^n y(s) + a_n s^{n-1} y(s) + \cdots + a_1 y(s) = c_{m+1} s^m u(s) + \cdots + c_1 u(s)$$

or

$$(s^n + a_n s^{n-1} + \cdots + a_1) y(s) = (c_{m+1} s^m + \cdots + c_1) u(s) \qquad (2.2-2)$$

The *transfer function*, $G_p(s)$, of a plant is defined as the ratio of the Laplace transform of the output, $y(s)$, to the Laplace transform of the input, $u(s)$, assuming that *all initial conditions are zero*. For the general example above, this becomes

$$\frac{y(s)}{u(s)} = G_p(s) = \frac{c_{m+1} s^m + c_m s^{m-1} + \cdots + c_1}{s^n + a_n s^{n-1} + \cdots + a_2 s + a_1} \qquad (2.2-3)$$

Note that the transfer function is the ratio of two polynomials in $s$, known as the numerator polynomial and the denominator polynomial. As we shall see in Chap. 4, the denominator polynomial plays the key role in determining the character of the behavior of the plant.

Several comments are in order at this time. First, the transfer function $G_p(s)$ completely characterizes the plant, since it contains all the information concerning the coefficients of the original differential equation (2.2-1) describing the plant. In other words, given the transfer function, it is possible to reconstruct the differential-equation description. In fact, this reconstruction may be done by inspection, by just cross-multiplying and letting $s^k$ equal the $k$th derivative.

Second, the transfer function is dependent only on the plant and not on the input or the initial conditions. The input does not enter into the transfer function, since the transfer function is defined as the ratio of the

Laplace transform of the output to the input. Note that the input does not appear on the right-hand side of Eq. (2.2-3). The statement that $G_p(s)$ is independent of the initial conditions is somewhat misleading since the definition of a transfer function requires that all initial conditions be zero. If the initial conditions are not zero, one must either return to the differential-equation description [Eq. (2.2-1)] or make use of the state-variable representation discussed later in this chapter. Since many properties of interest are completely determined by the plant, the lack of initial-condition information is not a serious problem, and transfer functions find a wide range of use. The reader is reminded, however, never to forget the basic assumption of zero initial conditions whenever he works with transfer functions.

The transfer function of Eq. (2.2-3) is said to have $n$ poles and $m$ zeros. The reason for this statement is more obvious if the transfer function $G_p(s)$ is represented in factored form as

$$G_p(s) = c_{m+1} \frac{(s + \delta_1)(s + \delta_2) \cdots (s + \delta_m)}{(s + \lambda_1)(s + \lambda_2) \cdots (s + \lambda_n)} = \frac{K_p N_p(s)}{D_p(s)}$$

where $N_p(s) = (s + \delta_1)(s + \delta_2) \cdots (s + \delta_m)$,

$$D_p(s) = (s + \lambda_1)(s + \lambda_2) \cdots (s + \lambda_n)$$

and $K_p = c_{m+1}$. Here $c_{m+1}$ is changed to $K_p$ to emphasize the fact that this is a gain inherently associated with the plant. The $m$ values of $s$, namely, $-\delta_1, -\delta_2, \ldots, -\delta_m$, which make the numerator polynomial of $G_p(s)$ zero, are known as the *zeros of* $G_p(s)$. The $n$ values of $s$, namely, $-\lambda_1, -\lambda_2, \ldots, -\lambda_n$, which make the denominator polynomial of $G_p(s)$ zero, or the overall transfer function infinite, are known as the *poles of* $G_p(s)$. Hence, we say that the transfer function has $n$ poles and $m$ zeros.

*Block diagrams.* It is often helpful to make a pictorial representation of a transfer function by means of a technique known as a *block diagram*. A block diagram of the plant described by the transfer function (2.2-3) is shown in Fig. 2.2-2. There we see that the quantity contained in the block is the transfer function relating the input and output of the block. It is common practice to assume that the block diagram represents information transmission in one direction only, from the input to the output, as indicated by the arrows in Fig. 2.2-2. In other words, if the output $y(t)$ is forced to have some behavior, from the block diagram the input is assumed to be unaffected. On the other hand, if the input is known, the output may be found by making use of the transfer function.

Although the block diagram of Fig. 2.2-2 is nothing more than a

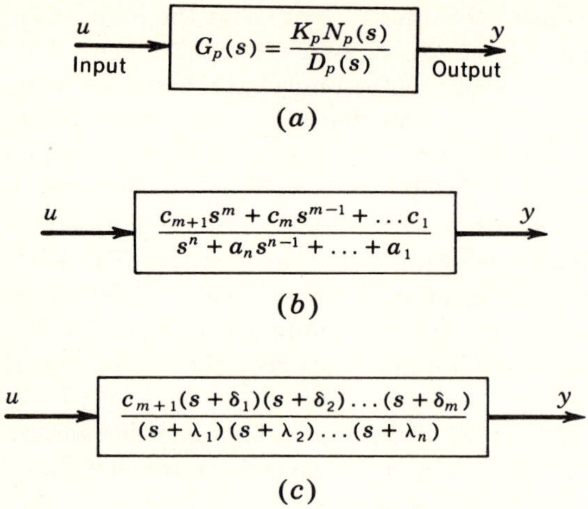

$$u \text{ (Input)} \longrightarrow \boxed{G_p(s) = \frac{K_p N_p(s)}{D_p(s)}} \longrightarrow y \text{ (Output)}$$

(a)

$$u \longrightarrow \boxed{\frac{c_{m+1}s^m + c_m s^{m-1} + \ldots c_1}{s^n + a_n s^{n-1} + \ldots + a_1}} \longrightarrow y$$

(b)

$$u \longrightarrow \boxed{\frac{c_{m+1}(s+\delta_1)(s+\delta_2)\ldots(s+\delta_m)}{(s+\lambda_1)(s+\lambda_2)\ldots(s+\lambda_n)}} \longrightarrow y$$

(c)

*Fig. 2.2-2*   Block diagrams.   (a) Compact form; (b) expanded form; (c) factored or pole-zero form.

pictorial representation of Eq. (2.2-3), the block-diagram representation often proves useful in gaining a clearer understanding of a control problem by making it possible to visualize the interrelations of the various parts of the problem.

The above development and, in particular, Fig. 2.2-2 illustrate a notation that is used throughout this book. The Laplace transform of a time function, such as $y(t)$, is indicated by simply replacing the time variable $t$ by the frequency variable $s$. The argument $s$ or $t$ is omitted only when there is no possibility of confusion or when there may be a dual meaning, as in the block diagrams of Fig. 2.2-2. There the variables may be interpreted as frequency-domain variables, $y(s)$, for example, and the blocks as transfer functions, or the variables may be interpreted as time-domain variables, $y(t)$, for example, and the blocks as linear time-domain operators, with $s$ indicating differentiation with respect to time, as usual.

With respect to the comments on notation, the reader is cautioned against the following pitfall. Although the equation

$$\frac{y(s)}{u(s)} = G_p(s) \tag{2.2-4}$$

is correct and unambiguous, the "analogous" time-domain statement, namely,

$$\frac{y(t)}{u(t)} = G_p(t)$$

is completely meaningless. Clearly the two are not analogous statements. The input-output relationship as expressed by the transfer-function relationship of Eq. (2.2-4) has meaning only for transformed quantities. This is, in fact, the basis for the definition of a transfer function: a ratio of the Laplace transform of the output to the Laplace transform of the input, with zero initial conditions.

Hence we see that the describing differential equation (2.2-1), the transfer function (2.2-3), and the block diagram of Fig. 2.2-2 are three entirely equivalent methods of representation. One may, by inspection, determine from any one of the three representations the other two. These three forms of representation constitute what we have loosely referred to as the *input-output method* of plant representation. To illustrate these various means of plant representation, two examples are presented.

*Example 2.2-1*   As the first example, let us consider the simple electrical circuit shown in Fig. 2.2-3. If we write a Kirchhoff loop equation for this circuit, we find that

$$L \frac{di(t)}{dt} + (R_1 + R_2)i(t) = u(t)$$

The output voltage $y(t)$ is given by

$$y(t) = R_2 i(t)$$

If these two equations are combined and $i(t)$ suppressed, the resulting differential equation is[1]

$$\frac{L}{R_2} \dot{y}(t) + \frac{R_1 + R_2}{R_2} y(t) = u(t)$$

[1] The dot convention for time derivatives is used whenever the order of the differentiation is low. Here $\dot{y}(t) = dy(t)/dt$ and $\ddot{y}(t) = d^2y(t)/dt^2$, etc.

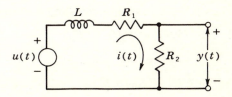

*Fig. 2.2-3*   Simple electrical circuit.

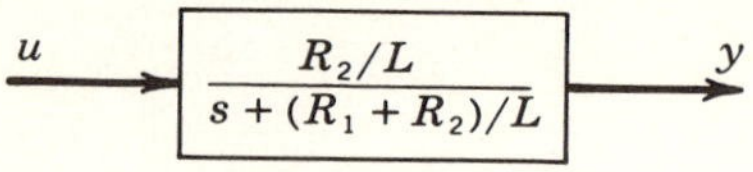

$$\frac{R_2/L}{s + (R_1 + R_2)/L}$$

*Fig. 2.2-4*  Block diagram of the circuit of Example 2.2-1.

Then the transfer function for this circuit is determined by taking the Laplace transform of this equation (assuming zero initial conditions) and solving for the ratio of $y(s)$ to $u(s)$, with the result that

$$\frac{y(s)}{u(s)} = G_p(s) = \frac{R_2}{Ls + (R_1 + R_2)} = \frac{R_2/L}{s + (R_1 + R_2)/L}$$

The block diagram for this system is shown in Fig. 2.2-4.

Note that in this example a cause-and-effect relationship is indicated by the arrows in the block diagram. Here $u(t)$ is the input and $y(t)$ is the output. If a current were given as flowing in the circuit, we could not predict the effect on $u(t)$. This is true in general. The block diagram presupposes the input as indicated, with the flow of information as shown by the arrow.

**Example 2.2-2**  The second example is a field-controlled dc motor driving an inertial and frictional load shown schematically in Fig. 2.2-5. The armature current is assumed to be constant. The input is the applied field voltage $e(t)$, the output is the shaft position $\theta_o(t)$, and

$i_f(t) =$ field current
$I_a =$ constant armature current
$T(t) =$ torque
$R_f =$ field resistance
$R_a =$ armature resistance
$L_f =$ field inductance
$\beta =$ viscous damping coefficient
$J =$ moment of inertia of motor and load
$K_T =$ torque constant

The Kirchhoff equation for the field circuit is

$$L_f \dot{i}_f(t) + R_f i_f(t) = e(t)$$

The newtonian equation for the mechanical load is

$$J \ddot{\theta}_o(t) + \beta \dot{\theta}_o(t) = T(t)$$

and the torque field-current relation is

$$T(t) = K_T i_f(t)$$

since the armature current is constant.   Let us transform each of these three equations to obtain

$$(sL_f + R_f)i_f(s) = e(s)$$
$$(Js^2 + \beta s)\theta_o(s) = T(s)$$

and

$$T(s) = K_T i_f(s)$$

These three algebraic frequency-domain equations can be solved simultaneously to obtain the desired transfer function $\theta_o(s)/e(s)$:

$$\frac{\theta_o(s)}{e(s)} = \frac{K_T}{(Js^2 + \beta s)(L_f s + R_f)} = \frac{K_T/JL_f}{s(s + \beta/J)(s + R_f/L_f)}$$

or

$$\frac{\theta_o(s)}{e(s)} = \frac{K_T}{JL_f s^3 + (L_f\beta + R_f J)s^2 + \beta R_f s}$$

The describing differential equation can now be written by inspection as

$$JL_f \ddot{\theta}_o(t) + (L_f\beta + R_f J)\ddot{\theta}_o(t) + \beta R_f \dot{\theta}_o(t) = K_T e(t)$$

or the block diagram can be drawn in either polynomial or pole-zero form as shown in Fig. 2.2-6.

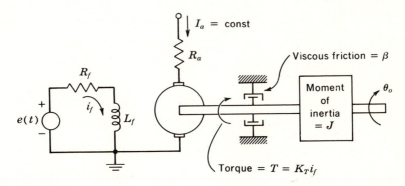

$$K_T = \text{Torque constant}$$

**Fig. 2.2-5**  Schematic diagram of a dc field-controlled motor.

$$\frac{K_T}{JL_f s^3 + (L_f \beta + R_f J)s^2 + \beta R_f s}$$

*e* →                                                                    → $\theta_o$

*(a)*

$$\frac{K_T / JL_f}{s(s + \beta/J)(s + R_f/L_f)}$$

*e* →                                                                    → $\theta_o$

*(b)*

**Fig. 2.2-6**   Block diagram of the field-controlled dc
motor: *(a)* polynomial form; *(b)* pole-
zero form.

*Block-diagram algebra.*   Closely associated with the use of block
diagrams to represent the input-output characteristics of a plant is a
set of procedures for block-diagram manipulation commonly referred
to as block-diagram algebra.   In essence these manipulations are nothing more than a graphical procedure for manipulating algebraic equations, such as the ones determined in Example 2.2-2.   Because of this
fact, these procedures are often helpful in finding the transfer function
of a complex plant by combining the block diagrams of various parts
of the plant in order to find the overall input-output block diagram and
hence the transfer function.   In addition, this block-diagram algebra may
be used in the design and analysis phases of a control problem to arrange
the system into some particularly advantageous form.

The simplest block-diagram manipulation involves the reduction
of two blocks in series or cascade, as shown in Fig. 2.2-7*a*, to one overall
block, as shown in Fig. 2.2-7*b*.   In order to verify this reduction, we

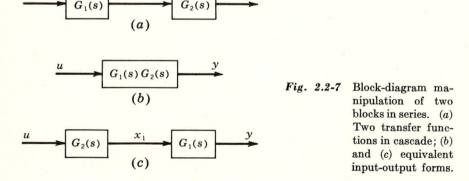

*u* → $G_1(s)$ → *x* → $G_2(s)$ → *y*

*(a)*

*u* → $G_1(s)\,G_2(s)$ → *y*

*(b)*

*u* → $G_2(s)$ → $x_1$ → $G_1(s)$ → *y*

*(c)*

**Fig. 2.2-7**   Block-diagram manipulation of two
blocks in series.   *(a)*
Two transfer func-
tions in cascade; *(b)*
and *(c)* equivalent
input-output forms.

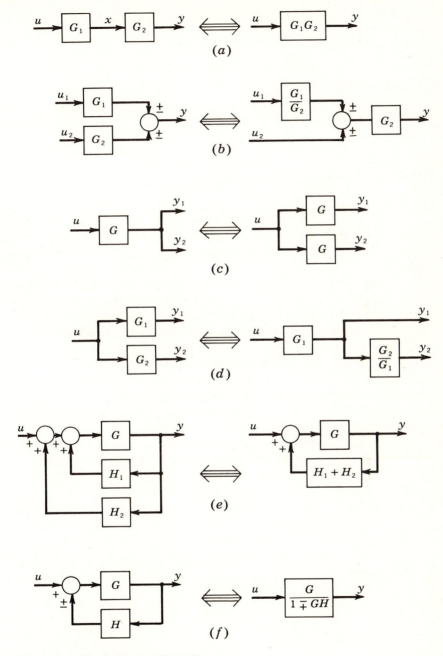

**Fig. 2.2-8** Block-diagram identities.

simply write the two algebraic frequency-domain equations represented by the two blocks

$$y(s) = G_2(s)x(s) \qquad x(s) = G_1(s)u(s)$$

and then combine them so that

$$y(s) = G_2(s)x(s) = G_2(s)G_1(s)u(s) = G_1(s)G_2(s)u(s)$$

The input-output transfer function becomes

$$\frac{y(s)}{u(s)} = G_1(s)G_2(s)$$

which is the desired result.

It should be noted that the order in which the two blocks occur does not alter the input-output result, so that the series combination of Fig. 2.2-7c also reduces to the single block of Fig. 2.2-7b. However, although the order of the blocks is unimportant from an input-output view, it does affect the internal variable found between the blocks. This important distinction will be discussed in detail in the section on state-variable representation (Sec. 2.4).

This block-diagram reduction may also be reversed in the sense that one may use it to divide one block into two blocks in series. In other words, the block-diagram equivalents indicated in Fig. 2.2-7 are reciprocal, and one may proceed from any one to any of the others.

It is obvious that by considering groups of two blocks at a time any number of blocks in series may be reduced to a single block. Similarly any single block may be broken into any number of blocks in series.

In addition to this simple block-diagram identity, there are a large number of others. Some of the more useful ones are summarized in Fig. 2.2-8. In addition to the familiar diagrams, this figure also contains a new symbol known as the error detector, the summing junction, or just the summer. Its properties are displayed in Fig. 2.2-9. This element plays an important role in almost all practical control systems.

All the identities shown in Fig. 2.2-8 may be verified by writing the output(s) of two plants. Consider the identity of Fig. 2.2-8f, for

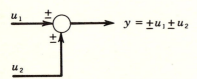

$$y = \pm u_1 \pm u_2$$

*Fig. 2.2-9* Summing junction.

example. The output of the left-hand representation is

$$y(s) = G(s)u(s) \pm G(s)H(s)y(s)$$

or

$$[1 \mp G(s)H(s)]y(s) = G(s)u(s)$$

The output is then

$$y(s) = \frac{G(s)}{1 \mp G(s)H(s)} u(s)$$

which is seen to be equivalent to the output of the representation on the right. Verification of the remaining identities is left to the reader as an exercise (see Exercise 2.2-1).

Since extensive use is made of this block-diagram algebra throughout the book, the reader is urged to study Fig. 2.2-8 carefully.

**_Example 2.2-3_** To illustrate the usefulness of the block-diagram identities of Fig. 2.2-8 in the representation of a plant, let us consider once again the field-controlled dc motor of Example 2.2-2. There three algebraic transformed equations were found for various portions of the plant. Those equations are repeated here for convenience:

$$(sL_f + R_f)i_f(s) = e(s)$$
$$(Js^2 + \beta s)\theta_o(s) = T(s)$$

and

$$T(s) = K_T i_f(s)$$

These three equations may be represented by three block diagrams, as shown in Fig. 2.2-10a. If the equivalent input and output variables of these three blocks are connected, the block diagram shown in Fig. 2.2-10b results. One may now make use of the series-combination rule of Fig. 2.2-8a to obtain the block diagram shown in Fig. 2.2-10c. Note that in this final block diagram the electrical and mechanical portions of the plant may still be separated.

**_Example 2.2-4_** As a second example, let us consider the representation of the armature-controlled dc motor shown schematically in Fig. 2.2-11. Once again the load is inertia plus friction; in this case, the field current is held constant. (This is often a much easier

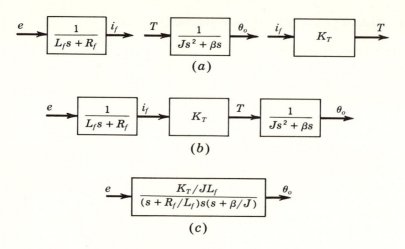

$$(a)$$

$$(b)$$

$$(c)$$

**Fig. 2.2-10**   Use of block-diagram algebra to find the transfer function for the field-controlled motor.   (*a*) Subsystems; (*b*) connected subsystems; (*c*) overall block diagram.

task to accomplish in practice than holding the armature current constant.)

Kirchhoff's equation for the armature circuit is

$$(sL_a + R_a)i_a(s) = e(s) - e_c(s)$$

where $e_c(s)$ is the back emf of the motor:

$$e_c(s) = K_v s\theta_o(s)$$

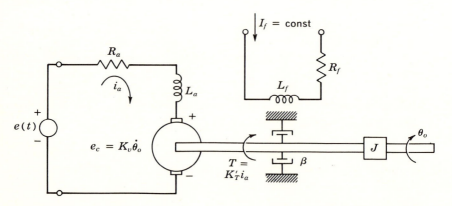

**Fig. 2.2-11**   Schematic diagram of a dc armature-controlled motor.

In the mechanical portion of the plant we have

$$(Js^2 + \beta s)\theta_o(s) = T(s)$$

and

$$T(s) = K'_T i_a(s)$$

If each of these four equations is represented by a block diagram, we have the result shown in Fig. 2.2-12. As the next step, we again connect the equivalent variables as shown in Fig. 2.2-13a. This result may be further reduced to the final form shown in Fig. 2.2-13c by making use of the identities given in Fig. 2.2-8a and f. From this final block diagram, we can easily write either the transfer function of the plant or the describing differential equation.

If the final block diagram of Fig. 2.2-13c is compared with the final block diagram of the field-controlled dc motor (Fig. 2.2-10c), it is seen that in the present case the electrical and mechanical portions of the plant cannot be separated. In practice, the fact that the electrical and mechanical portions of the plant cannot be separated is not important, and field-controlled motors are much more common than armature-controlled motors. For purposes of illustration, however, this separation into electrical and mechanical time constants is helpful and the armature-controlled motor is used extensively in examples throughout this text.

To be able to connect the subsystem block diagrams as we have done in the two above examples, it is necessary that the "impedance" levels match. In other words, if the subsystem block diagrams are determined on the basis of given termination and input conditions, these con-

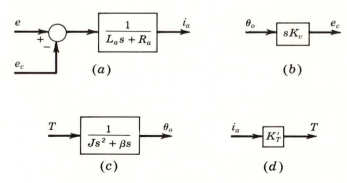

**Fig. 2.2-12**  Subsystems of the armature-controlled motor.

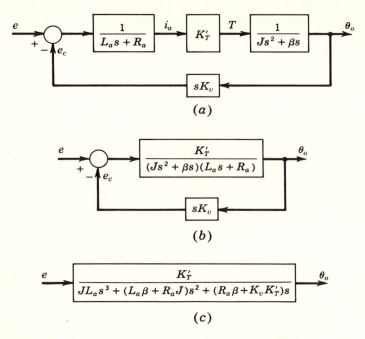

**Fig. 2.2-13**  Block diagrams of the armature-controlled motor: (*a*)
connected subsystems; (*b*) first reduction; (*c*) final
block diagram.

ditions must be met when the blocks are connected.  For example, if the
block diagram of an electrical circuit is computed on the basis of an open-
circuit termination, as was done in Example 2.2-1, then, in order to use
this result, it is necessary to ensure that the termination is effectively an
open circuit.

As the last example of this section, let us consider an electrical network slightly more complicated than the one of Example 2.2-1.  The
object of this final example is to illustrate that any set of coupled differential equations can be represented in block-diagram form.

**Example 2.2-5**  This example is based upon the two-loop network
of Fig. 2.2-14.  Here the describing integrodifferential equations are

$$R_1 i_1(t) + \frac{1}{C} \int_0^t i_1(t')\, dt' - \frac{1}{C} \int_0^t i_2(t')\, dt' = u(t)$$

$$R_2 i_2(t) + (L_1 + L_2) \frac{di_2(t)}{dt} + \frac{1}{C} \int_0^t i_2(t')\, dt' - \frac{1}{C} \int_0^t i_1(t')\, dt' = 0$$

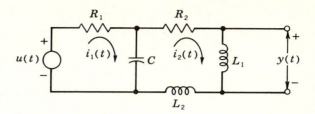

**Fig. 2.2-14**  Circuit for Example 2.2-5.

and

$$y(t) = L_1 \frac{di_2(t)}{dt}$$

These equations may be transformed, again assuming zero initial conditions, to yield

$$\left(R_1 + \frac{1}{Cs}\right) i_1(s) - \frac{1}{Cs} i_2(s) = u(s)$$

$$\frac{-1}{Cs} i_1(s) + \left[(L_1 + L_2)s + R_2 + \frac{1}{Cs}\right] i_2(s) = 0$$

and

$$y(s) = L_1 s i_2(s)$$

A picture of each of these equations is given in Fig. 2.2-15.  In Fig.

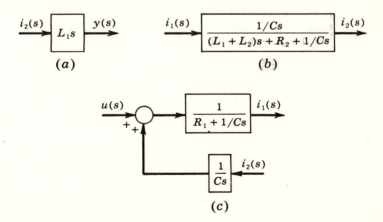

**Fig. 2.2-15**  Block-diagram representation of the transformed equations of Fig. 2.2-14.

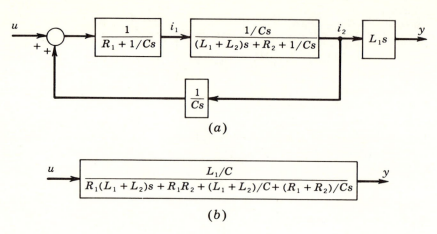

(a)

(b)

***Fig. 2.2-16***  Overall transfer functions for Example 2.2-5.  (*a*) Connected subsystems; (*b*) final block diagram.

2.2-15*a*, a transfer function for $y(s)/i_2(s)$ is indicated, as dictated by the last equation above.    The transfer-function relationship between $i_2(s)$ and $i_1(s)$ is determined from the middle equation directly above and is pictured in Fig. 2.2-15*b*.    The remaining equation is pictured in Fig. 2.2-15*c*.    In this last figure, $i_2(s)$ is assumed to be available, and the operation of addition is indicated by the summer.

The three isolated blocks of Fig. 2.2-15 are joined to form Fig. 2.2-16*a*.    The overall transfer function from $u(s)$ to $y(s)$ may be determined by block-diagram reduction from this diagram or from the transformed differential equations.    The final overall transfer function $y(s)/u(s)$ is indicated in Fig. 2.2-16*b*.

This example also illustrates an important assumption that should be emphasized.    It is assumed throughout this book that the physical system in question is modeled with the minimum number of variables necessary for a complete description.    For example, in this problem if a different current had been assigned to the inductance $L_2$ than that which flowed through $L_1$, a redundant equation would have resulted, that is, an equation that could be eliminated without affecting the accuracy of the plant description.    We always assume the plant is modeled with the minimum number of equations possible.

Before we leave this section it should be reemphasized that all the block-diagram identities presented in Fig. 2.2-8 are valid only in an input-output sense.    Because of this, internal variables may be and often are distorted or even obliterated.    Consider, for example, the three block-diagram representations of the armature-controlled motor shown in Fig. 2.2-13. In Fig. 2.2-13a, we can identify three meaningful physical, internal variables: $i_a$, $e_c$, and $T$. In Fig. 2.2-13c, however, none of these variables appears. The same is true in Fig. 2.2-16b.

This complete lack of regard for internal variables is a characteristic of the input-output representation of a plant. By its very name, this method is labeled as being strongly related to input-output properties. Such an approach is often sufficient, but there are also many situations in which a more complete description including the internal behavior of the plant is necessary. In these cases, one must use the state-variable representation. Before discussing this method, however, we shall introduce some of the concepts of matrix theory so that they may be used to simplify our development of the state-variable representation.

**Exercises 2.2**  *2.2-1.*  Verify the block-diagram identities of Fig. 2.2-8 and use them to represent the plant shown in Fig. 2.2-17a in the form given in Fig. 2.2-17b and c.

*answers:*

$$H(s) = 2s + 1 \qquad G(s) = \frac{10}{s(s + 21)}$$

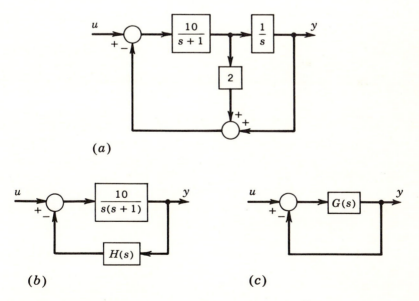

**Fig. 2.2-17**  Exercise 2.2-1.

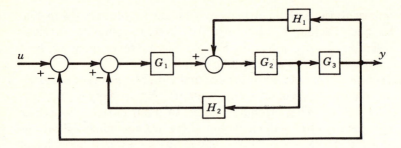

**Fig. 2.2-18**   Exercise 2.2-2.

**2.2-2.**   For the plant shown in Fig. 2.2-18, show that $y(s)/u(s)$ is

$$\frac{y(s)}{u(s)} = \frac{G_1 G_2 G_3}{1 + G_1 G_2 H_2 + G_2 G_3 H_1 + G_1 G_2 G_3}$$

**2.2-3.**   The plant shown in Fig. 2.2-19 consists of a field-controlled motor and a constant-field generator. The motor has a torque constant $= K_T$ newtons-m/amp. The generator constant $= K_v$ volts/(rad)/(sec); assume that the torque required to turn the generator under load $T_g$ is $K_v i_{ag}$. The connecting shaft has an inertia of $J$ and viscous friction $\beta$. Draw a block diagram for the plant and find $y(s)/u(s)$.

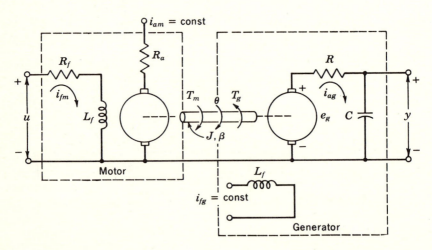

**Fig. 2.2-19**   Exercise 2.2-3.

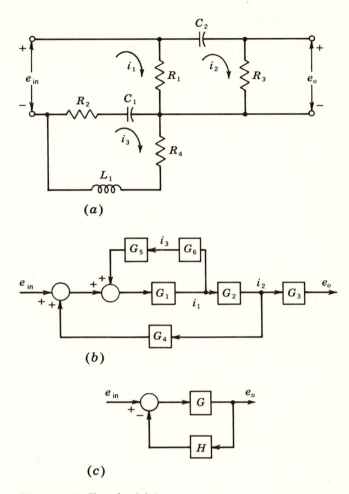

**(a)**

**(b)**

**(c)**

**Fig. 2.2-20**   Exercise 2.2-4.

*answer:*

$$\frac{y(s)}{u(s)} = \frac{K_T K_v}{(sL_f + R_f)[(Js + \beta)(sRC + 1) + sK_v\,^2C]}$$

*2.2-4.*   For the electrical network shown in Fig. 2.2-20a show that the block diagram for this network is as given in Fig. 2.2-20b and find each of the indicated transfer functions. Reduce the block diagram of Fig. 2.2-20b to that of Fig. 2.2-20c.

*answers:*

$$G_1 = \frac{C_1 s}{(R_1 + R_2)C_1 s + 1} \qquad G_2 = \frac{R_1 C_2 s}{(R_1 + R_3)C_2 s + 1}$$

$$G_3 = R_3 \qquad G_4 = R_1 \qquad G_5 = \frac{sR_2 C_1 + 1}{sC_1}$$

$$G_6 = \frac{R_2 C_1 s + 1}{(R_2 + R_4)C_1 s + L_1 C_1 s^2 + 1}$$

$$G = \frac{G_1 G_2 G_3}{1 - G_1 G_2 G_4} \qquad H = -\frac{G_5 G_6}{G_2 G_3}$$

or

$$G = \frac{G_1 G_2 G_3}{1 - G_1 G_5 G_6} \qquad H = -\frac{G_4}{G_3}$$

## 2.3   *Matrices and matrix algebra*

Since our use of matrices is for convenience rather than of necessity, the treatment is purposely kept brief.   The object here is not to be definitive but only to present the material that will be useful in the remainder of the book.

*Definition.*   A *matrix* is a rectangular array of elements, arranged in $m$ horizontal rows and $n$ vertical columns,[1] that obeys a matrix algebra discussed below.   The matrix **A** is therefore

$$\mathbf{A} = [a_{ij}] = \begin{bmatrix} a_{11} & a_{12} & \cdots & a_{1n} \\ a_{21} & a_{22} & \cdots & a_{2n} \\ \cdots & \cdots & \cdots & \cdots \\ a_{m1} & a_{m2} & \cdots & a_{mn} \end{bmatrix} \begin{matrix} \text{the second row} \\ \\ \\ \text{the } n\text{th column} \end{matrix} \qquad (2.3\text{-}1)$$

The notation used throughout the book is exemplified by Eq. (2.3-1). A matrix is designated by a boldface capital letter, such as **A**, as a shorthand notation for the entire array.   When the array itself is given, it is set off in brackets for emphasis.   The elements of the matrix are indicated by a small letter with a double subscript.   Thus the element of the matrix **A** that is located in the $i$th row and the $j$th column is $a_{ij}$.

It must be emphasized that **A** only symbolizes the matrix which is the entire array of $mn$ elements.   Whenever we wish to make a

---

[1] The $m$ and $n$ used here are positive integers and are not to be confused with the number of zeros and poles of $G_p(s)$.

specific computation, we must make use of the arrayed form.   Hence we see that *matrices do not solve problems;* they only provide a useful and systematic tool for representation and possible manipulation.

The elements of a matrix may be real or complex numbers or functions of either time or frequency variables.   Typical examples of matrices are

$$(a) \quad \begin{bmatrix} 1+j & 2 \\ 0 & 5 \\ 3+j & 9 \end{bmatrix} \quad (b) \quad [t^2 \quad 1 \quad 0]$$

$$(c) \quad \begin{bmatrix} \dfrac{s}{s+1} & 0 \\ \dfrac{1}{s} & s+2 \\ & s+4 \end{bmatrix} \quad (d) \quad \begin{bmatrix} x_1(t) \\ x_2(t) \\ x_3(t) \end{bmatrix} \qquad (2.3\text{-}2)$$

Here, for example, $a_{31}$, the element in the third row and first column of the first matrix in (2.3-2), is $3+j$.

*Classification.*   A matrix is classified according to the number of its rows and columns.   If a matrix has $m$ rows and $n$ columns, it is commonly referred to as an $m \times n$ *matrix* or of order $m \times n$.   A matrix that has $n$ rows and only one column is known as a *column matrix* or *vector.*[1]   To distinguish a vector from a general matrix, a lowercase boldface letter is used to denote a vector:

$$\mathbf{x} = \begin{bmatrix} x_1 \\ x_2 \\ \cdot \\ \cdot \\ \cdot \\ x_n \end{bmatrix} = \text{col } (x_1, x_2, \ \ldots \ , x_n)$$

Note that the elements of a column matrix have only one subscript, which indicates their location in the column.

A matrix with only one row and $n$ columns is known as a *row matrix* or row vector.   If $n = m$, the matrix is referred to as a *square matrix*, since its array is square.   A matrix that does not fit into any of the above categories is called simply a *rectangular matrix*.   The classi-

---

[1] The equating of a vector to a column matrix is not exactly correct, since a vector has a fixed geometric meaning whereas a column matrix is only a representation of a given vector in one coordinate system.   However, the usage is common and is followed here.

fications of the four matrices listed in (2.3-2) are as follows:

| Matrix | Order | Classification |
|--------|-------|----------------|
| a | $3 \times 2$ | Rectangular |
| b | $1 \times 3$ | Row |
| c | $2 \times 2$ | Square |
| d | $3 \times 1$ | Column |

A $1 \times 1$ matrix obeys all the normal algebraic rules of scalars and hence is considered a scalar.

In the class of square matrices, two important special cases arise. If a square matrix **A** is such that $a_{ij} = 0$ for all $i \neq j$, the matrix is referred to as a *diagonal matrix*. The matrix

$$\mathbf{A} = \begin{bmatrix} 2 & 0 & 0 \\ 0 & -1 & 0 \\ 0 & 0 & 3 \end{bmatrix}$$

is diagonal. If, in addition to being diagonal, the elements on the diagonal are all unity, that is, $a_{ii} = 1$ for all $i$, the matrix is known as the *identity matrix*, designated as **I**.

$$\mathbf{I} = \begin{bmatrix} 1 & 0 & \cdots & 0 \\ 0 & 1 & \cdots & 0 \\ & & \cdots & \\ 0 & 0 & \cdots & 1 \end{bmatrix} = \begin{bmatrix} 1 & & & 0 \\ & 1 & & \\ & & \ddots & \\ 0 & & & 1 \end{bmatrix}$$

*Transpose.* The *transpose* of any matrix is formed by interchanging the rows and columns of the matrix and is indicated by the superscript $T$, so that

$$\mathbf{A}^T = \begin{bmatrix} a_{11} & a_{21} & \cdots & a_{m1} \\ a_{12} & a_{22} & \cdots & a_{m2} \\ & & \cdots & \\ a_{1n} & a_{2n} & \cdots & a_{mn} \end{bmatrix}$$

For example, the transpose of the matrix

$$\mathbf{A} = \begin{bmatrix} 1 & 3 \\ 0 & 2 \\ 5 & 0 \end{bmatrix}$$

is

$$\mathbf{A}^T = \begin{bmatrix} 1 & 0 & 5 \\ 3 & 2 & 0 \end{bmatrix}$$

A row matrix may be thought of as the transpose of a column matrix, so that $\mathbf{x}^T$ is the row matrix:

$$\mathbf{x}^T = [x_1 \; x_2 \; \cdot \cdot \cdot \; x_n]$$

*Matrix algebra.* Closely associated with the above concepts, dealing with a single matrix, is a matrix algebra that governs the manipulation of two matrices. These procedures are summarized below.

*Equality.* A matrix $\mathbf{A}$ is equal to a matrix $\mathbf{B}$ if and only if their corresponding elements are equal. In other words, $\mathbf{A} = \mathbf{B}$ if and only if $a_{ij} = b_{ij}$ for all $i$ and $j$. Two matrices may be equated only if they are of the same order.

*Addition.* The sum of two matrices $\mathbf{A}$ and $\mathbf{B}$ is a matrix $\mathbf{C}$ whose elements are the sum of the respective elements of $\mathbf{A}$ and $\mathbf{B}$. Therefore, $\mathbf{A} + \mathbf{B} = \mathbf{C}$ if and only if $c_{ij} = a_{ij} + b_{ij}$ for all $i$ and $j$. Addition, like equality, is defined only for two matrices of the same order. As an example, the sum of the two matrices

$$\mathbf{A} = \begin{bmatrix} 1 & 1 \\ 0 & 2 \\ 1 & -1 \end{bmatrix} \quad \text{and} \quad \mathbf{B} = \begin{bmatrix} 0 & 3 \\ 1 & -2 \\ 5 & 2 \end{bmatrix}$$

is the matrix

$$\mathbf{C} = \begin{bmatrix} 1 & 4 \\ 1 & 0 \\ 6 & 1 \end{bmatrix}$$

Matrix addition is commutative, $\mathbf{A} + \mathbf{B} = \mathbf{B} + \mathbf{A}$, and associative, $\mathbf{A} + (\mathbf{B} + \mathbf{C}) = (\mathbf{A} + \mathbf{B}) + \mathbf{C}$.

*Multiplication by a scalar.* The multiplication of a matrix $\mathbf{A}$ by a scalar $h$ is accomplished by multiplying each element of $\mathbf{A}$ by $h$. Therefore, if $\mathbf{C} = h\mathbf{A}$, then $c_{ij} = ha_{ij}$ for all $i$ and $j$. For example, consider the following result:

$$h \begin{bmatrix} 1 & 0 & 1 \\ 2 & 5 & 1 \end{bmatrix} = \begin{bmatrix} h & 0 & h \\ 2h & 5h & h \end{bmatrix}$$

The operation of scalar multiplication distributes over vector addition so that $h(\mathbf{A} + \mathbf{B}) = h\mathbf{A} + h\mathbf{B}$.

Integration and differentiation as well as any other scalar operator, such as the Laplace transform, are also treated on this same element-by-element basis. For example,

$$\frac{d}{dt} \begin{bmatrix} t^2 & 1 \\ \sin t & t \end{bmatrix} = \begin{bmatrix} 2t & 0 \\ \cos t & 1 \end{bmatrix}$$

*Multiplication.* The matrix product **AB** is defined only if the *number of columns of the matrix* **A** *is equal to the number of rows of the matrix* **B**. If this requirement is satisfied, the matrices are said to be *conformal*. If the matrices **A** and **B** are conformal, the product **AB** of the $m \times r$ matrix **A** with the $r \times n$ matrix **B** is an $m \times n$ matrix **C** whose elements are defined by

$$c_{ij} = \sum_{k=1}^{r} a_{ik}b_{kj} \qquad (2.3\text{-}3)$$

Since this summation is over the columns of **A** and the rows of **B**, it is obvious why the matrices must be conformal. Note that the number of rows of **C** equals the number of rows of **A**, and the number of columns of **C** equals the number of columns of **B**. As an assistance in remembering this result, the following procedure may be used:

cancel

$$[m \times r] \text{ times } [r \times n] \text{ equals } [m \times n]$$

As an example, consider the matrix product

$$\begin{bmatrix} 1 & -1 & 2 \\ 3 & 0 & 1 \end{bmatrix} \begin{bmatrix} 1 & 0 \\ 1 & 2 \\ -1 & 3 \end{bmatrix} = \begin{bmatrix} -2 & 4 \\ 2 & 3 \end{bmatrix}$$

Here, for example, the $c_{21}$ term of the resulting matrix is determined as

$$c_{21} = a_{21}b_{11} + a_{22}b_{21} + a_{23}b_{31} = (3)(1) + (0)(1) + (1)(-1) = 2$$

From this example, we see that the element $c_{ij}$ is formed as the sum of the products of the corresponding elements of the $i$th row of **A** and the $j$th column of **B**. This computation is facilitated by visualizing the $j$th column of **B** as being removed from **B** and placed above the $i$th row of **A**. The corresponding elements of **A** and **B** are then multiplied vertically and summed horizontally. For example, let us use this procedure to find once again the $c_{21}$ element in the above illustration.

$$\begin{bmatrix} 1 & -1 & 2 \\ \boxed{1 \quad 1 \quad -1} \\ \times + \times + \times \\ 3 & 0 & 1 \end{bmatrix} \begin{bmatrix} 1 & 0 \\ 1 & 2 \\ -1 & 3 \end{bmatrix} = \begin{bmatrix} -2 & 4 \\ 2 & 3 \end{bmatrix}$$

With a small amount of practice, matrix multiplication may be accomplished quickly and accurately by using this method.

As a consequence of the requirement of conformality, the following results are observed:

1.  (Square matrix) $\times$ (square matrix) = (square matrix)
2.  (Row matrix) $\times$ (square matrix) = (row matrix)
3.  (Square matrix) $\times$ (column matrix) = (column matrix)
4.  (Row matrix) $\times$ (column matrix) = (scalar)
5.  (Column matrix) $\times$ (row matrix) = (square matrix)

In addition, the following properties of matrix multiplication are noted:

1.  Matrix multiplication is not necessarily commutative; that is, **AB** is not necessarily equal to **BA**.   Consider, for example, the matrices

    $$\mathbf{A} = \begin{bmatrix} 1 & 1 \\ 0 & 3 \end{bmatrix} \quad \text{and} \quad \mathbf{B} = \begin{bmatrix} 1 & 2 \\ 3 & 0 \end{bmatrix}$$

    Here the product **AB** is

    $$\mathbf{AB} = \begin{bmatrix} 4 & 2 \\ 9 & 0 \end{bmatrix}$$

    which is obviously not equal to the product **BA**:

    $$\mathbf{BA} = \begin{bmatrix} 1 & 7 \\ 3 & 3 \end{bmatrix}$$

    The product **BA** may not even be defined if **B** and **A** are not conformal.
    Because matrix multiplication is not commutative, one must be careful to preserve the order of the matrices when making any manipulations.   The term pre- or postmultiplication indicates whether the matrix is multiplied from the right or the left.   For example, in the triple matrix product **ABC**, **A** premultiplies **B** and **C** postmultiplies **B**.
2.  Matrix multiplication is associative.   In other words,

    $$\mathbf{A(BC)} = \mathbf{(AB)C} \tag{2.3-4}$$

3.  Matrix multiplication distributes with respect to matrix addition; that is,

    $$\mathbf{A(B + C)} = \mathbf{AB + AC} \tag{2.3-5}$$

4. If the matrix product $\mathbf{AB} = 0$, it cannot be concluded that either $\mathbf{A}$ or $\mathbf{B}$ is necessarily zero. As an example, consider the matrices

$$\mathbf{A} = \begin{bmatrix} 0 & 0 \\ 2 & 4 \end{bmatrix} \quad \text{and} \quad \mathbf{B} = \begin{bmatrix} 2 & 0 \\ -1 & 0 \end{bmatrix}$$

Here both $\mathbf{AB} = 0$ and $\mathbf{BA} = 0$ even though neither $\mathbf{A}$ nor $\mathbf{B}$ is zero. This is simply one indication that one must be careful in translating scalar concepts into the realm of matrices.

5. If $\mathbf{AB} = \mathbf{AC}$, it is not necessarily true that $\mathbf{B}$ and $\mathbf{C}$ are equal. This is a consequence of the above property.

*Inversion.* Division is not defined for matrices; it is replaced by an operation known as *matrix inversion* for *square* matrices. The square matrix $\mathbf{A}^{-1}$ is defined as the inverse of the square matrix $\mathbf{A}$ if

$$\mathbf{A}^{-1}\mathbf{A} = \mathbf{A}\mathbf{A}^{-1} = \mathbf{I} \tag{2.3-6}$$

where $\mathbf{I}$ is the identity matrix. Not all square matrices have an inverse, since in order to have an inverse the determinant[1] of the matrix must be nonzero. If the determinant is nonzero, the matrix possesses an inverse and is said to be *nonsingular*. If, on the other hand, the determinant is zero, the matrix has no inverse and is *singular*.

For nonsingular matrices, the inverse may be determined by the use of the adjoint matrix. The adjoint matrix, adj ($\mathbf{A}$), is related to the matrix $\mathbf{A}$ by means of the cofactors of the elements of $\mathbf{A}$. The *cofactor*, cof ($a_{ij}$), of the element $a_{ij}$ is $(-1)^{i+j}$ times the determinant of the matrix formed by deleting the $i$th row and the $j$th column of $\mathbf{A}$. The cofactors are often referred to as *signed* minors since they differ from the minor of the elements only by the $(-1)^{i+j}$ factor. The adjoint of the matrix $\mathbf{A}$ is then defined as the transpose of the cofactor matrix, that is, the matrix whose elements are the cofactors of $\mathbf{A}$. Thus the adjoint of $\mathbf{A}$ is

$$\text{adj }(\mathbf{A}) = [\text{cof }(a_{ij})]^T = \begin{bmatrix} \text{cof }(a_{11}) & \text{cof }(a_{21}) & \cdots & \text{cof }(a_{n1}) \\ \text{cof }(a_{12}) & \text{cof }(a_{22}) & \cdots & \text{cof }(a_{n2}) \\ \cdots\cdots\cdots\cdots\cdots\cdots\cdots\cdots\cdots \\ \text{cof }(a_{1n}) & \text{cof }(a_{2n}) & \cdots & \text{cof }(a_{nn}) \end{bmatrix} \tag{2.3-7}$$

In terms of the adjoint matrix, the inverse of $\mathbf{A}$ is given by

$$\mathbf{A}^{-1} = \frac{\text{adj }(\mathbf{A})}{\det (\mathbf{A})} \tag{2.3-8}$$

[1] A knowledge of determinants is assumed. See F. E. Hohn, "Elementary Matrix Algebra," The Macmillan Company, New York, 1958.

Here we see why the determinant of **A** must be nonzero, since otherwise the above expression would involve division by zero.

***Example 2.3-1***   To illustrate the above procedure, let us determine the inverse of the matrix

$$\mathbf{A} = \begin{bmatrix} 1 & 0 & 2 \\ 0 & 3 & 5 \\ 0 & 2 & 4 \end{bmatrix}$$

In order to check if the matrix is nonsingular, the determinant may be formed by expanding in terms of the minors of the first column.

$$\det(\mathbf{A}) = 1(12 - 10) + 0 + 0 = 2 \neq 0$$

and **A** is nonsingular and therefore possesses an inverse.

The next step is to determine the cofactor matrix which is given by

$$[\text{cof } (a_{ij})] = \begin{bmatrix} 2 & 0 & 0 \\ 4 & 4 & -2 \\ -6 & -5 & 3 \end{bmatrix}$$

Here, for example, the cofactor of $a_{23}$ is formed by eliminating the second row and the third column from the given matrix and multiplying the determinant of the resulting $2 \times 2$ matrix by $(-1)^{2+3}$. Thus cof $(a_{23})$ is formed as

$$[\text{cof } (a_{23})] = (-1)^{2+3} \det \begin{bmatrix} 1 & 0 & 2 \\ 0 & 3 & 5 \\ 0 & 2 & 4 \end{bmatrix} = (-1)^{2+3} \det \begin{bmatrix} 1 & 0 \\ 0 & 2 \end{bmatrix} = -2$$

The other cofactors are found by a similar process.

The adjoint matrix is easily found by transposing the cofactor matrix,

$$\text{adj } (\mathbf{A}) = [\text{cof } (a_{ij})]^T = \begin{bmatrix} 2 & 4 & -6 \\ 0 & 4 & -5 \\ 0 & -2 & 3 \end{bmatrix}$$

and the inverse of **A** is computed by means of Eq. (2.3-8):

$$\mathbf{A}^{-1} = \frac{\text{adj } (\mathbf{A})}{\det (\mathbf{A})} = \begin{bmatrix} 1 & 2 & -3 \\ 0 & 2 & -2.5 \\ 0 & -1 & 1.5 \end{bmatrix}$$

It is suggested that the reader verify that $\mathbf{A}^{-1}\mathbf{A} = \mathbf{A}\mathbf{A}^{-1} = \mathbf{I}$.

Before concluding this brief introduction to matrices, we give the two following identities for future reference:

1.  The inverse of the product of two *nonsingular* matrices is the reverse product of the inverse matrices; that is,

$$(\mathbf{AB})^{-1} = \mathbf{B}^{-1}\mathbf{A}^{-1} \qquad\qquad (2.3\text{-}9)$$

2.  The transpose of the product of two matrices is the reverse product of the transpose matrices; that is,

$$(\mathbf{AB})^T = \mathbf{B}^T\mathbf{A}^T \qquad\qquad (2.3\text{-}10)$$

With this matrix material as a background, let us turn our attention once again to the primary problem of plant representation.

**Exercises 2.3**   *2.3-1.*   Perform the following matrix multiplication:

(a) $\begin{bmatrix} 1 & 0 & 3 \\ 5 & 2 & 1 \end{bmatrix} \begin{bmatrix} 1 & 0 \\ 3 & 2 \\ 0 & 1 \end{bmatrix}$ 

(b) $\begin{bmatrix} a_{11} & a_{12} \\ a_{21} & a_{22} \end{bmatrix} \begin{bmatrix} x_1 \\ x_2 \end{bmatrix}$

(c) $[x_1 \ \ x_2] \begin{bmatrix} q_{11} & q_{12} \\ q_{21} & q_{22} \end{bmatrix} \begin{bmatrix} x_1 \\ x_2 \end{bmatrix}$

(d) $\begin{bmatrix} \alpha \\ \beta \end{bmatrix} [\alpha \ \ \beta]$

*answers:*

(a) $\begin{bmatrix} 1 & 3 \\ 11 & 5 \end{bmatrix}$

(b) $\begin{bmatrix} a_{11}x_1 + a_{12}x_2 \\ a_{21}x_1 + a_{22}x_2 \end{bmatrix}$

(c) $q_{11}x_1{}^2 + q_{12}x_1x_2 + q_{21}x_1x_2 + q_{22}x_2{}^2$

(d) $\begin{bmatrix} \alpha^2 & \alpha\beta \\ \alpha\beta & \beta^2 \end{bmatrix}$

*2.3-2.*   Determine the inverses of the following matrices:

(a) $\begin{bmatrix} 1 & 2 \\ 5 & 3 \end{bmatrix}$

(b) $\begin{bmatrix} 0 & 1 & 1 \\ 2 & 2 & 5 \\ 1 & 0 & 3 \end{bmatrix}$

(c) $\begin{bmatrix} s & -1 \\ s+2 & s+3 \end{bmatrix}$

*answers:*

(a) $\dfrac{-1}{7} \begin{bmatrix} 3 & -2 \\ -5 & 1 \end{bmatrix}$

(b) $\dfrac{-1}{3} \begin{bmatrix} 6 & -3 & 3 \\ -1 & -1 & 2 \\ -2 & 1 & -2 \end{bmatrix}$

(c) $\dfrac{1}{s^2 + 4s + 2} \begin{bmatrix} s+3 & 1 \\ -s-2 & s \end{bmatrix}$

*2.3-3.*  Show that the inverse of the general second-order matrix

$$\mathbf{A} = \begin{bmatrix} a_{11} & a_{12} \\ a_{21} & a_{22} \end{bmatrix}$$

is

$$\mathbf{A}^{-1} = \frac{1}{\det (\mathbf{A})} \begin{bmatrix} a_{22} & -a_{12} \\ -a_{21} & a_{11} \end{bmatrix} = \frac{1}{a_{11}a_{22} - a_{12}a_{21}} \begin{bmatrix} a_{22} & -a_{12} \\ -a_{21} & a_{11} \end{bmatrix}$$

*2.3-4.*  Verify Eqs. (2.3-9) and (2.3-10) for the matrices

$$\mathbf{A} = \begin{bmatrix} 1 & 0 \\ 2 & 3 \end{bmatrix} \qquad \mathbf{B} = \begin{bmatrix} 1 & 4 \\ 2 & 1 \end{bmatrix}$$

*answers:*

$$(\mathbf{AB})^{-1} = \mathbf{B}^{-1}\mathbf{A}^{-1} = \frac{-1}{21} \begin{bmatrix} 11 & -4 \\ -8 & 1 \end{bmatrix}$$

$$(\mathbf{AB})^T = \mathbf{B}^T\mathbf{A}^T = \begin{bmatrix} 1 & 8 \\ 4 & 11 \end{bmatrix}$$

## 2.4  State-variable representation

As mentioned at the end of Sec. 2.2, the state-variable method of plant representation is concerned not only with the input-output properties of the plant but also with its *complete internal behavior*.  This is the feature that distinguishes the state-variable representation from the input-output representation.

*The concept of state.*  Fundamental to the use of the state-variable means of representation is the concept of state.  It is important to stress at the outset that the concept of state is, first of all, a physical concept.  However, as discussed in Sec. 2.1, it is often convenient to describe the behavior of a physical object by means of a mathematical model. It is in terms of this mathematical model that we discuss the concept of state.

In particular, let us begin our discussion again with the $n$th-order describing differential equation (2.2-1).  This equation is repeated here for reference.

$$\frac{d^n y(t)}{dt^n} + a_n \frac{d^{n-1} y(t)}{dt^{n-1}} + \cdots + a_1 y(t)$$

$$= c_{m+1} \frac{d^m u(t)}{dt^m} + \cdots + c_1 u(t) \quad (2.4\text{-}1)$$

It is well known that to solve such an $n$th-order differential equation one must know, in addition to the input for $t \geq t_o$, a set of $n$ initial conditions.[1]   These initial conditions are usually chosen, because of the method of solution, as the value of the output $y$ and its $n - 1$ derivatives at the time $t_o$, that is, $y(t_o)$, $\dot{y}(t_o)$, $\ldots$ , $d^{n-1}y(t_o)/dt^{n-1}$.   However, this is not the only set of initial conditions that may be used.   The values at $t_o$ of any set of $n$ linearly independent variables of the plant, $x_1(t_o)$, $x_2(t_o)$, $\ldots$ , $x_n(t_o)$, are also sufficient.

A set of $n$ variables $x_1(t)$, $\ldots$ , $x_n(t)$ is said to be *linearly independent* if there is no set of constants, $\alpha_1$, $\alpha_2$, $\ldots$ , $\alpha_n$, other than all zero that satisfy the following equation for all $t \geq t_o$:

$$\alpha_1 x_1(t) + \alpha_2 x_2(t) + \cdots + \alpha_n x_n(t) = 0 \tag{2.4-2}$$

Otherwise the set is said to be *linearly dependent*.

The values at $t_o$ of any set of $n$ linearly independent plant variables are known as the state of the plant at $t_o$.   In more precise mathematical language, this definition takes the following form.   The *state at $t_o$* of an $n$th-order plant is described by a set of $n$ numbers, $x_1(t_o)$, $x_2(t_o)$, $\ldots$ , $x_n(t_o)$, which, along with the input to the plant for $t \geq t_o$, is sufficient to determine the behavior of the plant for all $t \geq t_o$.

In other words, the state of the plant represents the minimum amount of information about the plant at $t_o$ that is necessary so that its future behavior can be determined without reference to the input before $t_o$.   At the same time, the state of the plant at $t_o$ represents a complete description of the plant in the sense that no other information except the input is needed to determine its response.   In addition, any other plant variables may be determined from a knowledge of the state.

Consider, for example, games of checkers or chess; although these are discrete processes and hence not like the plants we are discussing, they illustrate the concept of state.   Here the state at the end of each move is simply the position of the pieces on the board.   If the game is interrupted, one could resume its play with this knowledge; it would not be necessary to know how the game had been played up to that point.

The variables $x_1(t)$, $x_2(t)$, $\ldots$ , $x_n(t)$ are known as state variables. In general, a *state variable* is any one of a set of $n$ variables, the knowledge of which is sufficient to describe completely the behavior of the plant, that is, a set of $n$ variables that determine the state of the plant at each instant of time.   In a third-order plant, for example, the variables

---

[1] A detailed discussion of the problem of solving such an $n$th-order differential equation will be presented in Chap. 4.

$y(t)$, $\dot{y}(t)$, and $\ddot{y}(t)$ form a set of state variables. On the other hand, $y(t)$, $\dot{y}(t)$, and $y(t) + \dot{y}(t)$ do not form a set of state variables since the last expression is not linearly independent of the first two and hence provides no new information.

From the preceding discussion it may appear that the state of a plant at any time is not unique. This, however, is not true; only the means of representing the state information is not unique. Any set of state variables that we may choose provides exactly the same information about the plant. This situation is similar to the process of specifying the size of a physical object in terms of different sets of units or in a different coordinate system. Assuredly the set of units used or the coordinate system does not change the physical object, although it may alter the description of the object.

**Example 2.4-1** As an illustration of the state-variable concepts discussed above, let us consider the problem of selecting a set of state variables for the field-controlled motor discussed in Examples 2.2-2 and 2.2-3. This plant is described by a third-order differential equation and the block diagrams of Fig. 2.2-10, both of which are repeated here for reference as Fig. 2.4-1 and

$$JL_f\dddot{\theta}_o(t) + (L_f\beta + R_fJ)\ddot{\theta}_o(t) + \beta R_f\dot{\theta}_o(t) = K_Te(t)$$

One possible set of state variables for this plant is $\theta_o(t)$, $\dot{\theta}_o(t)$, and $\ddot{\theta}_o(t)$. This set of state variables is mathematically sufficient to describe the plant completely, since these three variables are linearly independent. In addition, this set of state variables has physical

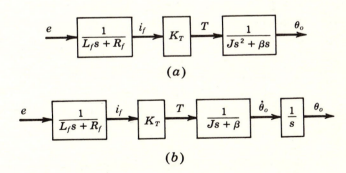

**Fig. 2.4-1** Block diagram of the field-controlled motor. (*a*) Connected subsystems; (*b*) alternative representation.

meaning since $\theta_o(t)$ is the output angular position, $\dot{\theta}_o(t)$ is the angular velocity, and $\ddot{\theta}_o(t)$ is the angular acceleration.

Let us suppose that for some reason the above set of state variables is not acceptable and therefore another set must be selected. An examination of Fig. 2.4-1$a$ reveals three plant variables: $i_f(t)$, $T(t)$, and $\theta_o(t)$. Perhaps these variables can serve as a set of state variables. This set of variables is not acceptable, however, since they are not linearly independent. This last statement is easily seen to be true since

$$T(t) = K_T i_f(t)$$

and therefore Eq. (2.4-2) can be satisfied with $\alpha_1 = 0$, $\alpha_2 = 1$, and $\alpha_3 = -K_T$, so that

$$(0)\theta_o(t) + (1)T(t) - (K_T)i_f(t) = 0$$

It is obvious that both $T(t)$ and $i_f(t)$ are not needed to describe the system since a knowledge of one completely determines the other. Therefore we must continue our search for an alternative set of state variables.

Since the block at the extreme right in Fig. 2.4-1$a$, which relates $T(t)$ and $\theta_o(t)$, is second order, perhaps we should consider the variables $\theta_o(t)$ and $\dot{\theta}_o(t)$. In order to do this, let us redraw Fig. 2.4-1$a$ as in Fig. 2.4-1$b$. The set of state variables $\theta_o(t)$, $\dot{\theta}_o(t)$, and $i_f(t)$ are now easily identified. This set of variables is linearly independent and therefore may be used as a set of state variables. Once again this set of state variables is also physically meaningful. This specification of the state involves a statement of the position, velocity, and field current. Other possible sets of state variables include the following:

1. $\theta_o(t)$, $\dot{\theta}_o(t)$, and $T(t)$
2. $\theta_o(t)$, $\dot{\theta}_o(t) + \alpha\dot{\theta}_o(t)$, and $T(t)$ or $i_f(t)$
3. $\theta_o(t)$, $\dot{\theta}_o(t) + \beta i_f(t)$, $T(t)$
   where $\alpha$ and $\beta$ are arbitrary constants and generate an infinite list of possible state variables.

*State-variable representation.* Plant representation in terms of state variables is accomplished by means of a set of $n$ first-order linear differential equations known as the *plant equations.* These equations take the general form

$$\dot{x}_1(t) = a_{11}x_1(t) + a_{12}x_2(t) + \cdots + a_{1n}x_n(t) + b_1u(t)$$
$$\dot{x}_2(t) = a_{21}x_1(t) + a_{22}x_2(t) + \cdots + a_{2n}x_n(t) + b_2u(t)$$
$$\cdots\cdots\cdots\cdots\cdots\cdots\cdots\cdots\cdots\cdots\cdots\cdots \quad (2.4\text{-}3)$$
$$\dot{x}_n(t) = a_{n1}x_1(t) + a_{n2}x_2(t) + \cdots + a_{nn}x_n(t) + b_nu(t)$$

and the $n$ variables $x_i(t)$ are the state variables. In addition, one needs an *output expression*, which is a linear algebraic equation relating the output of the plant to the state variables. This equation takes the form

$$y(t) = c_1x_1(t) + c_2x_2(t) + \cdots + c_nx_n(t) \qquad (2.4\text{-}4)$$

Use of the matrix concepts introduced in the preceding section permits the plant equations and the output expression to be written in a particularly simple and convenient form. For example, the plant equation may be written as

$$\dot{\mathbf{x}}(t) = \mathbf{A}\mathbf{x}(t) + \mathbf{b}u(t) \qquad \textbf{(Ab)}$$

Here $\mathbf{x}(t)$ is the $n$-dimensional vector known as the *state vector;* its elements are state variables so that

$$\mathbf{x}(t) = \text{col } (x_1(t), x_2(t), \ldots, x_n(t))$$

The vector $\dot{\mathbf{x}}(t)$ is the derivative of the state vector, and its elements are therefore the derivatives of the state variables, or

$$\dot{\mathbf{x}}(t) = \frac{d}{dt}\mathbf{x}(t) = \frac{d}{dt}\text{col } (x_1(t), x_2(t), \ldots, x_n(t))$$

$$= \text{col } (\dot{x}_1(t), \dot{x}_2(t), \ldots, \dot{x}_n(t))$$

The $n \times n$ matrix $\mathbf{A}$, which is defined as usual by

$$\mathbf{A} = \begin{bmatrix} a_{11} & a_{12} & \cdots & a_{1n} \\ a_{21} & a_{22} & \cdots & a_{2n} \\ \cdots & \cdots & \cdots & \cdots \\ a_{n1} & a_{n2} & \cdots & a_{nn} \end{bmatrix}$$

is known as the *plant matrix.* The $n$-dimensional vector

$$\mathbf{b} = \text{col } (b_1, b_2, \ldots, b_n)$$

is referred to as the *control vector* since it describes how the control input $u(t)$ affects the plant. If all these results are combined, the expanded form of Eq. (**Ab**) becomes

$$
\begin{bmatrix} \dot{x}_1(t) \\ \dot{x}_2(t) \\ \cdots \\ \dot{x}_n(t) \end{bmatrix} = \begin{bmatrix} a_{11} & a_{12} & \cdots & a_{1n} \\ a_{21} & a_{22} & \cdots & a_{2n} \\ \cdots & \cdots & \cdots & \cdots \\ a_{n1} & a_{n2} & \cdots & a_{nn} \end{bmatrix} \begin{bmatrix} x_1(t) \\ x_2(t) \\ \cdots \\ x_n(t) \end{bmatrix} + \begin{bmatrix} b_1 \\ b_2 \\ \cdots \\ b_n \end{bmatrix} u(t)
\tag{Ab}
$$

which, if the indicated matrix operations are completed, is equal to the initial plant equations (2.4-3).

In a similar fashion, the output expression may be written as

$$
y(t) = \mathbf{c}^T \mathbf{x}(t)
\tag{c}
$$

The *n*-dimensional vector $\mathbf{c} = \text{col}\,(c_1, c_2, \ldots, c_n)$ is known as the *output vector*. In expanded form this equation becomes

$$
y(t) = \begin{bmatrix} c_1 & c_2 & \cdots & c_n \end{bmatrix} \begin{bmatrix} x_1(t) \\ x_2(t) \\ \cdots \\ x_n(t) \end{bmatrix} = c_1 x_1(t) + c_2 x_2(t) + \cdots + c_n x_n(t)
\tag{c}
$$

Once again we see that this result is identical with the output expression of Eq. (2.4-4). Often we let $y = x_1$ so that all of the elements of **c** are zero except $c_1$, which is equal to 1.

Since Eqs. (**Ab**) and (**c**) play fundamental roles and therefore appear frequently in this book, they are designated by special and common symbols for the convenience of the reader. In addition, the equation symbols have been selected to assist the reader in remembering the form of the equation.

The use of the vector-matrix formulation therefore allows us to write the state-variable representation in terms of two simple equations:

$$
\dot{\mathbf{x}}(t) = \mathbf{A}\mathbf{x}(t) + \mathbf{b}u(t)
\tag{Ab}
$$

and

$$
y(t) = \mathbf{c}^T \mathbf{x}(t)
\tag{c}
$$

The simplicity of these equations as compared with the original form is obvious. Of course, these equations only symbolize Eqs. (2.4-3) and (2.4-4). Whenever an actual calculation is to be made for a specific plant, either the original equations (2.4-3) and (2.4-4) or the expanded form of the above matrix equations must be used, since only they contain the detailed information regarding the given plant. On the other hand, the compact symbolic representation is useful for general derivations and discussions.

The frequency-domain form of the state-variable representation of Eqs. (**Ab**) and (**c**) may be obtained by taking the Laplace transform of these equations. Since we will use the results to find a transfer function for the plant, let us assume that all initial conditions are zero.

$$s\mathbf{x}(s) = \mathbf{A}\mathbf{x}(s) + \mathbf{b}u(s) \qquad\qquad (2.4\text{-}5)$$

$$y(s) = \mathbf{c}^T\mathbf{x}(s) \qquad\qquad (2.4\text{-}6)$$

In terms of this frequency-domain form, we may draw a block diagram of the state-variable representation as shown in Fig. 2.4-2. This result may be compared with the above representation if both sides of Eq. (2.4-5) are divided by $s$ so that

$$\mathbf{x}(s) = \frac{1}{s}\,\mathbf{I}[\mathbf{A}\mathbf{x}(s) + \mathbf{b}u(s)]$$

The broad arrows in Fig. 2.4-2 are used to distinguish vectorial quantities, such as $\mathbf{x}$, from scalars such as $u$.

In addition to the general symbolic form of the block-diagram representation in Fig. 2.4-2, it is often useful to make a detailed block-diagram representation of the state-variable equations. Such a block diagram may be easily formed by beginning with $n$ $(1/s)$ blocks, each of whose input and output is labeled as $\dot{x}_i$ and $x_i$, $i = 1, 2, \ldots, n$, respectively. These blocks are then connected by making use of the state-variable equations to generate each of the $\dot{x}_i$'s. The output is formed from the output equation as the last step. A block diagram of this form, that is, one containing $n$ $(1/s)$ blocks, is referred to as the *elementary block diagram* of the state-variable representation.

As an illustration, Fig. 2.4-3 shows the elementary block diagram for the most general second-order plant. The state-variable representation of this plant is

$$\dot{\mathbf{x}} = \begin{bmatrix} a_{11} & a_{12} \\ a_{21} & a_{22} \end{bmatrix}\mathbf{x} + \begin{bmatrix} b_1 \\ b_2 \end{bmatrix}u \qquad \text{and} \qquad y = [c_1 \;\; c_2]\mathbf{x}$$

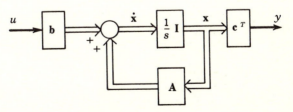

**Fig. 2.4-2**   General block diagram for the state-variable representation.

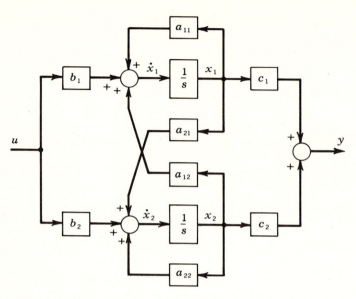

**Fig. 2.4-3**   Elementary block diagram of the general second-order plant.

In most practical problems, many of the elements of **A**, **b**, and **c** are zero so that many of the interconnections are not needed.

The exact form of the plant representation, that is, the elements of **A**, **b**, and **c**, depends on the set of state variables chosen for the plant. Once again, it must be emphasized that the plant does not change, only its representation.   There is, in fact, an infinite number of state-variable representations for a given plant, depending on which set of state variables has been selected for its representation.

> **Example 2.4-2**   To illustrate the state-variable means of representation presented above, let us consider once again the field-controlled motor.   Here the problem is to represent this plant by using two of the sets of state variables found in Example 2.4-1.
>
> In particular, let us begin with the set $\theta_o(t)$, $\dot{\theta}_o(t)$, and $\ddot{\theta}_o(t)$. Here we define
>
> $$x_1(t) = \theta_o(t)$$
> $$x_2(t) = \dot{\theta}_o(t)$$
> $$x_3(t) = \ddot{\theta}_o(t)$$
>
> and
>
> $$u(t) = e(t)$$

In terms of these definitions, the describing differential equation becomes[1]

$$JL_f \dot{x}_3(t) + (L_f \beta + R_f J)x_3(t) + \beta R_f x_2(t) = K_T u(t)$$

or

$$\dot{x}_3(t) = -\frac{\beta R_f}{JL_f} x_2(t) - \frac{L_f \beta + R_f J}{JL_f} x_3(t) + \frac{K_T}{JL_f} u(t)$$

The other two plant equations are provided by the definition of the $x_i$'s since

$$\dot{x}_1(t) = x_2(t) \qquad \text{and} \qquad \dot{x}_2(t) = x_3(t)$$

Since the output is $\theta_o(t)$, the output expression becomes

$$y(t) = \theta_o(t) = x_1(t)$$

Hence we have a complete state-variable representation in terms of the variables $\theta_o(t)$, $\dot{\theta}_o(t)$, and $\ddot{\theta}_o(t)$. In the matrix form, we have

$$\mathbf{A} = \begin{bmatrix} 0 & 1 & 0 \\ 0 & 0 & 1 \\ 0 & -\beta R_f/JL_f & -(L_f\beta + R_f J)/JL_f \end{bmatrix}$$

$$\mathbf{b} = \begin{bmatrix} 0 \\ 0 \\ K_T/JL_f \end{bmatrix} \qquad \mathbf{c} = \begin{bmatrix} 1 \\ 0 \\ 0 \end{bmatrix}$$

The elementary block diagram for this state-variable representation is shown in Fig. 2.4-4.

[1] See Example 2.2-2 or 2.4-1 for the describing differential equations.

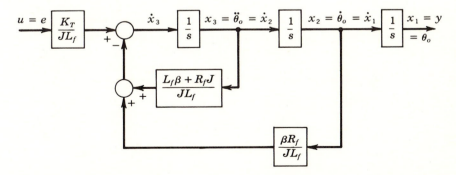

**Fig. 2.4-4**   Elementary block diagram for the state-variable representation of the field-controlled motor, using the variables $\theta_o(t)$, $\dot{\theta}_o(t)$, and $\ddot{\theta}_o(t)$.

Let us consider next the representation of this plant by means of the variables $\theta_o(t)$, $\dot{\theta}_o(t)$, and $i_f(t)$.   Here we define

$$x_1(t) = \theta_o(t)$$
$$x_2(t) = \dot{\theta}_o(t)$$
$$x_3(t) = i_f(t)$$
$$u(t) = e(t)$$

From these definitions, one plant equation is known, namely,

$$\dot{x}_1(t) = x_2(t)$$

In order to find the other two plant equations, we must consider the two original first- and second-order equations determined for this plant.   These are repeated here for reference as

$$L_f \frac{di_f(t)}{dt} + R_f i_f(t) = e(t)$$

and

$$J\ddot{\theta}_o(t) + \beta\dot{\theta}_o(t) = T(t)$$

Substituting the above definitions into these equations yields

$$L_f \dot{x}_3(t) + R_f x_3(t) = u(t)$$
$$J \dot{x}_2(t) + \beta x_2(t) = T(t)$$

Since $T(t) = K_T i_f(t)$, the last equation becomes

$$J\dot{x}_2(t) + \beta x_2(t) = K_T x_3(t)$$

so that the three plant equations are

$$\dot{x}_1(t) = x_2(t)$$
$$\dot{x}_2(t) = -\frac{\beta}{J} x_2(t) + \frac{K_T}{J} x_3(t)$$
$$\dot{x}_3(t) = -\frac{R_f}{L_f} x_3(t) + \frac{1}{L_f} u(t)$$

Once again the output expression is

$$y(t) = x_1(t)$$

The **A**, **b**, and **c** matrices have now become

$$\mathbf{A} = \begin{bmatrix} 0 & 1 & 0 \\ 0 & -\beta/J & K_T/J \\ 0 & 0 & -R_f/L_f \end{bmatrix} \qquad \mathbf{b} = \begin{bmatrix} 0 \\ 0 \\ 1/L_f \end{bmatrix} \qquad \mathbf{c} = \begin{bmatrix} 1 \\ 0 \\ 0 \end{bmatrix}$$

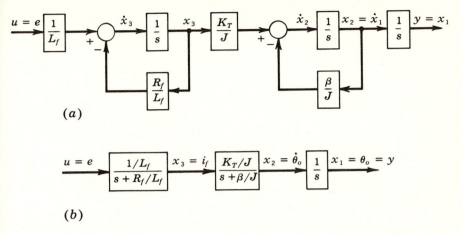

(a)

(b)

**Fig. 2.4-5**  Block diagrams of the field-controlled motor, using the variables $\theta_o(t)$, $\dot{\theta}_o(t)$, and $i_f$. (a) Elementary block diagram; (b) block diagram with some of the internal loops eliminated.

Note that only the **c** matrix has been unaffected by the change of variables, although this is not always the case.

The elementary block diagram of this state-variable representation is shown in Fig. 2.4-5a. In Fig. 2.4-5b, some of the internal loops have been removed to make the result more convenient. Note, however, that the state variables are still clearly labeled and have not been suppressed.

*Relation to transfer function.* The question that must naturally arise in the mind of the reader is: How are the state-variable representation and the input-output representation related? In particular, one would like to know how to find the transfer function of a plant, given its state-variable representation, and how to find its state-variable representation, given its transfer-function representation.

Since the transfer-function representation specifies only the input-output behavior, one can always make an arbitrary selection of state variables for a plant specified only by a transfer function. This means that, in general, an infinite number of state-variable representations exist for a given transfer function. The next two sections discuss two of the more common methods of choosing the state variables, beginning with a transfer-function representation.

On the other hand, if a state-variable representation of a plant is known, the transfer function of the plant is completely and uniquely specified. This fact is just one more manifestation that the state-variable representation is a more complete description. In order to determine the transfer function of the plant $y(s)/u(s)$ from the state-variable representation, we begin with the frequency-domain form of the state-variable representation as given by Eqs. (2.4-5) and (2.4-6). These equations are repeated here for reference:

$$sx(s) = Ax(s) + bu(s) \tag{2.4-5}$$
$$y(s) = c^T x(s) \tag{2.4-6}$$

It must be remembered that, when writing these equations, we assumed that all initial conditions were zero. This assumption is proper in this situation since we are searching for a transfer function.

Grouping the two $x(s)$ terms in Eq. (2.4-5), we have

$$(sI - A)x(s) = bu(s) \tag{2.4-7}$$

where the identity matrix has been introduced to maintain dimensionality and to allow the indicated factoring. If both sides of this equation are premultiplied by the matrix $(sI - A)^{-1}$, then Eq. (2.4-7) becomes

$$x(s) = (sI - A)^{-1}bu(s) \tag{2.4-8}$$

The matrix $(sI - A)^{-1}$ is referred to as the *resolvent matrix* and is designated by $\Phi(s)$.

$$\Phi(s) = (sI - A)^{-1} \tag{2.4-9}$$

In terms of this definition, Eq. (2.4-8) becomes

$$x(s) = \Phi(s)bu(s)$$

If this result is substituted into Eq. (2.4-6), $y(s)$ is given by

$$y(s) = c^T \Phi(s)bu(s)$$

so that the transfer function $y(s)/u(s)$ is

$$\frac{y(s)}{u(s)} = G_p(s) = c^T \Phi(s)b \tag{2.4-10}$$

Since the resolvent matrix is

$$\Phi(s) = (sI - A)^{-1} = \frac{\text{adj } (sI - A)}{\det (sI - A)}$$

$G_p(s)$ becomes

$$G_p(s) = \frac{c^T[\text{adj } (sI - A)]b}{\det (sI - A)} \tag{2.4-11}$$

Here the scalar quantity $c^T[\text{adj }(sI - A)]b$ is simply a polynomial in $s$ and forms the numerator polynomial of $G_p(s)$. The scalar quantity $\det(sI - A)$, which is also a polynomial in $s$, therefore forms the denominator polynomial of $G_p(s)$. In other words, if $G_p(s)$ is written as the ratio of a gain times a numerator polynomial $K_p N_p(s)$ to a denominator polynomial $D_p(s)$, that is,

$$G_p(s) = \frac{K_p N_p(s)}{D_p(s)} \tag{2.4-12}$$

then

$$K_p N_p(s) = c^T[\text{adj }(sI - A)]b \tag{2.4-13}$$

and

$$D_p(s) = \det(sI - A) \tag{2.4-14}$$

From the above result, we see that the values of $s$ that satisfy the equation

$$\det(sI - A) = 0 \tag{2.4-15}$$

also satisfy the equation

$$D_p(s) = 0$$

and therefore are poles of $G_p(s)$. In matrix terminology, these values of $s$ are known as *eigenvalues* of the matrix $A$, and Eq. (2.4-15) is referred to as the *characteristic equation* of the matrix $A$. Hence we see that the eigenvalues of $A$ correspond to the poles of $G_p(s)$.

***Example 2.4-3*** To illustrate the use of Eq. (2.4-10), let us consider the representation of the field-controlled motor in terms of the variables $\theta_o(t)$, $\dot{\theta}_o(t)$, and $i_f(t)$ found in Example 2.4-2. That representation is repeated here for reference.

$$\dot{x}(t) = \begin{bmatrix} 0 & 1 & 0 \\ 0 & -\beta/J & K_T/J \\ 0 & 0 & -R_f/L_f \end{bmatrix} x(t) + \begin{bmatrix} 0 \\ 0 \\ 1/L_f \end{bmatrix} u(t)$$

$$y(t) = [1 \quad 0 \quad 0]x(t)$$

For this problem, the matrix $(sI - A)$ becomes

$$(sI - A) = \begin{bmatrix} s & -1 & 0 \\ 0 & s+\beta/J & -K_T/J \\ 0 & 0 & s+R_f/L_f \end{bmatrix}$$

and its inverse is

$$\Phi(s) = (s\mathbf{I} - \mathbf{A})^{-1}$$
$$= \begin{bmatrix} 1/s & 1/s(s + \beta/J) & K_T/s(Js + \beta)(s + R_f/L_f) \\ 0 & 1/(s + \beta/J) & K_T/(Js + \beta)(s + R_f/L_f) \\ 0 & 0 & 1/(s + R_f/L_f) \end{bmatrix}$$

Therefore the transfer function is given by

$$G_p(s) =$$
$$[1 \quad 0 \quad 0] \begin{bmatrix} 1/s & 1/s(s + \beta/J) & K_T/s(Js + \beta)(s + R_f/L_f) \\ 0 & 1/(s + \beta/J) & K_T/(Js + \beta)(s + R_f/L_f) \\ 0 & 0 & 1/(s + R_f/L_f) \end{bmatrix} \begin{bmatrix} 0 \\ 0 \\ 1/L_f \end{bmatrix}$$
$$= \frac{K_T}{s(Js + \beta)(L_f s + R_f)} = \frac{K_T/JL_f}{s(s + \beta/J)(s + R_f/L_f)}$$

which is identical to the previous result of Example 2.2-2. The reader should carry out the above procedure for the other state-variable representation found in Example 2.4-2.

Although Eq. (2.4-10) provides a direct method for determining the transfer function of a plant from a state-variable representation of the plant, it is not an easy method, since it requires the inversion of the matrix $(s\mathbf{I} - \mathbf{A})$. The inversion of a matrix is never an easy task, and the job is even more difficult in this case since the elements are functions of $s$. Because of this problem, it is sometimes easier to obtain the transfer function by carrying out block-diagram reductions on the state-variable representation.[1] In fact, by using this approach, for certain types of state-variable representations the transfer function may be obtained by inspection.[2] For high-order systems, the use of a digital computer is probably the best approach.[3]

*Example 2.4-4* Consider, for example, the dc field-controlled-motor problem of Example 2.4-3. The block diagram of the state-variable representation is shown in Fig. 2.4-5b. A simple combination of the three series blocks provides the desired transfer function.

The problem of determining a state-variable representation of a plant whose transfer function is known is more complicated than the

[1] This subject will be discussed in more detail in Chap. 3.

[2] See, for example, the phase-variable representation of the next section.

[3] See J. L. Melsa, "Computer Programs for Computational Assistance in the Study of Linear Control Theory," McGraw-Hill Book Company, New York, 1970.

above problem, since there is an infinite number of state-variable representations for the same transfer function.    The next two sections discuss two of the more common means of selecting the state variables.

**Exercises 2.4**  *2.4-1.*  Represent the field-controlled motor by means of the variables

(a)  $\theta_o(t)$, $\dot{\theta}_o(t)$, and $T(t)$
(b)  $\theta_o(t)$, $\theta_o(t) + \dot{\theta}_o(t)$, and $i_f(t)$

In each case show that the input-output transfer function is unchanged.

*answers:*

(a)  $\dot{x} = \begin{bmatrix} 0 & 1 & 0 \\ 0 & -\beta/J & 1/J \\ 0 & 0 & -R_f/L_f \end{bmatrix} x + \begin{bmatrix} 0 \\ 0 \\ K_T/L_f \end{bmatrix} u \qquad y = x_1$

(b)  $\dot{x} = \begin{bmatrix} -1 & 1 & 0 \\ \beta/J - 1 & 1 - \beta/J & K_T/J \\ 0 & 0 & -R_f/L_f \end{bmatrix} x + \begin{bmatrix} 0 \\ 0 \\ 1/L_f \end{bmatrix} u \qquad y = x_1$

*2.4-2.*  Find the transfer function $y(s)/u(s)$ for the field-controlled motor by the use of the matrix equation (2.4-10) if the state variables $\theta_o$, $\dot{\theta}_o$, and $\ddot{\theta}_o$ are used.    (See Example 2.4-2.)

*answer:*

$$\frac{y(s)}{u(s)} = \frac{K_T/JL_f}{s(s + \beta/J)(s + R_f/L_f)}$$

*2.4-3.*  Determine the transfer function of the field-controlled motor by the use of block-diagram reductions on Fig. 2.4-4.    Label each step that you take with the appropriate portion of Fig. 2.2-8.

*2.4-4.*  Find the elementary block diagrams for each of the state-variable representations given below.

(a)  $\dot{x} = \begin{bmatrix} 0 & 1 & -1 \\ 0 & -2 & 1 \\ 0 & 2 & -3 \end{bmatrix} x + \begin{bmatrix} 0 \\ 2 \\ 1 \end{bmatrix} u \qquad y = [1 \quad 1 \quad 0]x$

(b)  $\dot{x} = \begin{bmatrix} 0 & 1 & 0 \\ 0 & -1 & 3 \\ -1 & 3 & -4 \end{bmatrix} x + \begin{bmatrix} 0 \\ 0 \\ 2 \end{bmatrix} u \qquad y = [2 \quad 1 \quad 0]x$

*2.4-5.*  Figure 2.4-6 shows a block-diagram representation of a set of state-variable equations for a third-order plant.    Find the ele-

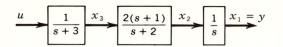

*Fig. 2.4-6* Exercise 2.4-5.

mentary block diagram for this same set of equations. HINT: It may be helpful to write the state-variable equations from Fig. 2.4-6.

## 2.5 Phase variables

Although there are an infinite number of means of selecting the state variables for any plant, and hence an infinite number of state-variable representations, only a limited number are in common use. These representations have either mathematical advantage or physical meaning. The method of phase variables presented in this section falls into the category of representations that possess mathematical advantage.

Phase variables are defined as the particular set of state variables that consists of one variable and its $n - 1$ derivatives. To introduce the method, let us consider the phase-variable representation of the $n$th-order plant whose block diagram is shown in Fig. 2.5-1. From the block diagram, the transfer function is seen to be

$$G_p(s) = \frac{y(s)}{u(s)} = \frac{1}{s^n + a_n s^{n-1} + \cdots + a_2 s + a_1}$$

This transfer function represents the following $n$th-order describing differential equation:

$$\frac{d^n y(t)}{dt^n} + a_n \frac{d^{n-1} y(t)}{dt^{n-1}} + \cdots + a_2 \frac{dy(t)}{dt} + a_1 y(t) = u(t) \qquad (2.5\text{-}1)$$

To express this equation in phase-variable form, let $x_1(t) = y(t)$, and then, according to the definition of phase variables, $x_2(t) = \dot{y}(t) = \dot{x}_1(t)$, $x_3(t) = \ddot{y}(t) = \dot{x}_2(t)$, . . . , $x_n(t) = d^{n-1}y(t)/dt^{n-1} = \dot{x}_{n-1}(t)$. If these definitions are substituted into Eq. (2.5-1), the result is

$$\dot{x}_n(t) + a_n x_n(t) + a_{n-1} x_{n-1}(t) + \cdots + a_2 x_2(t) + a_1 x_1(t) = u(t) \quad (2.5\text{-}2)$$

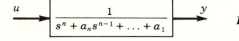

*Fig. 2.5-1* An $n$th-order plant with unity numerator.

In addition, we have the $n-1$ defining equations

$$\dot{x}_1(t) = x_2(t)$$
$$\dot{x}_2(t) = x_3(t)$$
$$\cdots \cdots \cdots$$
$$\dot{x}_{n-1}(t) = x_n(t)$$

and the output expression

$$y(t) = x_1(t)$$

These equations are of the state-variable form (**Ab**) and (**c**), where

$$
\mathbf{A} = \begin{bmatrix}
0 & 1 & 0 & \cdots & 0 & 0 \\
0 & 0 & 1 & \cdots & 0 & 0 \\
\cdots & \cdots & \cdots & \cdots & \cdots & \cdots \\
0 & 0 & 0 & \cdots & 0 & 1 \\
-a_1 & -a_2 & -a_3 & \cdots & -a_{n-1} & -a_n
\end{bmatrix}
\qquad
\mathbf{b} = \begin{bmatrix} 0 \\ 0 \\ \cdot \\ \cdot \\ \cdot \\ 0 \\ 1 \end{bmatrix}
\qquad
\mathbf{c} = \begin{bmatrix} 1 \\ 0 \\ \cdot \\ \cdot \\ \cdot \\ 0 \\ 0 \end{bmatrix}
$$

$$(2.5\text{-}3)$$

For example, the phase-variable representation for the third-order plant

$$G_p(s) = \frac{y(s)}{u(s)} = \frac{1}{s^3 + a_3 s^2 + a_2 s + a_1}$$

is

$$\dot{x}_1(t) = x_2(t) \qquad \dot{x}_2(t) = x_3(t)$$
$$\dot{x}_3(t) = -a_1 x_1(t) - a_2 x_2(t) - a_3 x_3(t) + u(t)$$

$$(2.5\text{-}4)$$

with

$$y(t) = x_1(t) \qquad\qquad\qquad\qquad\qquad\qquad\qquad (2.5\text{-}5)$$

The matrices **A**, **b**, and **c** are therefore

$$
\mathbf{A} = \begin{bmatrix}
0 & 1 & 0 \\
0 & 0 & 1 \\
-a_1 & -a_2 & -a_3
\end{bmatrix}
\qquad
\mathbf{b} = \begin{bmatrix} 0 \\ 0 \\ 1 \end{bmatrix}
\qquad
\mathbf{c} = \begin{bmatrix} 1 \\ 0 \\ 0 \end{bmatrix}
$$

Possible block diagrams for the phase-variable representation of both the $n$th-order plant and the above third-order example are shown in Fig. 2.5-2. A simple examination of these block diagrams, of the original transfer function or $n$th-order differential equation, and of the phase-variable representation reveals that any of these means of representation may be determined from the other *by inspection*. This is one of the

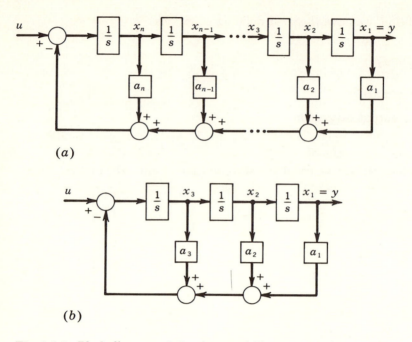

(a)

(b)

**Fig. 2.5-2**   Block diagrams of the phase-variable representation.   (a) An
nth-order plant; (b) a third-order plant.

important features of the phase-variable representation and is very valuable in situations where it is desired to transfer easily from one method of
representation to the other.

An alternative block diagram for the phase-variable representation
is shown in Fig. 2.5-3.   This block-diagram form is suggested by a knowl-

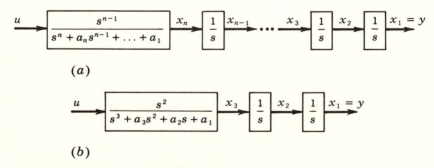

(a)

(b)

**Fig. 2.5-3**   Alternative block diagram of the phase-variable representation.   (a)
An nth-order plant; (b) a third-order plant.

edge of the transfer function $G_p(s)$. To some, the phase-variable representation is not as obvious in this alternative form. To others, this alternative representation in terms of series blocks alone is more appealing than the structure of Fig. 2.5-2, where inherent feedback within the plant might be implied. Inherent feedback is not intended in Fig. 2.5-2 nor is a series plant structure implied in Fig. 2.5-3, since phase variables are a mathematical convenience not meant to describe accurately the plant structure. The elementary block diagram form of Fig. 2.5-2 is adopted throughout the remainder of this book.

So far we have considered only the case where the transfer function $G_p(s)$ has a unity numerator. In order to include the general case, a slight modification must be made in the above approach. To see why this is necessary, let us consider the same third-order example except with a zero added, so that

$$G_p(s) = \frac{c_2 s + c_1}{s^3 + a_3 s^2 + a_2 s + a_1}$$

If we proceed as before by letting $x_1(t) = y(t)$, $x_2(t) = \dot{y}(t)$, and $x_3(t) = \ddot{y}(t)$, then Eq. (2.5-2) becomes

$$\dot{x}_3(t) + a_3 x_3(t) + a_2 x_2(t) + a_1 x_1(t) = c_2 \dot{u}(t) + c_1 u(t)$$

The phase-variable representation would now contain a $\dot{u}(t)$ term on the right-hand side, which is a violation of the assumed form of Eq. (**Ab**).

In order to avoid this problem, the transfer function is divided into two parts, as shown in Fig. 2.5-4:

$$G_p(s) = \frac{y(s)}{u(s)} = \frac{x_1(s)}{u(s)} \frac{y(s)}{x_1(s)}$$

where

$$\frac{x_1(s)}{u(s)} = \frac{1}{s^3 + a_3 s^2 + a_2 s + a_1}$$

and

$$\frac{y(s)}{x_1(s)} = c_2 s + c_1$$

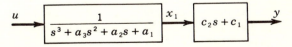

**Fig. 2.5-4**   Method of handling zeros.

The transfer function $x_1(s)/u(s)$ is identical to the original transfer function without the added zero, and therefore its phase-variable representation is still given by Eq. (2.5-4). From the second transfer function, however, we see that $y(t)$ is no longer equal to $x_1(t)$ but is now

$$y(t) = c_2\dot{x}_1(t) + c_1 x_1(t) = c_2 x_2(t) + c_1 x_1(t)$$

where the second expression has been written by making use of the fact that $\dot{x}_1(t) = x_2(t)$.

The complete phase-variable representation of this plant has the form

$$\dot{x}(t) = \begin{bmatrix} 0 & 1 & 0 \\ 0 & 0 & 1 \\ -a_1 & -a_2 & -a_3 \end{bmatrix} x(t) + \begin{bmatrix} 0 \\ 0 \\ 1 \end{bmatrix} u(t)$$
$$y(t) = \begin{bmatrix} c_1 & c_2 & 0 \end{bmatrix} x(t)$$

A comparison of this result with the representation of the original system as given by Eqs. (2.5-4) and (2.5-5) indicates that the only change that has been made is in the output expression. This is, in fact, always the case. Therefore the phase-variable representation for the general $n$th-order system with $m$ zeros, $m < n$,

$$G_p(s) = \frac{c_{m+1}s^m + c_m s^{m-1} + \cdots + c_2 s + c_1}{s^n + a_n s^{n-1} + \cdots + a_2 s + a_1} \tag{2.5-6}$$

is

$$\dot{x}(t) = \begin{bmatrix} 0 & 1 & 0 & \cdots & 0 & 0 \\ 0 & 0 & 1 & \cdots & 0 & 0 \\ \cdots & \cdots & \cdots & \cdots & \cdots & \cdots \\ 0 & 0 & 0 & \cdots & 0 & 1 \\ -a_1 & -a_2 & -a_3 & \cdots & -a_{n-1} & -a_n \end{bmatrix} x(t) + \begin{bmatrix} 0 \\ 0 \\ \cdot \\ \cdot \\ \cdot \\ 0 \\ 1 \end{bmatrix} u(t) \tag{2.5-7}$$

$$y(t) = \begin{bmatrix} c_1 & c_2 & \cdots & c_m & c_{m+1} & 0 & \cdots & 0 \end{bmatrix} x(t) \tag{2.5-8}$$

The block diagrams for the third-order example and the general $n$th-order case are shown in Fig. 2.5-5. Note that the phase-variable representation is still easily determined from the transfer function by inspection, and vice versa. Note also that the elements of $\mathbf{b}$ are all zero except for $b_n$, which is always equal to 1, and that the first $n - 1$ rows of $\mathbf{A}$ are of the same form.

Although the use of phase variables provides a simple means of representing a plant in state-variable form, the method has the disadvantage that the phase variables are not, in general, meaningful physical variables of the plant and therefore not available for measurement or

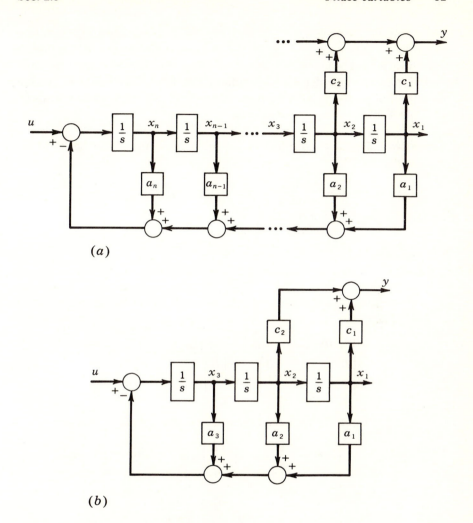

**Fig. 2.5-5**   General phase-variable representations.   (a) An $n$th-order plant; (b) a third-order plant with one zero.

manipulation. If the transfer function has a unity numerator, the phase variables are the output and its $n - 1$ derivatives. If $n$ is greater than 2 or 3, it is physically difficult to obtain these derivatives. If, on the other hand, $G_p(s)$ has one or more zeros, these phase variables bear little or no resemblance to real physical quantities in the plant.

Thus, although the phase-variable representation provides mathematical advantages, it is not a practical set of state variables from a con-

trol point of view, since we eventually wish to measure all the state
variables. This does not mean that phase variables are not a useful
and even valuable means of representation. On the contrary, we shall
make extensive use of them throughout this book. In the next section
a method of state-variable representation that is closely tied to measur-
able physical variables is considered.

**Exercises 2.5**  *2.5-1.*  Represent the following plants by means of phase
variables:

(a)  $G_p(s) = \dfrac{1}{s^2 + 2s + 3}$     (b)  $G_p(s) = \dfrac{s + 2}{s^3 + 3s^2 + 2s + 10}$

(c)  $G_p(s) = \dfrac{10(s + 1)(s + 2)}{s(s + 3)(s + 4)}$

*answers:*

(a)  $\dot{\mathbf{x}} = \begin{bmatrix} 0 & 1 \\ -3 & -2 \end{bmatrix} \mathbf{x} + \begin{bmatrix} 0 \\ 1 \end{bmatrix} u$     $y = [1 \quad 0]\mathbf{x}$

(b)  $\dot{\mathbf{x}} = \begin{bmatrix} 0 & 1 & 0 \\ 0 & 0 & 1 \\ -10 & -2 & -3 \end{bmatrix} \mathbf{x} + \begin{bmatrix} 0 \\ 0 \\ 1 \end{bmatrix} u$     $y = [2 \quad 1 \quad 0]\mathbf{x}$

(c)  $\dot{\mathbf{x}} = \begin{bmatrix} 0 & 1 & 0 \\ 0 & 0 & 1 \\ 0 & -12 & -7 \end{bmatrix} \mathbf{x} + \begin{bmatrix} 0 \\ 0 \\ 1 \end{bmatrix} u$     $y = [20 \quad 30 \quad 10]\mathbf{x}$

*2.5-2.*  Find the transfer functions for the following plants:

(a)  $\dot{\mathbf{x}}(t) = \begin{bmatrix} 0 & 1 \\ -4 & -5 \end{bmatrix} \mathbf{x}(t) + \begin{bmatrix} 0 \\ 1 \end{bmatrix} u(t)$     $y(t) = [5 \quad 0]\mathbf{x}(t)$

(b)  $\dot{\mathbf{x}}(t) = \begin{bmatrix} 0 & 1 & 0 \\ 0 & 0 & 1 \\ -6 & -3 & -1 \end{bmatrix} \mathbf{x}(t) + \begin{bmatrix} 0 \\ 0 \\ 1 \end{bmatrix} u(t)$

$y(t) = [1 \quad 2 \quad 0]\mathbf{x}(t)$

(c)  $\dot{\mathbf{x}}(t) = \begin{bmatrix} 0 & 1 & 0 \\ 0 & 0 & 1 \\ 0 & -1 & -3 \end{bmatrix} \mathbf{x}(t) + \begin{bmatrix} 0 \\ 0 \\ 1 \end{bmatrix} u(t)$

$y(t) = [5 \quad 10 \quad 0]\mathbf{x}(t)$

*answers:*

(a)  $G_p(s) = \dfrac{5}{s^2 + 5s + 4}$     (b)  $G_p(s) = \dfrac{2s + 1}{s^3 + s^2 + 3s + 6}$

(c)  $G_p(s) = \dfrac{10s + 5}{s^3 + 3s^2 + s}$

*2.5-3.* Find the phase-variable representation of the armature-controlled motor of Example 2.2-4.

*answer:*

$$\dot{x} = \begin{bmatrix} 0 & 1 & 0 \\ 0 & 0 & 1 \\ 0 & -(R_a\beta + K_vK_T')/JL_a & -(L_a\beta + R_aJ)/JL_a \end{bmatrix} x + \begin{bmatrix} 0 \\ 0 \\ 1 \end{bmatrix} u$$

$$y = [K_T'/JL_a \quad 0 \quad 0]x$$

## 2.6  Physical variables

The method of plant representation in terms of real physical variables relies heavily on an understanding of the physical character of the plant. It is an engineer's approach to the problem of state-variable representation, as opposed to the mathematician's approach of the preceding section. There no attempt was made to relate the state variables to real physical quantities. Here the method is almost intuitive, and the reader may feel that the approach is so straightforward that it should not be called a method.

Let us begin our discussion of physical variables by considering a third-order plant whose transfer function is

$$G_p(s) = \frac{c_1}{s^3 + a_3s^2 + a_2s + a_1}$$

This transfer function might represent either a field-controlled or an armature-controlled dc motor or, in fact, any other third-order plant with no zeros.

This is exactly the strength as well as the weakness of the transfer-function representation. All plants with the same input-output characteristics, that is, the same describing differential equations, are assigned the same transfer function. Therefore, if we are given only the transfer function, we may only speculate on the physical origins of the problem. Obviously a field-controlled motor is not the same physical device as an armature-controlled motor, yet the transfer function makes no distinction. Therefore, in order to represent a plant in terms of a set of meaningful physical variables, we must know more than just the transfer function; we must know the physical origin of the problem.

Since we have already used a set of physical variables, namely, $\theta_o(t)$, $\dot{\theta}_o(t)$, and $i_f(t)$, in Example 2.4-2 to represent the field-controlled motor, let us assume that the above transfer function represents an armature-controlled motor of the nature described in Example 2.2-4. There a set of four transformed equations were found to describe the plant.  These equations are presented here by means of their differential-equation counterparts.

$$L_a \frac{di_a(t)}{dt} + R_a i_a(t) = e(t) - e_c(t)$$

$$e_c(t) = K_v \dot{\theta}_o(t)$$

$$J\ddot{\theta}_o(t) + \beta \dot{\theta}_o(t) = T(t)$$

$$T(t) = K'_T i_a(t)$$

In addition the expanded block diagram of Fig. 2.2-13a is repeated here as Fig. 2.6-1.

In the above equations five variables are identified, $i_a(t)$, $\theta_o(t)$, $\dot{\theta}_o(t)$, $T(t)$, and $e_c(t)$, and at first glance it might appear that the plant should have five state variables.  However, we always assume that the plant is specified with the minimum number of state variables possible.  We see that both $T(t)$ and $i_a(t)$ may not be used, since they are not linearly independent.  Since $i_a(t)$ is perhaps the easier variable to measure, let us use it as one state variable.  The output $\theta_o(t)$ is a logical choice for a second variable.  Although we could choose $e_c(t)$ as the third state variable, we see that it is linearly dependent on $\dot{\theta}_o(t)$ and therefore $\dot{\theta}_o(t)$ could also be used.  Since the back emf does not exist in a physical sense, and since $\dot{\theta}_o(t)$ can be easily measured by means of a tachometer, we choose the following set of physical variables as the state variables for the system: $i_a(t)$, $\theta_o(t)$, and $\dot{\theta}_o(t)$.  Therefore, let

$$x_1(t) = i_a(t)$$
$$x_2(t) = \theta_o(t)$$
$$x_3(t) = \dot{\theta}_o(t)$$

and

$$u(t) = e(t)$$

If these definitions are substituted into the above equations for this

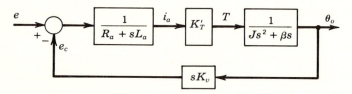

**Fig. 2.6-1**   Armature-controlled motor.

plant, we have

$$L_a\dot{x}_1(t) + R_a x_1(t) = u(t) - e_c(t)$$
$$e_c(t) = K_v x_3(t)$$
$$J\dot{x}_3(t) + \beta x_3(t) = T(t)$$
$$T(t) = K'_T x_1(t)$$

Combining the two algebraic equations with the differential equations, we obtain

$$L_a\dot{x}_1(t) + R_a x_1(t) = u(t) - K_v x_3(t)$$
$$J\dot{x}_3(t) + \beta x_3(t) = K'_T x_1(t)$$

or

$$\dot{x}_1(t) = -\frac{R_a}{L_a} x_1(t) - \frac{K_v}{L_a} x_3(t) + \frac{1}{L_a} u(t)$$
$$\dot{x}_3(t) = \frac{K'_T}{J} x_1(t) - \frac{\beta}{J} x_3(t)$$

In addition, we have the defining equation

$$\dot{x}_2(t) = x_3(t)$$

In the form of the matrix equation (**Ab**), this result becomes

$$\dot{\mathbf{x}}(t) = \begin{bmatrix} -R_a/L_a & 0 & -K_v/L_a \\ 0 & 0 & 1 \\ K'_T/J & 0 & -\beta/J \end{bmatrix} \mathbf{x}(t) + \begin{bmatrix} 1/L_a \\ 0 \\ 0 \end{bmatrix} u(t)$$

Since the output $y(t)$ is $\theta_o(t) = x_2(t)$, the output expression is

$$y(t) = x_2(t)$$

which in matrix form is

$$y(t) = [0 \quad 1 \quad 0]\mathbf{x}(t)$$

In order to make the $\theta_o(t) = x_3(t)$ variable more obvious, the block diagram of Fig. 2.6-1 has been redrawn as Fig. 2.6-2a. This block diagram has been further redrawn as Fig. 2.6-2b by joining the two places where $x_3(t)$ appears or, equivalently, by making use of the block-diagram identity of Fig. 2.2-8d. Note that the above physical-variable representation can be easily written from this block-diagram representation. Of course, Fig. 2.6-2b could be reduced to the overall block diagram of Fig. 2.2-13c. Obviously, physically meaningful state variables could not be chosen from that block diagram.

Note that in the resulting block diagram (Fig. 2.6-2) the output variable is labeled $x_2$ and the variables are not in any particular numerical

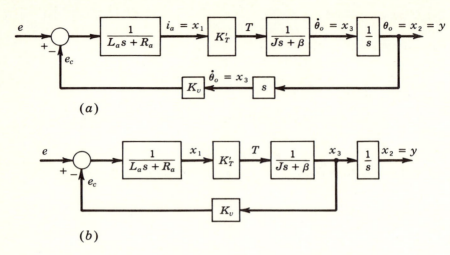

**Fig. 2.6-2**  Physical-variable representation of the armature-controlled dc motor. (a) Initial block diagram; (b) reduced block diagram.

order. Previously we have always designated the state variable that appears on the right in the block diagram as $x_1$, and the state variables appeared in numerical order. This ordering of the state variables is a natural consequence of the consideration of phase variables first. Here we wish to emphasize that no such ordering is necessary. The state variables may be called anything desired. In this example, for instance, had the output been designated as $x_1$, $\dot{\theta}_o$ as $x_2$, and $i_a$ as $x_3$, the state equations would have included the following matrices:

$$\mathbf{A^*} = \begin{bmatrix} 0 & 1 & 0 \\ 0 & -\beta/J & K_T'/J \\ 0 & -K_v/L_a & -R_a/L_a \end{bmatrix} \qquad \mathbf{b^*} = \begin{bmatrix} 0 \\ 0 \\ 1/L_a \end{bmatrix} \qquad \mathbf{c^*} = \begin{bmatrix} 1 \\ 0 \\ 0 \end{bmatrix}$$

Although $\mathbf{A^*}$, $\mathbf{b^*}$, and $\mathbf{c^*}$ are different from the matrices associated with the original choice of state variables, the transfer functions are identical; that is,

$$G_p(s) = \mathbf{c}^T(s\mathbf{I} - \mathbf{A})^{-1}\mathbf{b} = \mathbf{c}^{*T}(s\mathbf{I} - \mathbf{A^*})^{-1}\mathbf{b^*}$$

Thus far we have considered only cases in which the numerator of $G_p(s)$ has been a constant. The treatment when $G_p(s)$ has zeros calls for slight modifications, much as in the phase-variable case. To initiate the discussion, we consider the two third-order plants of Fig. 2.6-3a and b. (Here, although the transfer functions are the same, the two figures do not represent the same plant.) In Fig. 2.6-3a the zero does not appear in the left-hand-most block, and no difficulty arises in writing the state equations in the form (**Ab**). The state equations may be written directly from that block diagram as

$$\frac{x_1(s)}{x_2(s)} = \frac{1}{s} \quad \rightarrow \quad \dot{x}_1 = x_2$$

$$\frac{x_2(s)}{x_3(s)} = \frac{s+3}{s+1} \quad \rightarrow \quad \dot{x}_2 = -x_2 + \dot{x}_3 + 3x_3$$

$$\frac{x_3(s)}{u(s)} = \frac{1}{s+5} \quad \rightarrow \quad \dot{x}_3 = -5x_3 + u$$

After the $\dot{x}_3$ term is eliminated from the second equation, the result is

$$\dot{x}_1 = x_2$$
$$\dot{x}_2 = -x_2 - 2x_3 + u$$
$$\dot{x}_3 = -5x_3 + u$$

Note here that the vector **b** now has two nonzero elements since

$$\mathbf{b} = \begin{bmatrix} 0 \\ 1 \\ 1 \end{bmatrix}$$

The reader may recall that in phase variables the effect of zeros is to add additional nonzero terms to the output vector **c**. In physical variables, additional nonzero terms are normally added in **b**, whereas **c** remains unaffected. The **A** matrix is also different.

If the state equations are written for the plant of Fig. 2.6-3*b*, the resulting equations contain a $\dot{u}$ term, and hence they are not of the required form, (**Ab**). The $\dot{u}$ term arises from the block containing the zero, as may be seen by writing the differential equation corresponding to the transfer function $x_3'(s)/u(s)$. This equation is

$$\dot{x}_3' = -5x_3' + \dot{u} + 3u$$

This difficulty may be avoided by redrawing Fig. 2.6-3*b*, using the

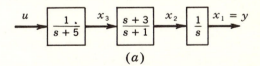

(a)

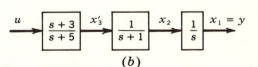

**Fig. 2.6-3** Two cases of a zero in third-order plants described in physical variables.

(b)

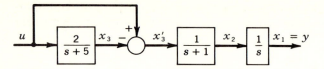

**Fig. 2.6-4**   Redrawing of Fig. 2.6-3b to eliminate the $u$ term in the describing equations.

feedforward form shown in Fig. 2.6-4.   This configuration is suggested by dividing the denominator into the numerator of $x_3'(s)/u(s)$, or

$$\frac{s+3}{s+5} = 1 - \frac{2}{s+5}$$

The consequences of this division are shown in Fig. 2.6-4, where it has become necessary to indicate a new state variable, $x_3$.   If the state equations are written in terms of $x_1$, $x_2$, and $x_3$, with $x_3'$ equal to $u - x_3$, the result is

$$\dot{x}_1 = x_2$$
$$\dot{x}_2 = -x_2 + x_3' = -x_2 - x_3 + u$$
$$\dot{x}_3 = -5x_3 + 2u$$

Here **b** again has an additional nonzero term, as **b** is now

$$\mathbf{b} = \begin{bmatrix} 0 \\ 1 \\ 2 \end{bmatrix}$$

It should be noted that, if $x_3'$ is a real physical variable, then, in all probability, $x_3$ is not.   [However, the configuration of Fig. 2.6-4 and unprimed variables are used hereafter to ensure that the describing equations are of the form (**Ab**) and (**c**).]   Answers determined in terms of $x_3$ therefore must be converted to $x_3'$ for physical interpretation.   This may be done by the inverse block-diagram manipulation required in going from Fig. 2.6-4 back to Fig. 2.6-3b, or it may be done by the use of the equation $x_3' = u - x_3$.

Before proceeding to the topic of linear changes of variables, let us make the following observations concerning phase variables vs. physical variables.   The phase-variable method provides a simple, direct, systematic, and unique means of translating the transfer-function information into state-variable form.   At the same time, phase variables in general lack physical significance, particularly if the plant transfer function has zeros.

In contrast, the physical-variable representation, because of its close relation to the physical plant, does not produce a unique form for the resulting state-variable representation; that is, the same transfer function may yield two or more different representations, depending on the physical plant involved. On the other hand, the state variables used in this approach are, by their very definition, real physically meaningful variables.

Although we shall make considerable use of both these approaches to state-variable representation throughout this book, we must never forget that our basic interest is in controlling real plants, and therefore we must ultimately deal with real physical variables.

**Exercises 2.6**    *2.6-1.*    Find the physical-variable representation of the plant shown in Fig. 2.6-5, using the variables $z_1(t)$, $\dot{z}_1(t)$, $z_2(t)$, $\dot{z}_2(t)$, with $z_2(t) = y(t)$ and $f(t) = u(t)$. Draw the elementary block diagram of the plant from this state-variable representation. Find the transfer function $y(s)/u(s)$.

*answers:*

If $x_1 = z_1$, $x_2 = \dot{z}_1$, $x_3 = z_2$, and $x_4 = \dot{z}_2$, then

$$\dot{x} = \begin{bmatrix} 0 & 1 & 0 & 0 \\ -K_1/m_1 & -\beta_1/m_1 & K_1/m_1 & 0 \\ 0 & 0 & 0 & 1 \\ K_1/m_2 & 0 & -(K_1+K_2)/m_2 & 0 \end{bmatrix} x + \begin{bmatrix} 0 \\ 1/m_1 \\ 0 \\ 0 \end{bmatrix} u$$

$$y = x_3$$

$$\frac{y(s)}{u(s)} = \frac{K_1}{(m_1 s^2 + \beta_1 s + K_1)(m_2 s^2 + K_1 + K_2) - K_1{}^2}$$

*2.6-2.*    Find the physical-variable representation for the plant given in Exercise 2.2-3, using the variables $i_{fm}$, $i_{ag}$, $\theta$, and $y$.

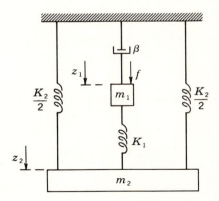

**Fig. 2.6-5**    Exercise 2.6-1.

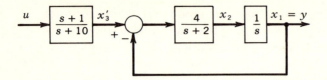

**Fig. 2.6-6**   Exercise 2.6-3.

*answer:*

$$\dot{\mathbf{x}} = \begin{bmatrix} 0 & 1/C & 0 & 0 \\ 0 & -1/RC - K_v^2/JR & -\beta/JR & K_vK_T/JR \\ 0 & -K_v^2/J & -\beta/J & K_vK_T/J \\ 0 & 0 & 0 & -R_f/L_f \end{bmatrix} \mathbf{x} + \begin{bmatrix} 0 \\ 0 \\ 0 \\ 1/L_f \end{bmatrix} u$$

$$y = x_1$$

*2.6-3.*   Find a physical-variable–state-variable representation of the system shown in Fig. 2.6-6 in the form (**Ab**) and (**c**) by defining an artificial state variable $x_3 = x_3' - u$.

*answer:*

$$\dot{x}_1 = x_2$$
$$\dot{x}_2 = -4x_1 - 2x_2 + 4x_3 + 4u$$
$$\dot{x}_3 = -10x_3 - 9u$$
$$y = x_1$$

*2.6-4.*   Show that the two block diagrams in Fig. 2.6-7 have the same input-output transfer function.

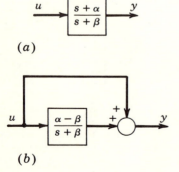

(*a*)

(*b*)                                        **Fig. 2.6-7**   Exercise 2.6-4.

## 2.7  *Linear transformation of variables*

In the three preceding sections we have discussed the fact that a given plant has an infinite number of state-variable representations. In this section, we examine more closely the relationship between these various state-variable representations by means of a technique known as a linear transformation.

To introduce this technique, let us consider once again the field-controlled motor. In Example 2.4-2 two different state-variable representations for this plant were found. The first was in terms of the variables $\theta_o$, $\dot{\theta}_o$, and $\ddot{\theta}_o$.

$$\dot{\mathbf{x}}(t) = \begin{bmatrix} 0 & 1 & 0 \\ 0 & 0 & 1 \\ 0 & -\beta R_f/JL_f & -(L_f\beta + R_fJ)/JL_f \end{bmatrix} \mathbf{x}(t) + \begin{bmatrix} 0 \\ 0 \\ K_T/L_fJ \end{bmatrix} u(t)$$

$$y(t) = \begin{bmatrix} 1 & 0 & 0 \end{bmatrix} \mathbf{x}(t)$$

where $x_1(t) = \theta_o(t)$, $x_2(t) = \dot{\theta}_o(t)$, $x_3(t) = \ddot{\theta}_o(t)$, and $u(t) = e(t)$.

The second representation was in terms of physical variables and took the form

$$\dot{\mathbf{x}}^*(t) = \begin{bmatrix} 0 & 1 & 0 \\ 0 & -\beta/J & K_T/J \\ 0 & 0 & -R_f/L_f \end{bmatrix} \mathbf{x}^*(t) + \begin{bmatrix} 0 \\ 0 \\ 1/L_f \end{bmatrix} u(t)$$

$$y(t) = \begin{bmatrix} 1 & 0 & 0 \end{bmatrix} \mathbf{x}^*(t)$$

where $x_1^*(t) = \theta_o(t)$, $x_2^*(t) = \dot{\theta}_o(t)$, $x_3^*(t) = i_f(t)$, and $u(t) = e(t)$. Here we have starred the physical-variable representation to distinguish it from the other state-variable representation.

In terms of these two representations, there are two questions that we investigate. First, how the two sets of state variables are related; and, second, how the two plant representations are related. Both questions have basically the same answer.

Let us begin with the first question. From the definitions of the two sets of state variables, we easily see that

$$x_1(t) = x_1^*(t) = \theta_o(t)$$
$$x_2(t) = x_2^*(t) = \dot{\theta}_o(t)$$

The relationship between the third variables in each set is not so obvious. To find this relationship, let us consider the second equation of the physical-variable representation

$$\dot{x}_2^*(t) = -\frac{\beta}{J} x_2^*(t) + \frac{K_T}{J} x_3^*(t)$$

Since $x_2^*(t) = x_2(t)$, $\dot{x}_2^*(t) = \dot{x}_2(t) = x_3(t)$, and this equation may be rewritten as

$$x_3(t) = -\frac{\beta}{J} x_2^*(t) + \frac{K_T}{J} x_3^*(t)$$

Therefore, we have the following three equations which relate the two sets of variables, and the first question has been answered.

$$x_1(t) = x_1^*(t)$$
$$x_2(t) = x_2^*(t)$$
$$x_3(t) = -\frac{\beta}{J} x_2^*(t) + \frac{K_T}{J} x_3^*(t)$$

Note, however, that each of these three equations is a linear equation, and therefore this result can be put into a matrix form as

$$\mathbf{x}(t) = \mathbf{P}\mathbf{x}^*(t) \tag{2.7-1}$$

where

$$\mathbf{P} = \begin{bmatrix} 1 & 0 & 0 \\ 0 & 1 & 0 \\ 0 & -\beta/J & K_T/J \end{bmatrix} \tag{2.7-2}$$

The relationship between the two sets of variables represented by Eq. (2.7-1) is known as a *linear transformation*, and the matrix $\mathbf{P}$ is referred to as the *transformation matrix*. Any two sets of state variables for a linear plant must be related by a linear transformation.

In order to answer the second question concerning the relationship of the two plant representations, let us consider a general problem. Suppose that we have two sets of state variables: a set $\mathbf{x}(t)$ whose plant representation is

$$\dot{\mathbf{x}}(t) = \mathbf{A}\mathbf{x}(t) + \mathbf{b}u(t) \tag{Ab}$$
$$y(t) = \mathbf{c}^T\mathbf{x}(t) \tag{c}$$

and a set $\mathbf{x}^*(t)$ with the associated representation

$$\dot{\mathbf{x}}^*(t) = \mathbf{A}^*\mathbf{x}^*(t) + \mathbf{b}^*u(t) \tag{A*b*}$$
$$y(t) = \mathbf{c}^{*T}\mathbf{x}^*(t) \tag{c*}$$

The variables $\mathbf{x}(t)$ and $\mathbf{x}^*(t)$ are related by the linear transformation

$$\mathbf{x}(t) = \mathbf{P}\mathbf{x}^*(t) \tag{2.7-1}$$

where $\mathbf{P}$ is a general *constant nonsingular matrix*. The problem is to find the relationship between the matrices $\mathbf{A}$, $\mathbf{b}$, and $\mathbf{c}$ and the matrices $\mathbf{A}^*$, $\mathbf{b}^*$, and $\mathbf{c}^*$.

Let us begin by taking the derivative of both sides of Eq. (2.7-1). Since **P** has constant elements, it is not involved in the differentiation, and

$$\dot{x}(t) = P\dot{x}^*(t) \tag{2.7-3}$$

The substitution of Eqs. (2.7-1) and (2.7-3) into Eqs. (**Ab**) and (**c**) yields

$$P\dot{x}^*(t) = APx^*(t) + bu(t) \tag{2.7-4}$$
$$y(t) = c^T Px^*(t)$$

The premultiplication of Eq. (2.7-4) by $P^{-1}$ provides the following result:

$$\dot{x}^*(t) = P^{-1}APx^*(t) + P^{-1}bu(t)$$
$$y(t) = c^T Px^*(t)$$

A comparison of these equations with Eqs. (**A\*b\***) and (**c\***) above reveals that

$$A^* = P^{-1}AP \qquad b^* = P^{-1}b \qquad \text{and} \qquad c^* = P^T c \tag{2.7-5}$$

Hence we see that the matrix **P** not only relates the two sets of state variables by means of Eq. (2.7-1) but also relates their associated representation by means of Eqs. (2.7-5).

The above multiplication by $P^{-1}$ indicates why the **P** matrix must be nonsingular. [Note that the **P** matrix given by Eq. (2.7-2) is nonsingular.] However, **P** may be any nonsingular matrix. Therefore a plant initially expressed by one set of state variables may be transformed into an infinite number of alternative representations.

In order to use Eqs. (2.7-5) to find **P** if **A**, **b**, **c**, **A\***, **b\***, and **c\*** are known, one premultiplies the first two parts of the equation by **P** to obtain

$$PA^* = AP \qquad Pb^* = b \qquad \text{and} \qquad c^* = P^T c \tag{2.7-6}$$

These three matrix equations generate a set of $n^2 + 2n$ linear equations in the elements of **P** which may be solved to determine **P**. Since there are only $n^2$ elements in the matrix **P**, the reader may wonder how it is possible to satisfy $n^2 + 2n$ equations. The simple answer is that exactly $2n$ of these equations are redundant. Therefore for any two compatible state-variable representations, that is, representations that truly describe the same plant, these equations may always be solved to determine **P**.

***Example 2.7-1*** Let us determine the transformation matrix **P** relating the two state-variable descriptions of the field-controlled motor. The answer is already known to be Eq. (2.7-2), and we simply wish to establish the validity of this result through the use of Eqs. (2.7-6). The plant matrix **A**, associated with the first set of state variables, is already known to be

$$
\mathbf{A} = \begin{bmatrix} 0 & 1 & 0 \\ 0 & 0 & 1 \\ 0 & -\beta R_f/JL_f & -(L_f\beta + R_fJ)/JL_f \end{bmatrix}
$$

and $\mathbf{A}^*$ associated with the physical variables is

$$
\mathbf{A}^* = \begin{bmatrix} 0 & 1 & 0 \\ 0 & -\beta/J & K_T/J \\ 0 & 0 & -R_f/L_f \end{bmatrix}
$$

If these matrices are substituted into the equation $\mathbf{PA}^* = \mathbf{AP}$, the expanded result is

$$
\begin{bmatrix} p_{11} & p_{12} & p_{13} \\ p_{21} & p_{22} & p_{23} \\ p_{31} & p_{32} & p_{33} \end{bmatrix} \begin{bmatrix} 0 & 1 & 0 \\ 0 & -\beta/J & K_T/J \\ 0 & 0 & -R_f/L_f \end{bmatrix} =
$$
$$
\begin{bmatrix} 0 & 1 & 0 \\ 0 & 0 & 1 \\ 0 & -\beta R_f/JL_f & -(L_f\beta + R_fJ)/JL_f \end{bmatrix} \begin{bmatrix} p_{11} & p_{12} & p_{13} \\ p_{21} & p_{22} & p_{23} \\ p_{31} & p_{32} & p_{33} \end{bmatrix}
$$

This matrix equation represents nine equations in nine unknowns. By equating elements of the matrices that result from the indicated multiplications, these nine equations are found to be

$$
0 = p_{21}
$$

$$
p_{11} - \frac{\beta}{J} p_{12} = p_{22}
$$

$$
\frac{K_T}{J} p_{12} - \frac{R_f}{L_f} p_{13} = p_{23}
$$

$$
0 = p_{31}
$$

$$
p_{21} - \frac{\beta}{J} p_{22} = p_{32}
$$

$$
\frac{K_T}{J} p_{22} - \frac{R_f}{L_f} p_{23} = p_{33}
$$

$$
0 = -\frac{\beta R_f}{JL_f} p_{21} - \frac{1}{JL_f} (L_f\beta + R_fJ)p_{31}
$$

$$
p_{31} - \frac{\beta}{J} p_{32} = -\frac{\beta R_f}{JL_f} p_{22} - \frac{1}{JL_f} (L_f\beta + R_fJ)p_{32}
$$

$$
\frac{K_T}{J} p_{32} - \frac{R_f}{L_f} p_{33} = -\frac{\beta R_f}{JL_f} p_{23} - \frac{1}{JL_f} (L_f\beta + R_fJ)p_{33}
$$

Because of the first and fourth equations, the seventh equation is redundant.   From the equation $\mathbf{Pb}^* = \mathbf{b}$, we see that

$$
\begin{bmatrix} p_{11} & p_{12} & p_{13} \\ p_{21} & p_{22} & p_{23} \\ p_{31} & p_{32} & p_{33} \end{bmatrix} \begin{bmatrix} 0 \\ 0 \\ 1/L_f \end{bmatrix} = \begin{bmatrix} 0 \\ 0 \\ K_T/L_f J \end{bmatrix}
$$

or, after multiplication, the three resulting equations are

$$\frac{p_{13}}{L_f} = 0$$

$$\frac{p_{23}}{L_f} = 0$$

$$\frac{p_{23}}{L_f} = \frac{K_T}{L_f J}$$

These equations provide considerable help in simplifying the nine preceding ones.   Now we know that not only are $p_{21}$ and $p_{31}$ equal to zero but $p_{13}$ and $p_{23}$ are also zero, with $p_{33}$ equal to $K_T/J$.   Our knowledge of $\mathbf{P}$ thus far is summed up in the following matrix:

$$
\mathbf{P} = \begin{bmatrix} p_{11} & p_{12} & 0 \\ 0 & p_{22} & 0 \\ 0 & p_{32} & K_T/J \end{bmatrix}
$$

The nine equations have been reduced to

$$p_{11} = p_{22}$$

$$\frac{K_T}{J} p_{12} = 0$$

$$-\frac{\beta}{J} p_{22} = p_{32}$$

$$\frac{K_T}{J} p_{22} = \frac{K_T}{J}$$

Now $\mathbf{P}$ is completely known and is as given in Eq. (2.7-2).   The eighth and ninth equations of the original nine equations are also redundant, bringing the total thus far to three redundant equations. The three equations that result from the remaining portion of Eqs. (2.7-6), namely, $\mathbf{c}^* = \mathbf{P}^T \mathbf{c}$, produce three more redundant equations. Thus, although we originally had 15 equations, 6 were redundant, or, as stated above, $2n$ are redundant, and the remaining $n^2$ equations uniquely specify $\mathbf{P}$.

Although the above example indicated the use of Eqs. (2.7-6) to obtain the transformation matrix $\mathbf{P}$, this is not necessarily the recommended procedure.   The method always works, but the algebra is often tedious.   The example was considered earlier in this section, and there we found six elements in the first two rows of **P** from the block diagram or by physical reasoning, whichever seems preferable.   If we had used these six elements of **P** to find the remaining elements, we would have had to solve only three simultaneous equations rather than nine.   A common case when it is necessary to know the transformation matrix is when the problem is worked in variables other than the physical variables and it is desired to transform the results to physical variables for interpretation.

The use of Eqs. (2.7-5) to transform one state-variable representation into another is simply a matter of selecting a nonsingular **P** matrix and then substituting **A**, **b**, **c**, and **P** into Eqs. (2.7-5) to obtain **A***, **b***, and **c***.   The only difficult part of this operation is the inversion of the matrix **P**.

*Example 2.7-2*   To illustrate the use of Eqs. (2.7-5) to transform a state-variable representation, let us assume that the physical-variable representation of the field-controlled motor is not known but that the transformation matrix **P** [Eq. (2.7-2)] is known as well as the representation in terms of $\theta_o$, $\dot{\theta}_o$, and $\ddot{\theta}_o$.   As a first step we obtain the inverse of **P**:

$$\mathbf{P}^{-1} = \begin{bmatrix} 1 & 0 & 0 \\ 0 & 1 & 0 \\ 0 & \beta/K_T & J/K_T \end{bmatrix}$$

Then **A*** is

$$\mathbf{A}^* = \begin{bmatrix} 1 & 0 & 0 \\ 0 & 1 & 0 \\ 0 & \beta/K_T & J/K_T \end{bmatrix} \begin{bmatrix} 0 & 1 & 0 \\ 0 & 0 & 1 \\ 0 & -\beta R_f/JL_f & -(L_f\beta + R_fJ)/JL_f \end{bmatrix}$$
$$\begin{bmatrix} 1 & 0 & 0 \\ 0 & 1 & 0 \\ 0 & -\beta/J & K_T/J \end{bmatrix}$$

$$= \begin{bmatrix} 0 & 1 & 0 \\ 0 & -\beta/J & K_T/J \\ 0 & 0 & -R_f/L_f \end{bmatrix}$$

The matrix **b*** is given by

$$\mathbf{b}^* = \begin{bmatrix} 1 & 0 & 0 \\ 0 & 1 & 0 \\ 0 & \beta/K_T & J/K_T \end{bmatrix} \begin{bmatrix} 0 \\ 0 \\ K_T/L_fJ \end{bmatrix} = \begin{bmatrix} 0 \\ 0 \\ 1/L_f \end{bmatrix}$$

and the matrix $\mathbf{c}^*$ becomes

$$\mathbf{c}^* = \begin{bmatrix} 1 & 0 & 0 \\ 0 & 1 & -\beta/J \\ 0 & 0 & K_T/J \end{bmatrix} \begin{bmatrix} 1 \\ 0 \\ 0 \end{bmatrix} = \begin{bmatrix} 1 \\ 0 \\ 0 \end{bmatrix}$$

If these results are compared with the physical-variable representation given at the beginning of this section, the answers for $\mathbf{A}^*$, $\mathbf{b}^*$, and $\mathbf{c}^*$ are seen to be correct.

The transformation of one state-variable representation into another representation may also be accomplished by means of block diagrams and block-diagram algebra. Consider, for example, Fig. 2.7-1, which shows both state-variable representations of the field-controlled motor. The physical variables were determined from the transformation matrix of Eq. (2.7-2). In order to transform the block diagram of Fig. 2.7-1 into the conventional block diagram for the physical-variable representation, we need only suppress the variable $x_3(t)$.

As a first step, we transfer both blocks connected to the $x_2^*(t)$ variable to the left side of the $1/s$ block by making use of the block-diagram identity of Fig. 2.2-8d. The result is shown in Fig. 2.7-2a.

Next, we combine the two sets of blocks that are in parallel by the use of the identity of Fig. 2.2-8e to obtain Fig. 2.7-2b. At this point, we

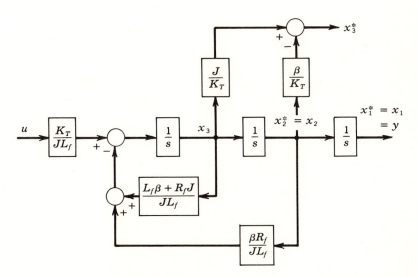

**Fig. 2.7-1**   Block diagram of the field-controlled motor, showing two sets of state variables.

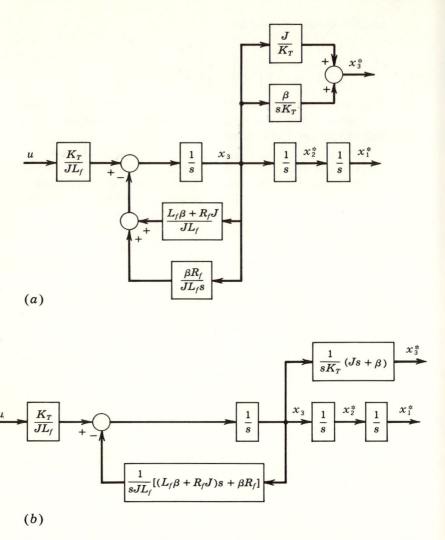

**Fig. 2.7-2**  Block-diagram manipulations of Fig. 2.7-1.

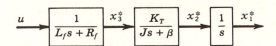

**Fig. 2.7-3**  Block diagram of the field-controlled
motor in physical variables.

make use of the block-diagram identities of Fig. 2.2-8$d$ and $f$ to redraw the
diagram as shown in Fig. 2.7-2$c$.   A final combination of the series blocks
gives the desired result, shown in Fig. 2.7-3.

Before concluding our discussion of linear transformation, let us
consider briefly two of its properties:

1.   Since the input-output transfer function is insensitive to the state
     variables used to represent the plant, therefore

$$\mathbf{c}^{*T}(s\mathbf{I} - \mathbf{A}^*)^{-1}\mathbf{b}^* = \mathbf{c}^T(s\mathbf{I} - \mathbf{A})^{-1}\mathbf{b} \tag{2.7-7}$$

2.   Since det $(s\mathbf{I} - \mathbf{A})$ is equal to the denominator of $G_p(s)$, which is also
     insensitive to the choice of state variables, then

$$\det (s\mathbf{I} - \mathbf{A}^*) = \det (s\mathbf{I} - \mathbf{A}) \tag{2.7-8}$$

This result also indicates that the eigenvalues of the plant are not
affected by a linear transformation of the state variables which
represent the plant.

**Exercise 2.7**   *2.7-1*.   Find the transformation matrix **P** that relates the
     phase-variable representation found in Exercise 2.5-3 and the physi-
     cal-variable representation of Sec. 2.6 for the armature-controlled
     motor by using physical reasoning rather than Eqs. (2.7-6).   Verify
     that this **P** matrix satisfies Eqs. (2.7-6).

   *answer*:

   If $\mathbf{x}$ = physical variables and $\mathbf{x}^*$ = phase variables, then

$$\mathbf{x} = \begin{bmatrix} 0 & \beta/JL_a & 1/L_a \\ K'_T/JL_a & 0 & 0 \\ 0 & K'_T/JL_a & 0 \end{bmatrix} \mathbf{x}^*$$

## 2.8  *Conclusions*

In this chapter the concepts and techniques of plant representation have
been developed and discussed.   Two basic methods of plant representa-

tion were presented.   The first, known as the input-output representation, deals only with the terminal characteristics of the plant and ignores all internal behavior.   The second, the state-variable representation, provides a complete description of the internal as well as the terminal behavior of the plant.

In addition to the introduction of these two methods, considerable effort has been made to provide the reader with manipulative techniques appropriate and useful for each method, such as block diagrams and matrix algebra as well as the linear transformation.   The interrelations of the two means of plant representation have also been emphasized, often by means of the block diagram.

The block diagram was introduced at the same time that the concept of the transfer function was introduced.   However, the careful reader no doubt observed that we continued to use the block diagram to picture a variety of different plants described in state-variable form. This is possible because the block diagram is a picture of transformed differential equations, regardless of the form in which they appear.   Block diagrams will continue to be used throughout the book.

In all the third-order examples considered in this section, the original describing differential equations were neither one third-order equation nor three first-order equations.   In models of electrical, mechanical, and electromechanical plants, it often happens that the original plant description is in terms of sets of coupled first- and second-order equations.   As convenient, these coupled equations are arranged as one nth-order differ-

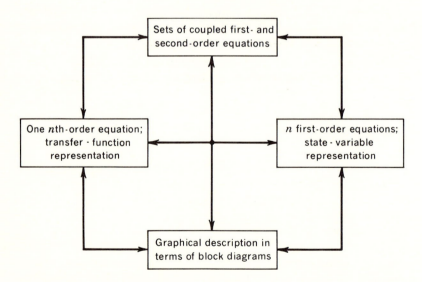

*Fig. 2.8-1*   Methods of plant representation.

ential equation, associated with the transfer function, or $n$ first-order equations, associated with the state-variable representation, as shown graphically in Fig. 2.8-1. Regardless of how the plant is represented, a block diagram may always be used to display the relationships of the plant variables.

## 2.9 Problems

*2.9-1.* The describing differential equations for the system shown in Fig. 2.9-1 are

$$m_1\ddot{x}_1 + \beta_1\dot{x}_1 + K_1x_1 - \beta_1\dot{x}_2 - K_1x_2 = f(t)$$
$$- \beta_1\dot{x}_1 - K_1x_1 + m_2\ddot{x}_2 + \beta_1\dot{x}_2 + (K_1 + K_2)x_2 = 0$$

Draw a block diagram for this plant and find the transfer function $x_2(s)/f(s)$. Write the state-variable representation for the plant, using the variables $z_1 = x_1$, $z_2 = \dot{x}_1$, $z_3 = x_2$, and $z_4 = \dot{x}_2$.

*2.9-2.* The schematic diagram for a feedback amplifier is shown in Fig. 2.9-2. Draw a block diagram for this plant and find the input-output transfer function for this circuit. The input impedance of the amplifier is assumed to be infinite, and the output impedance is zero.

*2.9-3.* Find the phase-variable and physical-variable representations of the amplifier of Prob. 2.9-2. Use the node voltages $v_1$ and $e_{out}$ as state variables in the physical-variable representation. Find the transformation matrix that relates these two state-variable representations.

*2.9-4.* Shown in Fig. 2.9-3 is a plant for measuring high-frequency displacements of a shake table. The potentiometer measures the difference in displacement between $M_1$ and $M_2$. The driving function is a position

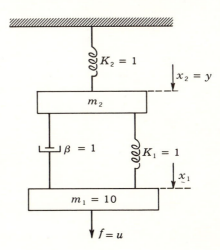

**Fig. 2.9-1**    Problem 2.9-1.

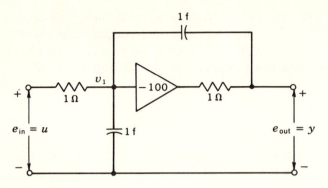

**Fig. 2.9-2**   Problem 2.9-2.

$x(t)$.   Ignore the acceleration produced by gravity, and assume infinite input and zero output impedance of the amplifier.   Assume that $v_2 = 0$ in the equilibrium condition $x = y = 0$.   The maximum travel of the linear-motion potentiometer is $a$.   Find the transfer function $v_0(s)/x(s)$. The transfer function of the amplifier is

$$\frac{v_1(s)}{v_2(s)} = \frac{k}{1 + Ts}$$

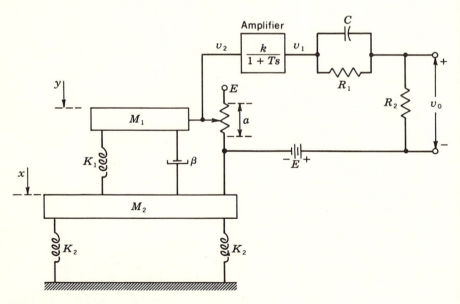

**Fig. 2.9-3**   Problem 2.9-4.

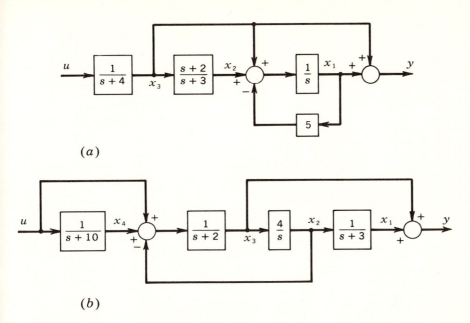

(a)

(b)

**Fig. 2.9-4**  Problems 2.9-5 and 2.9-6.

*2.9-5.*  Find the state-variable representations for each of the plants shown in block-diagram form in Fig. 2.9-4.  Use the variables indicated in the diagram.  Draw the elementary block diagrams for each of these plants.  Find the input-output transfer function $y(s)/u(s)$ for each plant.

*2.9-6.*  Find the phase-variable representation for each of the plants of Fig. 2.9-4.  Determine the linear-transformation matrices that relate the phase-variable representation and the representation found in Prob. 2.9-5.

*2.9-7.*  In order to represent the plant shown in Fig. 2.9-5 in state-variable form as Eqs. (**Ab**) and (**c**), it is convenient to define an artificial state variable $x_3$ to replace the $x_3'$ shown.  Find the state-variable representation in terms of the variables $x_1$, $x_2$, and $x_3$.  Following some design computations, it is decided that $u$ should be equal to

$$u = x_1 + 2x_2 + 0.5x_3$$

In order to use this result it is necessary to express it in terms of $x_1$, $x_2$, and $x_3'$.  Show how this may be done, and find the expression of $u$ in terms of $x_1$, $x_2$, and $x_3'$.

*2.9-8.*  Suppose that a given state-variable representation for an $n$th-

**Fig. 2.9-5**  Problem 2.9-7.

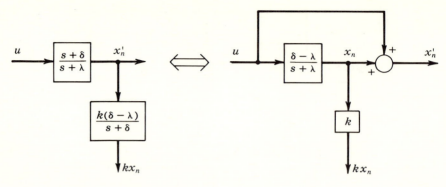

**Fig. 2.9-6**  Problem 2.9-10.

order plant

$$\dot{x} = Ax + bu$$
$$y = c^T x$$

and the phase-variable representation for the same plant

$$\dot{x}^* = A^* x^* + b^* u$$
$$y = (c^*)^T x^*$$

are related by the linear transformation

$$x(t) = Px^*(t)$$

Show that $P$ may be determined by the recursive relation

$$p^n = b$$

and

$$p^{n-i} = A p^{n-i+1} + a_{n-i+1} b \qquad i = 1, 2, \ldots, n - 1$$

where the $p^i$'s are the columns of $P$, that is, $P = [p^1 | p^2 | \cdots | p^n]$, and $a_i$ is the coefficient of $s^{i-1}$ in the characteristic polynomial of the plant. Use this technique to find the transformation matrices requested in Prob. 2.9-6.

**2.9-9.**  For the plant

$$\dot{x} = \begin{bmatrix} -1 & 1 & -1 \\ 1 & 0 & 1 \\ 0 & 0 & 1 \end{bmatrix} x + \begin{bmatrix} 0 \\ 0 \\ 2 \end{bmatrix} u \qquad y = [1 \quad 0 \quad 1]x$$

(a)  Find the elementary block diagram.
(b)  Find the phase-variable representation.

**2.9-10.**  Verify the block-diagram identity shown in Fig. 2.9-6.

# *three*  *closed-loop-system representation*

## *3.1  Introduction*

In Chap. 2 we discussed the representation of a given plant, under the assumption that the unalterable plant is described by linear differential equations with constant coefficients.   Often the describing equations appear as sets of simultaneous equations of first and second order.   With a little manipulation, these equations may be expressed as either one $n$th-order equation or as a set of $n$ first-order equations.   But regardless of how the plant is described, it always proved convenient to picture the governing equations by drawing a block diagram.

The reader will recall that the block diagram

involves not the differential equations themselves but the Laplace transform form of the given differential equations. This is an important point. Pictures prove to be useful, both academically and in practice, and hence in this chapter we continue to utilize the Laplace transform, block-diagram approach. In addition, we adopt the state-variable representation as being the most general and turn our attention to means of utilizing information concerning the various state variables as a means of controlling the given unalterable plant.

It is important at the outset to realize the difference between the terms *plant* and *system*. The plant is described by Eqs. (**Ab**) and (**c**), or

$$\dot{\mathbf{x}} = \mathbf{A}\mathbf{x} + \mathbf{b}u \tag{Ab}$$
$$y = \mathbf{c}^T\mathbf{x} \tag{c}$$

Here the input $u$ is termed the *control input*. The closed-loop control system, which is the concern of this chapter, is described by Eqs. (**Ab**) and (**c**), plus an additional equation that specifies the control signal. The most general form of the control signal considered in this book is given as Eq. (**k**), or

$$u = K(r - \mathbf{k}^T\mathbf{x}) = K(r - k_1x_1 - k_2x_2 - \cdots - k_nx_n) \tag{k}$$

The input $r$ is designated as the *reference input*, that is, the input that the control system is attempting to follow. In the general case each of the state variables $x_i$ is weighted by a linear gain or attenuation factor $k_i$ and subtracted from the reference input.

We begin our treatment of closed-loop-system representation by discussing some of the more common effects of feedback in Sec. 3.2. The following section presents the methods and techniques of closed-loop-system representation in terms of a specific-example system. In Sec. 3.4 the results and methods that have been developed in terms of a specific example in Sec. 3.3 are generalized. This generalization is based upon the utilization of either the plant resolvent matrix, as previously mentioned in Chap. 2, or alternatively in terms of the system resolvent matrix. The matrix approach to the system resolvent matrix is also discussed in Sec. 3.4 and an alternative block-diagram approach in Sec. 3.5.

## 3.2 The effects of feedback

The basic concept of feedback was introduced in Chap. 1. Some of the effects of feedback were mentioned in addition to a discussion of systems for which feedback could be used advantageously. Before we discuss the representation of closed-loop, feedback systems, it seems advisable to examine the effects of feedback in more detail.

Although the effects of feedback are too numerous to list in their entirety, four effects are usually of primary interest:

1.  The reduction of sensitivity to plant-parameter variations
2.  The reduction of sensitivity to output disturbances
3.  The ability to control the system bandwidth
4.  The ability to control the system transient response

Although these results are not universally obtained, one or more of these effects is usually the primary goal of a feedback control system.   Let us consider each of these items in more detail.

*Reduction of sensitivity to plant-parameter variations.*   To illustrate the effect of feedback on plant-parameter sensitivity most clearly, let us consider the simple cascade system shown in Fig. 3.2-1a.   Each of the two blocks represents a positive-gain, *frequency-independent*, linear amplifier. This same system is shown in Fig. 3.2-1b with the addition of a feedback loop around the first amplifier.

The transfer function (gain) of these two systems may be easily found by making use of the block-diagram reductions of Chap. 2.   For the open-loop system (Fig. 3.2-1a) the result is

$$\frac{y(s)}{r(s)} = M_o(s) = M_o = GG_1 \qquad (3.2\text{-}1)$$

and for the closed-loop system (Fig. 3.2-1b) we have

$$\frac{y(s)}{r(s)} = M_c(s) = M_c = \frac{GG_1}{1 + GH} \qquad (3.2\text{-}2)$$

Notice that the input-output transfer functions are specifically indicated as being not dependent on $s$, since $G$, $G_1$, and $H$ are assumed to be frequency-independent.

As a preliminary indication of the effect of feedback on the sensitiv-

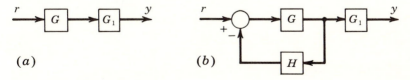

*Fig. 3.2-1*   Two simple frequency-independent systems: (a) open-loop system; (b) closed-loop system.

ity picture, let us examine the closed-loop gain $M_c$ when the amplifier gain $G$ becomes large. In order to do this, let us rewrite $M_c$ in the following form:

$$M_c = \frac{G_1}{1/G + H}$$

Now if $G$ becomes very large while $H$ remains fixed, eventually the $1/G$ term becomes small compared with $H$, and $M_c$ is approximately

$$M_c \approx \frac{G_1}{H}$$

In this case the closed-loop gain has become effectively independent of the amplifier gain $G$ as long as $G$ is large, which is an obvious case of insensitivity. It is clearly not possible to obtain this result in the case of the open-loop gain function $M_o$.

If we assume, for the moment, that $G_1 = 1$, the closed-loop gain $M_c$ depends only on $H$ whenever $1/G \ll H$ or $GH \gg 1$, where $GH$ is the negative of the loop gain. If the gain $H$ can be fixed precisely, as, for example, when $H$ is formed by a precision passive voltage divider, then the closed-loop gain is independent of $G$ as long as $G$ is large enough. This is exactly the approach used to stabilize an amplifier gain in a large number of practical situations.

In order to place the above discussion of sensitivity on a more mathematical, quantitative basis, it is necessary to introduce a concept known as a *sensitivity function*. Although there are a large number of ways in which such a concept may be defined, one normally refers to the *percentage* variation in some specific system quality, such as gain, with respect to a *percentage* variation in the system parameter in question. Therefore the sensitivity function $S_\alpha^M$ of the gain $M$ with respect to the parameter $\alpha$ is written as

$$S_\alpha^M = \frac{\%\text{ change in } M}{\%\text{ change in } \alpha} = \frac{dM/M}{d\alpha/\alpha} = \frac{\alpha}{M}\frac{dM}{d\alpha} \qquad (3.2\text{-}3)$$

The effect of changes in the parameter $\alpha$ on $M$ is minimized if the sensitivity function $S_\alpha^M$ is minimized.

Let us examine the effect of feedback on the gain sensitivity for the systems shown in Fig. 3.2-1 by making use of the above sensitivity-function definition. In particular, let us consider first the sensitivity situation with regard to the amplifier gain $G$. For the open-loop system, the sensitivity function is given by

$$S_G^{M_o} = \frac{G}{M_o}\frac{dM_o}{dG} = \frac{1}{G_1}G_1 = 1$$

For the closed-loop system, the sensitivity function is

$$S_G^{M_c} = \frac{G}{M_c}\frac{dM_c}{dG} = \frac{1}{1 + GH} \tag{3.2-4}$$

These results confirm the conclusions in the previous discussion. The sensitivity of the gain of the closed-loop system, $M_c$, to $G$ is less than the sensitivity of open-loop gain to $G$. In addition, the sensitivity of the closed-loop system can be made as small as desired by increasing the loop gain function $GH$.

The sensitivity functions for the other amplifier gains may be found in a similar fashion. The results are given in Table 3.2-1. From this table, we note that the addition of the feedback loop has no effect with respect to the sensitivity functions related to $G_1$. This is not surprising, since $G_1$ is not included in the feedback loop.

**Table 3.2-1   *Sensitivity Functions***

| Parameter | Open-loop System | Closed-loop System |
|:---:|:---:|:---:|
| $G$ | $S_G^{M_o} = 1$ | $S_G^{M_c} = \dfrac{1}{1 + GH}$ |
| $G_1$ | $S_{G_1}^{M_o} = 1$ | $S_{G_1}^{M_c} = 1$ |
| $H$ | $\cdots$ | $S_H^{M_c} = \dfrac{-GH}{1 + GH}$ |

Although it is not possible to discuss the effect of feedback on the sensitivity function related to $H$, since $H$ does not exist in the open-loop system, it is interesting to note that the sensitivity of $M_c$ to $H$ increases as the loop gain $GH$ is increased. This result, which is just the opposite of the result obtained for the gain $G$, is not difficult to explain. As the loop gain $GH$ is increased, the closed-loop gain $M_c$ begins to depend more and more on $H$ and less and less on $G$. Hence, as the sensitivity to variations in the gain $G$ is reduced, the sensitivity to variations in $H$ must increase.

In effect, then, one can conclude that the addition of feedback has not really reduced the sensitivity but has merely allowed the sensitivity to be transferred, in part, to the feedback elements. This effect, however, is of significant practical importance since the feedback-path elements are often under the complete control of the designer and may therefore be rigidly specified.

On the other hand, the forward-path elements may be beyond the direct control of the designer, since they compose the fixed plant being controlled which is assumed to be unalterable.   In addition, the forward-path elements may not be known exactly and may vary.   Hence the ability, which feedback provides, to transfer the sensitivity dependence to elements in the feedback path may be very helpful in a practical situation. This general topic was also discussed in Chap. 1 where it was indicated that the use of feedback does not eliminate the need for calibration but simply allows the calibration to be done at a more convenient and accepta-ble part of the system.   The need for such a transference of calibration and the existence of an acceptable alternative location for the calibration to be executed are two of the basic factors in deciding the usefulness of a feedback system.

In terms of this simple example, we can draw the following tentative conclusions regarding the effects of feedback on sensitivity:

1.   Feedback has no effect on elements which are not included in the feedback loop.
2.   Feedback may be used to reduce the sensitivity of the closed-loop transfer function to forward-path elements by transferring the sensi-tivity to feedback-path elements.

Although based only on our simple example, these results remain essen-tially correct for even the most general case.

Of course, if the elements involved are frequency-dependent transfer functions, as they generally are, these effects of feedback on sensitivity are also frequency-dependent.   In particular, if $G$ were actually $G(s)$ and $H$ were actually $H(s)$, Eq. (3.2-4) would remain unaltered, since differentia-tion is not with respect to $s$.   To indicate the resulting dependence on $s$, Eq. (3.2-4) is rewritten as

$$S_{G(s)}{}^{M_c(s)} = \frac{1}{1 + G(s)H(s)} \qquad (3.2\text{-}5)$$

In order for this sensitivity function to remain less than unity so that feedback reduces the sensitivity, *the magnitude of the denominator of Eq. (3.2-5) must remain greater than* 1 for the range of $s$ of interest, or the following condition must be satisfied

$$|1 + G(s)H(s)| > 1 \qquad (3.2\text{-}6)$$

In later chapters on design, considerable effort will be expended to achieve this result.

There is also a large number of other ways in which sensitivity may

be defined.    In all these cases, the conclusions listed above remain more
or less unaltered.

*Reduction of sensitivity to output disturbances.*    As a basis for this
discussion, we consider the control system pictured in Fig. 3.2-2.    In
addition to the usual reference input $r$, this figure includes an additional
input $d$ which is a disturbance input.    As a practical example, we might
consider the control of a radar antenna.    Then $G_p(s)$ would include the
motor and gearing necessary to control the antenna, and $H(s)$ would
include any components that were added in the feedback path to accomplish control.    The disturbance $d$ would be any influence that tends to
disturb the output, such as the wind blowing on the antenna.    To maintain control, we wish to minimize the effect of these disturbances.

If, for the moment, the reference input $r(t)$ is allowed to be zero, the
block diagram of Fig. 3.2-2$a$ may be redrawn with the disturbance as the
input.    This is done in Fig. 3.2-2$b$.    The transfer function $y(s)/d(s)$ then
becomes

$$\frac{y(s)}{d(s)} = \frac{1}{1 + KG_p(s)H(s)} \tag{3.2-7}$$

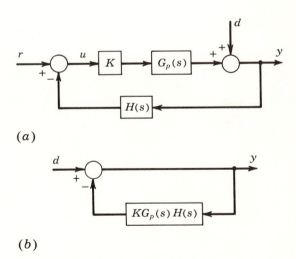

(a)

(b)

***Fig. 3.2-2***    Illustration of the effect of feedback on
output disturbances.    ($a$) Typical con-
trol system including an output disturbance;    ($b$) closed-loop system redrawn with the disturbance as the input.

This result is very similar to that expressed in Eq. (3.2-5), and we see that, in order to minimize the effect of output disturbances, once again it is necessary to ensure that *one plus the loop gain is greater than one for all frequencies.*

*Control of bandwidth.*    The bandwidth of a system is usually defined in terms of the response of the system to sinusoidal inputs.    Here let us define bandwidth as the frequency at which the magnitude of the output is $1/\sqrt{2} = 0.707$ times the magnitude of the output at very low frequency.    It will be shown later in Chap. 5 in the discussion of frequency response that the magnitude of the output in response to a sinusoidal input may be determined by evaluating the system transfer function at the frequency of interest.    Let us use the two particularly first-order simple examples of Fig. 3.2-3 to illustrate the effect of feedback on system bandwidth.

The transfer function $y(s)/r(s)$ for the open-loop system (Fig. 3.2-3a) is given by

$$\frac{y(s)}{r(s)} = M_o(s) = \frac{1}{s+1}$$

When $s = j\omega = j1$, the magnitude of $M_o(s)$ is

$$|M_o(j1)| = \frac{1}{|j1+1|} = \frac{1}{\sqrt{2}}$$

and this is 0.707 times the magnitude of $M_o(s)$ at $s = 0$.

Contrast this result with the closed-loop system of Fig. 3.2-3b. Here $M_c(s)$ is $1/(s+1+\beta)$, and the low-frequency value of $|M_c(j\omega)|$ is found by setting $s = j0$ so that

$$|M_c(j0)| = \frac{1}{1+\beta}$$

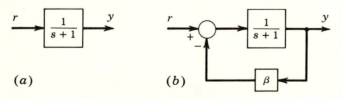

(a)                              (b)

**Fig. 3.2-3**    Two systems illustrating the effect of feedback on bandwidth.    (a) Open-loop system;    (b) closed-loop system.

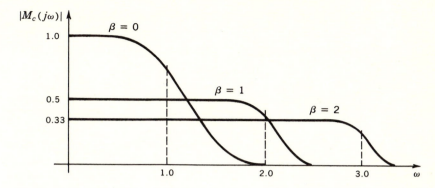

**Fig. 3.2-4**   Frequency response of a closed-loop system for $\beta = 0$, 1, and 2.

At $\omega = 1 + \beta$, the magnitude of $M_c(j\omega)$ is

$$|M_c[j(1 + \beta)]| = \frac{1}{|j(1 + \beta) + 1 + \beta|} = \frac{1}{(1 + \beta)\sqrt{2}}$$

or it also is 0.707 of its dc value. Hence the bandwidth of the closed-loop system is greater than that of the open-loop system as long as $\beta$ is positive (negative feedback).

However, in addition to increasing the bandwidth, the use of feedback has *reduced* the low-frequency gain since the low-frequency gain of the open-loop system is 1, whereas that of the closed-loop system is $1/(1 + \beta)$. A graphical representation of this situation is shown in Fig. 3.2-4 for $\beta = 0$ (open-loop system), $\beta = 1$, and $\beta = 2$.

From this illustration, it is obvious that the increase in bandwidth is achieved only at the expense of reducing the low-frequency gain. Here we see that feedback does not provide a magic wand for increasing bandwidth but provides only a means of trading gain for bandwidth.

*Control of the system transient response.* One of the most striking effects of feedback is its influence on the time response of a system. With linear state-variable feedback, $y(s)/r(s)$ may be set equal to *any* transfer function with the same number of poles and zeros as the plant transfer function $G_p(s)$. This is a very general statement, and it indicates the power of the state-variable-feedback approach. The assumption is that each of the state variables is available and may be measured accurately. If these assumptions are violated, and they often are, then one must be satisfied with a closed-loop transfer function that is not exactly as desired. The ramifications of these statements will be discussed in detail in later chapters, particularly in terms of synthesis methods.

In the discussion of the four principal effects of feedback that forms the bulk of this section, two examples involved amplifiers. This was done to make a point in as simple a fashion as possible. Feedback is used extensively in amplifier design for exactly the same reasons that have been specified above. The principal differences between amplifier design and control-system design are in the frequency ranges covered and the level of power involved. These two differences are closely related. If one is to control a radar antenna, as mentioned previously, it is possible that the antenna might weigh tons. On the basis of intuition alone, one does not expect an object of such weight to respond to high frequencies. In addition, it is clear that the control of such a heavy object would require huge amounts of power, as compared with a typical amplifier design. In general, automatic control systems involve the manipulation of relatively large amounts of power, and as the power being controlled increases, the frequency spectrum of interest decreases. Rarely is a control engineer concerned with frequencies above 1,000 rad/sec.

One effect of feedback that has not been mentioned is the tendency of a closed-loop system toward instability. This tendency may be considered as the cost of achieving the advantageous effects described earlier in this section. Much of the control engineer's task is to ensure adequate performance and at the same time to maintain the necessary degree of stability.

With these brief comments concerning the effects of feedback as background, let us consider the procedure for the representation of closed-loop systems.

### 3.3   Linear state-variable feedback—an example

As an aid in orienting our thinking on closed-loop-system representation, let us return to the general-system configuration discussed in the early part of Chap. 2 and illustrated in Fig. 2.2-1. At that point, the idea of state variables had not been introduced. Let us now redraw Fig. 2.2-1 as Fig. 3.3-1, this time on the basis of state-variable equations. The plant is described by Eqs. (**Ab**) and (**c**); hence the plant portion of the closed-loop system is identical to that portrayed in Fig. 2.4-2. Again, the broad arrows indicate vectorial quantities, and the scalar quantities $y$ and $u$ are indicated by a single line.

The closed-loop nature of the system of Fig. 3.3-1 is indicated by the presence of the controller to generate the control signal $u$ from the knowledge of the state variables. In contrast to Fig. 2.2-1, here the

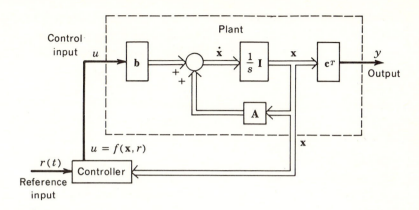

**Fig. 3.3-1**   General closed-loop system with state-variable feedback.

input to the controller is specifically indicated as the state of the plant $\mathbf{x}$ rather than the vague statement previously used, "information about the plant." It is important to note that, except for the reference input, the state of the plant $\mathbf{x}$ is the only information needed by the controller. This must obviously be true since the state, by definition, provides a complete description of the plant, and any other information concerning the plant can be synthesized from the state. The output $y$, for example, can always be obtained from a knowledge of the state by means of the output expression

$$y = \mathbf{c}^T\mathbf{x}$$

and therefore we need not show the controller as being dependent on the output.

The output of the controller, namely, the control input $u$, is shown in this general state-variable feedback system as a general function $f(\mathbf{x},r)$ of the state and the reference input. However, rather than this general, vague form of controller, we consider a more specific case known as *linear state-variable feedback.*

In the linear state-variable-feedback case, the control input is computed by multiplying by a gain $K$ the difference between the reference input and a weighted (linear) sum of the state variables. As a mathematical expression, this controller relation takes the form

$$u = f(\mathbf{x},r) = K[r - (k_1x_1 + k_2x_2 + \cdots + k_nx_n)] \tag{k}$$

Here the $k_i$'s are referred to as *feedback coefficients*, since they are the coefficients of the various state variables in our linear feedback expression for the control input. The gain $K$ is referred to as the *controller gain*.

Equation (**k**) may be simplified by making use of matrix notation:

$$u = K(r - \mathbf{k}^T\mathbf{x}) \tag{k}$$

where

$$\mathbf{k} = \text{col}\,(k_1, k_2, \ldots, k_n) \tag{3.3-1}$$

A graphical representation of the linear-state-variable-feedback-system configuration is shown in Fig. 3.3-2. Except for some minor modifications made in Chap. 9, this system configuration will be used throughout the remainder of the book. The linear state-variable-feedback system has been chosen for consideration for three basic reasons.

First, the configuration results in a linear system, thereby allowing us to make use of the powerful transform techniques for analysis and synthesis. Second, this configuration is sufficiently general to obtain satisfactory performance in many *practical* control problems. The minor modifications made in Chap. 9 involve the addition of linear dynamics to the controller. This change allows even a broader class of systems to be effectively handled.

Third, the linear state-variable-feedback approach serves as a suitable introduction to a broad array of topics usually referred to as *modern control theory*. In fact, the use of state-variable feedback is one of the

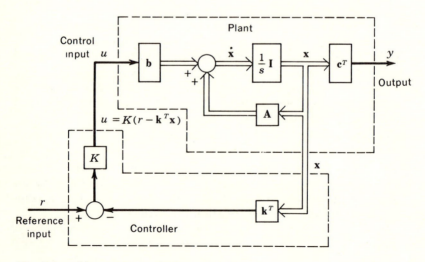

*Fig. 3.3-2* Linear state-variable-feedback system.

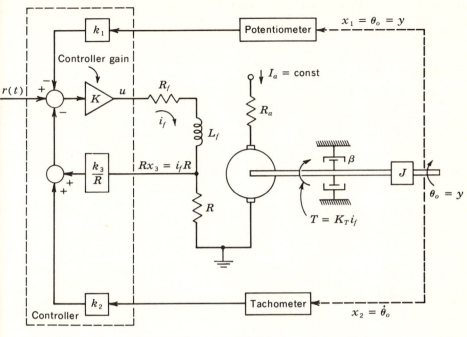

*Fig. 3.3-3* Simple positioning system using a field-controlled dc motor.

most important practical contributions that has resulted from this modern approach.

The above statements and conclusions may not be completely obvious at this time; this should not be a cause for concern. As the reader gains a greater understanding of and appreciation for the linear-state-variable-feedback configuration in this and the following chapters, the above comments will become more meaningful.

Instead of dwelling further on generalities concerning linear state-variable feedback, let us consider a specific example of a simple positioning system in order to illustrate its practical implications. The next section demonstrates the generality of results obtained from this specific example.

Let us consider the field-controlled dc motor discussed in Chap. 2. This dc motor is to be used to supply the torque to position an inertial load, such as a telescope. Control of the plant, that is, the motor and the load, is to be achieved through state-variable feedback, as in Fig. 3.3-3. The state variables used to describe the system are the output, $x_1 = \theta_o$; the derivative of the output, $x_2 = \dot{\theta}_o$; and the motor field cur-

rent, $x_3 = i_f$.   The output is converted into a voltage signal by means
of a potentiometer, and a tachometer is used to measure the angular
velocity.   (Many control motors contain a tachometer as an integral
part of the motor.)   The motor field current is obtained by measuring
the voltage across a small series resistor.

A block diagram of this simple positioning system is shown in Fig.
3.3-4a, with the controller and the plant clearly indicated.   In order to
simplify our further discussion of this example, let us assume specific
numerical values for the parameters so that the system shown in Fig.
3.3-4b results.   Here we have not specifically isolated the controller from
the plant.   This is the convention that we adopt for future use.   The

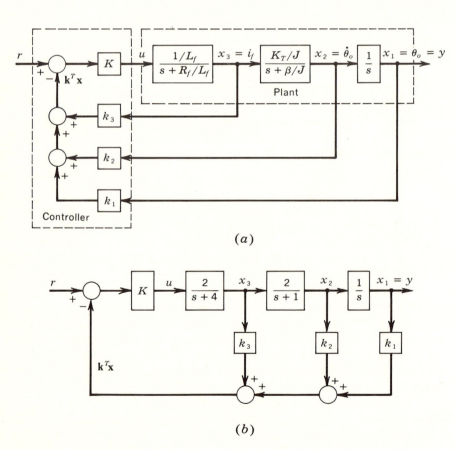

*(a)*

*(b)*

**Fig. 3.3-4**   Block diagram of the system of Fig. 3.3-3.   (a) General system; (b) the
system if the following numerical values are used for the parameters:
$L_f = 0.5$, $R_f = 2$, $K_T/J = 2$, and $\beta/J = 1$.

summers, the controller gain $K$, and the feedback coefficients **k** are always part of the controller and in the physical system are not necessarily located together. As indicated in Fig. 3.3-4*b*, there are four parameters to be chosen in the system: the three feedback coefficients $k_1$, $k_2$, $k_3$ and the controller gain $K$.

The state-variable representation of the *plant* may be easily obtained by inspection from Fig. 3.3-4*b* as

$$\dot{\mathbf{x}} = \begin{bmatrix} 0 & 1 & 0 \\ 0 & -1 & 2 \\ 0 & 0 & -4 \end{bmatrix} \mathbf{x} + \begin{bmatrix} 0 \\ 0 \\ 2 \end{bmatrix} u \tag{3.3-2}$$

$$y = \begin{bmatrix} 1 & 0 & 0 \end{bmatrix} \mathbf{x} \tag{3.3-3}$$

Similarly, the input-output transfer function of the plant may be obtained directly from Fig. 3.3-4*b* or from the state-variable representation by making use of Eq. (2.4-10). The result in either case is

$$\frac{y(s)}{u(s)} = G_p(s) = \frac{4}{s(s+1)(s+4)} \tag{3.3-4}$$

In future work the quantity $KG_p(s)$ will prove to be important, and here we define the transfer function $KG_p(s)$ as the *forward transfer function*. From Eq. (3.3-4) the forward transfer function is simply

$$KG_p(s) = K\frac{y(s)}{u(s)} \tag{3.3-5}$$

The forward transfer function can also be thought of as the transfer function from the reference input $r$ to the output $y$ with all the feedback paths open, or $\mathbf{k} = 0$. Thus an alternative expression for the forward transfer function is

$$KG_p(s) = \frac{y(s)}{r(s)}\bigg|_{\mathbf{k}=0} \tag{3.3-6}$$

The notation in Eq. (3.3-6) implies that $y(s)/r(s)$ is evaluated with all the $k_i$'s set equal to zero, that is, with no feedback.

In order to find the state-variable representation of the entire closed-loop system, it is necessary only to substitute the expression (**k**) for $u$ into the state-variable representation of the plant. If this is done, one obtains

$$\dot{\mathbf{x}} = \begin{bmatrix} 0 & 1 & 0 \\ 0 & -1 & 2 \\ 0 & 0 & -4 \end{bmatrix} \mathbf{x} + \begin{bmatrix} 0 \\ 0 \\ 2 \end{bmatrix} K(r - \begin{bmatrix} k_1 & k_2 & k_3 \end{bmatrix} \mathbf{x})$$

or

$$\dot{x} = \begin{bmatrix} 0 & 1 & 0 \\ 0 & -1 & 2 \\ 0 & 0 & -4 \end{bmatrix} x + \begin{bmatrix} 0 \\ 0 \\ 2 \end{bmatrix} Kr - \begin{bmatrix} 0 \\ 0 \\ 2 \end{bmatrix} K[k_1 \quad k_2 \quad k_3] x$$

If the indicated matrix multiplication is carried out and the two terms involving x are grouped, the result is

$$\dot{x} = \left( \begin{bmatrix} 0 & 1 & 0 \\ 0 & -1 & 2 \\ 0 & 0 & -4 \end{bmatrix} + \begin{bmatrix} 0 & 0 & 0 \\ 0 & 0 & 0 \\ -2Kk_1 & -2Kk_2 & -2Kk_3 \end{bmatrix} \right) x + \begin{bmatrix} 0 \\ 0 \\ 2 \end{bmatrix} Kr$$

which may be reduced to

$$\dot{x} = \begin{bmatrix} 0 & 1 & 0 \\ 0 & -1 & 2 \\ -2Kk_1 & -2Kk_2 & -4 - 2Kk_3 \end{bmatrix} x + \begin{bmatrix} 0 \\ 0 \\ 2 \end{bmatrix} Kr \qquad (3.3\text{-}7)$$

The output expression is still

$$y = [1 \quad 0 \quad 0] x \qquad (3.3\text{-}8)$$

Equations (3.3-7) and (3.3-8) are the desired state-variable representation of the closed-loop system. Note that there are two differences between this system representation and the original plant representation of Eqs. (3.3-2) and (3.3-3). First, the reference input times $K$, or $Kr$, has replaced the control input $u$; second, the **A** matrix has been modified by the feedback of the state variables. The **b** and **c** vectors have remained unchanged. Although we could obtain the closed-loop transfer function $y(s)/r(s)$ from the above state-variable representation, let us approach the problem in a different manner.

Our approach is to force the system into one of the two configurations shown in Fig. 3.3-5. These configurations are referred to as the $H_{eq}$ and the $G_{eq}$ reductions. The transfer functions $H_{eq}(s)$ and $G_{eq}(s)$ are read as $H$ equivalent and $G$ equivalent. This terminology was chosen

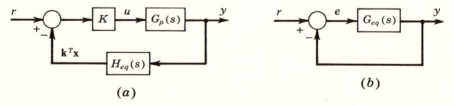

*(a)*      *(b)*

**Fig. 3.3-5**  Two basic configurations for the closed-loop system: (*a*) the *H*-equivalent form; (*b*) the *G*-equivalent form.

because neither $H_{eq}(s)$ nor $G_{eq}(s)$ is a physical transfer function. Although $H_{eq}(s)$ and $G_{eq}(s)$ are completely artificial, these methods of system representation are convenient ways to describe a large variety of closed-loop systems. In addition, these configurations serve to display the effects of state-variable feedback in different ways.

If the system of Fig. 3.3-4*b* is reduced to either configuration of Fig. 3.3-5, the closed-loop transfer function may be easily determined from the identity of Fig. 2.2-8*f*:

$$\frac{y(s)}{r(s)} = \frac{KG_p(s)}{1 + KG_p(s)H_{eq}(s)} = \frac{G_{eq}(s)}{1 + G_{eq}(s)} \tag{3.3-9}$$

Because of the frequent use made of both these configurations in later work, let us consider the reduction of the system of Fig. 3.3-4*b* to each of these two forms. We begin with the $H_{eq}$ reduction of Fig. 3.3-5*a*.

The $H_{eq}$ configuration may be achieved by making alternative use of the identities of Fig. 2.2-8*c* and *e* to move the inner state-variable feedback loops outward. This procedure begins with the innermost loop and is successively applied until the desired $H_{eq}$ form is obtained. This procedure is illustrated in a step-by-step fashion in Fig. 3.3-6 for the system of Fig. 3.3-4*b*.

In Fig. 3.3-6*a* the path containing the $k_3$ feedback coefficient has been moved to the right past the $2/(s + 1)$ and $1/s$ blocks and combined with the $k_1$ feedback coefficient. The $k_2$ feedback is next moved to the right and also combined with the $k_1$ block in order to achieve the desired $H_{eq}$ configuration shown in Fig. 3.3-6*b*. From Fig. 3.3-6*b* it is easily seen that $H_{eq}(s)$ is given by

$$H_{eq}(s) = \frac{k_3 s^2 + (k_3 + 2k_2)s + 2k_1}{2} \tag{3.3-10}$$

Note that the numerator of $H_{eq}(s)$ is a quadratic in $s$, of the form $As^2 + Bs + C$, and that the location of the zeros may be determined as desired by an appropriate choice of $k_1$, $k_2$, and $k_3$. It is interesting to note also that the zero locations are independent of the controller gain $K$.

The closed-loop transfer function of the system may now be obtained by making use of Eq. (3.3-9). The result, as shown in Fig. 3.3-6*c*, is

$$\frac{y(s)}{r(s)} = \frac{4K}{s^3 + (5 + 2Kk_3)s^2 + (4 + 2Kk_3 + 4Kk_2)s + 4Kk_1} \tag{3.3-11}$$

The denominator of $y(s)/r(s)$ is a cubic in $s$, and the three coefficients may be adjusted by varying the feedback coefficients and the gain $K$. Thus the poles of the closed-loop system may be placed at any desired location. By selecting the three coefficients in the closed-loop denomi-

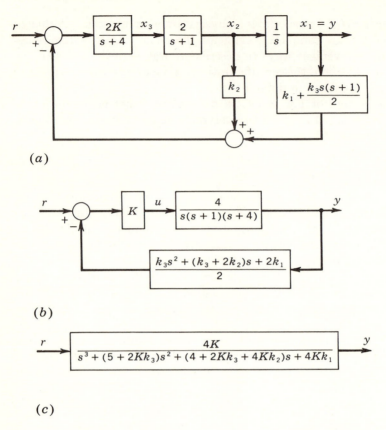

(a)

(b)

(c)

**Fig. 3.3-6**  Step-by-step reduction to the $H_{eq}$ form.     (a) Movement
of $k_3$ forward to combine with $k_1$; (b) $H_{eq}$ form; (c) resulting closed-loop transfer function.

nator, only three equations are established for the determination of the
four variable elements: the three feedback coefficients and the gain.
Hence one of these elements may be selected at random, or one additional relationship such as zero steady-state error may be enforced.   We
shall see later that both methods are of practical importance.

Let us consider next the reduction of the system of Fig. 3.3-4b to
the $G_{eq}$ form of Fig. 3.3-5b.   This may be accomplished by beginning
with the inner loop and applying the reduction of Fig. 2.2-8f on each
successive feedback loop.   The outermost loop must be broken into two
loops containing $k_1 - 1$ and $1$ as feedback coefficients, so that the desired
unity-gain feedback form may be achieved.   This procedure is illustrated in a step-by-step manner in Fig. 3.3-7.

In Fig. 3.3-7$a$ the $k_3$ feedback loop has been reduced.  Note that $k_3$ affects only the position of the pole that had been at $s = -4$.  If the transfer function of the first block had been $K(s + a)/(s + 4)$ rather than just $2K/(s + 4)$, the zero at $s = -a$ would be unaffected.  Thus the effect of feedback through pure-gain elements is to alter pole locations but to leave zeros, if any exist, unaltered.  *No new zeros are ever created.*  These statements are true regardless of the complexity of the plant transfer function, as we shall show in the next section.

The reduction of the $k_2$ loop is shown in Fig. 3.3-7$b$ along with the division of the $k_1$ loop into two parts as required.  The final reduction of the $k_1 - 1$ loop yields the desired single, unity-gain-feedback-loop,

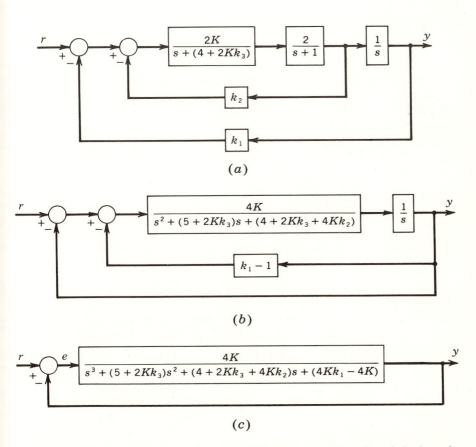

(a)

(b)

(c)

**Fig. 3.3-7**   Step-by-step reduction to the $G_{eq}$ form.   (a) Absorption of the $k_3$ path; (b) absorption of $k_2$ and the division of $k_1$; (c) $G_{eq}$ form.

$G_{eq}$ form as shown in Fig. 3.3-7c, where $G_{eq}$ is easily recognized as

$$G_{eq}(s) = \frac{4K}{s^3 + (5 + 2Kk_3)s^2 + (4 + 2Kk_3 + 4Kk_2)s + (4Kk_1 - 4K)}$$

$$(3.3-12)$$

The closed-loop transfer function may be determined by applying Eq. (3.3-9) and is identical to the result obtained by using the $H_{eq}$ approach. Note that the input applied to $G_{eq}(s)$ is the error signal $e(s) = r(s) - y(s)$, that is, the difference between the input $r(s)$ and the output $y(s)$.

On the basis of this third-order example, it seems rather a large step to assume that all single-input, single-output linear control systems can be represented in terms of either the $G_{eq}(s)$ or $H_{eq}(s)$ form. However, this is the case, and it is difficult to overemphasize the utility of these two forms of system representation. In the next section the quantities $G_{eq}(s)$ and $H_{eq}(s)$ are determined in terms of the matrices that make up the system description, that is, in terms of $\mathbf{A}$, $\mathbf{b}$, $\mathbf{c}$, and $\mathbf{k}$.

With regard to this third-order example, we were also able to conclude that the zeros of the closed-loop system are identical to those of the open-loop plant. Perhaps more important is the fact that the closed-loop-system poles may be positioned anywhere. This result is also true in general, as is indicated in the next section.

**Exercises 3.3**    *3.3-1.*    Find the $H_{eq}(s)$ and $G_{eq}(s)$ representations of the feedback control system of Fig. 3.3-8.

*answers:*

$$H_{eq}(s) = \frac{0.1s^2 + 2.3s + 2}{2}$$

$$G_{eq}(s) = \frac{2K}{s^3 + (13 + 0.1K)s^2 + (30 + 2.3K)s}$$

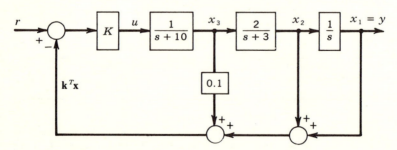

**Fig. 3.3-8**    Exercise 3.3-1.

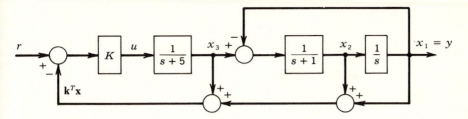

**Fig. 3.3-9**   Exercise 3.3-2.

*3.3-2.*   Determine the closed-loop transfer function $y(s)/r(s)$ for the system of Fig. 3.3-9 by using both forms of Eq. (3.3-9).

*answer:*

$$\frac{y(s)}{r(s)} = \frac{K}{s^3 + (6 + K)s^2 + (6 + 2K)s + (5 + 2K)}$$

## 3.4   Linear state-variable feedback—general case

In many situations the block-diagram methods of the preceding section provide an easy means of reducing a given closed-loop system to either the $G_{eq}(s)$ or the $H_{eq}(s)$ form.   In this section we do not consider a specific example as we did in Sec. 3.3, and therefore there are no block diagrams with which to work.   Hence we are forced to return to the more universal matrix approach.   By starting with the system equations it is a relatively simple matter to determine general expressions for both $G_{eq}(s)$ and $H_{eq}(s)$.   In addition, expressions for the loop transfer function $KG_p(s)H_{eq}(s)$ and the closed-loop transfer function $y(s)/r(s)$ are also derived, both in terms of the resolvent matrix of the plant and in terms of the resolvent matrix of the system.   The reader may recall that in Chap. 2 we established that the plant transfer function can be written as

$$G_p(s) = \frac{y(s)}{u(s)} = \mathbf{c}^T \boldsymbol{\Phi}(s)\mathbf{b}$$

Here $G_p(s)$ is given in terms of the resolvent matrix of the plant.

In order to treat the most general case, let us assume an $n$th-order plant described as always by

$$\dot{\mathbf{x}} = \mathbf{A}\mathbf{x} + \mathbf{b}u \tag{Ab}$$

with

$$y = \mathbf{c}^T\mathbf{x} \tag{c}$$

Next we assume that $u$ is generated by a linear state-variable-feedback relationship of the form

$$u = K(r - \mathbf{k}^T \mathbf{x}) \tag{k}$$

If this expression is substituted into Eq. (**Ab**), the desired closed-loop representation is obtained:

$$\dot{\mathbf{x}} = \mathbf{A}\mathbf{x} - K\mathbf{b}\mathbf{k}^T\mathbf{x} + K\mathbf{b}r$$

Grouping the two terms involving $\mathbf{x}$, we have a complete description of the closed-loop system.

$$\dot{\mathbf{x}} = (\mathbf{A} - K\mathbf{b}\mathbf{k}^T)\mathbf{x} + K\mathbf{b}r \tag{3.4-1}$$
$$y = \mathbf{c}^T\mathbf{x}$$

In order to put this result into a more compact form, let us make the following definition:

$$\mathbf{A}_k = \mathbf{A} - K\mathbf{b}\mathbf{k}^T \tag{3.4-2}$$

Then the closed-loop-system representation becomes

$$\dot{\mathbf{x}} = \mathbf{A}_k\mathbf{x} + K\mathbf{b}r \tag{$\mathbf{A}_k\mathbf{b}$}$$
$$y = \mathbf{c}^T\mathbf{x} \tag{c}$$

Once again we note that the closed-loop-system representation is identical to the original plant representation except that the $\mathbf{A}$ matrix has been changed to the *closed-loop-system matrix* $\mathbf{A}_k$ and $u$ becomes $Kr$.

Before considering the $G_{eq}$ and $H_{eq}$ approach to closed-loop-system representation, let us determine the closed-loop transfer function directly from the above state-variable representation. This result may be obtained by employing the same procedure that was used in Sec. 2.4 to find the plant transfer function. We begin by taking the Laplace transform of Eqs. ($\mathbf{A}_k\mathbf{b}$) and (**c**), assuming all initial conditions are zero.

$$s\mathbf{x}(s) = \mathbf{A}_k\mathbf{x}(s) + K\mathbf{b}r(s) \tag{3.4-3}$$
$$y(s) = \mathbf{c}^T\mathbf{x}(s) \tag{3.4-4}$$

If we group the two terms involving $\mathbf{x}(s)$ in Eq. (3.4-3) and solve for $\mathbf{x}(s)$, the result is

$$\mathbf{x}(s) = K(s\mathbf{I} - \mathbf{A}_k)^{-1}\mathbf{b}r(s) \tag{3.4-5}$$

Following the previous definition of $\mathbf{\Phi}(s)$, we define the *closed-loop resolvent matrix* as[1]

$$\mathbf{\Phi}_k(s) = (s\mathbf{I} - \mathbf{A}_k)^{-1} = (s\mathbf{I} - \mathbf{A} + K\mathbf{b}\mathbf{k}^T)^{-1} \tag{3.4-6}$$

---

[1] See Prob. 3.7-8 for a relationship between $\mathbf{\Phi}(s)$ and $\mathbf{\Phi}_k(s)$.

In terms of this definition, Eq. (3.4-5) becomes

$$\mathbf{x}(s) = K\boldsymbol{\Phi}_k(s)\mathbf{b}r(s) \tag{3.4-7}$$

If this result is substituted into Eq. (3.4-4), $y(s)$ is given by

$$y(s) = K\mathbf{c}^T\boldsymbol{\Phi}_k(s)\mathbf{b}r(s)$$

so that the closed-loop transfer function is

$$\frac{y(s)}{r(s)} = K\mathbf{c}^T\boldsymbol{\Phi}_k(s)\mathbf{b} \tag{3.4-8}$$

This is the desired result.

By making use of the fact that

$$\boldsymbol{\Phi}_k(s) = (s\mathbf{I} - \mathbf{A}_k)^{-1} = \frac{\text{adj } (s\mathbf{I} - \mathbf{A}_k)}{\det (s\mathbf{I} - \mathbf{A}_k)}$$

the above result may be written as

$$\frac{y(s)}{r(s)} = K\frac{\mathbf{c}^T[\text{adj } (s\mathbf{I} - \mathbf{A}_k)]\mathbf{b}}{\det (s\mathbf{I} - \mathbf{A}_k)} \tag{3.4-9}$$

Here $\mathbf{c}^T[\text{adj } (s\mathbf{I} - \mathbf{A}_k)]\mathbf{b}$ forms the numerator polynomial of the closed-loop transfer function, and $\det (s\mathbf{I} - \mathbf{A}_k)$ forms the denominator, or characteristic, polynomial. The characteristic equation is formed by setting the characteristic polynomial equal to zero:

$$\det (s\mathbf{I} - \mathbf{A}_k) = 0$$

The values of $s$ that satisfy the above equation are then the poles or *eigenvalues of the closed-loop system*. The characteristic polynomial and the characteristic equation derive their names from the fact that the nature of the closed-loop response is characterized by the values or locations of the closed-loop poles. It is important to distinguish between the open- and closed-loop poles. The open-loop poles are associated with $G_p(s)$ and are often known from the knowledge of the given plant. The poles of the closed-loop system are a consequence of the effects of feedback, and they must be found by factoring the characteristic polynomial.[1]

Now let us find matrix expressions for $G_{eq}(s)$ and $H_{eq}(s)$. As previously noted, the $H_{eq}$ and $G_{eq}$ configurations may always be obtained by the use of the block-diagram identities of Fig. 2.2-8. However, the appropriate manipulations are not always as obvious or as simple as they were in the example of Sec. 3.3. The procedures of that section are easy to apply only if the plant consists of a simple cascade of first-order blocks with no internal feedback.

---

[1] Methods of factoring are discussed in Appendix C.

Although there are many control systems that fit this category and for which block-diagram manipulations are simple and effective, there are also many systems that are not of this form. For such systems, general matrix expressions for $H_{eq}(s)$ and $G_{eq}(s)$ can often be helpful. In addition, such matrix expressions are very useful in establishing general conclusions concerning the effects of linear state-variable feedback. These matrix expressions are also useful in the development of computer programs for the analysis and design of state-variable-feedback systems.[1]

Lest the reader be confused, it should be noted that block-diagram manipulations are usually the most efficient method for determining $H_{eq}(s)$ and $G_{eq}(s)$ even for highly interacting plants. However, no general rules can be given for these manipulations except in the special pure cascade case, and this has been covered in detail in Sec. 3.3. Matrices are therefore the only universally applicable approach to the determination of $H_{eq}(s)$ and $G_{eq}(s)$.

For easy reference, the $H_{eq}$ and $G_{eq}$ configurations are repeated in Fig. 3.4-1. Let us begin with the determination of $H_{eq}(s)$. From Fig. 3.4-1a, we see that $H_{eq}(s)$ is given by

$$H_{eq}(s) = \frac{\mathbf{k}^T \mathbf{x}(s)}{y(s)}$$

Since $y(s) = \mathbf{c}^T \mathbf{x}(s)$, this expression becomes

$$H_{eq}(s) = \frac{\mathbf{k}^T \mathbf{x}(s)}{\mathbf{c}^T \mathbf{x}(s)}$$

If we next use Eq. (2.4-8) to write $\mathbf{x}(s)$ as

$$\mathbf{x}(s) = \mathbf{\Phi}(s)\mathbf{b}u(s)$$

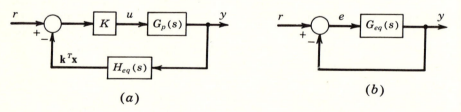

$$(a) \qquad\qquad\qquad (b)$$

**Fig. 3.4-1**  Two basic configurations for the linear state-variable-feedback system: (a) $H_{eq}$ form; (b) $G_{eq}$ form.

---

[1] See J. L. Melsa, "Computer Programs for Computational Assistance in the Study of Linear Control Theory," McGraw-Hill Book Company, New York, 1970.

the expression for $H_{eq}(s)$ becomes

$$H_{eq}(s) = \frac{\mathbf{k}^T \mathbf{\Phi}(s)\mathbf{b}}{\mathbf{c}^T \mathbf{\Phi}(s)\mathbf{b}} \tag{3.4-10}$$

This is the desired matrix expression for $H_{eq}(s)$. The reader is urged to determine $H_{eq}(s)$ for the system discussed in Sec. 3.3 by using this expression. (See Exercise 3.4-1.)

In terms of this result, the *open-loop* or just *loop transfer function* is defined as $KG_p(s)H_{eq}(s)$ takes on a particularly simple form since, as the reader will remember $G_p(s)$ is given by

$$G_p(s) = \mathbf{c}^T \mathbf{\Phi}(s)\mathbf{b} \tag{3.4-11}$$

and the forward transfer function $KG_p(s)$ is therefore

$$KG_p(s) = K\mathbf{c}^T \mathbf{\Phi}(s)\mathbf{b} \tag{3.4-12}$$

Therefore the open-loop transfer function $KG_p(s)H_{eq}(s)$ becomes simply

$$KG_p(s)H_{eq}(s) = K\mathbf{c}^T \mathbf{\Phi}(s)\mathbf{b} \frac{\mathbf{k}^T \mathbf{\Phi}(s)\mathbf{b}}{\mathbf{c}^T \mathbf{\Phi}(s)\mathbf{b}} = K\mathbf{k}^T \mathbf{\Phi}(s)\mathbf{b} \tag{3.4-13}$$

where the cancellation is possible because the quantities involved are scalars.

Since the closed-loop transfer function is given by Eq. (3.3-9) as

$$\frac{y(s)}{r(s)} = \frac{KG_p(s)}{1 + KG_p(s)H_{eq}(s)}$$

the above results may be used to write $y(s)/r(s)$:

$$\frac{y(s)}{r(s)} = \frac{K\mathbf{c}^T \mathbf{\Phi}(s)\mathbf{b}}{1 + K\mathbf{k}^T \mathbf{\Phi}(s)\mathbf{b}} \tag{3.4-14}$$

This expression affords an alternative method of determining $y(s)/r(s)$ by matrix means. Note that this result depends on the plant resolvent matrix $\mathbf{\Phi}(s)$ rather than the closed-loop resolvent matrix $\mathbf{\Phi}_k(s)$.

In order to find the matrix expression for $G_{eq}(s)$, we begin by solving Eq. (3.3-9) for $G_{eq}(s)$ in terms of $G_p(s)$ and $H_{eq}(s)$.

$$G_{eq}(s) = \frac{KG_p(s)}{1 + KG_p(s)H_{eq}(s) - KG_p(s)}$$

If we substitute the above expressions for $KG_p(s)$ and $KG_p(s)H_{eq}(s)$, then $G_{eq}(s)$ becomes

$$G_{eq}(s) = \frac{K\mathbf{c}^T \mathbf{\Phi}(s)\mathbf{b}}{1 + K(\mathbf{k} - \mathbf{c})^T \mathbf{\Phi}(s)\mathbf{b}} \tag{3.4-15}$$

Alternatively we can solve Eq. (3.3-9) for $G_{eq}(s)$ in terms of $y(s)/r(s)$:

$$G_{eq}(s) = \frac{y(s)/r(s)}{1 - y(s)/r(s)}$$

Obviously if we use Eq. (3.4-14) for $y(s)/r(s)$, we obtain Eq. (3.4-15) for $G_{eq}(s)$ once again. However, if we substitute Eq. (3.4-8) for $y(s)/r(s)$, we obtain a new expression for $G_{eq}(s)$ in terms of the closed-loop resolvent matrix:

$$G_{eq}(s) = \frac{K\mathbf{c}^T\mathbf{\Phi}_k(s)\mathbf{b}}{1 - K\mathbf{c}^T\mathbf{\Phi}_k(s)\mathbf{b}} \tag{3.4-16}$$

So far we have determined two expressions for $y(s)/r(s)$ and $G_{eq}(s)$, one depending on $\mathbf{\Phi}(s)$ and one on $\mathbf{\Phi}_k(s)$. We may express $H_{eq}(s)$ in terms of $\mathbf{\Phi}_k(s)$ by simply substituting Eq. (3.4-7) for $\mathbf{x}(s)$ rather than Eq. (2.4-8) to obtain

$$H_{eq}(s) = \frac{\mathbf{k}^T\mathbf{\Phi}_k(s)\mathbf{b}}{\mathbf{c}^T\mathbf{\Phi}_k(s)\mathbf{b}} \tag{3.4-17}$$

As yet we have no expression for the plant in terms of $\mathbf{\Phi}_k(s)$, the system resolvent matrix. It may seem somewhat unnecessary to express $G_p(s)$ in terms of $\mathbf{\Phi}_k(s)$, but since this is easily accomplished, it is done here for completeness. Since $y(s)/r(s)$ and $H_{eq}(s)$ are both known in terms of $\mathbf{\Phi}_k(s)$ from Eqs. (3.4-8) and (3.4-17), Eq. (3.3-9) may be solved directly for $KG_p(s)$:

$$\frac{KG_p(s)}{1 + KG_p(s)[\mathbf{k}^T\mathbf{\Phi}_k(s)\mathbf{b}/\mathbf{c}^T\mathbf{\Phi}_k(s)\mathbf{b}]} = K\mathbf{c}^T\mathbf{\Phi}_k(s)\mathbf{b}$$

or

$$KG_p(s) = \frac{K\mathbf{c}^T\mathbf{\Phi}_k(s)\mathbf{b}}{1 - K\mathbf{k}^T\mathbf{\Phi}_k(s)\mathbf{b}} \tag{3.4-18}$$

Thus $G_p(s)$ is just

$$G_p(s) = \frac{\mathbf{c}^T\mathbf{\Phi}_k(s)\mathbf{b}}{1 - K\mathbf{k}^T\mathbf{\Phi}_k(s)\mathbf{b}} \tag{3.4-19}$$

and the loop transfer function $KG_p(s)H_{eq}(s)$ is also known to be

$$KG_p(s)H_{eq}(s) = \frac{K\mathbf{k}^T\mathbf{\Phi}_k(s)\mathbf{b}}{1 - K\mathbf{k}^T\mathbf{\Phi}_k(s)\mathbf{b}} \tag{3.4-20}$$

All the above matrix expressions are summarized for easy reference in Table 3.4-1.

## Table 3.4-1    *Summary of Matrix-relation Results*

| | $\Phi(s)$ *Expression* | $\Phi_k(s)$ *Expression* |
|---|---|---|
| $G_p(s)$<br>Plant transfer<br>function | $\mathbf{c}^T\Phi(s)\mathbf{b}$ <br> Eq. (3.4-11) | $\dfrac{\mathbf{c}^T\Phi_k(s)\mathbf{b}}{1 - Kk^T\Phi_k(s)\mathbf{b}}$ <br> Eq. (3.4-19) |
| $KG_p(s)$<br>Forward transfer<br>function | $K\mathbf{c}^T\Phi(s)\mathbf{b}$ <br> Eq. (3.4-12) | $\dfrac{K\mathbf{c}^T\Phi_k(s)\mathbf{b}}{1 - Kk^T\Phi_k(s)\mathbf{b}}$ <br> Eq. (3.4-18) |
| $H_{eq}(s)$<br>Equivalent feedback<br>transfer function | $\dfrac{k^T\Phi(s)\mathbf{b}}{\mathbf{c}^T\Phi(s)\mathbf{b}}$ <br> Eq. (3.4-10) | $\dfrac{k^T\Phi_k(s)\mathbf{b}}{\mathbf{c}^T\Phi_k(s)\mathbf{b}}$ <br> Eq. (3.4-17) |
| $KG_p(s)H_{eq}(s)$<br>Open-loop transfer<br>function | $Kk^T\Phi(s)\mathbf{b}$ <br> Eq. (3.4-13) | $\dfrac{Kk^T\Phi_k(s)\mathbf{b}}{1 - Kk^T\Phi_k(s)\mathbf{b}}$ <br> Eq. (3.4-20) |
| $G_{eq}(s)$<br>Equivalent forward<br>transfer function | $\dfrac{K\mathbf{c}^T\Phi(s)\mathbf{b}}{1 + K(k - \mathbf{c})^T\Phi(s)\mathbf{b}}$ <br> Eq. (3.4-15) | $\dfrac{K\mathbf{c}^T\Phi_k(s)\mathbf{b}}{1 - K\mathbf{c}^T\Phi_k(s)\mathbf{b}}$ <br> Eq. (3.4-16) |
| $y(s)/r(s)$<br>Closed-loop transfer<br>function | $\dfrac{K\mathbf{c}^T\Phi(s)\mathbf{b}}{1 + Kk^T\Phi(s)\mathbf{b}}$ <br> Eq. (3.4-14) | $K\mathbf{c}^T\Phi_k(s)\mathbf{b}$ <br> Eq. (3.4-8) |

***Example 3.4-1***    As an example of the use of the expressions of Table 3.4-1, let us return to the field-controlled motor of Sec. 3.3, as given in the block diagram of Fig. 3.3-4.    In order to make the example more specific, let us choose $k_1 = 1$, $k_2 = \frac{13}{40}$, and $k_3 = \frac{7}{20}$.

Before any use can be made of Table 3.4-1, the equations describing either the plant or the system must be written.    From these equations, $\mathbf{A}$ or $\mathbf{A}_k$ may be used to find $\Phi(s)$ or $\Phi_k(s)$, as required in the expressions for the various transfer functions.    Here use is made of the plant equations, which are

$$\dot{x}_1 = x_2$$
$$\dot{x}_2 = -x_2 + 2x_3$$
$$\dot{x}_3 = -4x_3 + 2u$$

so that $\mathbf{A}$ is

$$\mathbf{A} = \begin{bmatrix} 0 & 1 & 0 \\ 0 & -1 & 2 \\ 0 & 0 & -4 \end{bmatrix}$$

and **b** and **c** are

$$\mathbf{b} = \begin{bmatrix} 0 \\ 0 \\ 2 \end{bmatrix} \quad \text{and} \quad \mathbf{c} = \begin{bmatrix} 1 \\ 0 \\ 0 \end{bmatrix}$$

Use of the formula $\boldsymbol{\Phi}(s) = (s\mathbf{I} - \mathbf{A})^{-1}$ yields

$$\boldsymbol{\Phi}(s) = (s\mathbf{I} - \mathbf{A})^{-1}$$
$$= \frac{1}{s(s+1)(s+4)} \begin{bmatrix} (s+1)(s+4) & s+4 & 2 \\ 0 & s(s+4) & 2s \\ 0 & 0 & s(s+1) \end{bmatrix}$$

and we may now use the $\boldsymbol{\Phi}(s)$ column of Table 3.4-1 to calculate the various transfer functions.

From Fig. 3.3-4 it is evident that the plant transfer function $G_p(s) = y(s)/u(s)$ is

$$G_p(s) = \frac{4}{s(s+1)(s+4)}$$

This same result may be obtained from the first entry in the $\boldsymbol{\Phi}(s)$ column of Table 3.4-1 [Eq. (3.4-11)], which states

$$G_p(s) = \mathbf{c}^T \boldsymbol{\Phi}(s) \mathbf{b} \tag{3.4-11}$$

In this example, therefore,

$$G_p(s) = \begin{bmatrix} 1 & 0 & 0 \end{bmatrix} \left\{ \frac{1}{s(s+1)(s+4)} \right.$$
$$\left. \begin{bmatrix} (s+1)(s+4) & s+4 & 2 \\ 0 & s(s+4) & 2s \\ 0 & 0 & s(s+1) \end{bmatrix} \begin{bmatrix} 0 \\ 0 \\ 2 \end{bmatrix} \right\}$$
$$= 2\phi_{13}(s) = \frac{4}{s(s+1)(s+4)}$$

as required.

Let us also calculate $H_{eq}(s)$, given in Table 3.4-1 as

$$H_{eq}(s) = \frac{\mathbf{k}^T \boldsymbol{\Phi}(s) \mathbf{b}}{\mathbf{c}^T \boldsymbol{\Phi}(s) \mathbf{b}}$$

The denominator of $H_{eq}(s)$ is already known, as it is just $G_p(s)$. The numerator is

$$\mathbf{k}^T \mathbf{\Phi}(s)\mathbf{b} = \begin{bmatrix} 1 & \frac{13}{40} & \frac{7}{20} \end{bmatrix} \begin{bmatrix} \phi_{11} & \phi_{12} & \phi_{13} \\ \phi_{21} & \phi_{22} & \phi_{23} \\ \phi_{31} & \phi_{32} & \phi_{33} \end{bmatrix} \begin{bmatrix} 0 \\ 0 \\ 2 \end{bmatrix}$$

$$= \begin{bmatrix} 1 & \frac{13}{40} & \frac{7}{20} \end{bmatrix} \begin{bmatrix} 2\phi_{13} \\ 2\phi_{23} \\ 2\phi_{33} \end{bmatrix} = \frac{4 + 13s/10 + 7s(s+1)/10}{s(s+1)(s+4)}$$

The feedback transfer function $H_{eq}(s)$ is found by dividing the above by $\mathbf{c}^T \mathbf{\Phi}(s)\mathbf{b}$, or

$$H_{eq}(s) = \frac{4 + 13s/10 + (7s/10)(s+1)}{4}$$

or equivalently

$$H_{eq}(s) = \frac{7s^2 + 20s + 40}{40}$$

The previous expression for $H_{eq}(s)$ from Eq. (3.3-10) was

$$H_{eq}(s) = \frac{k_3 s^2 + (k_3 + 2k_2)s + 2k_1}{2}$$

and the two results are seen to agree exactly for the given feedback coefficients.

It may seem somewhat ridiculous in the above example to use the matrix expressions to find $G_p(s)$ and $H_{eq}(s)$. They may be determined almost by inspection from Fig. 3.3-4, whereas the matrix approach requires the inversion of a $3 \times 3$ matrix. The point of this section is generality and a demonstration of the equivalence of the block-diagram and the matrix approaches. In an actual application, the easier method is used. In situations where the plant being controlled is not simply a group of transfer functions in series, the matrix approach may be easier. In any event, the generality of the matrix approach is convenient to illustrate the properties of linear state-variable feedback.

Exercise 3.4-1 requires the completion of Example 3.4-1 to find all the transfer functions indicated in Table 3.4-1.

*Properties of linear state-variable feedback.* In order to establish some of the properties of linear state-variable feedback, let us consider a general plant having $n$ poles and $m$ zeros whose transfer function is therefore

$$G_p(s) = \frac{c_{m+1}s^m + c_m s^{m-1} + \cdots + c_1}{s^n + a_n s^{n-1} + \cdots + a_1}$$

Since we are interested in properties of transfer functions, the specific state-variable representation is immaterial. Therefore, for simplicity, we make use of phase variables so that the plant representation becomes

$$\dot{\mathbf{x}} = \begin{bmatrix} 0 & 1 & \cdots & 0 \\ 0 & 0 & \cdots & 0 \\ \cdots & \cdots & \cdots & \cdots \\ -a_1 & -a_2 & \cdots & -a_n \end{bmatrix} \mathbf{x} + \begin{bmatrix} 0 \\ 0 \\ \cdots \\ 1 \end{bmatrix} u$$

and

$$y = [c_1 \quad c_2 \quad \cdots \quad c_{m+1} \quad 0 \quad \cdots \quad 0]\mathbf{x}$$

Let us examine $H_{eq}(s)$ first by considering Eq. (3.4-10). Since $\mathbf{\Phi}(s)$ is given by

$$\mathbf{\Phi}(s) = (s\mathbf{I} - \mathbf{A})^{-1} = \frac{\text{adj }(s\mathbf{I} - \mathbf{A})}{\det (s\mathbf{I} - \mathbf{A})} \tag{3.4-21}$$

Eq. (3.4-10) may be written

$$\begin{aligned} H_{eq}(s) &= \frac{\mathbf{k}^T[\text{adj }(s\mathbf{I} - \mathbf{A})]\mathbf{b}/\det (s\mathbf{I} - \mathbf{A})}{\mathbf{c}^T[\text{adj }(s\mathbf{I} - \mathbf{A})]\mathbf{b}/\det (s\mathbf{I} - \mathbf{A})} \\ &= \frac{\mathbf{k}^T[\text{adj }(s\mathbf{I} - \mathbf{A})]\mathbf{b}}{\mathbf{c}^T[\text{adj }(s\mathbf{I} - \mathbf{A})]\mathbf{b}} \end{aligned} \tag{3.4-22}$$

If we recall that $G_p(s)$ can be expressed by Eq. (2.4-11) or Eq. (3.4-11) as

$$G_p(s) = \mathbf{c}^T\mathbf{\Phi}(s)\mathbf{b} = \frac{\mathbf{c}^T[\text{adj }(s\mathbf{I} - \mathbf{A})]\mathbf{b}}{\det (s\mathbf{I} - \mathbf{A})}$$

we see that

$$\det (s\mathbf{I} - \mathbf{A}) = s^n + a_n s^{n-1} + \cdots + a_1 \tag{3.4-23}$$

and

$$\mathbf{c}^T[\text{adj }(s\mathbf{I} - \mathbf{A})]\mathbf{b} = c_{m+1}s^m + c_m s^{m-1} + \cdots + c_1 \tag{3.4-24}$$

By analogous reasoning, $\mathbf{k}^T[\text{adj }(s\mathbf{I} - \mathbf{A})]\mathbf{b}$ must therefore be equal to

$$\mathbf{k}^T[\text{adj }(s\mathbf{I} - \mathbf{A})]\mathbf{b} = k_n s^{n-1} + k_{n-1}s^{n-2} + \cdots + k_1 \tag{3.4-25}$$

since the only change from Eq. (3.4-24) is that $\mathbf{k}$ has been substituted for $\mathbf{c}$. Note, however, that $c_i$ is zero for $i > m + 1$, whereas normally this is not true for the elements of $\mathbf{k}$.

If Eqs. (3.4-24) and (3.4-25) are substituted into Eq. (3.4-22) for $H_{eq}(s)$, we have

$$H_{eq}(s) = \frac{k_n s^{n-1} + k_{n-1} s^{n-2} + \cdots + k_1}{c_{m+1} s^m + c_m s^{m-1} + \cdots + c_1} \tag{3.4-26}$$

Here it is seen that the numerator of $H_{eq}(s)$ is a $(n-1)$st-order polynomial in $s$ and that all the coefficients are adjustable by proper selection of **k**. In other words, $H_{eq}(s)$ has $n-1$ arbitrary zeros whose locations are under the direct control of the designer.

In addition, we note the poles of $H_{eq}(s)$ are exactly equal to the zeros of $G_p(s)$; that is, the denominator polynomial of $H_{eq}(s)$ is equal to the numerator polynomial of $G_p(s)$.

The loop transfer function $KG_p(s)H_{eq}(s)$ is therefore given by

$$KG_p(s)H_{eq}(s) = \frac{K(k_n s^{n-1} + k_{n-1} s^{n-2} + \cdots + k_1)}{s^n + a_n s^{n-1} + \cdots + a_2 s + a_1} \tag{3.4-27}$$

The arbitrary zeros of $H_{eq}(s)$ are the only zeros of the loop transfer function, and the poles of $G_p(s)$ are the only poles. In other words, the loop transfer function has poles where $G_p(s)$ has poles and $n-1$ zeros whose locations are determined by **k**.

Let us consider next the closed-loop transfer function $y(s)/r(s)$ by making use of Eq. (3.4-14) to write $y(s)/r(s)$ as

$$\frac{y(s)}{r(s)} = \frac{K\mathbf{c}^T \mathbf{\Phi}(s)\mathbf{b}}{1 + K\mathbf{k}^T \mathbf{\Phi}(s)\mathbf{b}}$$

If we now substitute for $\mathbf{\Phi}(s)$ by using Eq. (3.4-21), $y(s)/r(s)$ becomes

$$\begin{aligned}
\frac{y(s)}{r(s)} &= \frac{K\mathbf{c}^T[\text{adj }(s\mathbf{I} - \mathbf{A})]\mathbf{b}/\det(s\mathbf{I} - \mathbf{A})}{1 + K\mathbf{k}^T[\text{adj }(s\mathbf{I} - \mathbf{A})]\mathbf{b}/\det(s\mathbf{I} - \mathbf{A})} \\
&= \frac{K\mathbf{c}^T[\text{adj }(s\mathbf{I} - \mathbf{A})]\mathbf{b}}{\det(s\mathbf{I} - \mathbf{A}) + K\mathbf{k}^T[\text{adj }(s\mathbf{I} - \mathbf{A})]\mathbf{b}}
\end{aligned} \tag{3.4-28}$$

By making use of Eqs. (3.4-23) to (3.4-25), this result may be rewritten as

$$\frac{y(s)}{r(s)} = \frac{K(c_{m+1} s^m + c_m s^{m-1} + \cdots + c_1)}{(s^n + a_n s^{n-1} + \cdots + a_1) + K(k_n s^{n-1} + k_{n-1} s^{n-2} + \cdots + k_1)}$$

Grouping the like powers of $s$ in the denominator, we have

$$\frac{y(s)}{r(s)} = \frac{K(c_{m+1} s^m + c_m s^{m-1} + \cdots + c_1)}{s^n + (a_n + Kk_n)s^{n-1} + \cdots + (a_1 + Kk_1)} \tag{3.4-29}$$

This result could also have been obtained by writing the phase-variable representation of the closed-loop system:

$$\dot{\mathbf{x}} = \begin{bmatrix} 0 & 1 & \cdots & 0 \\ 0 & 0 & \cdots & 0 \\ \cdots & \cdots & \cdots & \cdots \\ -(a_1 + Kk_1) & -(a_2 + Kk_2) & \cdots & -(a_n + Kk_n) \end{bmatrix} \mathbf{x} + \begin{bmatrix} 0 \\ 0 \\ \cdot \\ \cdot \\ \cdot \\ 1 \end{bmatrix} Kr$$

$$y = [c_1 \quad c_2 \quad \cdots \quad c_{m+1} \quad 0 \quad \cdots \quad 0]\mathbf{x}$$

From this result, the closed-loop transfer function may be determined by inspection. The result is identical to Eq. (3.4-29).

From Eq. (3.4-29) we see that the coefficients of the denominator polynomial of $y(s)/r(s)$ may be adjusted at will by proper selection of **k** and $K$. Once again, since there are only $n$ coefficients and $n + 1$ adjustable parameters, it is possible to fix arbitrarily one of the parameters to enforce another condition on the system.

Note also the fact that except for $K$ the numerator of $y(s)/r(s)$ is identical to the numerator of $G_p(s)$; that is, the closed-loop zeros are equal to the open-loop zeros. In other words, linear state-variable feedback has no effect on the zeros of $y(s)/r(s)$. However, the freedom of choice of the poles of $y(s)/r(s)$ means that closed-loop poles may be positioned anywhere that is desired. A means of removing unwanted zeros is to place a closed-loop pole at the same location as the zero, effectively canceling it.

To summarize, the following conclusions concerning the effects of linear state-variable feedback may be made.

1. Feedback transfer function $H_{eq}(s)$:
   a. The poles of $H_{eq}(s)$ are the zeros of $G_p(s)$.
   b. $H_{eq}(s)$ has $n - 1$ arbitrary zeros.
2. Loop transfer function $KG_p(s)H_{eq}(s)$:
   a. The poles of $KG_p(s)H_{eq}(s)$ are the poles of $G_p(s)$.
   b. The zeros of $KG_p(s)H_{eq}(s)$ are the zeros of $H_{eq}(s)$; that is, there are $n - 1$ arbitrary zeros.
3. Closed-loop transfer function $y(s)/r(s)$:
   a. The poles of $y(s)/r(s)$ may be arbitrarily positioned by proper selection of **k** and $K$.
   b. The zeros of $y(s)/r(s)$ are the zeros of $G_p(s)$.

One might view the above conclusions jointly and say that the ability to select the locations of the loop-transfer-function zeros, which are the zeros of $H_{eq}(s)$, ensures that the closed-loop poles can be located at will.

If one is not careful, two possible situations can arise that may appear to violate the properties of state-variable feedback developed above. First, if the system description is not in the form of Eq. (**Ab**), the properties stated above are usually violated.[1] This situation most often occurs because of a zero in the input block which causes a $\dot{u}$ term to appear in the equations, as discussed in Sec. 2.6. The solution to this problem is to redefine the state variables, using, for example, the feedforward scheme suggested in Sec. 2.6 so that the system is represented by Eqs. (**Ab**) and (**c**).

The second situation is a more insidious one. If $H_{eq}(s)$ is found by using block-diagram methods, it is possible that poles and zeros of $H_{eq}(s)$ may be unknowingly canceled, resulting in an $H_{eq}(s)$ that does not contain as poles all the zeros of $G_p(s)$. Consider, for example, the system shown in Fig. 3.4-2. Using block-diagram manipulations, one may easily find that

$$H_{eq}(s) = k_1 \quad \text{and} \quad G_p(s) = \frac{s+4}{(s+1)(s+2)(s+5)}$$

Note that the zero of $G_p(s)$ at $s = -4$ does *not* appear as a pole of $H_{eq}(s)$, as required. However, if the matrix equation (3.4-10) is used, the correct $H_{eq}(s)$ is found:

$$H_{eq}(s) = \frac{k_1(s+4)}{s+4}$$

In order to avoid this problem, whenever block-diagram methods are used to find $H_{eq}(s)$, one should always check to see if all the zeros of $G_p(s)$ appear as poles of $H_{eq}(s)$. If there are some zeros that do not appear, then $H_{eq}(s)$ should be multiplied in the numerator and denominator by these factors in order to ensure that all the zeros of $G_p(s)$ appear as poles of $H_{eq}(s)$. This may seem an unnecessary complication of the problem; however, by ensuring that the properties of $H_{eq}(s)$ are always satisfied, the later developments may be considerably simplified by excluding special cases. After the reader gains familiarity with the

[1] See Probs. 3.7-6 and 3.7-7.

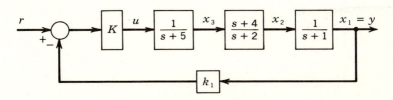

***Fig. 3.4-2***   Example illustrating how $H_{eq}(s)$ can be incorrectly found by block-diagram methods.

methods, he will find several shortcuts that he can use. But, at the beginning, the reader is urged to follow the rules carefully in order to avoid difficulty.

One point that the reader should note from the above discussion is that the poles and zeros of $H_{eq}(s)$ should never be canceled lest the properties of $H_{eq}(s)$ be destroyed. In addition, the poles of $G_p(s)$ should never be canceled by the zeros of $H_{eq}(s)$ in forming the open-loop transfer function although the poles of $H_{eq}(s)$ and the zeros of $G_p(s)$, which are always identical, may be canceled.

In this section we have attempted to generalize the means of closed-loop-system representation discussed in the preceding section and to establish some of the properties of linear state-variable feedback. We shall make repeated use of these results in the following chapters for the analysis and eventual design of closed-loop systems.

**Exercises 3.4**  *3.4-1.* Complete Example 3.4-1 by using the matrix expressions given in Table 3.4-1 to find $G_p(s)$, $H_{eq}(s)$, $G_{eq}(s)$, $KG_p(s)H_{eq}(s)$, and $y(s)/r(s)$ for the field-controlled motor shown in Fig. 3.3-4. As in Example 3.4-1, let $k_1 = 1$, $k_2 = \frac{13}{40}$, and $k_3 = \frac{7}{20}$. Both the $\mathbf{\Phi}(s)$ and the $\mathbf{\Phi}_k(s)$ expressions should be used. Compare the answers with those obtained by block-diagram manipulation in Sec. 3.3.

*3.4-2.* Find $H_{eq}(s)$, $G_{eq}(s)$, and $y(s)/r(s)$ for the positioning system of Fig. 3.4-3 by means of block-diagram manipulations if $R_a = 5$,

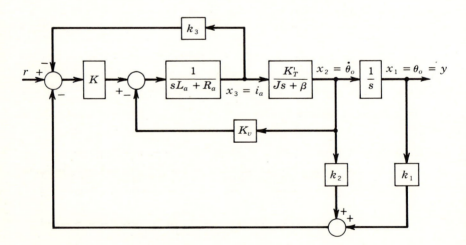

**Fig. 3.4-3**  Exercise 3.4-2.

$L_a = 1$, $K_v = 2$, $K_T' = 2$, $J = 1$, and $\beta = 1$.   Verify the results by the use of the matrix expressions given in Table 3.4-1.

*answers:*

$$H_{eq}(s) = \frac{k_3 s^2 + (k_3 + 2k_2)s + 2k_1}{2}$$

$$G_{eq}(s) = \frac{2K}{s^3 + (6 + Kk_3)s^2 + (9 + Kk_3 + 2Kk_2)s + (2Kk_1 - 2K)}$$

$$\frac{y(s)}{r(s)} = \frac{2K}{s^3 + (6 + Kk_3)s^2 + (9 + Kk_3 + 2Kk_2)s + 2Kk_1}$$

*3.4-3.*   Indicate each correct answer.   More than one answer may be correct.

(a)   The poles of $y(s)/r(s)$ are as follows:
   (*i*)   The poles of $G_p(s)$
   (*ii*)   The zeros of $H_{eq}(s)$
   (*iii*)   Arbitrarily located by $K$ and $\mathbf{k}$
   (*iv*)   The poles of $G_{eq}(s)$
   (*v*)   None of the above
(b)   The zeros of $y(s)/r(s)$ are as follows:
   (*i*)   The zeros of $H_{eq}(s)$
   (*ii*)   The poles of $G_{eq}(s)$
   (*iii*)   The poles of $H_{eq}(s)$
   (*iv*)   Arbitrarily located by $K$ and $\mathbf{k}$
   (*v*)   None of the above
(c)   The loop gain $KG_p(s)H_{eq}(s)$ has the following:
   (*i*)   $n - 1$ arbitrary poles
   (*ii*)   $n$ zeros
   (*iii*)   Arbitrary poles and zeros
   (*iv*)   Zeros identical to those of $G_p(s)$
   (*v*)   None of the above
(d)   A linear transformation of variables does the following:
   (*i*)   It does not alter the eigenvalues.
   (*ii*)   It changes the transfer function.
   (*iii*)   It does not change the $\mathbf{c}$ matrix.
   (*iv*)   It requires a nonsingular transformation matrix.
   (*v*)   It does none of the above.

*answers:*

(*a*)   *iii*; (*b*)   *iii*; (*c*)   *v*; (*d*)   *i, iv*

### 3.5   *The resolvent matrix—a block-diagram approach*

In order to use any of the matrix expressions for the transfer functions in Table 3.4-1, it is necessary to obtain either the open- or the closed-loop resolvent matrix; that is, it is necessary to find $\mathbf{\Phi}(s)$ or $\mathbf{\Phi}_k(s)$. This may be accomplished by using the definitions [Eqs. (2.4-9) and (3.4-6)], which are repeated here:

$$\mathbf{\Phi}(s) = (s\mathbf{I} - \mathbf{A})^{-1} \tag{3.5-1}$$
$$\mathbf{\Phi}_k(s) = (s\mathbf{I} - \mathbf{A} + K\mathbf{b}\mathbf{k}^T)^{-1} = (s\mathbf{I} - \mathbf{A}_k)^{-1} \tag{3.5-2}$$

If these matrix definitions are used, one obtains the entire resolvent matrix.

Often one is not actually interested in the entire resolvent matrix. For instance, suppose that it is desired to find the closed-loop transfer function for the system of Fig. 3.3-4$a$, through the use of the matrix methods of Sec. 3.4. For that figure, the state-variable equations are

$$\dot{x}_1 = x_2$$
$$\dot{x}_2 = -\frac{\beta}{J} x_2 + \frac{K_T}{J} x_3$$
$$\dot{x}_3 = -\frac{R_f}{L_f} x_3 - \frac{1}{L_f} K(r - \mathbf{k}^T\mathbf{x})$$

so that the **b** vector is just

$$\mathbf{b} = \text{col } (0,0,1/L_f)$$

and, since $y = x_1$, **c** is simply

$$\mathbf{c} = \text{col } (1,0,0)$$

From Eq. (3.4-8), $y(s)/r(s)$ is therefore

$$\frac{y(s)}{r(s)} = K\mathbf{c}^T\mathbf{\Phi}_k(s)\mathbf{b}$$

or

$$\frac{y(s)}{r(s)} = \begin{bmatrix} 1 & 0 & 0 \end{bmatrix} \begin{bmatrix} \phi_{k(11)} & \phi_{k(12)} & \phi_{k(13)} \\ \phi_{k(21)} & \phi_{k(22)} & \phi_{k(23)} \\ \phi_{k(31)} & \phi_{k(32)} & \phi_{k(33)} \end{bmatrix} \begin{bmatrix} 0 \\ 0 \\ 1/L_f \end{bmatrix}$$

In expanded form this expression for $y(s)/r(s)$ involves only one element of $\mathbf{\Phi}_k(s)$, since $y(s)/r(s)$ reduces to[1]

$$\frac{y(s)}{r(s)} = \frac{1}{L_f} \phi_{k(13)}(s)$$

---

[1] Parentheses are used on the subscripts of the elements of $\mathbf{\Phi}_k(s)$ to avoid confusion.

If one uses the definition to find $\boldsymbol{\Phi}_k(s)$, unnecessary work is done since only one element of $\boldsymbol{\Phi}_k(s)$ is needed in this case. For this reason it is convenient to have a method for determining individual elements of either $\boldsymbol{\Phi}(s)$ or $\boldsymbol{\Phi}_k(s)$.

The individual elements of $\boldsymbol{\Phi}(s)$ or $\boldsymbol{\Phi}_k(s)$ may be obtained from the matrix equations by determining only the required elements of the adj $(s\mathbf{I} - \mathbf{A}_k)$. In the above example, we need only $\phi_{k(13)}(s)$ so that only that element of adj $(s\mathbf{I} - \mathbf{A}_k)$ needs to be determined. Since the elements of the adjoint matrix are determined separately, this element can be found without computing any of the others. In particular, $\phi_{k(13)}$ is given by

$$\phi_{k(13)} = (-1)^{1+3} \frac{\text{cof } (s\mathbf{I} - \mathbf{A}_k)_{31}}{\det (s\mathbf{I} - \mathbf{A}_k)}$$

In addition to this matrix approach, the individual elements of $\boldsymbol{\Phi}(s)$ or $\boldsymbol{\Phi}_k(s)$ may be determined by the use of block-diagram techniques. This is the approach discussed in this section. The method used to find the elements of the resolvent matrix is based upon the *elementary block-diagram description* of the state-variable equations. The state-variable equations may describe either the open-loop plant, in which case the open-loop resolvent matrix results, or the closed-loop system, in which case the elements of the closed-loop resolvent matrix are found. Attention is focused upon the determination of the elements of $\boldsymbol{\Phi}_k(s)$, since the closed-loop resolvent matrix is necessary in the time-response development of the next chapter.

As a first step, let us show that the elements of $\boldsymbol{\Phi}_k(s)$ are related to initial-condition responses of the various state variables. We begin by assuming that the input to the system is zero, $r(t) = 0$, so that Eq. $(\mathbf{A}_k\mathbf{b})$ becomes simply

$$\dot{\mathbf{x}} = \mathbf{A}_k\mathbf{x} \tag{$\mathbf{A}_k$}$$

Next we take the Laplace transform of this equation as before, except that we no longer assume that the initial conditions on $\mathbf{x}$ are zero. The resulting transformed equation therefore has $-\mathbf{x}(0)$ on the left side so that

$$s\mathbf{x}(s) - \mathbf{x}(0) = \mathbf{A}_k\mathbf{x}(s)$$

or

$$(s\mathbf{I} - \mathbf{A}_k)\mathbf{x}(s) = \mathbf{x}(0)$$

If we solve for $\mathbf{x}(s)$, the result is

$$\mathbf{x}(s) = (s\mathbf{I} - \mathbf{A}_k)^{-1}\mathbf{x}(0)$$

or, in terms of the closed-loop resolvent matrix,

$$\mathbf{x}(s) = \mathbf{\Phi}_k(s)\mathbf{x}(0) \tag{3.5-3}$$

In expanded form the $i$th equation of Eq. (3.5-3) is

$$x_i(s) = \phi_{k(i1)}(s)x_1(0) + \phi_{k(i2)}(s)x_2(0) + \cdots + \phi_{k(in)}(s)x_n(0) \tag{3.5-4}$$

If the initial conditions are assumed to be zero except for the $j$th variable, Eq. (3.5-4) becomes

$$x_i(s) = \phi_{k(ij)}(s)x_j(0)$$

or

$$\phi_{k(ij)}(s) = \frac{x_i(s)}{x_j(0)} \tag{3.5-5}$$

From Eq. (3.5-5) we observe that the element $\phi_{k(ij)}(s)$ is equal to the ratio of the Laplace transform of the $i$th state variable to the initial condition of the $j$th state variable with *all other initial conditions and inputs equal to zero.*

In order to use block-diagram manipulations to find the elements of $\mathbf{\Phi}_k(s)$ by means of Eq. (3.5-5), it is necessary to have some procedure for handling initial conditions on the block diagram. The reader will remember that all initial conditions were assumed to be zero when the transfer function and hence the block-diagram representation were developed. One way of solving this problem is to replace the initial conditions by equivalent impulse inputs that are added to the system through fictitious summers. This is illustrated in Fig. 3.5-1, where we have pictured only one element of a supposedly complicated closed-loop system. If it is assumed for the moment that $\delta_j(s)$ is the only input to the $j$th integrator shown, the transfer-function relationship between this input and the output is simply

$$\frac{x_j(s)}{\delta_j(s)} = \frac{1}{s}$$

If it is assumed that the magnitude of the impulse is $x_j(0)$, since the

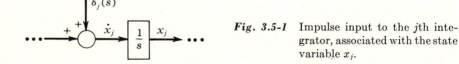

**Fig. 3.5-1** Impulse input to the $j$th integrator, associated with the state variable $x_j$.

Laplace transform of an impulse is just the magnitude of the impulse, $\delta_j(s)$ is just

$$\delta_j(s) = x_j(0)$$

so that

$$x_j(s) = \frac{x_j(0)}{s}$$

To find the initial value of $x_j(t)$, or $x_j(t)$ evaluated at $t = 0$, the initial-value theorem may be used.    (See Appendix A.)    Thus $x_j(t = 0)$ is

$$\lim_{t \to 0} x_j(t) = \lim_{s \to \infty} sx_j(s) = x_j(0)$$

This is just the required initial condition, and we have shown that an impulse input may be used to establish initial conditions on an integrator.
    It is now necessary to adapt this ability to represent initial conditions on integrators by impulses to the problem of finding the elements of the resolvent matrix.    The elementary block-diagram representation of state-variable equations as discussed in Sec. 2.4 requires the use of $n$ integrators with a unity numerator, much like the $j$th integrator considered in Fig. 3.5-1.    Therefore let us return to this elementary block-diagram representation and spell out the way in which the element $\phi_{k(ij)}(s)$ may be determined.    Because the procedure presupposes only that the dynamic process being modeled is described in $n$ state-variable equations, it is equally suited to finding the elements of $\Phi(s)$.    Hence the procedure is stated as a means of finding the elements of the resolvent matrix, either that of the open- or the closed-loop system.    The basic assumption, however, is that the physical system being modeled is described by $n$ first-order differential equations.    On the basis of this assumption, the following block-diagram procedure may be used to find the elements of the resolvent matrix.

1.  Start the block diagram by drawing $n$ blocks, each of which includes the element $1/s$.    Call the output of the $j$th block $x_j$ and the input $\dot{x}_j$, $j = 1, 2, \ldots, n$.
2.  At the input to each box, include a fictitious summer, through which an impulse input $\delta_j(s)$ may be applied.
3.  Complete the block diagram according to the describing state-variable equations.    Assume that the input is zero.
4.  In order to find the element of the resolvent matrix that lies in the $i$th row and the $j$th column, that is, $\phi_{k(ij)}(s)$ or $\phi_{ij}(s)$, redraw the

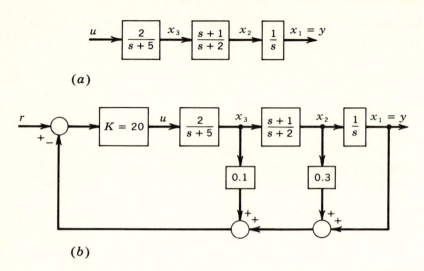

(a)

(b)

**Fig. 3.5-2**  Open-loop plant and closed-loop system of Example 3.5-1.  (a) Open-loop plant; (b) closed-loop system.

block diagram with $\delta_j(s)$ as the input and $x_i(s)$ as the output, with *all other inputs equal to zero.*

5.  The transfer function

$$\frac{x_i(s)}{\delta_j(s)} = \frac{x_i(s)}{x_j(0)} = \phi_{k(ij)}(s) \tag{3.5-6}$$

is the desired element of the resolvent matrix.  [Note that Eq. (3.5-6) is identical with Eq. (3.5-5) and that both are stated in terms of the closed-loop resolvent matrix.]

The following examples illustrate the use of this five-step procedure to find different elements of both $\Phi(s)$ and $\Phi_k(s)$.  This is done by first considering the third-order plant of Fig. 3.5-2a and determining elements of $\Phi(s)$ associated with this plant.  When this plant is controlled through the use of state-variable feedback, the closed-loop configuration of Fig. 3.5-2b results.  Various elements of $\Phi_k(s)$ associated with this closed-loop system are then found by the block-diagram approach outlined above.

**Example 3.5-1**  It is desired to find the resolvent matrix associated with the open-loop plant of Fig. 3.5-2a.  Even though this problem is given with the block diagram specified, it is necessary to begin by describing the dynamics of the plant in three first-order differential equations.  These equations may be written directly from the block diagram as

$$\frac{x_1(s)}{x_2(s)} = \frac{1}{s} \quad \longrightarrow \quad \dot{x}_1 = x_2$$

$$\frac{x_2(s)}{x_3(s)} = \frac{s+1}{s+2} \quad \longrightarrow \quad \dot{x}_2 = -2x_2 + x_3 + \dot{x}_3$$

$$\frac{x_3(s)}{u(s)} = \frac{2}{s+5} \quad \longrightarrow \quad \dot{x}_3 = -5x_3 + 2u$$

If a substitution is made for $\dot{x}_3$ in the second equation, the standard form of the state-variable equations, that is, the form of Eq. (**Ab**), results:

$$\dot{x}_1 = x_2$$
$$\dot{x}_2 = -2x_2 - 4x_3 + 2u \qquad\qquad (3.5\text{-}7)$$
$$\dot{x}_3 = -5x_3 + 2u$$

These equations are now in the form necessary for the application of the five-step procedure. Step 1 is accomplished in Fig. 3.5-3$a$, and in this case $n = 3$, so that three integrator blocks are needed. Step 2 is completed in Fig. 3.5-3$b$, and three unit impulses

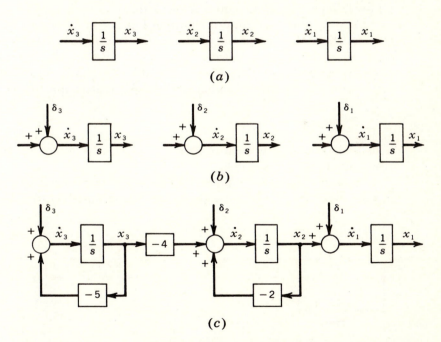

$(a)$

$(b)$

$(c)$

**Fig. 3.5-3**   First three steps in the block-diagram approach to finding elements of $\mathbf{\Phi}(s)$.   $(a)$ First step in the five-step procedure; $(b)$ second step; $(c)$ diagram completed according to the state-variable equations.

are indicated as inputs to the three integrator boxes. This step was not included in the discussion of elementary block-diagram representation in Sec. 2.4. Otherwise the first three steps of this procedure are identical to those needed to complete the elementary block diagram.

In Fig. 3.5-3c the three isolated integrators are connected according to the state-variable equations (3.5-7). Note that $u$ has been omitted since here it is the input and has been set equal to zero. We now have the block-diagram configuration needed to compute any of the elements of $\Phi(s)$. Before proceeding, let us state that $\Phi(s)$ as determined by matrix manipulations is

$$\Phi(s) = \begin{bmatrix} 1/s & 1/s(s+2) & -4/s(s+2)(s+5) \\ 0 & s/(s+2) & -4/(s+2)(s+5) \\ 0 & 0 & 1/(s+5) \end{bmatrix} \tag{3.5-8}$$

We use this result to check the answers that are generated by the block-diagram method. As a start, let us find $\phi_{12}(s)$. According to step 5, $\phi_{12}(s)$ may be found by finding the transfer function from the unit impulse input $\delta_2(s)$ to the state variable $x_1(s)$ with all other inputs equal to zero. A block diagram with $\delta_2(s)$ as the input and $x_1(s)$ as the output is shown in Fig. 3.5-4a, as required by step 4. Note that $u$ and the delta functions other than $\delta_2(s)$ were assumed to be zero in drawing Fig. 3.5-4a. By a simple block-diagram reduction of Fig. 3.5-4a, Fig. 3.5-4b results. From the latter figure the transfer function $x_1(s)/\delta_2(s)$ is written directly as

$$\frac{x_1(s)}{\delta_2(s)} = \frac{1}{s(s+2)}$$

This is the required element of $\Phi(s)$, or $\phi_{12}(s)$, and, of course, the result agrees with Eq. (3.5-8).

It is seen from Eq. (3.5-8) that $\phi_{23}(s)$ involves a minus sign in the numerator. Let us use the block-diagram method to deter-

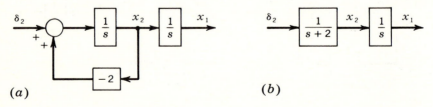

(a)     (b)

**Fig. 3.5-4**  Reduction of Fig. 3.5-3c necessary to determine $\phi_{12}$.  (a) Redrawing with $\delta_2$ as input and $x_1$ as output; (b) reduced version of (a).

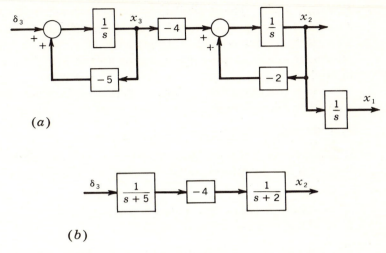

(a)

(b)

**Fig. 3.5-5**   Reduction of the block diagram of Fig. 3.5-3c necessary to determine $\phi_{23}$.   (a) Redrawing with $\delta_3$ as the input and $x_3$ as the output; (b) reduced version.

mine $\phi_{23}(s)$ and see how this negative sign comes about. We may start immediately with step 4, by proceeding from Fig. 3.5-3c. This time $\delta_3(s)$ is to be the input and $x_2(s)$ is to be the output, with all other inputs zero. Figure 3.5-5a indicates this situation, and this is further reduced to Fig. 3.5-5b, from which the required element of $\mathbf{\Phi}(s)$ may be written immediately:

$$\phi_{23}(s) = \frac{x_2(s)}{\delta_3(s)} = \frac{-4}{(s+2)(s+5)}$$

One additional fact is evident from the block-diagram approach that may not be quite as obvious from the matrix definition. This is the fact that $\phi_{21}(s)$, $\phi_{31}(s)$, and $\phi_{32}(s)$ are all zero. From Fig. 3.5-3c it is evident that $x_2(s)$ is not affected by the input $\delta_1(s)$. The arrows in the block diagram indicate the direction of the transmission of the signal, so that both $x_2(s)$ and $x_3(s)$ are unaltered by the input $\delta_1(s)$. Hence both $\phi_{21}(s)$ and $\phi_{31}(s)$ are zero. Similarly the state variable $x_3$ is not affected by the input $\delta_2(s)$, and so $\phi_{32}(s)$ is also zero.

It is not difficult to determine the remaining elements of $\mathbf{\Phi}(s)$ by this block-diagram method. However, if the complete resolvent matrix is required, it is probably simpler to use the matrix definition of Eq. (3.5-1).

***Example 3.5-2***    This example is based on the block diagram of Fig. 3.5-2b and involves the determination of two elements of the closed-loop resolvent matrix $\Phi_k(s)$.    The two elements found in this example are $\phi_{k(12)}(s)$ and $\phi_{k(23)}(s)$, analogous to the two elements found in $\Phi(s)$.

As a starting point, return to Eqs. (3.5-7), where $u$ may be read from Fig. 3.5-2b as

$$u = 20(-x_1 - 0.3x_2 - 0.1x_3)$$

If this value of $u$ is substituted into Eqs. (3.5-7), the state-variable equations for the closed-loop system are

$$\dot{x}_1 = x_2$$
$$\dot{x}_2 = -40x_1 - 14x_2 - 8x_3 + 40r(t) \qquad\qquad (3.5\text{-}9)$$
$$\dot{x}_3 = -40x_1 - 12x_2 - 9x_3 + 40r(t)$$

The equations are pictured in elementary block-diagram form in Fig. 3.5-6, with the addition of the delta-function inputs at the input to each integrator.    Now $r$ is the input and is set equal to zero and therefore is not shown.    This block diagram is considerably more complicated than Fig. 3.5-3c, associated with the plant, because of the generous use of feedback in the closed-loop system. We see immediately that every delta-function input affects all the state variables so that we cannot expect any of the elements of $\Phi_k(s)$ to be zero.

In order to find $\Phi_{k(12)}(s)$ it is necessary to redraw Fig. 3.5-6 with $\delta_2(s)$ as the input and $x_1(s)$ as the output.    As a preliminary

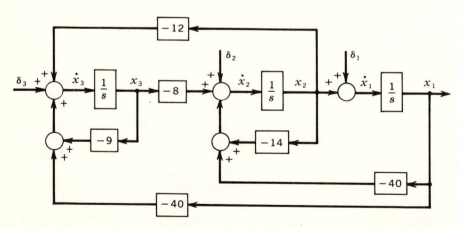

***Fig. 3.5-6***    Elementary block-diagram representation of the state-variable equations (3.5-9) with provision for initial-condition inputs.

step, Fig. 3.5-7*a* is drawn with $r$, $\delta_1$, and $\delta_3$ equal to zero. Note that the feedback from $x_2$ to $x_3$ has been moved to $x_1$ by multiplying by $s$. The idea in mind here is eventually to achieve a block diagram in $H_{eq}(s)$ form. This is made clearer in Fig. 3.5-7*b*, and finally the $H_{eq}(s)$-type form is realized in Fig. 3.5-7*c*. From Fig.

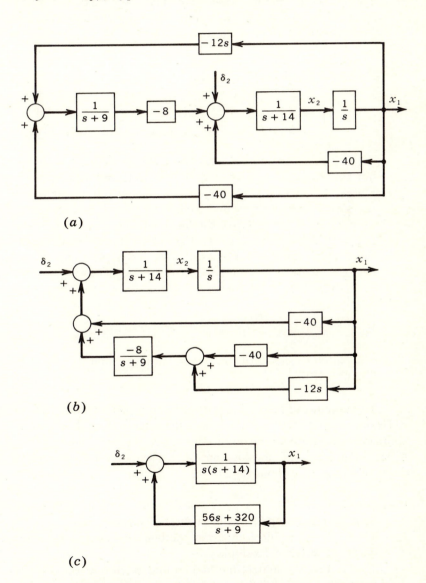

(*a*)

(*b*)

(*c*)

**Fig. 3.5-7** Block-diagram reductions to find $\phi_{k(12)}(s)$. (*a*) Preliminary redrawing; (*b*) further reduction; (*c*) final reduction.

3.5-7c the transfer function $x_1(s)/\delta_2(s)$ is readily found to be

$$\frac{x_1(s)}{\delta_2(s)} = \frac{s + 9}{s^3 + 23s^2 + 70s + 40} = \phi_{k(12)}(s)$$

This is the desired $\phi_{k(12)}(s)$. This may be checked with the corresponding element of $\boldsymbol{\Phi}_k(s)$ as determined by the matrix definition of Eq. (3.5-2) by which $\boldsymbol{\Phi}_k(s)$ is found to be

$$\boldsymbol{\Phi}_k(s) = \frac{1}{s^3 + 23s^2 + 70s + 40}$$
$$\begin{bmatrix} s^2 + 23s + 30 & s + 9 & -8 \\ -40(s + 1) & s(s + 9) & -8s \\ -40(s + 2) & -(12s + 40) & s^2 + 14s + 40 \end{bmatrix} \quad (3.5\text{-}10)$$

In order to find $\phi_{k(23)}(s)$, we once again start with Fig. 3.5-6 but this time we assume $\delta_3(s)$ is the input and $x_2(s)$ is the output. Figure 3.5-8a shows the required input-output configuration. In this figure, feedback around each integrator has been removed for simplicity. In Fig. 3.5-8b the feedback from $x_1$ has been moved to $x_2$ by multiplying by $1/s$. In Fig. 3.5-8c the feedback path around the block $1/(s + 14)$ has been eliminated, and the two parallel feedback paths from the output to the input have been combined.

From Fig. 3.5-8c the transfer function $x_2(s)/\delta_3(s)$ is found to be

$$\frac{x_2(s)}{\delta_3(s)} = \frac{-8s}{s^3 + 23s^2 + 70s + 40} = \phi_{k(23)}(s)$$

This value checks with the corresponding element of $\boldsymbol{\Phi}_k(s)$ given in Eq. (3.5-10).

The amount of labor involved in finding just two elements of $\boldsymbol{\Phi}_k(s)$ in Example 3.5-2 serves to emphasize once again that the block-diagram approach is not advantageous if the entire resolvent matrix is needed. At the same time, the development of this section relates the matrix and the block-diagram approaches.

**Exercises 3.5**    *3.5-1.*    Use the block-diagram method of this section to find the remaining elements of $\boldsymbol{\Phi}(s)$ of Example 3.5-1.

*3.5-2.*    Use the block-diagram method to find the remaining elements of $\boldsymbol{\Phi}_k(s)$ of Example 3.5-2.

*3.5-3.*    Use the matrix method to find $\phi_{k(12)}$ and $\phi_{k(23)}$ for the system of Example 3.5-2.

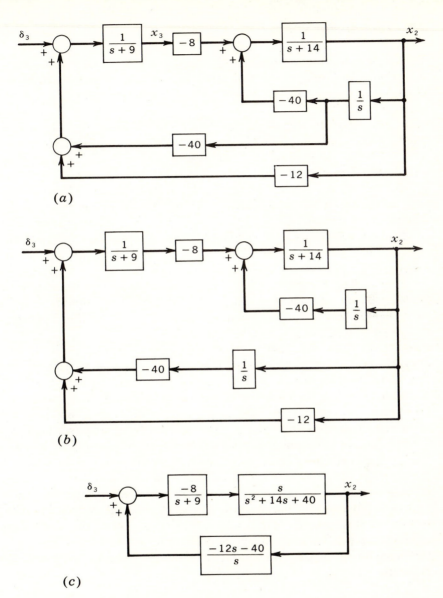

**Fig. 3.5-8** Block-diagram reduction to find $\phi_{k(23)}(s)$.  (a) Preliminary redrawing; (b) further reduction; (c) final reduction.

## 3.6    Conclusions

This section concludes not only this chapter but the discussion of representation. In Chap. 2 the emphasis was on plant representation and description, whereas in this chapter we have concentrated solely on the representation and description of the closed-loop system. It is important once again to emphasize the fundamental difference between the plant and the system. The plant is the physical entity being controlled. The system is created through the use of feedback of the plant state variables. The differential equations that describe the resulting system are different from those of the plant, or, alternatively, the closed-loop or system transfer function is different from the plant transfer function. The point of this chapter is to emphasize that this difference exists and to illustrate ways in which this difference can be expressed quantitatively.

As in Chap. 2, we have once again stressed the dual approach to system representation through the use of matrix methods and transfer functions. The $H$-equivalent and $G$-equivalent representations prove to be particularly important in future work; Table 3.4-1 summarizes the relationships between these quantities and the state-variable equations, expressed in terms of the matrix notation.

At the beginning of this chapter the basic effects of feedback were pointed out. Feedback reduces the effects of plant variations and decreases the sensitivity of the system to output disturbances. In addition, feedback allows control over bandwidth and the system transient response. Although these points were mentioned, no problems involving these ideas were given. Just *how* feedback affects sensitivity, for example, is a matter of analysis, and we are now prepared to begin the analysis portion of the book. Analysis is concerned primarily with the closed-loop system. We wish to be able to answer any question concerning the closed-loop system that has been created through the use of feedback. Because of our interest in the closed-loop system, the final section of this chapter was concerned primarily with the closed-loop resolvent matrix $\mathbf{\Phi}_k(s)$. The closed-loop resolvent matrix is the key to the behavior of all the system state variables for any initial conditions or inputs.

## 3.7    Problems

*3.7-1.*   Indicate the correct answer or answers to the following:
(a)   The zeros of $H_{eq}(s)$ are as follows:
  (i)   The poles of $G_p(s)$

        (*ii*)    The zeros of $G_p(s)$
      (*iii*)    The poles of $y(s)/r(s)$
      (*iv*)    The zeros of $y(s)/r(s)$
       (*v*)    None of the above
(*b*)    The poles of $H_{eq}(s)$ are as follows:
       (*i*)    The poles of $G_p(s)$
      (*ii*)    The zeros of $G_p(s)$
     (*iii*)    The poles of $y(s)/r(s)$
     (*iv*)    The zeros of $y(s)/r(s)$
      (*v*)    None of the above
(*c*)    The zeros of $y(s)/r(s)$ are as follows:
       (*i*)    The zeros of $G_{eq}(s)$
      (*ii*)    The poles of $G_{eq}(s)$
     (*iii*)    The zeros of $G_p(s)$
     (*iv*)    The poles of $G_p(s)$
      (*v*)    None of the above
(*d*)    The following may be positioned arbitrarily by the proper selection of $K$ and **k**:
       (*i*)    The poles of $G_p(s)$
      (*ii*)    The zeros of $G_p(s)$
     (*iii*)    The poles of $H_{eq}(s)$
     (*iv*)    The zeros of $H_{eq}(s)$
      (*v*)    The poles of $y(s)/r(s)$
     (*vi*)    The zeros of $y(s)/r(s)$

*3.7-2.*  The following closed-loop-system state-variable equations are known.

$$\dot{x}_1 = x_2 \qquad \dot{x}_2 = -x_1 - 2.2x_3 - 3.5x_2 + r$$
$$\dot{x}_3 = -0.5x_2 - x_1 - 5.2x_3 + r \qquad y = x_1$$

Find (*a*) a block diagram of the system and (*b*) the state-variable representation of the plant if $u$ is known to be

$$u = r - x_1 - 0.5x_2 - 0.2x_3$$

*3.7-3.*  A particular control system is described by the following vectors and matrices:

$$\mathbf{A} = \begin{bmatrix} -2 & -1.5 \\ -1 & -4.5 \end{bmatrix} \qquad \mathbf{c} = \begin{bmatrix} 1 \\ 0 \end{bmatrix}$$

$$\mathbf{b} = \begin{bmatrix} 1 \\ 1 \end{bmatrix} \qquad \mathbf{k} = \begin{bmatrix} 1 \\ 0.5 \end{bmatrix}$$

$$K = 10$$

Use matrix methods to find the following:

(a)  $G_p(s)$
(b)  $y(s)/r(s)$

*3.7-4.*  Given the state-variable representation

$$\dot{\mathbf{x}} = \begin{bmatrix} 0 & 1 \\ -1 & -1 \end{bmatrix} \mathbf{x} + \begin{bmatrix} 0 \\ 1 \end{bmatrix} u \qquad y = \begin{bmatrix} 1 & 1 \end{bmatrix} \mathbf{x}$$

and $u = r - \mathbf{k}^T \mathbf{x}$ where

$$\mathbf{k} = \begin{bmatrix} 1 \\ 0.5 \end{bmatrix} \qquad \text{and} \qquad K = 1$$

(a)  Draw the block-diagram representation for the plant and for the system.
(b)  Using matrix techniques, find

     (i)   $\boldsymbol{\Phi}(s)$
     (ii)  $\boldsymbol{\Phi}_k(s)$
     (iii) $H_{eq}(s)$
     (iv)  $G_p(s)$
     (v)   $y(s)/r(s)$

*3.7-5.*  Given the plant shown in Fig. 3.7-1 to be controlled, where $u$ is defined as $u = K(r - \mathbf{k}^T \mathbf{x})$ with $K = 1$ and $\mathbf{k} = \text{col}\,(1, \alpha, 0)$,

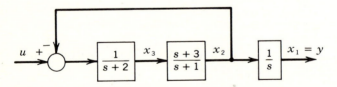

**Fig. 3.7-1**  Problem 3.7-5.

(a)  Find the state-variable representation, that is, $n$ first-order equations, for the plant and for the closed-loop system.
(b)  Using block-diagram manipulations, find $G_p(s)$, $H_{eq}(s)$, $G_{eq}(s)$, $y(s)/r(s)$, $\phi_{13}(s)$.

*3.7-6.*  Find the $H_{eq}(s)$ representation of the feedback control systems of Fig. 3.7-2.

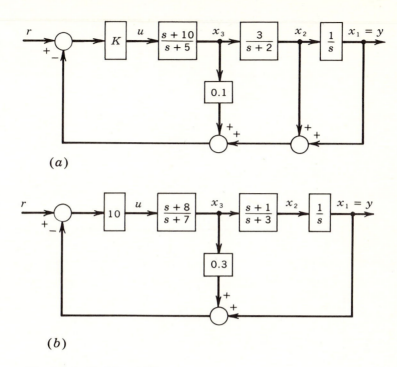

*(a)*

*(b)*

**Fig. 3.7-2**  Problem 3.7-6.

*3.7-7.*  Both the systems given in Fig. 3.7-2 result in an $H_{eq}(s)$ whose poles are *not* the zeros of $G_p(s)$.  This is true because the state-variable equations contain a $\dot{u}$ term and are not of the form (**Ab**).  Rework Prob. 3.7-6 with the state variables indicated in Fig. 3.7-3.  (See Prob. 2.9-10 for a useful block-diagram identity.)

*3.7-8.*  Use the matrix identity

$$(\mathbf{M}\mathbf{N}^T + \mathbf{I})^{-1} = \mathbf{I} - \mathbf{M}(\mathbf{N}^T\mathbf{M} + \mathbf{I})^{-1}\mathbf{N}^T$$

to show that $\boldsymbol{\Phi}_k(s)$ may be written as

$$\boldsymbol{\Phi}_k(s) = \boldsymbol{\Phi}(s)[\mathbf{I} - K\mathbf{b}\mathbf{k}^T\boldsymbol{\Phi}(s)/(1 + K\mathbf{k}^T\boldsymbol{\Phi}(s)\mathbf{b})]$$

*3.7-9.*  For the system

$$\dot{\mathbf{x}} = \begin{bmatrix} 0 & 1 & 0 \\ 0 & -2 & 2 \\ -1 & 0 & -4 \end{bmatrix}\mathbf{x} + \begin{bmatrix} 0 \\ 0 \\ 1 \end{bmatrix}u$$

$$u = 10(r - \begin{bmatrix} 1 & 1 & 1 \end{bmatrix}\mathbf{x}) \qquad y = x_1$$

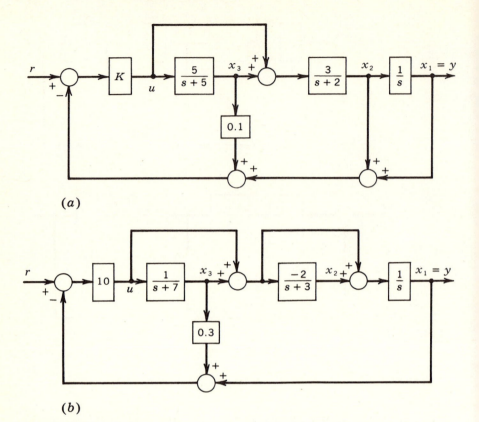

(a)

(b)

**Fig. 3.7-3**   Problem 3.7-7.

find the following:

(a)    Elementary block diagram
(b)    $G_p(s)$
(c)    $H_{eq}(s)$
(d)    $G_{eq}(s)$
(e)    $y(s)/r(s)$

# *four*   *time response*

## 4.1   *Introduction*

This is the first chapter concerned primarily with analysis. Chapters 2 and 3 dealt with means of representing a plant and a closed-loop system. Now we are interested in how the state variables associated with the final closed-loop system behave for a variety of inputs and/or initial conditions.

Although the reader is assumed to have a basic knowledge of Laplace transform methods, this chapter begins with a review of partial-fraction-expansion methods. The algebraic and residue methods that are useful in finding inverse Laplace transforms are treated along with the closely related graphical approach.   Graphical procedures are a powerful tool in control engineering, since the accuracies obtained are usually compatible with the accuracies with which the given plant or system is described.

The review of inverse Laplace transforms is followed by a discussion of second-order-system response.   The specific interest is the response to a step-function input and the time-domain specifications associated with the step response.   The following section treats the response of the general $n$th-order system to general inputs with initial conditions on all the state variables.   The general case is treated from the matrix point of view, and alternative approaches related to the system block diagrams are also discussed.

The last section of this chapter deals with time-domain methods of determining total system response.   Again the general case is treated, this time in terms of the state transition matrix and the convolution integral.

Although the advent of the digital computer has to some extent eliminated the need for the hand computation of time response,[1] this does not mean that the material of this chapter is useless.   Quite the contrary; this material still plays an important role in control theory, especially in the area of system design.   Of particular importance is the ability to achieve a reasonably accurate knowledge of the *qualitative* nature of the response without extensive computation.   This is the basic goal of this chapter.

## 4.2   *Partial-fraction-expansion methods*

It has been assumed thus far that the reader has a working knowledge of Laplace transform methods.   In fact, the whole underlying philosophy of this book is that it is simpler to treat dynamic systems, those described by differential equations, in the frequency or $s$ domain than in the time domain.

In working in the complex frequency domain, two problems arise: first, the transformation of the given differential equations to the $s$ domain and, second, the transformation of the problem solution back to the time domain.   The two previous chapters described a variety of methods of representing both open- and closed-loop systems as sets of differential equations, $n$ first-order differential equations, or, equivalently, in terms of block diagrams or transfer functions.   Here we consider in detail the

---

[1] See J. L. Melsa, "Computer Programs for Computational Assistance in the Study of Linear Control Theory," McGraw-Hill Book Company, New York, 1970.

inverse problem, that is, the problem of going from the complex $s$ domain back to the time domain.

The basic assumption is that the function whose inverse Laplace transform is desired is given as a ratio of polynomials in $s$ and that the denominator of this function appears in factored form. For purposes of discussion, we assume a general function $f(s)$ given as

$$f(s) = \frac{N(s)}{D(s)} = \frac{N(s)}{(s + s_1)(s + s_2) \cdots (s + s_n)} \tag{4.2-1}$$

Here the numerator of $f(s)$ is a polynomial in $s$ of order less than $n$. In the factored form of $D(s)$, the $s_i$'s may be zero, real, or complex, and two or more values of $s_i$ may be equal. The nature of the time response depends only on $D(s)$; hence $N(s)$ is left in the unfactored form. To emphasize the different types of the roots of $D(s)$, Eq. (4.2-1) is simplified and written as

$$f(s) = \frac{N(s)}{D(s)} = \frac{N(s)}{(s + s_1)(s + s_2)^k (s + \alpha + j\beta)(s + \alpha - j\beta)} \tag{4.2-2a}$$

$$f(s) = \frac{N(s)}{D(s)} = \frac{N(s)}{(s + s_1)(s + s_2)^k [(s + \alpha)^2 + \beta^2]} \tag{4.2-2b}$$

Here the $s_1$ and the $s_2$ may be any real numbers, and one of these roots is assumed to be repeated $k$ times. A single set of complex conjugate roots is indicated either separately, as in Eq. (4.2-2a), or as a product, as in Eq. (4.2-2b). Although Eqs. (4.2-2) are less general than Eq. (4.2-1), the latter form is sufficient to illustrate the partial-fraction-inversion method to be used here. The different forms of Eqs. (4.2-2a) and (4.2-2b) indicate alternative means used to handle the complex conjugate terms.

Let us assume initially that no complex conjugate terms are present so that Eqs. (4.2-2) may be further reduced to

$$f(s) = \frac{N(s)}{D(s)} = \frac{N(s)}{(s + s_1)(s + s_2)^k} \tag{4.2-3}$$

Once again the order of $N(s)$ is assumed to be less than that of $D(s)$. Partial-fraction expansion requires that Eq. (4.2-3) be expressed as

$$f(s) = \frac{N(s)}{D(s)} = \frac{R_{11}}{s + s_1} + \frac{R_{2k}}{(s + s_2)^k} + \frac{R_{2(k-1)}}{(s + s_2)^{k-1}} + \cdots + \frac{R_{21}}{s + s_2} \tag{4.2-4}$$

The problem is to find the numerator constants $R_{ij}$. Here the $i$ subscript is used to identify the index of the root, and $j$ indicates the power of the associated denominator term. Once the numerator constants are known, the inversion of the individual terms is a simple matter that may be carried out with the aid of Table 4.2-1 or Appendix B. The actual

**Table 4.2-1    Table of Laplace Transform Pairs**

|     | $f(s)$ | $f(t)$ |
|-----|--------|--------|
| 1.  | $1$ | Impulse, $\delta(t)$ |
| 2.  | $\dfrac{1}{s}$ | Step, $\mu(t)$ |
| 3.  | $\dfrac{1}{s^2}$ | Ramp, $t\mu(t)$ |
| 4.  | $\dfrac{n!}{s^{n+1}}$ | $t^n\mu(t)$ |
| 5.  | $\dfrac{1}{s+\alpha}$ | $e^{-\alpha t}\mu(t)$ |
| 6.  | $\dfrac{1}{(s+\alpha)^2}$ | $te^{-\alpha t}\mu(t)$ |
| 7.  | $\dfrac{\beta}{s^2+\beta^2}$ | $\sin\beta t\mu(t)$ |
| 8.  | $\dfrac{s}{s^2+\beta^2}$ | $\cos\beta t\mu(t)$ |
| 9.  | $\dfrac{\beta}{(s+\alpha)^2+\beta^2}$ | $e^{-\alpha t}\sin\beta t\mu(t)$ |
| 10. | $\dfrac{s+\alpha}{(s+\alpha)^2+\beta^2}$ | $e^{-\alpha t}\cos\beta t\mu(t)$ |
| 11. | $\dfrac{s+a_o}{(s+\alpha)^2+\beta^2}$ | $\dfrac{1}{\beta}[(a_o-\alpha)^2+\beta^2]^{\frac{1}{2}}e^{-\alpha t}\sin(\beta t+\psi)\mu(t)$ where $\psi=\arctan\dfrac{\beta}{a_o-\alpha}$ |

partial-fraction expansion may be accomplished by three methods: completely algebraic procedures, the method of residues, or graphical methods from a pole-zero plot of $f(s)$.  These three approaches complement each other, and all are discussed below.

*Algebraic approach.*  The algebraic approach involves equating $N(s)/D(s)$ as given by Eq. (4.2-3) to the right-hand side of Eq. (4.2-4), forming a common denominator for the partial-fraction expansion, and finally equating the coefficients of equal powers of $s$ in the result with those of $N(s)$.  This procedure is illustrated in the following example.

***Example 4.2-1***   Let the closed-loop transfer function of a particular system be

$$\frac{y(s)}{r(s)} = \frac{2.5(s + 2)}{(s + 5)(s + 1)^2}$$

Let us assume that the input is a unit step, or $r(s) = 1/s$, so that $y(s)$ becomes

$$y(s) = \frac{2.5(s + 2)}{s(s + 5)(s + 1)^2}$$

This $y(s)$ corresponds to the $f(s)$ in the previous discussion, and it is desired to expand $y(s)$ in partial-fraction form as

$$y(s) = \frac{N(s)}{D(s)} = \frac{R_{11}}{s} + \frac{R_{21}}{s + 5} + \frac{R_{32}}{(s + 1)^2} + \frac{R_{31}}{s + 1}$$

If the right-hand side of the above equation is written as a ratio of polynomials in $s$ by use of the common denominator $[s(s + 5)(s + 1)^2]$, it becomes

$$y(s) = \frac{1}{s(s + 5)(s + 1)^2} [s^3(R_{11} + R_{21} + R_{31})$$
$$+ s^2(7R_{11} + 2R_{21} + R_{32} + 6R_{31})$$
$$+ s(11R_{11} + R_{21} + 5R_{32} + 5R_{31}) + 5R_{11}]$$

Since the numerator of $y(s)$ is just $2.5(s + 2)$, the following four simultaneous linear equations result by equating the coefficients of equal powers of $s$ in the two numerator expressions for $y(s)$:

$$\begin{aligned} R_{11} + R_{21} + R_{31} &= 0 \\ 7R_{11} + 2R_{21} + R_{32} + 6R_{31} &= 0 \\ 11R_{11} + R_{21} + 5R_{32} + 5R_{31} &= 2.5 \\ 5R_{11} &= 5 \end{aligned} \qquad (4.2\text{-}5)$$

These equations may be solved to yield

$$R_{11} = 1 \qquad R_{21} = \tfrac{3}{32}$$
$$R_{32} = -\tfrac{5}{8} \qquad R_{31} = -\tfrac{35}{32}$$

Therefore the partial-fraction form for $y(s)$ becomes

$$y(s) = \frac{1}{s} + \frac{\tfrac{3}{32}}{s + 5} - \frac{\tfrac{5}{8}}{(s + 1)^2} - \frac{\tfrac{35}{32}}{s + 1}$$

and from Table 4.2-1, $y(t)$ is found to be[1]

$$y(t) = 1 + \tfrac{3}{32}e^{-5t} - \tfrac{5}{8}te^{-t} - \tfrac{35}{32}e^{-t}$$

[1] Since $y(t)$ is zero for $t < 0$, this result should be multiplied by $\mu(t)$ but we shall omit $\mu(t)$ where there is no chance of confusion.

The above example illustrates the conceptual simplicity of the algebraic approach, as well as the computational difficulties that are involved. In order to find all the unknown constants in the partial-fraction expansion, it is necessary to solve four simultaneous equations. In general it is necessary to solve $n$ equations for an $n$th-order $f(s)$.

*Residue method.* The residue method avoids this difficulty but adds the difficulty of differentiation if multiple roots are involved. To illustrate this method, let us consider again Eqs. (4.2-3) and (4.2-4). The right-hand sides of these two equations are set equal to yield

$$\frac{N(s)}{(s + s_1)(s + s_2)^k} = \frac{R_{11}}{s + s_1} + \frac{R_{2k}}{(s + s_2)^k} + \frac{R_{2(k-1)}}{(s + s_2)^{k-1}}$$

$$+ \cdots + \frac{R_{21}}{s + s_2} \quad (4.2\text{-}6)$$

If Eq. (4.2-6) is multiplied by $s + s_1$ on both sides of the equal sign and then evaluated at $s = -s_1$, all terms on the right, except $R_{11}$, become zero, and the resulting expression completely specifies $R_{11}$:

$$R_{11} = \frac{N(s)}{(s + s_2)^k} \bigg|_{s = -s_1}$$

A similar procedure may be used to evaluate $R_{2k}$. In this case, however, both sides of Eq. (4.2-6) must be multiplied by $(s + s_2)^k$ and then evaluated at $s = -s_2$. The result of this procedure is the following expression for $R_{2k}$:

$$R_{2k} = \frac{N(s)}{s + s_1} \bigg|_{s = -s_2}$$

Unfortunately, the only coefficient associated with the multiple pole at $s = -s_2$ that may be evaluated in this simple fashion is $R_{2k}$. The remaining $R_{2j}$'s, $j < k$, must be evaluated by the use of differentiation. The constant $R_{2(k-1)}$ may be isolated from Eq. (4.2-6) by multiplying both sides by $(s + s_2)^k$, then differentiating with respect to $s$, and finally evaluating at $s = -s_2$. This is done in two steps below, to give the following expression

$$\frac{N(s)}{s + s_1} = \frac{R_{11}(s + s_2)^k}{s + s_1} + R_{2k} + R_{2(k-1)}(s + s_2)$$

$$+ \cdots + R_{21}(s + s_2)^{k-1}$$

Then

$$\left[ \frac{d}{ds} \frac{N(s)}{s + s_1} \right] \bigg|_{s = -s_2} = R_{2(k-1)}$$

In an entirely similar manner, the following expression for the $R_{ij}$ coefficient for a $k$th-order root can be established by successive differentiation to be

$$R_{ij} = \left\{ \frac{1}{(k-j)!} \frac{d^{k-j}}{ds^{k-j}} \left[ (s + s_i)^k \frac{N(s)}{D(s)} \right] \right\} \Big|_{s=-s_i} \tag{4.2-7}$$

If $k$ is equal to 1, that is, if the root is simple, this expression leads to the expression found previously:

$$R_{i1} = \frac{(s + s_i)N(s)}{D(s)} \Big|_{s=-s_i} \tag{4.2-8}$$

The general expression of Eq. (4.2-7) involves tedious differentiation if $k$ is larger than 2. The coefficients $R_{i1}$ are called *residues* in complex-variable theory; hence the name: the residue method.

To demonstrate the residue method, let us rework the problem of Example 4.2-1.

**Example 4.2-2**   Once again we wish to express $y(s)$ in the form

$$y(s) = \frac{2.5(s+2)}{s(s+5)(s+1)^2} = \frac{R_{11}}{s} + \frac{R_{21}}{s+5} + \frac{R_{32}}{(s+1)^2} + \frac{R_{31}}{s+1}$$

The residues associated with the poles at $s = 0$ and $s = -5$ are easily determined from Eq. (4.2-8) as

$$R_{11} = \frac{2.5(s+2)}{(s+5)(s+1)^2} \Big|_{s=0} = \frac{5}{5} = 1 \tag{4.2-9}$$

$$R_{21} = \frac{2.5(s+2)}{s(s+1)^2} \Big|_{s=-5} = \frac{-7.5}{-5(16)} = \frac{3}{32} \tag{4.2-10}$$

Application of Eq. (4.2-7) is illustrated in the evaluation of both $R_{32}$ and $R_{31}$. Since the root at $s = -1$ is repeated twice, the index $k = 2$. The determination of $R_{32}$ involves no differentiation, and substitution into Eq. (4.2-7) results in

$$R_{32} = \frac{2.5(s+2)}{s(s+5)} \Big|_{s=-1} = \frac{2.5}{-4} = -\frac{5}{8} \tag{4.2-11}$$

The evaluation of $R_{31}$, on the other hand, does require differentiation and is given by

$$R_{31} = \left[ \frac{1}{1!} \frac{d}{ds} \frac{2.5(s+2)}{s(s+5)} \right] \Big|_{s=-1}$$

$$= \left[ \frac{-2.5(s+2)(2s+5)}{s^2(s+5)^2} + \frac{2.5}{s(s+5)} \right] \Big|_{s=-1} = -\frac{35}{32}$$

These are the same values for the partial-fraction-expansion coefficients that were determined in Example 4.2-1.

Although the differentiation involved in Example 4.2-2 is not difficult, it would have been considerably more complicated if $k$ had been 3, rather than 2.   By combining the algebraic and residue approaches, the disadvantages of both methods are largely overcome.   Through use of the residue method, three of the four unknown constants are evaluated with ease.   With three of the four constants in Eqs. (4.2-5) known, the solution of the four simultaneous equations becomes trivial.   Thus the two methods of partial-fraction expansion complement each other.

Thus far, attention has been focused upon cases involving repeated roots.   Such cases are much more likely to appear as textbook examples and problems than to appear in practice, particularly in closed-loop systems.   The reader may not appreciate the truth of this statement until the root-locus method is discussed in Chap. 7, but it is nevertheless a fact. A much more common occurrence than repeated roots is the presence of complex conjugate roots.   Here, in order to avoid unnecessary computational difficulties, we assume that complex conjugate roots are always simple, an assumption that the authors have never seen violated in practice.

To initiate the discussion of complex conjugate roots, let us return to Eqs. (4.2-2a) and (4.2-2b).   In order to deemphasize repeated roots, let us assume that $k = 0$, so that Eqs. (4.2-2a) and (4.2-2b) become

$$f(s) = \frac{N(s)}{D(s)} = \frac{N(s)}{(s + s_1)(s + \alpha + j\beta)(s + \alpha - j\beta)} \qquad (4.2\text{-}12a)$$

$$f(s) = \frac{N(s)}{D(s)} = \frac{N(s)}{(s + s_1)[(s + \alpha)^2 + \beta^2]} \qquad (4.2\text{-}12b)$$

These two equations indicate the two methods by which the complex conjugate roots can be handled.   If Eq. (4.2-12a) is rewritten and expanded into partial-fraction form, it becomes

$$f(s) = \frac{N(s)}{D(s)} = \frac{N(s)}{(s + s_1)(s + s_2)(s + s_3)} = \frac{R_{11}}{s + s_1} + \frac{R_{21}}{s + s_2} + \frac{R_{31}}{s + s_3}$$
$$(4.2\text{-}13)$$

where $s_2 = \alpha + j\beta$ and $s_3 = \alpha - j\beta$.

Here the fact that two of the roots are complex conjugates is not important, and the residues at each simple pole, whether complex or not, may be evaluated in the usual way. Thus the three residues are

$$R_{11} = \frac{N(s)}{(s + s_2)(s + s_3)}\bigg|_{s = -s_1} = \frac{N(s)}{(s + \alpha + j\beta)(s + \alpha - j\beta)}\bigg|_{s = -s_1}$$

$$R_{21} = \frac{N(s)}{(s + s_1)(s + s_3)}\bigg|_{s = -s_2} = \frac{N(s)}{(s + s_1)(s + \alpha - j\beta)}\bigg|_{s = -\alpha - j\beta}$$

$$R_{31} = \frac{N(s)}{(s + s_1)(s + s_2)}\bigg|_{s = -s_3} = \frac{N(s)}{(s + s_1)(s + \alpha + j\beta)}\bigg|_{s = -\alpha + j\beta}$$

It will be shown later in this section that $R_{21}$ and $R_{31}$ are always complex conjugates so that Eq. (4.2-13) is often written

$$f(s) = \frac{N(s)}{D(s)} = \frac{R_{11}}{s + s_1} + \frac{R}{s + \alpha + j\beta} + \frac{\bar{R}}{s + \alpha - j\beta}$$

where the overbar indicates the complex conjugate. Once either $R$ or $\bar{R}$ is found, the other is known. The use of the residue method to obtain the partial-fraction expansion of an $f(s)$ involving complex conjugate roots is illustrated in Example 4.2-3.

**Example 4.2-3** Let the closed-loop transfer function $y(s)/r(s)$ be given as

$$\frac{y(s)}{r(s)} = \frac{16(s + 5)}{(s + 10)[(s + 2)^2 + 2^2]} = \frac{16(s + 5)}{(s + 10)(s + 2 + j2)(s + 2 - j2)}$$

It is desired to find the impulse response of this system, in which case $r(s)$ is just 1, and $y(s)$ is

$$y(s) = \frac{16(s + 5)}{(s + 10)(s + 2 + j2)(s + 2 - j2)}$$

$$= \frac{R_{11}}{s + 10} + \frac{R}{s + 2 + j2} + \frac{\bar{R}}{s + 2 - j2}$$

$R_{11}$ is easily found to be $-80/68$, and $R$ is

$$R = \frac{16(s + 5)}{(s + 10)(s + 2 - j2)}\bigg|_{s = -2 - j2} = \frac{16(3 - j2)}{(8 - j2)(-4j)}$$

$$= 1.75/\underline{70.4°} = 1.75e^{j(70.4°)} = 0.588 + j1.65$$

Therefore $\bar{R}$ is

$$\bar{R} = 1.75/\underline{-70.4°} = 1.75e^{-j(70.4°)} = 0.588 - j1.65$$

and $y(s)$ in partial-fraction form is

$$y(s) = \frac{-\frac{80}{68}}{s + 10} + \frac{1.75e^{j(70.4°)}}{s + 2 + j2} + \frac{1.75e^{-j(70.4°)}}{s + 2 - j2} \tag{4.2-14}$$

(Note that, as is common in engineering, the exponential here is shown in degrees rather than in radians.)

Now $y(t)$ may be found directly from Table 4.2-1, although the answer is not in a convenient form, since both the pole and the residue are complex numbers. To express the answer in terms of a sinusoidal function, it is convenient to add the last two terms of the partial-fraction expansion of $y(s)$ before taking the inverse transform. If this is done in this example, $y(s)$ becomes

$$y(s) = \frac{-1.175}{s + 10} + \frac{1.175(s + 7.60)}{(s + 2)^2 + 2^2} \qquad (4.2\text{-}15)$$

The inverse transforms may be written directly from Table 4.2-1 as

$$y(t) = -1.175e^{-10t} + \frac{1.175}{2}[(7.60 - 2)^2 + 2^2]^{\frac{1}{2}}e^{-2t} \sin (2t + \psi)$$

where $\psi = \arctan [2/(7.60 - 2)]$ or

$$y(t) = -1.175e^{-10t} + 3.50e^{-2t} \sin (2t + 19.6°) \qquad (4.2\text{-}16)$$

The algebraic approach to partial-fraction expansion in which complex conjugate roots are involved is based upon an $f(s)$ written as in Eq. (4.2-12b). In expanded form, Eq. (4.2-12b) is written

$$y(s) = \frac{R_{11}}{s + s_1} + \frac{As + B}{(s + \alpha)^2 + \beta^2} \qquad (4.2\text{-}17)$$

This is the same form as Eq. (4.2-15) of Example 4.2-3, and the obvious goal is to avoid the manipulation of the complex numbers and to realize the convenient sinusoidal form of the answer directly. The algebraic approach to Example 4.2-3 is illustrated below.

**Example 4.2-4**   Let $y(s)$ be written in the desired partial-fraction-expansion form of Eq. (4.2-17) as

$$y(s) = \frac{16(s + 5)}{(s + 10)[(s + 2)^2 + 2^2]} = \frac{R_{11}}{s + 10} + \frac{As + B}{(s + 2)^2 + 2^2}$$

$$= \frac{s^2(R_{11} + A) + s(4R_{11} + 10A + B) + 8R_{11} + 10B}{(s + 10)[(s + 2)^2 + 2^2]}$$

When the numerator terms above have been equated, the three equations that must be solved simultaneously for $R_{11}$, $A$, and $B$ are

$$R_{11} + A = 0 \qquad 4R_{11} + 10A + B = 16 \qquad 8R_{11} + 10B = 80$$

Of course, $R_{11}$ may be found more easily from the residue formula, and if this is done, both $A$ and $B$ follow with little effort. The values of these unknown coefficients are

$$R_{11} = -1.175 \qquad A = 1.175 \qquad B = 8.94 = 1.175(7.60)$$

The resulting $y(s)$ is exactly that of Eq. (4.2-15), and $y(t)$ is once again given by Eq. (4.2-16).

Thus far in this section, we have worked two example problems, both by the residue method and by the algebraic method. In each case the resulting $y(t)$ is a sum of time functions rather than a plot of $y(t)$ versus time. In one sense, a plot of $y(t)$ versus time is a more satisfactory answer than a sum of time functions, since by looking at the plot we know the behavior of the output for all time. In order to obtain the same information from a series of time functions, each function must be evaluated at a number of different values of time and the results added and plotted. Often we accept the analytic expression for $y(t)$, even though the exact nature of the output for all time can be determined only from the analytic expression after considerable computation.

In attempting to ascertain the nature of the output from its component parts, one automatically looks to see which terms have the largest coefficients, the longest time constants, etc. In short, one often makes a mental approximation of the output. This section is concluded by a discussion of graphical means of determining residues. The object is not necessarily to replace the analytic methods of calculating residues but rather to provide a means by which the approximate contribution of each pole to the total output may be gauged.

*Graphical approach.* As an introduction to the graphical approach, we consider the $s$-plane drawing of Fig. 4.2-1$a$. There one pole is located

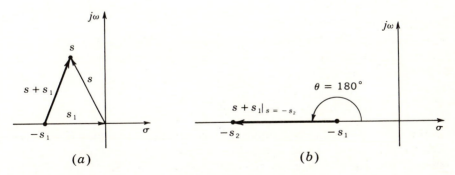

**Fig. 4.2-1** Graphical evaluation of residues. ($a$) Illustration of the distance $s + s_1$ for an arbitrary $s$; ($b$) the distance $s + s_1$ evaluated at $s = -s_2$.

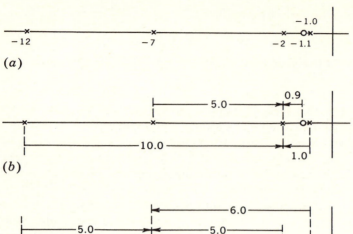

**Fig. 4.2-2** $f(s) = K(s + 1.1)/[(s + 1)(s + 2)(s + 7)(s + 12)]$. (a) Pole-zero plot; (b) residue at $s = -2$; (c) residue at $s = -7$.

at the point $-s_1$, and an arbitrary $s$ location is also indicated. The distances $s_1$ and $s$ are shown, as well as their sum. In Fig. 4.2-1b the arbitrary $s$ location is replaced by the specific value of $s$, $s = -s_2$. Thus the term $s + s_1$ evaluated at $s = -s_2$ is just the distance indicated, along with its phase angle.

Of course, terms such as this one are needed in evaluating residues. Therefore for simple poles the values of the residues, and hence the contribution to the total output, may be determined by measuring or estimating distances on the $s$ plane.

To be specific, consider the $s$−plane plot of Fig. 4.2-2a. There four poles are indicated at $s = -1, -2, -7$, and $-12$, and a zero at $s = -1.1$. The corresponding $f(s)$ is

$$f(s) = \frac{K(s + 1.1)}{(s + 1)(s + 2)(s + 7)(s + 12)}$$

If the exact inverse transformation is to be taken, the gain $K$ must be known. However, if one wishes to estimate only the relative contribution of each of the poles to the time response $f(t)$, the value of $K$ is unimportant. The distances involved in calculating the residue at

$s = -2$ are indicated in Fig. 4.2-2*b*, and the distances involved in calculating the residue at $s = -7$ are indicated in Fig. 4.2-2*c*. From Fig.
4.2-2*b*, the residue at $s = -2$ is

$$R_{21} = \frac{-0.9K}{(-1)(5)(10)} = \frac{0.9K}{50}$$

For a rough approximation, the distances to the pole and zero near 1 are
almost the same, and they might be ignored. This approximation is even
better in the evaluation of the residue at $s = -7$, which from Fig. 4.2-2*c*
is given by

$$R_{31} = \frac{-5.9K}{(-6)(-5)(+5)} \approx \frac{-K}{25}$$

By comparing Fig. 4.2-2*b* and *c*, it is easily seen that the residue for the
pole at $s = -2$ is smaller than that for the pole at $s = -7$. However,
because the time constant is larger, $\frac{1}{2}$ compared with $\frac{1}{7}$ sec, the transient
term associated with the residue at $s = -2$ contributes to the output for
a longer time.

Because of the very short distance between the pole at $-1$ and the
zero at $-1.1$, the residue of the pole at $s = -1$ is very small, because this
distance appears in the numerator when evaluating residues. Therefore,
even though this pole has the smallest time constant, it does not affect the
output to any extent because its residue is so small.

The utilization of graphical procedures to determine the complete
system response is illustrated in Example 4.2-5 for a case involving complex conjugate poles.

***Example 4.2-5*** As an example, consider the partial-fraction
expansion of the $y(s)$ used in Examples 4.2-3 and 4.2-4.

$$y(s) = \frac{16(s + 5)}{(s + 10)[(s + 2)^2 + 2^2]} = \frac{R_{11}}{s + 10}$$
$$+ \frac{R}{s + 2 + j2} + \frac{\bar{R}}{s + 2 - j2}$$

The pole-zero plot of this function is shown in Fig. 4.2-3. As
before, $R$ is associated with the pole located at $s = -\alpha - j\beta$, or,
in this case, $s = -2 - j2$. Note that this is the pole in the lower
half plane. The distances involved in the calculation of the residue
of the pole at $s = -10$ are indicated in Fig. 4.2-4*a*. This residue,

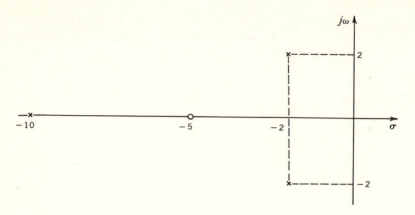

**Fig. 4.2-3**  Pole-zero plot of $y(s) = 16(s + 5)/\{(s + 10)[(s + 2)^2 + 2^2]\}$.

$R_{11}$, is readily found to be

$$R_{11} = \frac{(16)(-5)}{\sqrt{68} \; /-\phi \; \sqrt{68} \; /\phi} = -\frac{80}{68}$$

Since we are trying to find $y(t)$, the gain $K = 16$ is necessary.

The vector distances, or the distances and angles, associated with the calculation of the residue at the pole $s = -2 - j2$ are indicated in Fig. 4.2-4$b$.  Thus $R$ is

$$R = \frac{16 \sqrt{13} \; /-33.7°}{4/-90° \; \sqrt{68} \; /-14.1°} = 1.75/70.4° = 0.588 + j1.65$$

Consider now the evaluation of $\bar{R}$, the residue associated with the pole at $s = -2 + j2$.  All the distances are the same as those in the calculation of $R$, and all the angles are equal but opposite in sign.  As a consequence, the real part of each vector distance remains the same but the complex part is opposite, or the resulting $\bar{R}$ is just the conjugate of $R$.  Thus $y(s)$ is completely known and is identical to that of Eq. (4.2-14) of Example 4.2-3.

Once $R$ is known, $\bar{R}$ is also known, and one might suspect that this is enough information to find the inverse transform associated with a pair of complex conjugate poles without working with complex numbers and without forcing the transformed function to look like the $f(s)$ of entry 11 in Table 4.2-1.  This is indeed the case, and if $R$ is a complex number, $a + jb$, then the inverse transform of the pair

$$\frac{R}{s + \alpha + j\beta} + \frac{\bar{R}}{s + \alpha - j\beta}$$

is just

$$2|R|e^{-\alpha t} \sin\left(\beta t + \arctan \frac{a}{b}\right)$$

or

$$2\sqrt{a^2 + b^2} \; e^{-\alpha t} \sin\left(\beta t + \arctan \frac{a}{b}\right)$$

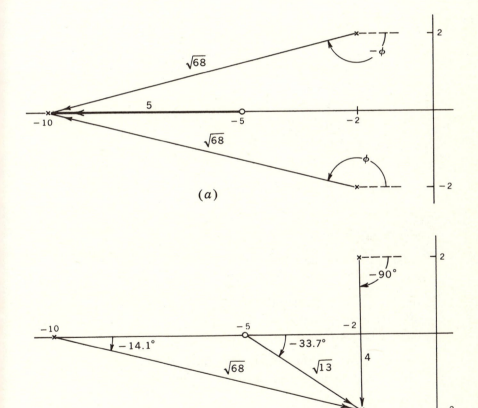

(a)

(b)

**Fig. 4.2-4** Distances and angles involved in residue evaluation. (a) Pole at $s = -10$; (b) pole at $s = -2 - j2$.

The proof of this statement is formulated as an exercise at the end of this section.   Note that if we are interested simply in the magnitude of the sinusoidal components then the angle need not be measured since the $|R|$ is all that is needed.

On the basis of this short review of partial-fraction-expansion methods, we are ready to consider the problem of finding the total system response.   As a start, in the next section the time response of a second-order system is considered in detail.

**Exercises 4.2**    *4.2-1.*   In a partial-fraction expansion involving complex conjugate roots, the following terms appear in $y(s)$:

$$y(s) = \frac{R}{s + \alpha + j\beta} + \frac{\bar{R}}{s + \alpha - j\beta}$$

If $R = a + jb$, use entry 11 of Table 4.2-1 to show that $y(t)$ is

$$y(t) = 2 \sqrt{a^2 + b^2} \, e^{-\alpha t} \sin \left( \beta t + \arctan \frac{a}{b} \right)$$

*4.2-2.*   If $y(s)$ is given as

$$y(s) = \frac{3(s + 4)}{(s + 1)(s + 2)(s + 6)}$$

find $y(t)$ in three different ways: by the algebraic approach, by residues, and through graphical means.

*answer:*

$$y(t) = \tfrac{9}{5}e^{-t} - \tfrac{3}{2}e^{-2t} - \tfrac{3}{10}e^{-6t}$$

*4.2-3.*   Use the most convenient approach to determine the time response of the given system to the following inputs:

(a)   $r(t) = \delta(t)$, an impulse
(b)   $r(t) = \mu(t)$, a step
(c)   $r(t) = t\mu(t)$, a ramp

In each case use the initial-value theorem to show that $y(0) = 0$. The given system is described by the closed-loop transfer function

$$\frac{y(s)}{r(s)} = \frac{12(s + 3)}{(s + 2)^2[(s + 3)^2 + 3^2]}$$

*answer:*

(c)   $y(t) = \tfrac{1}{2}t - \tfrac{1}{2} + \tfrac{3}{10}te^{-2t} + \tfrac{27}{50}e^{-2t} + \tfrac{1}{15}e^{-3t} \sin (3t - 36.7°)$

## 4.3   Step-function response of second-order systems

In this section the time response of a second-order system to a step-function input is discussed in detail.   The second-order system is important because it is easy to analyze and understand.   For this reason closed-loop-system specifications are often given in terms of behavior typical of the second-order case.   In other words, the performance of a high-order system is discussed in terms of a dominant set of second-order poles; that is, the high-order system is approximat₂d by a second-order system.   Thus a thorough understanding of the second-order case is important before the total time response for the general case is discussed in the following section.

The discussion here is based upon the second-order differential equation

$$\frac{d^2y(t)}{dt^2} + 2\zeta\omega_n\frac{dy(t)}{dt} + \omega_n{}^2y(t) = \omega_n{}^2r(t) \qquad (4.3\text{-}1)$$

where $r(t)$ is a step-function input and the initial conditions are assumed to be zero.   Here the variable $\zeta$ is called the *damping ratio*, and $\omega_n$ is referred to as the *undamped natural frequency*.   Equation (4.3-1) is purposely written in terms of these parameters in order to express the solution conveniently.   In addition, a number of aspects of the solution are most easily written in terms of $\zeta$ and $\omega_n$.

The physical origins of the second-order differential equation (4.3-1) are many and varied.   For example, a series electrical circuit containing resistive, inductive, and capacitive elements is typical.   A system consisting of a spring, mass, and viscous friction is a mechanical analog.   Here attention is directed to the state-variable-feedback system shown in Fig. 4.3-1.   This may be thought of as a simple positioning servo-

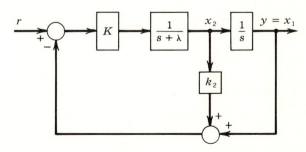

**Fig. 4.3-1**   Second-order system utilizing state-variable feedback.

mechanism in which the power element is either a field- or armature-controlled dc motor. In either case, the inductance of the control winding is assumed to be negligible so that the plant may be approximated by the second-order form shown.

The closed-loop transfer function $y(s)/r(s)$ associated with Fig. 4.3-1 may be found by using either the $H_{eq}$ or $G_{eq}$ reduction:

$$\frac{y(s)}{r(s)} = \frac{K}{s^2 + s(\lambda + Kk_2) + K} \tag{4.3-2}$$

The Laplace transform of Eq. (4.3-1), with zero initial conditions, on the other hand, is

$$\frac{y(s)}{r(s)} = \frac{\omega_n^2}{s^2 + 2\zeta\omega_n s + \omega_n^2} \tag{4.3-3}$$

and Eqs. (4.3-2) and (4.3-3) are equal when

$$K = \omega_n^2 \qquad \lambda + Kk_2 = 2\zeta\omega_n \tag{4.3-4}$$

Therefore by proper selection of $K$ and $k_2$ it is possible to cause $\zeta$ and $\omega_n$ to have any desired values. In other words, by varying $K$ and $k_2$ it is possible to adjust $\zeta$ and $\omega_n$ and therefore the closed-loop response of the system of Fig. 4.3-1. We shall consider this process in more detail later, but first let us investigate the basic nature of the time response.

The closed-loop transfer function (4.3-3) has poles at

$$s = -\zeta\omega_n \pm \sqrt{(\zeta^2 - 1)\omega_n^2}$$

The nature of the closed-loop response depends almost completely on the damping ratio. The value of $\omega_n$ simply adjusts the time scale, as we shall see later. If $\zeta$ is greater than 1, both closed-loop poles are real and negative; if $\zeta$ is equal to 1, both closed-loop poles are equal to $-\zeta\omega_n$; and if $\zeta$ is less than 1, the closed-loop poles are complex conjugates, given as

$$s = -\zeta\omega_n \pm j\omega_n \sqrt{1 - \zeta^2} \tag{4.3-5}$$

The latter situation is the case of practical interest, and we shall concentrate our discussion on it.

The location of the closed-loop poles with respect to the damping ratio and the undamped natural frequency is indicated in Fig. 4.3-2. In order to investigate the response for the closed-loop system, it is necessary to determine the inverse Laplace transform of $y(s)$ for some $r(s)$. In this chapter the input is chosen as a step since this is a typical and convenient test input. For $r(s) = 1/s$, $y(s)$ is given by

$$y(s) = \frac{\omega_n^2}{s[(s + \zeta\omega_n)^2 + (\omega_n \sqrt{1 - \zeta^2})^2]} \tag{4.3-6}$$

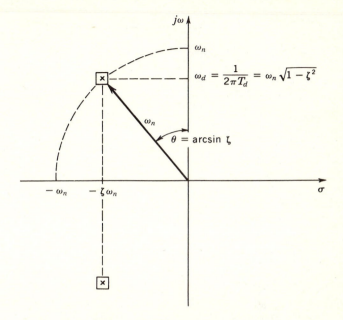

**Fig. 4.3-2**  Location of the closed-loop poles with respect to $\zeta$ and $\omega_n$.

Equation (4.3-6) has been written in this form to emphasize the complex conjugate structure of the closed-loop system. To find $y(t)$ it is convenient to expand Eq. (4.3-6) into partial-fraction form and use the short table of transforms (Table 4.2-1), as advocated in the preceding section. Thus $y(s)$ is

$$y(s) = \frac{1}{s} - \frac{s + 2\zeta\omega_n}{(s + \zeta\omega_n)^2 + (\omega_n\sqrt{1 - \zeta^2})^2}$$

With the aid of the short table of transform pairs, $y(t)$ is written directly as

$$y(t) = 1 - \frac{1}{\sqrt{1 - \zeta^2}}\, e^{-\zeta\omega_n t} \sin(\omega_n\sqrt{1 - \zeta^2}\, t + \psi) \qquad (4.3\text{-}7)$$

where

$$\psi = \arctan \frac{\sqrt{1 - \zeta^2}}{\zeta} \qquad (4.3\text{-}8)$$

Here the actual frequency of oscillation in radians per second is $\omega_n\sqrt{1 - \zeta^2}$ and is known as the *damped frequency* $\omega_d$. A typical oscil-

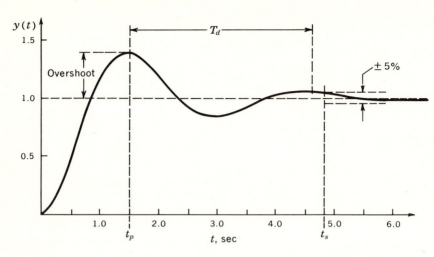

**Fig. 4.3-3**  Typical underdamped $(0 < \zeta < 1)$ second-order-system response for a unit step imput.

latory response is given in Fig. 4.3-3, and pertinent characteristics of the curve are labeled.    These are discussed below.

The period of oscillation, $T_d$, associated with the damped frequency $\omega_d$ is

$$T_d = \frac{2\pi}{\omega_d} = \frac{2\pi}{\omega_n \sqrt{1 - \zeta^2}} \qquad (4.3\text{-}9)$$

The time to the maximum, or peak, value of the output is $t_p$.    The maximum value of $y(t)$ may be found by taking the derivative of $y(t)$ with respect to $t$ and equating the result to zero.    If this is done, $dy(t)/dt$ is

$$\frac{dy(t)}{dt} = \frac{\zeta\omega_n}{\sqrt{1 - \zeta^2}} e^{-\zeta\omega_n t} \sin(\omega_d t + \psi) - \omega_n e^{-\zeta\omega_n t} \cos(\omega_d t + \psi) = 0$$

$$(4.3\text{-}10)$$

This is zero when $t$ is

$$t = \frac{n\pi}{\omega_d} = \frac{n\pi}{\omega_n \sqrt{1 - \zeta^2}} \qquad n = 0, 1, 2, \ldots$$

The peak value of $y(t)$ occurs when $n = 1$ or

$$t_p = \frac{\pi}{\omega_n \sqrt{1 - \zeta^2}} = \frac{T_d}{2} \qquad (4.3\text{-}11)$$

The percent overshoot (PO) is 100 times the peak value of $y(t)$ minus the step size, divided by the step size.    This is a normalized quantity which is independent of the step size, although here we have assumed a unit step. The peak value of the response, $y(t)_{max}$, can be found by substituting $t_p$ into the expression for $y(t)$. The result can be reduced to

$$y(t)_{max} = 1 + \exp - \frac{\zeta\pi}{\sqrt{1 - \zeta^2}} \qquad (4.3\text{-}12)$$

Therefore the percent overshoot is simply

$$PO = 100 \exp - \frac{\zeta\pi}{\sqrt{1 - \zeta^2}} \qquad (4.3\text{-}13)$$

The overshoot for a range of values of $\zeta$ is shown in Fig. 4.3-4.

The settling time $t_s$ is normally defined as the time required for the response to remain within 5 percent of its final value. Since the magnitude of the sinusoidal portion of Eq. (4.3-7) is always less than or equal to 1, for convenience the settling time is often approximated as the time beyond which the overshoot (or undershoot) is less than 0.05 or

$$\exp - \frac{\zeta n\pi}{\sqrt{1 - \zeta^2}} < 0.05$$

or when

$$\frac{\zeta n\pi}{\sqrt{1 - \zeta^2}} > 3$$

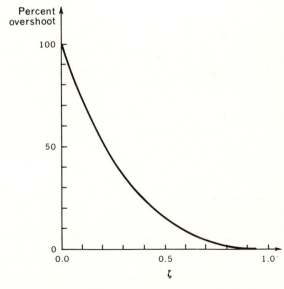

Fig. 4.3-4   Percent overshoot as a function of damping ratio $\zeta$.

In terms of $t_p$ the latter expression is

$$\zeta\omega_n n t_p > 3$$

or $t_s$ is that multiple of $t_p$ for which

$$t_s = n t_p > \frac{3}{\zeta\omega_n}$$

Often $t_s$ is taken simply as

$$t_s = \frac{3}{\zeta\omega_n} \tag{4.3-14}$$

Note that $T_d$, $t_p$, and $t_s$ are all inversely proportional to $\omega_n$. In other words, the quantities $\omega_n T_d$, $\omega_n t_p$, and $\omega_n t_s$ are independent of $\omega_n$. The percent overshoot is completely independent of $\omega_n$. If the time response $y(t)$ is plotted against the dimensionless normalized variable $\omega_n t$, the response is solely a function of $\zeta$. This justifies the comment made earlier that the nature of the response is almost completely determined by the value of $\zeta$ and that $\omega_n$ serves merely to time-scale the response. To illustrate more clearly the influence of $\zeta$ and $\omega_n$ on the time response, let us consider three special cases: (1) vary $\zeta$ with $\omega_n$ constant, (2) vary $\omega_n$ with $\zeta$ constant, and (3) vary $\zeta$ and $\omega_n$ with the product $\zeta\omega_n$ constant. Each of these cases is discussed separately below.

The variation of $\zeta$ while holding $\omega_n$ constant can be accomplished in the state-variable-feedback system of Fig. 4.3-1 by varying $k_2$ with $K$ fixed. Figure 4.3-5a indicates the position of the closed-loop poles for a number of values of $\zeta$. The locus of the closed-loop poles as $\zeta$ is varied from 0 to 1 is a semicircle of radius $\omega_n$ shown by the dashed line in Fig. 4.3-5a. The corresponding step responses are shown in Fig. 4.3-5b. Note that $T_d$, $t_s$, and $t_p$ as well as the percent overshoot vary as $\zeta$ is changed.

In order to vary $\omega_n$ while keeping $\zeta$ constant, it is convenient to solve Eqs. (4.3-4) for $k_2$ in terms of $\zeta$, $\omega_n$, and $K$ as

$$k_2 = \frac{2\omega_n\zeta - \lambda}{K} \tag{4.3-15}$$

Since $\omega_n{}^2 = K$, Eq. (4.3-15) may be written

$$k_2 = \frac{2\sqrt{K}\,\zeta - \lambda}{K} \tag{4.3-16}$$

Now, as $\omega_n$ is varied by changing $K$, $k_2$ may be determined so that $\zeta$ remains constant. Figure 4.3-6a shows the locus of the closed-loop poles in this case; it is a straight line passing through the origin. The step

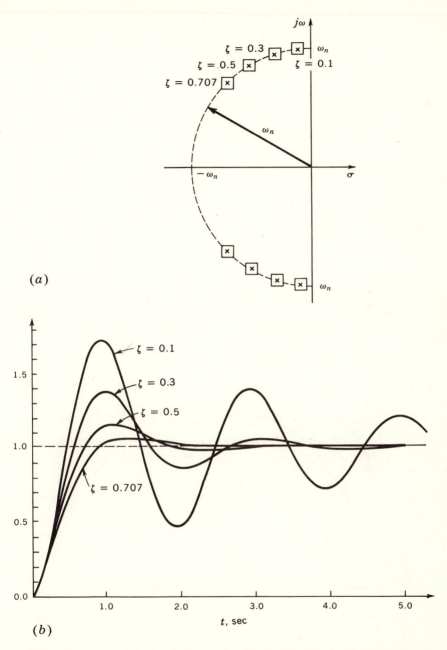

**Fig. 4.3-5**   Variation of $\zeta$ with $\omega_n = 1.0$.    (a) Plot of the closed-loop poles; (b) step response.

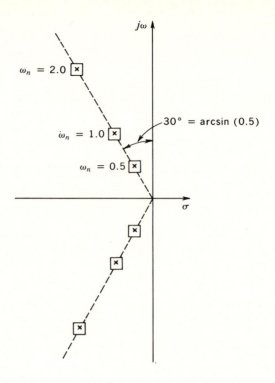

(a)

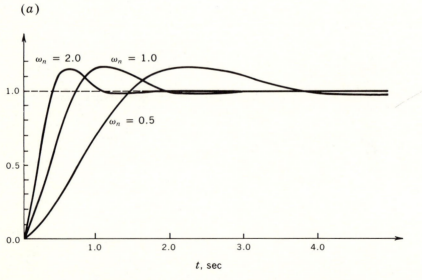

(b)

**Fig. 4.3-6** Variation of $\omega_n$ with $\zeta = 0.5$. (a) Plot of the closed-loop poles; (b) step response.

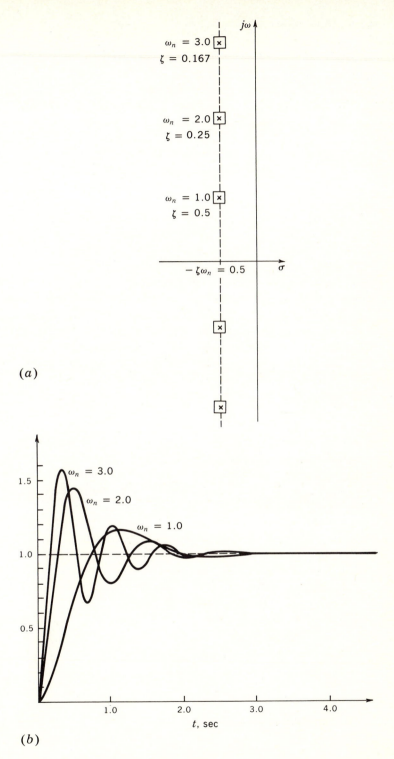

$(a)$

$(b)$

**Fig. 4.3-7**  Variation of $\zeta$ and $\omega_n$ with $\zeta\omega_n = 0.5$.   $(a)$ Plot of the closed-loop poles; $(b)$ step response.

responses for these same values of $\omega_n$ are shown in Fig. 4.3-6b. In this case, we see that the basic character of the response, in particular the overshoot, is constant since $\zeta$ is constant but that $T_d$, $t_s$, and $t_p$ decrease as $\omega_n$ is increased.

When $\zeta$ and $\omega_n$ are both varied while the product $\zeta\omega_n$ is kept constant, the locus of the closed-loop poles becomes the vertical straight line $s = -\zeta\omega_n$ as shown in Fig. 4.3-7a. The practical implementation in this case is achieved by varying $k_2$ and $K$ while holding the product $Kk_2$ constant. The step responses are shown in Fig. 4.3-7b. Note that, in this case, $t_s$ remains constant although $T_d$, $t_p$, and the overshoot vary.

The quantities $t_s$, $t_p$, and PO can be expressed in terms of the second-order parameters $\zeta$ and $\omega_n$. In addition, there are a number of other terms used to describe the system response for which no analytic expression exists. These quantities are the rise time $t_r$ and the delay time $t_d$. Rise time is defined as the time required by the system to rise from 10 to 90 percent of its final value. Delay time is defined as the time required for the system to reach 50 percent of its final value. These two means of description of the closed-loop response are mentioned here for completeness and will be discussed in detail in Chap. 8.

The discussion above has centered about the case in which $0 < \zeta < 1$, or the case of the underdamped system. If $\zeta = 1$, the step-function response is said to be critically damped and is of the form

$$y(t) = 1 - \omega_n t e^{-\omega_n t} - e^{-\omega_n t}$$

In the overdamped case, the damping ratio is greater than 1, and the response contains two exponential terms with different time constants. As the damping ratio approaches 1 from either above or below, it becomes more and more difficult to establish the actual damping ratio by input-output measurements of the closed-loop system.

**Exercises 4.3**  *4.3-1.*  Verify Eqs. (4.3-10) and (4.3-11).

*4.3-2.*  Compare the system of Fig. 4.3-8 with that of Fig. 4.3-1. Under what conditions can the two systems have the same response to a given input signal? Under these conditions is the initial-condition response the same?

*4.3-3.*  In the system of Fig. 4.3-8, let $\lambda = 2$ and plot the closed-

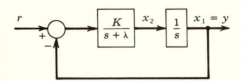

**Fig. 4.3-8**  Exercises 4.3-2 and 4.3-3.

loop poles for $K = 0, \frac{1}{4}, \frac{1}{2}, 1, 2, 5,$ and 10.   If these points are connected by a straight line, this represents the locus of roots of the closed-loop system as $K$ varies from 0 to a large value.

## 4.4   Total time response

In the preceding section the output response was determined as a function of time for a second-order system subject to a unit step input and no initial conditions.   In the general case treated in this section, the order of the system is not restricted nor is the nature of the input. Initial conditions are allowed on any of the state variables and, in addition to the output response, here we determine the behavior of all the state variables as a function of time.

As noted in Chaps. 2 and 3, it is easiest to formulate a general problem in terms of the matrix notation, although this does not make the work of obtaining the answer any easier, just more systematic. Assume that the plant to be controlled is described by Eq. (**Ab**),

$$\dot{\mathbf{x}} = \mathbf{A}\mathbf{x} + \mathbf{b}u \tag{Ab}$$

and that the control is achieved by state-variable feedback so that $u$ is given by

$$u = K(r - \mathbf{k}^T\mathbf{x})$$

The closed-loop system is once again

$$\dot{\mathbf{x}} = (\mathbf{A} - K\mathbf{b}\mathbf{k}^T)\mathbf{x} + K\mathbf{b}r = \mathbf{A}_k\mathbf{x} + K\mathbf{b}r \tag{$\mathbf{A}_k\mathbf{b}$}$$

where $\mathbf{A}_k$ is the describing matrix of the closed-loop system defined as before by

$$\mathbf{A}_k = \mathbf{A} - K\mathbf{b}\mathbf{k}^T$$

The output $y$ is given by

$$y = \mathbf{c}^T\mathbf{x} \tag{c}$$

Thus the two equations that concern us are Eqs. ($\mathbf{A}_k\mathbf{b}$) and (**c**).   These must be solved for $\mathbf{x}(t)$ and $y(t)$.   The procedure that we use here is based on the Laplace transform.   In the next section an alternative time-domain solution will be considered.

The Laplace transform of each side of Eq. ($\mathbf{A}_k\mathbf{b}$) is

$$\mathcal{L}[\dot{\mathbf{x}}(t)] = s\mathbf{x}(s) - \mathbf{x}(0)$$

and

$$\mathcal{L}[\mathbf{A}_k\mathbf{x} + K\mathbf{b}r] = \mathbf{A}_k\mathbf{x}(s) + K\mathbf{b}r(s)$$

so that the transformed version of Eq. $(\mathbf{A}_k\mathbf{b})$ is just

$$s\mathbf{x}(s) - \mathbf{x}(0) = \mathbf{A}_k\mathbf{x}(s) + K\mathbf{b}r(s)$$

or

$$\mathbf{x}(s) = (s\mathbf{I} - \mathbf{A}_k)^{-1}\mathbf{x}(0) + K(s\mathbf{I} - \mathbf{A}_k)^{-1}\mathbf{b}r(s) \qquad (4.4\text{-}1)$$

where the identity matrix $\mathbf{I}$ has been added for dimensional compatibility. In terms of the resolvent matrix, Eq. (4.4-1) becomes

$$\mathbf{x}(s) = \mathbf{\Phi}_k(s)\mathbf{x}(0) + K\mathbf{\Phi}_k(s)\mathbf{b}r(s) \qquad (4.4\text{-}2)$$

since

$$\mathbf{\Phi}_k(s) = (s\mathbf{I} - \mathbf{A}_k)^{-1} \qquad (4.4\text{-}3)$$

The total time response $\mathbf{x}(t)$ is therefore

$$\mathbf{x}(t) = \mathcal{L}^{-1}[\mathbf{\Phi}_k(s)]\mathbf{x}(0) + K\mathcal{L}^{-1}[\mathbf{\Phi}_k(s)\mathbf{b}r(s)] \qquad (4.4\text{-}4)$$

Notice that the solution for $\mathbf{x}(t)$ is divided into two parts: that associated with initial conditions and that associated with the forcing function.

The apparent simplicity of the solution for $\mathbf{x}(t)$, that is, Eq. (4.4-4), is deceiving. In order to obtain the solution, one must first find the resolvent matrix of the closed-loop system, as discussed in Chap. 3. Then the indicated multiplications must be carried out, which is a simple job but one that produces, in the general case, $2n^2$ terms. The state variables are then found as functions of time by taking the inverse Laplace transform of these $2n^2$ terms. Once $\mathbf{x}(t)$ is known, $y(t)$ is easily found.

As a second-order example, let us consider the system shown in Fig. 4.4-1. This system might be a dc positioning system with tachometer feedback to produce a voltage proportional to the derivative of the out-

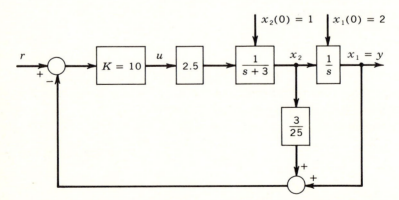

**Fig. 4.4-1**  Second-order system.

put. On the basis of the block diagram, the system equations are written as

$$\dot{x}_1 = x_2$$
$$\dot{x}_2 = -3x_2 + 25(r - \tfrac{3}{25}x_2 - x_1)$$

or

$$\dot{x}_1 = x_2$$
$$\dot{x}_2 = -25x_1 - 6x_2 + 25r$$

so that $\mathbf{A}_k$ and $K\mathbf{b}$ are

$$\mathbf{A}_k = \begin{bmatrix} 0 & 1 \\ -25 & -6 \end{bmatrix} \quad \text{and} \quad K\mathbf{b} = \begin{bmatrix} 0 \\ 25 \end{bmatrix}$$

Also

$$(s\mathbf{I} - \mathbf{A}_k) = \mathbf{\Phi}_k^{-1} = \begin{bmatrix} s & -1 \\ 25 & s+6 \end{bmatrix}$$

The resolvent matrix is therefore

$$\mathbf{\Phi}_k(s) = \frac{1}{(s+3)^2 + 4^2} \begin{bmatrix} s+6 & 1 \\ -25 & s \end{bmatrix}$$

Note particularly that the determinant of the system matrix $(s\mathbf{I} - \mathbf{A}_k)$ appears in the denominator of every element of $\mathbf{\Phi}_k(s)$. This determinant is a polynomial in $s$ and, as previously mentioned, is commonly referred to as the characteristic polynomial of the closed-loop system. The *form* of the behavior of each state variable is the same, since they have the same characteristic equation and hence the same pole locations.

In the above development $\mathbf{\Phi}_k(s)$ was determined completely by matrix manipulations. The elements of $\mathbf{\Phi}_k(s)$ could also have been found by block-diagram manipulations as discussed in Chap. 3.

Once $\mathbf{\Phi}_k(s)$ is known, we are ready to substitute into Eq. (4.4-2). In expanded form this equation is now

$$\mathbf{x}(s) = \frac{1}{(s+3)^2 + 4^2} \begin{bmatrix} s+6 & 1 \\ -25 & s \end{bmatrix} \begin{bmatrix} x_1(0) \\ x_2(0) \end{bmatrix}$$
$$+ \frac{1}{(s+3)^2 + 4^2} \begin{bmatrix} s+6 & 1 \\ -25 & s \end{bmatrix} \begin{bmatrix} 0 \\ 25 \end{bmatrix} r(s)$$

For the initial conditions of $\mathbf{x}(0) = \text{col}\,(2,1)$, $\mathbf{x}(s)$ is given by

$$\mathbf{x}(s) = \frac{1}{(s+3)^2 + 4^2} \begin{bmatrix} 2(s+6) + 1 + 25r(s) \\ s - 50 + 25sr(s) \end{bmatrix}$$
$$= \frac{1}{(s+3)^2 + 4^2} \begin{bmatrix} 2s+13 \\ s-50 \end{bmatrix} + \frac{1}{(s+3)^2 + 4^2} \begin{bmatrix} 25 \\ 25s \end{bmatrix} r(s)$$

For the moment let us discuss initial-condition response only by assuming that $r(s)$ is zero. In this case the expressions for $x_1(s)$ and $x_2(s)$ reduce to the following, where the subscript "ic" is used to indicate response to initial conditions only.

$$x_1(s)_{ic} = \frac{2(s + 6.5)}{(s + 3)^2 + 4^2}$$

$$x_2(s)_{ic} = \frac{s - 50}{(s + 3)^2 + 4^2}$$

After inverse Laplace transformation, $x_1(t)_{ic}$ and $x_2(t)_{ic}$ are

$$x_1(t)_{ic} = 2.65e^{-3t} \sin (4t + 48.8°)$$
$$x_2(t)_{ic} = 13.3e^{-3t} \sin (4t + 175.7°)$$

The reader is urged to show that these are related by the derivative, even though direct differentiation of $x_1(t)$ results in an answer that contains both sine and cosine terms. Of course, $x_2(t)$ must be the derivative of $x_1(t)$, as these two state variables are defined in this way by the given differential equations and/or the block diagram of Fig. 4.4-1.

In this example the roots of the characteristic equation, that is, the closed-loop poles, are located at $s = -3 \pm j4$, and the damping ratio is $\zeta = 0.6$. Both $x_1$ and $x_2$ are plotted as functions of time in Fig. 4.4-2. An alternative state-space representation of the initial-condition response is given in Fig. 4.4-3. Here the coordinates are the state variables, and time is a parameter along the curve and does not appear explicitly. Common points are labeled in Figs. 4.4-2 and 4.4-3 to aid in correlation of the two plots. Much of modern control theory is oriented in terms of state variables and their behavior in state space. Unfortunately it is difficult to draw a state space of more than two dimensions.

The forced response is contributed by the last term in Eq. (4.4-4). If $r(s)$ is assumed to be a unit step function, the problem of finding the inverse transform is identical to that treated in Sec. 4.2, with the result that

$$x_1(t)_f = 1 - 1.25e^{-3t} \sin (4t + 53.2°)$$
$$x_2(t)_f = 6.25e^{-3t} \sin 4t$$

Here the subscript $f$ indicates that only the forced response is being considered. Although in this example we consider the input forcing function to be a unit step, it should be emphasized that the solution given in the frequency domain in Eq. (4.4-4) is valid for any $r(t)$ having a Laplace transform.

Since the system with which we are dealing is linear, the principle of superposition applies, and the total response is found by simply adding the initial-condition and forced responses. Thus in this problem the

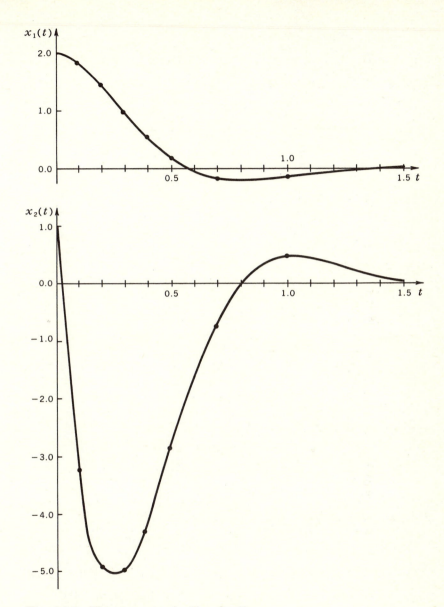

***Fig. 4.4-2*** Time response of $x_1(t)$ and $x_2(t)$.

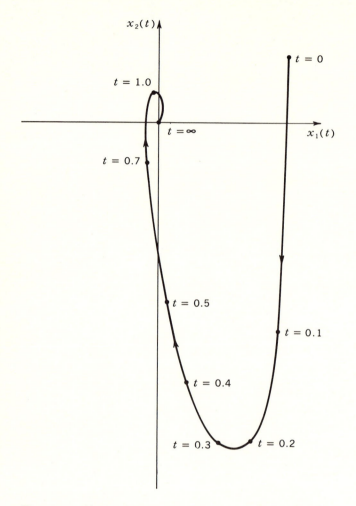

**Fig. 4.4-3**  Alternative state-space representation of $x_1(t)$ and
$x_2(t)$.

total response is

$$x_1(t) = 1 - 1.25e^{-3t} \sin(4t + 53.2°) + 2.65e^{-3t} \sin(4t + 48°)$$
$$= 1 + 1.414e^{-3t} \sin(4t + 45°)$$
$$x_2(t) = 6.25e^{-3t} \sin 4t + 13.3e^{-3t} \sin(4t + 175.7°)$$
$$= 7.08e^{-3t} \sin(4t + 171.9°)$$

$$(4.4\text{-}5)$$

Once again it is important to emphasize that, even though $\mathbf{x}(t)$ may
be completely determined in an analytic sense, we still have little idea of

the actual behavior of the various state variables unless we plot them. The preceding example serves to illustrate this point.   If in that example we were to plot Eq. (4.4-5), the response curves for any initial conditions other than those specifically assumed at the beginning of the example would be unknown.   This is always the case, and for this reason it is very common to assume that initial conditions are zero.   This leads to thinking in terms of transfer functions, as a transfer function is defined in terms of zero initial conditions.   One danger of so thinking is that one tends to concentrate on overall input-output relationships, rather than on the transfer functions that relate the individual state variables.

In the above development, the response of the individual state variables was carried out as separate scalar problems of inverse Laplace transforms.   In other words, the residues for each $x_i(s)$ were computed separately.   An alternative procedure is to determine the inverse transform in terms of matrix notation.   Although the latter approach probably does not provide a significant computational advantage, it makes the labor somewhat more systematic.

Let us suppose, for example, that $\mathbf{x}(s)$ has been determined from Eq. (4.4-2) and written in the following form:[1]

$$\mathbf{x}(s) = \frac{\mathbf{n}(s)}{D(s)} = \frac{\mathbf{n}(s)}{(s + s_1)(s + s_2) \cdots (s + s_n)} \qquad (4.4\text{-}6)$$

Here $\mathbf{n}(s)$ is a vector of polynomials in $s$ which are the numerator polynomials of each of the state variables.   The coefficients of the elements of $\mathbf{n}(s)$ depend on the $\mathbf{x}(0)$ and $r(s)$ being used.   In addition, $D(s)$ contains any poles that have been added by the forcing function, if one is present.

Now let us expand $\mathbf{x}(s)$ in the usual fashion as

$$\mathbf{x}(s) = \frac{\mathbf{r}_{11}}{s + s_1} + \frac{\mathbf{r}_{21}}{s + s_2} + \cdots + \frac{\mathbf{r}_{n1}}{s + s_n} \qquad (4.4\text{-}7)$$

where the residue vectors are given by

$$\mathbf{r}_{i1} = \left. \frac{(s + s_i)\mathbf{n}(s)}{D(s)} \right|_{s = -s_i} \qquad (4.4\text{-}8)$$

In terms of the partial-fraction expansion of Eq. (4.4-7), the time response $\mathbf{x}(t)$ may be written

$$\mathbf{x}(t) = \mathbf{r}_{11}e^{-s_1 t} + \mathbf{r}_{21}e^{-s_2 t} + \cdots + \mathbf{r}_{n1}e^{-s_n t} \qquad (4.4\text{-}9)$$

The advantage of this procedure is that it groups the evaluation of the residues associated with each pole into one operation.

[1] Only simple poles are represented here; the case of multiple poles follows directly.

*Example 4.4-1*    To illustrate the use of the above procedure, let us determine again the total time response of the system shown in Fig. 4.4-1. Once again we assume that $\mathbf{x}(0) = \text{col }(2,1)$ and $r(s) = 1/s$ so that $\mathbf{x}(s)$ from Eq. (4.4-2) becomes

$$\mathbf{x}(s) = \frac{1}{s[(s + 3)^2 + 4^2]} \begin{bmatrix} 2s^2 + 13s + 25 \\ s^2 - 25s \end{bmatrix} \qquad (4.4\text{-}10)$$

Rather than treat the complex conjugate poles as separate complex terms, let us expand $\mathbf{x}(s)$ in the following form, after the development of Sec. 4.2:

$$\mathbf{x}(s) = \frac{\mathbf{r}_{11}}{s} + \frac{\boldsymbol{\alpha} s + \boldsymbol{\beta}}{(s + 3)^2 + 4^2} \qquad (4.4\text{-}11)$$

Here $\mathbf{r}_{11}$ is easily found to be

$$\mathbf{r}_{11} = \frac{1}{(s + 3)^2 + 4^2} \begin{bmatrix} 2s^2 + 13s + 25 \\ s^2 - 25s \end{bmatrix} \Bigg|_{s=0} = \begin{bmatrix} 1 \\ 0 \end{bmatrix}$$

Therefore Eq. (4.4-11) becomes

$$\begin{aligned}
\mathbf{x}(s) &= \frac{1}{s} \begin{bmatrix} 1 \\ 0 \end{bmatrix} + \frac{\boldsymbol{\alpha} s + \boldsymbol{\beta}}{(s + 3)^2 + 4^2} \\
&= \frac{1}{s[(s + 3)^2 + 4^2]} \begin{bmatrix} (1 + \alpha_1)s^2 + (13 + \beta_1)s + 25 \\ \alpha_2 s^2 + \beta_2 s \end{bmatrix} \qquad (4.4\text{-}12)
\end{aligned}$$

If the coefficients of equal powers of $s$ in Eqs. (4.4-10) and (4.4-12) are equated, we find that $\boldsymbol{\alpha} = \text{col }(1,1)$ and $\boldsymbol{\beta} = \text{col }(7, -25)$ so that the partial-fraction expansion of Eq. (4.4-11) becomes

$$\mathbf{x}(s) = \frac{1}{s} \begin{bmatrix} 1 \\ 0 \end{bmatrix} + \frac{1}{(s + 3)^2 + 4^2} \left( s \begin{bmatrix} 1 \\ 1 \end{bmatrix} + \begin{bmatrix} 7 \\ -25 \end{bmatrix} \right)$$

Now, by using the short table of transform pairs, we find once again that $\mathbf{x}(t)$ is

$$\mathbf{x}(t) = \begin{bmatrix} 1 \\ 0 \end{bmatrix} + \begin{bmatrix} 1.414e^{-3t} \sin (4t + 45°) \\ 7.08e^{-3t} \sin (4t + 171.9°) \end{bmatrix}$$

Before leaving this section, let us examine one additional way of looking at the problem of finding the total system response. The use of the resolvent matrix is a very systematic approach, and if one is interested in the response for all initial conditions and any input, that is probably the optimum way to approach the problem. This is particularly true if all the elements of the vector $\mathbf{b}$ are nonzero. Often, as mentioned above, initial conditions are assumed to be zero. Also, we have often observed that in cascade-type systems, that is, systems in which the transfer-function blocks appear in series, only one element of **b** is non-zero; this is the $n$th element. Under these circumstances it may be easier just to find $x_1(t)$ and then find all the other state-variable time descriptions from $x_1(t)$. Example 4.4-2 illustrates the procedure.

**Example 4.4-2**  This example is concerned with the determination of the total time response of the system of Fig. 4.4-4. The initial conditions on all the state variables are zero; the input is a step function occurring at $t = 0$. The output $y$ is just $x_1(t)$, and since $y(s)/r(s)$ is

$$\frac{y(s)}{r(s)} = \frac{KG_p(s)}{1 + KG_p(s)H_{eq}(s)}$$

$x_1(s)$ is easily found from a knowledge of $H_{eq}(s)$ and the plant transfer function. Here $H_{eq}(s)$ is

$$H_{eq} = \tfrac{1}{80}(7s^2 + 38s + 80)$$

so that $y(s)/r(s)$ becomes

$$\frac{y(s)}{r(s)} = \frac{80}{s^3 + 14s^2 + 48s + 80} = \frac{80}{[(s + 2)^2 + 2^2](s + 10)}$$

For a step-function input, $x_1(s)$ is

$$y(s) = x_1(s) = \frac{80}{s[(s + 2)^2 + 2^2](s + 10)}$$

Therefore $x_1(t)$ is

$$x_1(t) = 1 - \tfrac{2}{17}e^{-10t} - 1.72e^{-2t}\sin(2t + 30.9°)$$

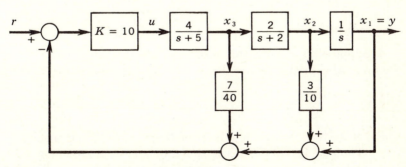

**Fig. 4.4-4**  System of Example 4.4-2.

From the block diagram or, alternatively, from the defining differential equations, it is seen that

$$\frac{x_1(s)}{x_2(s)} = \frac{1}{s} \quad \text{or} \quad x_2(t) = \frac{dx_1(t)}{dt}$$

Thus $x_2(t)$ may be found by just differentiating $x_1(t)$. Similarly $x_3(t)$ is just

$$x_3(t) = \frac{1}{2}\left[\frac{dx_2(t)}{dt} + 2x_2(t)\right]$$

and the complete solution is known in analytic form.

As yet we have made no mention of why anyone should desire the total system response. The answer is that one must be sure that the system under consideration will indeed behave as the analysis predicts. The analysis predicts the behavior of a linear system subject to a particular input. One must be sure that the linear model of the system is valid by verifying that none of the state variables has exceeded its range of linear operation. In addition, it may be necessary to verify that certain physical limitations or specifications of performance have not been violated. To illustrate the point, let us reexamine the system of Fig. 4.4-1. Again we think of this system as a positioning servomechanism with a dc motor for a power element and tachometer feedback. The dc motor is a velocity-limited device. If one increases the input voltage to a dc motor, its output velocity increases proportionately until iron saturation occurs. The output velocity then remains essentially constant regardless of how much voltage is applied.

Assume in the initial discussion of the system of Fig. 4.4-1 that the largest input that could be expected is a unit step input. The problem was worked for a unit step input, and the velocity $x_2(t)$ for this input was found to be

$$x_2(t) = 6.5e^{-3t} \sin 4t$$

Suppose we rework this problem with the same plant but change the forward gain and the feedback coefficient so that $y(s)/r(s)$ is now

$$\frac{y(s)}{r(s)} = \frac{100}{(s+6)^2 + 8^2}$$

The new system requires a gain $K$ of 40 and a feedback coefficient $k_2$ of $\frac{9}{100}$. The pole locations for the closed-loop system are indicated in Fig. 4.4-5 and compared with the original closed-loop poles. The damping

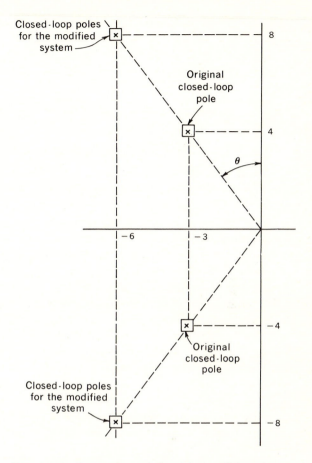

**Fig. 4.4-5** Comparison of closed-loop-pole locations.

ratio has been kept the same, although the poles are now located twice as far from the origin of the $s$ plane.

In order to find the response of both state variables to a unit step input, let us proceed as in Example 4.4-2. Again $y(s)/r(s)$ is equal to $x_1(s)/r(s)$, or

$$x_1(s) = \frac{100}{s[(s + 6)^2 + 8^2]} = \frac{1}{s} - \frac{s + 12}{(s + 6)^2 + 8^2}$$

and $x_1(t)$ is

$$x_1(t) = 1 - \tfrac{5}{4}e^{-6t} \sin (8t + 53.2°)$$

Here the time constant is shorter, and the damped frequency of oscillation is higher. The system responds faster and hence might be considered to be better.

The velocity is easily found, since

$$x_2(t) = \frac{dx_1(t)}{dt}$$

so that $x_2(t)$ is

$$x_2(t) = 12.5e^{-6t} \sin 8t$$

The maximum velocity is twice the maximum velocity found previously. Assume that the system had originally been designed so that a unit step input required the maximum effort from the power element; that is, the step input drove the motor at its maximum velocity. Then, by altering the gain and feedback coefficient, it only *appears* that the resulting system is faster and better. For the same unit step input, the modified system saturates, and we have no idea of whether this system is better or worse than the previous one, because the linear model is no longer adequate. For steps of $\frac{1}{2}$ the system is clearly faster, but it was initially postulated that steps of unit magnitude could be expected.

As yet we have discussed only analysis of given control-system configurations. Ultimately we shall be concerned with synthesis; however, the last step of any synthesis procedure is always one of analysis.

**Exercises 4.4** *4.4-1.* Consider the system described by the following two first-order differential equations:

$$\dot{x}_1 = x_2$$
$$\dot{x}_2 = -2x_1 - 3x_2$$

Find the response of this system to the initial conditions $x_1(0) = 1$, $x_2(0) = 1$ by using (*a*) the resolvent-matrix approach and (*b*) classical Laplace transform methods.

*answer:*

$$\mathbf{x}(t) = \begin{bmatrix} 3e^{-t} - 2e^{-2t} \\ -3e^{-t} + 4e^{-2t} \end{bmatrix}$$

*4.4-2.* Consider the block diagram shown in Fig. 4.4-6. Determine the resolvent matrix for this closed-loop system and find $x_1(t)$, $x_2(t)$, and $x_3(t)$ if $r(t)$ is a unit impulse and $\mathbf{x}(0) = 0$.

*answer:*

$$x_1(t) = \tfrac{9}{5}e^{-t} - \tfrac{3}{2}e^{-2t} - \tfrac{3}{10}e^{-6t}$$

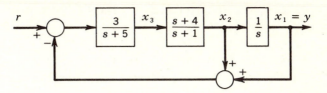

**Fig. 4.4-6**   Exercise 4.4-2.

*4.4-3.*   Repeat Exercise 4.4-2 but use the closed-loop transfer function $y(s)/r(s)$ to find $x_1(s) = y(s)$. From $x_1(t)$ find $x_2(t)$ and $x_3(t)$ by using the state-variable equations.

## 4.5   Time-domain methods

The object of this section is to present an alternative time-domain technique for determining the total system response $\mathbf{x}(t)$. The method makes use of the inverse Laplace transform of the resolvent matrix, which is referred to as the *state transition matrix* $\boldsymbol{\Phi}_k(t)$, and requires the utilization of the convolution integral. The discussion of the convolution integral serves to give further justification for our previous interpretation of the elements of $\phi_{k(ij)}(s)$ as impulse-response functions. The evaluation of the convolution integral proves to be difficult, and there is no shortcut approach analogous to partial-fraction-expansion techniques. Hence this section also serves to justify our almost exclusive preference for Laplace transform methods as the most convenient mathematical medium for the discussion of analysis and synthesis procedures in automatic control. The next chapter on frequency response will serve as further justification for our concentration on $s$-plane methods.

As an introduction, let us return to the solution of Eq. ($\mathbf{A}_k\mathbf{b}$), Eq. (4.4-4), which is repeated below:

$$\mathbf{x}(t) = \mathcal{L}^{-1}[\boldsymbol{\Phi}_k(s)]\mathbf{x}(0) + K\mathcal{L}^{-1}[\boldsymbol{\Phi}_k(s)\mathbf{b}r(s)] \qquad (4.5\text{-}1)$$

To find the contribution to $\mathbf{x}(t)$ due to the forcing function $r(t)$, it is necessary to find the inverse Laplace transform

$$\mathbf{x}_f(t) = K\mathcal{L}^{-1}[\boldsymbol{\Phi}_k(s)\mathbf{b}r(s)] \qquad (4.5\text{-}2)$$

This inverse Laplace transformation involves the inverse of the product of two Laplace transforms, $r(s)$ and $\boldsymbol{\Phi}_k(s)$.

Rather than find the inverse transform indicated in Eq. (4.5-2), let

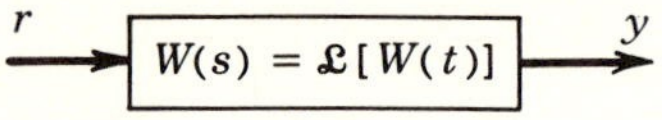

$$r \longrightarrow \boxed{W(s) = \mathcal{L}[W(t)]} \longrightarrow y$$

*Fig. 4.5-1*   Linear system with a transfer function $W(s)$ driven by an input $r(t)$.

us consider a more general case pictured in Fig. 4.5-1. Here $W(s)$ is the transfer function of any linear system, and the output $y(s)$ is simply

$$y(s) = W(s)r(s)$$

so that $y(t)$ is

$$y(t) = \mathcal{L}^{-1}[W(s)r(s)]$$

From the definition of the Laplace transform, we may write $r(s)$ as

$$r(s) = \int_0^\infty r(\tau)e^{-s\tau} \, d\tau \tag{4.5-3}$$

Here the variable $\tau$ was chosen as the integration variable in order to avoid later confusion. In terms of Eq. (4.5-3), $y(t)$ becomes

$$y(t) = \mathcal{L}^{-1}\left[ W(s) \int_0^\infty r(\tau)e^{-s\tau} \, d\tau \right] \tag{4.5-4}$$

Under very mild mathematical restrictions[1] the inverse Laplace operation and the integration in Eq. (4.5-4) may be interchanged to yield

$$y(t) = \int_0^\infty r(\tau) \, d\tau \, \mathcal{L}^{-1}[W(s)e^{-s\tau}]$$

But from the shifting theorem (see Appendix A),

$$\mathcal{L}^{-1}[W(s)e^{-s\tau}] = W(t - \tau)\mu(t - \tau)$$

and therefore $y(t)$ is

$$y(t) = \int_0^\infty r(\tau)W(t - \tau)\mu(t - \tau) \, d\tau \tag{4.5-5}$$

A case of special interest occurs when $r(t)$ is an impulse so that $r(s)$ is just 1. Then $y(s)$ is equal to $W(s)$, and $y(t)$ is exactly $W(t)$. This impulse response is often referred to as the *weighting function* of the system. In other words, $W(t)$ is the response of the system due to an impulse at $t = 0$. In terms of this interpretation, it is possible to make a simplification in Eq. (4.5-5). If $W(t)$ is the response due to an impulse applied at $t = 0$, it is clear that $W(t)$ must be zero for $t < 0$ since the system cannot respond before an input is applied. Therefore $W(t - \tau)$ is zero for negative values of the argument or for $\tau > t$, and the upper limit on the integral of Eq. (4.5-5) may be changed from $\infty$

[1] W. R. LePage, "Complex Variables and the Laplace Transform for Engineers," McGraw-Hill Book Company, New York, 1961.

to $t$ so that $y(t)$ becomes

$$y(t) = \mathcal{L}^{-1}[W(s)r(s)] = \int_0^t r(\tau)W(t - \tau)\,d\tau \qquad (4.5\text{-}6)$$

Equation (4.5-6) is known as the *convolution integral*. Note that multiplication in the $s$ domain becomes convolution in the time domain.

  Utilization of the concept of the impulse response or the weighting function makes possible a physical interpretation of the convolution integral. Assume that the impulse response or the weighting function of a given system is as indicated in Fig. 4.5-2$a$. Here the weighting function is associated with an impulse applied at 0. At any time $t$, $W(t)$ is the value of the system response due to the impulse at 0. If we are stationed at $t$ on the time axis, it would appear that an impulse had been applied $t$ sec in the past. Figure 4.5-2$b$ indicates the contribution to the output at time $t$ due to an impulse that occurred $\tau$ sec in the past. The ordinate here is $W(t - \tau)$. In determining the system output at any time $t$, we must be concerned with all the impulse inputs that may have occurred in the past. Since the system is linear, the responses due to these impulse inputs may then simply be added.

  With this in mind, let us return to the convolution integral and look briefly at $r(\tau)$. Suppose that $r(\tau)$ is an arbitrary input, as in Fig. 4.5-3$a$. Let us approximate this input by a series of impulses. This may be done

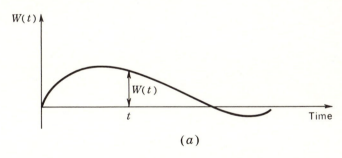

(a)

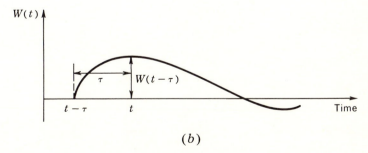

(b)

**Fig. 4.5-2** Impulse response. (a) Impulse applied at time $= 0$; (b) impulse applied at time $= t - \tau$.

by breaking $r(\tau)$ into a number of segments each of which is $d\tau$ wide. Here $d\tau$ is a differential quantity, and the area of each segment is approximately $r(\tau_i)\, d\tau$. As long as $d\tau$ is very small compared with the time constants of the system to which $r(t)$ is applied, each of these areas may be approximated with an impulse of magnitude equal to the area of the segment, as in Fig. 4.5-3$b$. The response due to each of these infinitesimal impulses then has the shape of the weighting function and a size proportional to the strength of the impulse.

Consider the contribution to the output at time $t$ of one of these differential inputs that occurs time $\tau$ in the past. Since the input is a differential quantity, the output is also expressed as a differential quantity, so that at time $t$ the output is

$$dy(t) = r(\tau)W(t - \tau)\, d\tau$$

Recall that $\tau$ ranges from 0 to $t$, and as $d\tau$ goes to zero, the contribution of each infinitesimal impulse is expressed as an integration as

$$y(t) = \int_0^t dy(t) = \int_0^t r(\tau)W(t - \tau)\, d\tau$$

Let us return now to the use of these concepts to evaluate $\mathbf{x}(t)$. The general expression for $\mathbf{x}(t)$ is given in terms of the required inverse Laplace

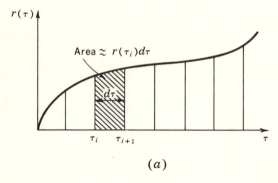

(*a*)

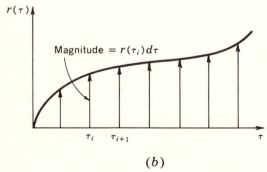

(*b*)

**Fig. 4.5-3**  An arbitrary input. (*a*) Represented in segments; (*b*) represented as a series of impulses.

transforms by Eq. (4.5-1).   The *state transition matrix* $\mathbf{\Phi}_k(t)$ of the closed-loop system is defined as the inverse Laplace transform of the resolvent matrix, or

$$\mathbf{\Phi}_k(t) = \mathcal{L}^{-1}[\mathbf{\Phi}_k(s)]$$

Then we may take the inverse transform required by Eq. (4.5-1) by replacing the multiplication in the $s$ domain by the convolution integral. The result, still in matrix notation, is then

$$\mathbf{x}(t) = \mathbf{\Phi}_k(t)\mathbf{x}(0) + K \int_0^t \mathbf{\Phi}_k(t - \tau)\mathbf{b}r(\tau)\, d\tau \tag{4.5-7}$$

In expanded form the equation for the $i$th state variable is just

$$x_i(t) = x_1(0)\phi_{k(i1)}(t) + x_2(0)\phi_{k(i2)}(t) + \cdots + x_n(0)\phi_{k(in)}(t)$$
$$+ K \int_0^t \phi_{k(i1)}(t - \tau)b_1 r(\tau)\, d\tau + \cdots + K \int_0^t \phi_{k(in)}(t - \tau)b_n r(\tau)\, d\tau$$

The convolution integrals here are the time-domain equivalents of the expressions noted in Eq. (4.5-2).   It is evident that in general one must evaluate $n^2$ convolution integrals in order to find the total system response, just as one must take $2n^2$ inverse Laplace transforms in Eq. (4.5-1).   The very practical question then arises: Which is easier, finding inverse transforms or evaluating convolution integrals?

To answer this question, let us take a particular case related to Fig. 4.5-1 in which $W(s)$ is second order, or

$$W(s) = \frac{1}{(s + \alpha)^2 + \beta^2}$$

or

$$W(t) = \frac{1}{\beta}\, e^{-\alpha t} \sin \beta t$$

and $W(t - \tau)$ is

$$W(t - \tau) = \frac{1}{\beta}\, e^{-\alpha(t-\tau)} \sin \beta(t - \tau)$$

For zero initial conditions and an arbitrary input $r(t)$, the output is

$$y(t) = \int_0^t r(\tau) \left[ \frac{1}{\beta}\, e^{-\alpha(t-\tau)} \sin \beta(t - \tau) \right] d\tau$$

Only in the case when $r(t)$ is a step function or an exponential is this integral easy to evaluate.   In that case evaluation is done by looking up the integral in a set of standard integral tables.   If $r(t)$ were allowed to be $t$, or a sinusoidal driving force, the usual tables are no longer adequate. However, the alternative approach through partial-fraction expansion of the transformed quantity would be no more difficult than problems already solved in this chapter.

In defense of the time-domain approach through the use of convolution, it should be mentioned that graphical and digital-computer methods are available to evaluate the convolution integral. These graphical methods are particularly suited to cases in which the input and/or weighting functions are not known analytically. Of course, before these methods may be used, it is necessary to know the weighting function. If an analytical description of the system is known, then $W(t)$ may be determined analytically. In addition, $W(t)$ may be determined experimentally in a number of ways even if a mathematical model for the system is not completely known.

The most obvious method of determining $W(t)$ is to apply an impulse to the system and measure the output. Of course, it is impossible actually to apply an impulse, but a pulse of very short duration is often an acceptable substitute. A second method for finding $W(t)$ is to differentiate, usually numerically, the step response of the system. The step response can normally be obtained directly and without difficulty. The weighting function can also be obtained by measuring the response of the system to sinusoidal inputs of a number of different frequencies. This frequency-response approach will be discussed in detail in the next chapter.

**Exercise 4.5**   *4.5-1.*   Consider Exercise 4.4-1 with the addition of a forcing term $r(t)$. The system equations become

$$\dot{x}_1 = x_2$$
$$\dot{x}_2 = -2x_1 - 3x_2 + 2r(t)$$

If the initial conditions are once again $x_1(0) = 1$ and $x_2(0) = 1$, use time-domain methods to find $x_1(t)$ and $x_2(t)$ if

(a)   $r(t) = $ step function, $\mu(t)$
(b)   $r(t) = t\mu(t)$
(c)   $r(t) = t^2\mu(t)$

## 4.6   Conclusions

This chapter is the first one that has been concerned with analysis, in particular, with the determination of the time response of a closed-loop control system. The assumptions made at the beginning of the chapter are worth repeating. It is assumed that the system is completely specified. That is, not only is the plant assumed to be known, but the gains and feedback coefficients are assumed to be specified as well.

In a sense, this chapter could have been the last one in the book. Often the plant is known, and the synthesis problem is to select the forward gains and the feedback coefficients so that a desired overall system results. Once synthesis has been completed, the final design must be analyzed to ensure that the final system performs as required. However, the synthesis procedure that we advocate depends upon a specification of the closed-loop poles and zeros. Hence one is unable to proceed to the synthesis technique unless one has a certain familiarity with the meaning of pole and zero locations on the $s$ plane.

As we proceeded through this chapter it was pointed out that it is often difficult to obtain the time response for any input and/or initial conditions exactly. In many cases we might be quite content with much less information, assuming that less information would be easier to obtain. That is the point of view that we shall adopt in the next two chapters. In fact, the next chapter begins with the assumption that the plant to be controlled exists physically but not as a transfer function written on a piece of paper. Before control can be attempted, the plant must first be identified. Once the plant is known, its stability must be examined before the system can be given any inputs or initial conditions. This is commonly done in the frequency domain, the topic of the following chapter.

## 4.7  Problems

*4.7-1.* The pole-zero plot of a given transfer function $y(s)/r(s)$ is shown in Fig. 4.7-1. If the input $r$ is an impulse, the output $y(t)$ is of the form

$$y(t) = K[Ae^{-t} + Be^{-\alpha t} + Ce^{-\gamma t} \sin (\omega t + \phi)]$$

What are the values of $A$, $B$, $C$, $\alpha$, $\gamma$, and $\omega$?

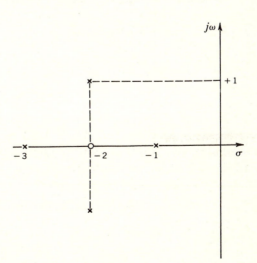

*Fig. 4.7-1*  Problem 4.7-1.

**4.7-2.**  If $y(s)$ is

$$y(s) = \frac{s + 2}{s(s + 1)[(s + 3)^2 + 4^2]}$$

find the partial-fraction expansion of $y(s)$.  Do not determine $y(t)$.

**4.7-3.**  The resolvent matrix $\Phi(s)$ for a given plant is

$$\Phi(s) = \begin{bmatrix} 1/s & 1/s(s + 1) & 3/[s(s + 1)(s + 5)] \\ 0 & 1/(s + 1) & 3/[(s + 1)(s + 5)] \\ 0 & 0 & 1/(s + 5) \end{bmatrix}$$

and                    $\mathbf{b} = \text{col } (0, 0, \tfrac{7}{3})$

Find $x_2(s)$ if $u$ is a step input and $\mathbf{x}(0) = \text{col } (1, 2, 3)$.

**4.7-4.**  Find the resolvent matrix $\Phi_k(s)$ for the closed-loop system of Fig. 4.7-2.  Check your result by first showing that $\phi_{k(12)}$ as determined by block-diagram manipulation agrees with your result and, second, that $\Phi_k(s)\,\Phi_k^{-1}(s) = \mathbf{I}$.  Recall that $\Phi_k(s)^{-1}$ is given as $(s\mathbf{I} - \mathbf{A}_k)$.

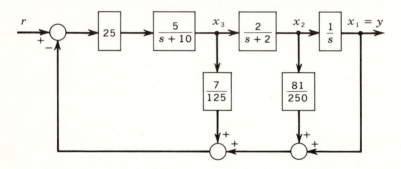

**Fig. 4.7-2**   Problems 4.7-4 to 4.7-6.

**4.7-5.**  For the system considered in Prob. 4.7-4, find $\mathbf{x}(t)$ from the closed-loop transfer function $y(s)/r(s)$ if $r(t)$ is a step-function input of magnitude $\tfrac{1}{2}$.

**4.7-6.**  Again consider the system of Fig. 4.7-2, and this time use the resolvent matrix found in Prob. 4.7-4 to find the total time response if the initial conditions are $\mathbf{x}(0) = \text{col } (0, 1, 0)$ and $r(s) = \tfrac{1}{2}/s$.

**4.7-7.**  Find the total time response of the plant shown in Fig. 4.7-3 if $u(s) = 1/s$ and $\mathbf{x}(0) = 0$.

(*a*)  Use the resolvent matrix to find the answer by defining a new state variable $x_3 = x_3' - u$.

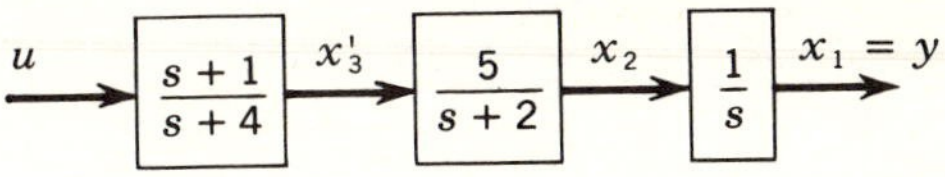

**Fig. 4.7-3**    Problem 4.7-7.

(b)    Find the answer by using the plant transfer function $G_p(s)$ to find $y(s) = x_1(s)$ and then applying the state-variable relations to find $x_2(s)$ and $x_3'(s)$.

*4.7-8.*    If the input to a plant is zero, the total time response may be represented as

$$\mathbf{x}(t) = \mathbf{\Phi}(t)\mathbf{x}(0)$$

For $t = 0$, this result becomes

$$\mathbf{x}(0) = \mathbf{\Phi}(0)\mathbf{x}(0)$$

and we see that $\mathbf{\Phi}(0)$ must equal the identity matrix $\mathbf{I}$ if this equation is to be satisfied for every $\mathbf{x}(0)$. This fact provides a simple check on the validity of $\mathbf{\Phi}(t)$. In addition, by using the initial-condition theorem, we may write

$$\mathbf{\Phi}(t)\Big|_{t=0} = \lim_{s \to \infty} [s\mathbf{\Phi}(s)] = \mathbf{I}$$

and use this expression as a check on $\mathbf{\Phi}(s)$.

Show that the above expressions are satisfied for $\mathbf{\Phi}(t)$ and $\mathbf{\Phi}(s)$ associated with the plant

$$\dot{\mathbf{x}} = \begin{bmatrix} 0 & 1 \\ -2 & -3 \end{bmatrix} \mathbf{x} + \begin{bmatrix} 0 \\ 1 \end{bmatrix} u$$

*4.7-9.*    Show that the state-transition matrix satisfies the homogeneous plant equation, i.e.,

$$\dot{\mathbf{\Phi}}(t) = \mathbf{A}\mathbf{\Phi}(t)$$

Verify the expression for the $\mathbf{\Phi}(t)$ of Prob. 4.7-8.

# *five*  *frequency response*

## *5.1  Introduction*

This is the second chapter concerned with analysis and is intended to complement and extend the results of Chap. 4.  The reader may be somewhat surprised that any further discussion of analysis is necessary, as we have given the complete solution for any initial conditions and/or forcing functions for either an open- or a closed-loop system.  Thus the complete time response for all the state variables may be calculated, and the behavior of the system for all time is known.

The complete solution for a single-input, single-output closed-loop system was given as Eq. (4.5-7) and is repeated here for reference.

$$\mathbf{x}(t) = \mathbf{\Phi}_k(t)\mathbf{x}(0) + K \int_0^t \mathbf{\Phi}_k(t - \tau)\mathbf{b}r(\tau) \; d\tau \tag{5.1-1}$$

Equation (5.1-1) is a matrix equation, and although the time solution is completely indicated, the actual determination of the solution is tedious. Even if this equation is solved so that the response $\mathbf{x}(t)$ is known analytically, in order to plot $\mathbf{x}(t)$ versus time it is necessary to proceed in a point-by-point fashion. This is even more work.

Although Eq. (5.1-1) is a general solution for any input, the discussion in Chap. 4 was restricted mainly to impulse and step inputs. Another important class of inputs that was not mentioned in the preceding chapter is the sinusoidal function. This chapter is devoted entirely to a discussion of the response to sinusoidal inputs. Initially the discussion is in terms of a general transfer function, $W(s)$. Particular transfer functions that are of interest are the transfer functions of the plant, $G_p(s)$; the closed-loop system, $y(s)/r(s)$; and the internal transfer functions relating the various state variables with the input $u(s)$ and with themselves. That is, we are interested in $x_i(s)/u(s)$ and $x_i(s)/x_j(s)$, $i \neq j$. The response of the latter transfer functions to a sinusoidal input is particularly important if the plant is unknown and one is seeking to determine by measurement its transfer function.

Section 5.2 is devoted to a discussion of the frequency-response function $W(j\omega)$. Of particular interest are its magnitude and phase angle. Section 5.3 discusses means of sketching the magnitude of this function. Specifically, it is important to develop means of plotting the frequency-response function that are less tedious than the point-by-point method required to plot $\mathbf{x}(t)$. Section 5.3 uses the straight-line approximations to make the plot of the magnitude of the frequency-response function almost trivial to accomplish.

Section 5.4 is a preliminary discussion of the phase plot; in particular, the conditions under which it is possible to determine the phase-angle diagram directly from the amplitude diagram are stressed. A semianalytical means of representing the phase angle at any point is discussed in Sec. 5.5 dealing with the arctangent approximation.

Sections 5.3 to 5.5 fall into the category of "how to do it" sections; that is, they explain the mechanics of handling the frequency-response function but give little indication as to why the frequency-response function is important. Its importance is emphasized in Sec. 5.6 on plant identification. The assumption of the preceding sections is that the transfer function is known, and we are asked to represent the frequency function associated with the given transfer function. In Sec. 5.6 the inverse problem is discussed; that is, we have a plot of an unknown frequency-response function, and it is desired to find the transfer function. This problem occurs in practice when a plant whose transfer function is unknown is to be controlled. As a first step in the design, the plant must be identified; that is, its transfer function must be determined so that calculations regarding the means of control can be made. Here the transfer functions relating the various state variables and the input to the plant $x_i(s)/u(s)$ and the transfer functions between the various state variables $x_i(s)/x_j(s)$ prove to be important.

## 5.2   *Frequency-response function*

Let us consider the time response of a rather general linear system driven by the sinusoidal time function $r(t) = A \sin \omega t$, as pictured in Fig. 5.2-1. The output is $y(t)$, and the transfer function of the system is $W(s)$, where $W(s)$ is assumed to be a ratio of polynomials in $s$. The output $y(t)$ contains, among other things, a sinusoidal component of the same frequency as the input of the form $AR(\omega) \sin [\omega t + \phi(\omega)]$. Here $R(\omega)$ is the ratio of the magnitude of the sinusoidal component of the output to the input, and the phase angle $\phi(\omega)$ is the phase difference between the input and the output. Both $R$ and $\phi$ have been designated as functions of $\omega$ since, in general, they vary as the input frequency is varied. The object of this section is to show that

$$R(\omega) = |W(s)|\Big|_{s=j\omega} = |W(j\omega)|$$

$$\phi(\omega) = \text{phase angle of } W(s)\Big|_{s=j\omega} = \arg W(j\omega)$$

This means that the magnitude and phase of the output for a sinusoidal input may be found by simply determining the magnitude and phase of $W(j\omega)$. The complex function $W(j\omega)$ is referred to as the *frequency-response function.*

To demonstrate the truth of this statement, let us examine the Laplace transform of $y(t)$ for the indicated input. Since the Laplace transform of $A \sin \omega t$ is $A\omega/(s^2 + \omega^2)$, $y(s)$ is

$$y(s) = W(s)r(s) = \frac{A\omega W(s)}{s^2 + \omega^2} = \frac{A\omega W(s)}{(s + j\omega)(s - j\omega)}$$

If a partial-fraction expansion of $y(s)$ is made, using any of the approaches

**Fig. 5.2-1**   General linear system with a sinusoidal input.

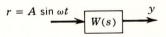

of Sec. 4.2, we obtain

$$y(s) = \frac{R_{11}}{s + j\omega} + \frac{R_{21}}{s - j\omega} + \text{other terms} \tag{5.2-1}$$

$$\underbrace{\phantom{\frac{R_{11}}{s + j\omega} + \frac{R_{21}}{s - j\omega}}}_{\substack{\text{Particular solution} \\ \text{(steady-state)}}} \quad \underbrace{\phantom{\text{other terms}}}_{\substack{\text{Complementary} \\ \text{solution (transient)}}}$$

The "other terms" arise from the poles of $W(s)$. In mathematical terminology, the first two terms on the right side of Eq. (5.2-1) are due to the sinusoidal forcing function and are therefore related to the particular solution; the "other terms" are the unforced or complementary solution.

One often refers to the particular solution as the steady-state solution and the complementary solution as the transient solution. Although these designations are often appropriate, there are situations in control-system applications where such labeling is inappropriate. In some plants, for example, which we shall refer to later as unstable, it is possible for the complementary solution to grow without bound as time increases. In such cases the use of the term transient response to describe the complementary solution is meaningless.

For the majority of systems, however, the complementary solution is, in fact, transient in nature and the particular solution becomes the usual steady-state response. In these cases, the frequency-response function may be given a particularly useful physical meaning; we shall discuss this interpretation later.

Independent of whether the complementary solution is transient or not, our present interest is only in the particular solution. We therefore ignore the complementary portion of the solution. To find the sinusoidal portion of $y(t)$, which we designate as $y(t)_s$ for sinusoidal, it is necessary to find $R_{11}$ and $R_{21}$. This is easily done by the method of residues so that

$$R_{11} = \frac{A\omega W(s)}{s - j\omega}\bigg|_{s=-j\omega} = \frac{-AW(-j\omega)}{2j}$$

and

$$R_{21} = \frac{A\omega W(s)}{s + j\omega}\bigg|_{s=j\omega} = \frac{AW(j\omega)}{2j}$$

In terms of these values for $R_{11}$ and $R_{21}$, $y(t)_s$ may now be written by taking the inverse transform of the first two terms on the right side of Eq. (5.2-1), or

$$y(t)_s = R_{11}e^{-j\omega t} + R_{21}e^{+j\omega t}$$

$$= \frac{-AW(-j\omega)}{2j} e^{-j\omega t} + \frac{AW(j\omega)}{2j} e^{+j\omega t} \tag{5.2-2}$$

We now let the complex function $W(j\omega)$ be expressed in rectangular form as

$$W(j\omega) = X(\omega) + jY(\omega) \qquad (5.2\text{-}3)$$

where both $X(\omega)$ and $Y(\omega)$ are real functions of frequency corresponding to the real and imaginary parts of $W(j\omega)$. Note that $|W(j\omega)|$ is

$$|W(j\omega)| = \sqrt{X(\omega)^2 + Y(\omega)^2}$$

and the phase angle or argument of $W(j\omega)$ is

$$\arg W(j\omega) = \arctan \frac{Y(\omega)}{X(\omega)}$$

For any particular frequency, $W(j\omega)$ is a complex number so that $W(-j\omega)$ is just the complex conjugate of $W(j\omega)$, or

$$W(-j\omega) = X(\omega) - jY(\omega)$$

The substitution of these expressions for $W(j\omega)$ and $W(-j\omega)$ into Eq. (5.2-2) for $y(t)_s$ yields

$$y(t)_s = \frac{-A}{2j}[X(\omega) - jY(\omega)]e^{-j\omega t} + \frac{A}{2j}[X(\omega) + jY(\omega)]e^{+j\omega t}$$

$$= AX(\omega)\frac{e^{j\omega t} - e^{-j\omega t}}{2j} + AY(\omega)\frac{e^{j\omega t} + e^{-j\omega t}}{2}$$

Making use of the Euler formula, we may write $y(t)_s$ as

$$y(t)_s = AX(\omega)\sin \omega t + AY(\omega)\cos \omega t$$

If $y(t)_s$ is written as a sine function with a phase angle, $y(t)_s$ becomes

$$y(t)_s = A[X(\omega)^2 + Y(\omega)^2]^{\frac{1}{2}} \sin [\omega t + \alpha(\omega)]$$

where $\alpha(\omega) = \arctan [Y(\omega)/X(\omega)]$. But this is exactly the form for the sinusoidal component of $y(t)$ postulated at the beginning of this section. Hence

$$R(\omega) = [X(\omega)^2 + Y(\omega)^2]^{\frac{1}{2}} = |W(j\omega)| \qquad (5.2\text{-}4)$$

and

$$\alpha(\omega) = \phi(\omega) = \arctan \frac{Y(\omega)}{X(\omega)} = \arg W(j\omega) \qquad (5.2\text{-}5)$$

Hence we have established that the frequency-response function may be obtained from the transfer function by letting $s = j\omega$.

This is an important result. Given the transfer function, the frequency-response function follows immediately. On the other hand, if the frequency-response function is known, as the result of experiment, for instance, then the transfer function may be found by replacing $\omega$ by $s/j$. That is, the frequency-response function and the transfer function are directly related to one another. If one is known, the other is also known. Of course, if the transfer function is known, the response to any input can be found; hence we may make the following statement: *The frequency-response function of a linear system uniquely determines the time response of the system to any known input.*

The converse of this statement is also true. In Chap. 4 one method discussed for finding the weighting function of a system was the use of an impulse input. Then the Laplace transform of the weighting function is the transfer function of the system in question. An alternative approach is to use a step input and differentiate the resulting output to find the impulse response. For the system of Fig. 5.2-1, $W(t)$, the weighting function, is

$$W(t) = \mathcal{L}^{-1}\left[\frac{y(s)}{r(s)}\right]$$

and this is true regardless of the input. It is often convenient to use a step and impulse as test inputs to determine $W(t)$. With $W(t)$ known, $W(s)$ may be found and hence the frequency-response function $W(j\omega)$. On the basis of these observations, we note the following property: *The frequency-response function of a linear system is uniquely determined by its time response to any known input.*

The combination of these two statements establishes the complete interdependence of the time- and frequency-response methods. The importance of this fact is that it allows one to translate time-domain specifications, such as rise time and overshoot, into frequency-domain specifications. One may often, therefore, carry out a complete design in the frequency domain where the procedures are usually simpler and more systematic.

There was some discussion of the interdependence of the time and frequency domains in Chap. 4 with regard to second-order systems in particular. In addition, the reader is undoubtedly already familiar with the relationship of such properties as bandwidth, a frequency-domain concept, and rise time, a time-domain characteristic. In this and the following chapters, this interplay of the time and frequency domains will be further developed in order to give the reader an appreciation for the usefulness of this duality of the time and frequency domains.

At present, however, our interest is simply in the meaningful representation of the frequency-response information. In particular, we are interested in graphical means of representation, since this allows us to examine the frequency-response data most easily.   Initially, the amplitude and phase characteristics are plotted separately; in the next chapter, the representation of both items on a single polar plot with frequency as a parameter will be considered.

To introduce the concept of graphical representation of the frequency-response information, let us consider the simple electrical network shown in Fig. 5.2-2.   In discussing this example, we illustrate the problems associated with a usual graphical representation.

The transfer function for this network may be easily found to be

$$\frac{y(s)}{u(s)} = G_p(s) = \frac{1}{1 + R_1 C_1 s}$$

and in this simple example $W(s)$ is a plant transfer function, $G_p(s)$.   The frequency-response function for this network may be obtained by setting $s = j\omega$ so that

$$G_p(j\omega) = \frac{1}{1 + jR_1 C_1 \omega}$$

The magnitude and phase of the frequency-response function are therefore

$$R(\omega) = |G_p(j\omega)| = \frac{1}{\sqrt{1 + (\omega R_1 C_1)^2}}$$

and

$$\phi(\omega) = \arg G_p(j\omega) = -\arctan \omega R_1 C_1$$

A plot of these two functions vs. frequency is shown in Fig. 5.2-3.

In addition to the asymptotic behavior of the plots for very low and very high frequencies, the value of the magnitude and phase functions can be determined *by inspection* at only one finite value of frequency, namely, $\omega = 1/R_1 C_1$.   Therefore, in order to obtain reasonably accurate plots, it is necessary to substitute various values of $\omega$ and carry out the computations in detail.   Even though in this case the computations are not difficult and usually five to ten appropriately selected points are sufficient, the effort involved is not trivial.

The situation becomes much worse when systems are cascaded. Consider, for example, the problem of plotting the magnitude and phase

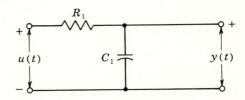

*Fig. 5.2-2*   Simple electrical network.

of the frequency-response function for the network shown in Fig. 5.2-4.
This network is formed by cascading a second $RC$ network with the $RC$
network shown in Fig. 5.2-2.

If we assume that the impedance levels of the two networks are such
that no loading occurs, the transfer function of the cascade network is

$$G_p(s) = \frac{1}{(1 + sR_1C_1)(1 + sR_2C_2)}$$

and the frequency-response function is therefore

$$G_p(j\omega) = \frac{1}{(1 + j\omega R_1C_1)(1 + j\omega R_2C_2)}$$

Even with the no-loading assumption to simplify the task, plotting
the magnitude and phase of $G_p(j\omega)$ is not easy.   Although the asymptotic
character of the two plots may still be easily determined, there is no longer

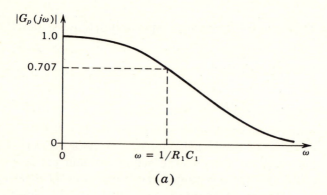

(*a*)

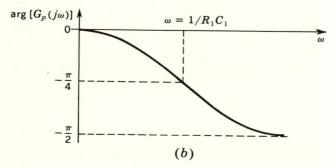

(*b*)

**Fig. 5.2-3**   Frequency-response function for the network of
Fig. 5.2-2.   (*a*) Magnitude; (*b*) phase.

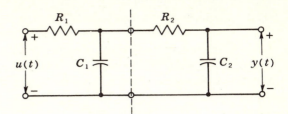

**Fig. 5.2-4**  Cascade network.

even a single frequency for which the value of the magnitude or phase may be determined by inspection.  In addition, since the character of the cascade frequency-response function may not be as obvious as its simple components, more points may be necessary to achieve an accurate representation.

Even if the magnitude plots of the two networks were known exactly, the job of finding the cascade plot would not be trivial, since point-by-point multiplication would be necessary.  Since the cascading of various component blocks is commonplace in control-system design and analysis, the above direct method is obviously inadequate.  The approach is cumbersome in the handling of simple networks, and its inability to facilitate the cascading of component systems makes the technique completely unacceptable.

Hence we are led to seek alternative procedures for plotting the magnitude and phase of the frequency-response function.  Such a procedure should have one or more of the following characteristics.  First, the procedure should allow a rapid and reasonably accurate plot to be made.  Second, composite systems should be handled with relative ease. Third, it should be possible to make the plots as accurate as needed in a given frequency range or ranges by a reasonably small amount of additional work.

Although it may appear that the above requirements are too much to hope for, in the following sections we shall develop procedures that more or less meet all these requirements.  We begin by considering the representation of the magnitude plot by means of a straight-line approximation.

## 5.3   *Magnitude plot—straight-line approximation*

In the preceding section we showed that the ratio of the magnitude of the sinusoidal component of the output to the magnitude of the sinusoidal input is simply the magnitude of the frequency-response function, or $|W(j\omega)|$. In addition, the phase angle of the sinusoidal component of the output relative to the phase angle of the sinusoidal input is just the argument or angle of $W(j\omega)$. Thus for any particular frequency it is possible to plot $y(t)_s$, since the amplitude, frequency, and phase angle of the response are known. However, because the plot of sinusoidal functions is so familiar, this is usually not done; rather, the amplitude and phase angle of $W(j\omega)$ are plotted for all frequencies. These curves supply all the information necessary to draw $y(t)_s$ for any frequency if such a plot is ever required.

In this chapter the amplitude and phase diagrams are considered separately; the polar plot will be discussed in the following chapter in terms of the Nyquist diagram. In this section we concentrate on the means by which the magnitude of the frequency-response function may be sketched with a minimum of difficulty. The phase plot will be discussed in Sec. 5.4. The procedure developed in this section involves the approximation of the magnitude plot as a sequence of straight lines. In order to develop the method, a rather simple-example transfer function of the form

$$W(s) = \frac{K(s + \omega_1)}{s(s + \omega_2)} \tag{5.3-1}$$

is considered.

This specific example is used to derive the procedure, since it illustrates the principal features of the approach without extensive notational problems. The extension of the approach to the general situation is relatively direct. In addition, several numerical examples are used to demonstrate further the application of the technique.

At the outset, notice that $W(s)$ in Eq. (5.3-1) has been factored and written in pole-zero form. That is, the coefficient of $s$ in each factor is 1, as the poles and zeros are each written as $s + \omega_i$. Let us rewrite Eq. (5.3-1) in an alternative form, so that the coefficient of $s$ in each factor is $1/\omega_i$. This is referred to as the *time-constant form* since each of the quantities $1/\omega_1$ and $1/\omega_2$ has the dimension of time. Although it is not necessary to write $W(j\omega)$ in this form, the results obtained below are more convenient if this time-constant form is used. In time-constant form Eq. (5.3-1) becomes

$$W(s) = \frac{K\omega_1}{\omega_2} \frac{1 + s/\omega_1}{s(1 + s/\omega_2)}$$

The frequency-response function in time-constant form is then

$$W(j\omega) = \frac{K\omega_1}{\omega_2} \frac{1 + j\omega/\omega_1}{j\omega(1 + j\omega/\omega_2)}$$

The magnitude of the frequency-response function is seen to be

$$|W(j\omega)| = \left|\frac{K\omega_1}{\omega_2}\right| \frac{|1 + j\omega/\omega_1|}{|j\omega||1 + j\omega/\omega_2|} \tag{5.3-2}$$

where use has been made of the fact that the magnitude of a product is the product of the magnitudes.

The difficulty of evaluating $|W(j\omega)|$ for various values of $\omega$ is that we must multiply (or divide) the various components of the magnitude expression of Eq. (5.3-2) for each value of $\omega$. Note that the same problem occurs when system components are cascaded. Since multiplication is one of the problems, let us consider the logarithm of $|W(j\omega)|$ so that the multiplication involved is replaced by the simpler job of adding logarithms. If this were the only advantage gained by this change, it would hardly be worthwhile. However, several other advantageous features are also associated with this logarithmic approach.

If we now take the logarithm of the magnitude of $W(j\omega)$, we obtain

$$\log|W(j\omega)| = \log\left|\frac{K\omega_1}{\omega_2}\right| + \log\left|1 + \frac{j\omega}{\omega_1}\right| - \log|j\omega| - \log\left|1 + \frac{j\omega}{\omega_2}\right| \tag{5.3-3}$$

Let us examine in detail one term of Eq. (5.3-3), the term $\log|1 + j\omega/\omega_1|$. In particular, let us examine the asymptotic behavior for both large and small $\omega$. For small $\omega$, the magnitude of the quantity $j\omega/\omega_1$ is small relative to 1, and since these terms add vectorially at right angles, the following approximation is made, namely, that

$$\log\left|1 + \frac{j\omega}{\omega_1}\right| \approx \log 1 = 0 \qquad \text{for } \omega \ll \omega_1 \tag{5.3-4}$$

Rather than plot this quantity against $\omega$, as done in Sec. 5.2, we assume that the quantity is to be plotted with the horizontal variable as $\log \omega$ rather than just $\omega$. Then the slope of the asymptotic approximation expressed in Eq. (5.3-4) is

$$\frac{d(0)}{d(\log \omega)} = 0 \tag{5.3-5}$$

Similarly for large $\omega$, 1 is small compared with $j\omega/\omega_1$, and hence

$$\log\left|1 + \frac{j\omega}{\omega_1}\right| \approx \log\left|\frac{j\omega}{\omega_1}\right| \qquad \text{for } \omega \gg \omega_1$$

Since we are concerned only with positive values of frequency, this result becomes

$$\log\left|1 + \frac{j\omega}{\omega_1}\right| \approx \log\frac{\omega}{|\omega_1|} = \log \omega - \log|\omega_1| \qquad \text{for } \omega \gg \omega_1 \tag{5.3-6}$$

The slope of this asymptotic approximation for large $\omega$ when plotted against log $\omega$ is

$$\frac{d}{d(\log \omega)} (\log \omega - \log |\omega_1|) = +1$$

The exact logarithmic plot of $|1 + j\omega/\omega_1|$ is plotted versus log $\omega$ as the dashed line in Fig. 5.3-1. The straight-line asymptotic approximations for large and small $\omega$ are also indicated in that figure. Note that the two straight-line approximations that form the high- and low-frequency asymptotes for the exact plot intersect at $\omega = |\omega_1|$. For this reason, $\omega_1$ is often referred to as a *break frequency*, since it is the point at which the corner, or break, of the straight-line approximation occurs. From Fig. 5.3-1, we observe that, whenever $\omega < |\omega_1|/10$ or $\omega > 10|\omega_1|$, the exact and approximate plots are almost identical.

For most control applications, the straight-line asymptotic approximations are often of sufficient accuracy. However, if a more accurate approximation is needed, an exact calculation for one or two points near the break point can be made. Note, in particular, that for $\omega = |\omega_1|$ the exact value of $|1 + j\omega/\omega_1|$ is $\sqrt{2}$. If correction of the straight-line approximation is necessary, this point is often adequate.

In practical applications, it is often more convenient to plot the actual values of $|1 + j\omega/\omega_1|$ and $\omega$ on log-log paper, as in Fig. 5.3-2, rather than the logarithms of these quantities on linear paper, as in Fig. 5.3-1. In this way, one may work directly with the quantities of interest and still gain the advantages of the logarithmic plot. This approach is used in all numerical examples.

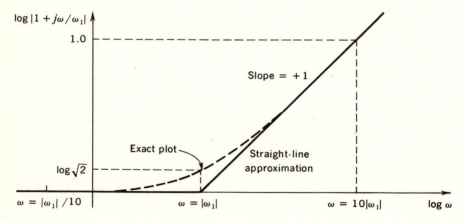

**Fig. 5.3-1**   Logarithmic plot of $|1 + j\omega/\omega_1|$.

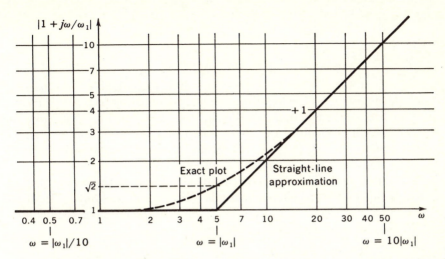

**Fig. 5.3-2**   Asymptotic straight-line approximation and exact plot of $|1 + j\omega/\omega_1|$ on log-log paper.

Note that in this case the reference for the vertical scale becomes the $|1 + j\omega/\omega_1| = 1$ line, and the low-frequency asymptote is coincident with this line. The slope of the high-frequency asymptote is indicated by placing a $+1$ on the high-frequency asymptote. On log-log paper, a slope of $\pm k$ implies that the magnitude changes by $\pm k$ decades as the frequency changes by one decade. For example, in Fig. 5.3-2 the magnitude of $1 + j\omega/\omega_1$ changes from 1 to 10 as $\omega$ increases from 5 to 50, thereby indicating a $+1$ slope.

If we examine the entire expression for the log $|W(j\omega)|$ as given by Eq. (5.3-3), the last term is seen to be identical in form to the term log $|1 + j\omega/\omega_1|$ except that it is preceded by a minus sign. Hence, at the break frequency of $\omega_2$, this term breaks downward with a slope of $-1$. Otherwise, it is the same as the $1 + j\omega/\omega_1$ term.

The other two terms of Eq. (5.3-3), namely, the log $|K\omega_1/\omega_2|$ and the log $|j\omega|$ terms, have even simpler representations on the logarithmic plot. The constant term, since it is frequency-independent, describes just a horizontal line passing through log $|W(j\omega)| = \log |K\omega_1/\omega_2|$. If the plot is made on log-log paper, the line coincides with the $|W(j\omega)| = |K\omega_1/\omega_2|$ line.

In order to treat the $j\omega$ term, observe that

$$\log |j\omega| = \log \omega$$

and the slope is

$$\frac{d(\log \omega)}{d(\log \omega)} = +1$$

which is just the equation of a straight line with slope of $+1$ passing through the origin on the logarithmic plot. On log-log paper, this means that the line passes through the $|W(j\omega)| = 1$, $\omega = 1$ point. Note that no approximation is involved in this case and that the exact plot is a single straight line. In Eq. (5.3-3), this term appears with a negative sign so that the slope becomes $-1$ rather than $+1$.

Let us plot each of the four constituent terms of the log $|W(j\omega)|$ as given in Eq. (5.3-3) on one logarithmic plot, as in Fig. 5.3-3a. In order to make the discussion more specific, numerical values have been chosen for $K$, $\omega_1$, and $\omega_2$ so that $W(j\omega)$ becomes

$$W(j\omega) = \frac{10(1 + j\omega/5)}{(j\omega)(1 + j\omega/50)}$$

In other words, $K = 100$, $\omega_1 = 5$, and $\omega_2 = 50$.

All that remains to be done is to add the contributions of the four terms in order to find the magnitude plot of $W(j\omega)$, since addition of logarithms is equivalent to multiplication of magnitudes. The resulting asymptotic plot of the magnitude of $W(j\omega)$ is shown in Fig. 5.3-3b, along with the exact plot for comparison. Once again, note that the approximation is reasonably accurate.

It must be remembered that when adding the effects of the basic terms we wish to add the logarithms. On the log-log plot where the axes are labeled with actual magnitude, we must add in terms of the geometric distance from the $|W(j\omega)| = 1$ axis, that is, the log $|W(j\omega)| = 0$ line. In other words, for the purpose of adding the effects of the four constituent terms, the vertical scale must be thought of as a linear scale with the zero reference at the $|W(j\omega)| = 1$ line.

Even the job of finding the composite plot of $|W(j\omega)|$ can be considerably simplified, since each of the four terms involves only straight lines of integer slopes. The $|W(j\omega)|$ plot must be similarly composed. Hence the plot of the $|W(j\omega)|$ is determined by fixing a single point on the plot by adding the effects of the four terms and then computing the slope of each straight-line segment by adding the slopes of the four terms.

Since the slope of the composite plot of the $|W(j\omega)|$ can change only at the break points of the constituent terms, it is a very simple job to find the entire plot. For example, for the frequency range $5 \leq \omega \leq 50$, the slope of the $|W(j\omega)|$ is equal to $-1 + 1 = 0$.

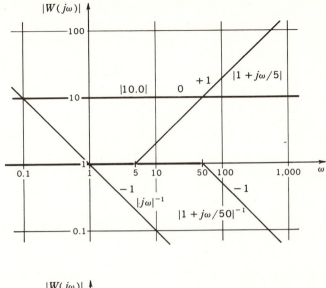

(a)

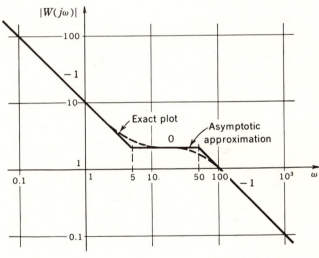

(b)

***Fig. 5.3-3*** Log-log plot of $|W(j\omega)|$.    (a) Constituent terms; (b) resultant plot.

The reason for selecting the time-constant form for $W(j\omega)$ rather than the original form of Eq. (5.3-1) should now be clear. If the latter representation had been selected, the low-frequency asymptote would have been of the form

$$\log |j\omega + \omega_1| = \log |\omega_1|$$

Therefore, rather than being a line coincident with the line $\log |W(j\omega)| = 0$, it would have been a line coincident with $\log |W(j\omega)| = \log |\omega_1|$. In this case, each of the factors of this form would have had a different low-frequency asymptote, thereby making the addition of their effects more difficult. The time-constant form is used whenever we plot asymptotic diagrams.

*Example 5.3-1* As another illustration of the use of the asymptotic magnitude diagram, let us determine the magnitude plot for the following transfer function:

$$\frac{x_i(s)}{u(s)} = \frac{0.1(s + 10)^2}{s(s + 100)}$$

This is the transfer function of the so-called lead-lag equalizer that is often put in series with the plant to modify the closed-loop frequency response and to ensure stability. The use of such series-compensation networks will be discussed in Chap. 10.

After putting $x_i(s)/u(s)$ into time-constant form, we have

$$\frac{x_i(s)}{u(s)} = \frac{0.1(1 + s/10)^2}{s(1 + s/100)}$$

The logarithm of the magnitude of $x_i(j\omega)/u(j\omega)$ is therefore

$$\log \left|\frac{x_i(j\omega)}{u(j\omega)}\right| = \log 0.1 + 2 \log \left|1 + \frac{j\omega}{10}\right| - \log |j\omega| - \log \left|1 + \frac{j\omega}{100}\right|$$

The only difference between the form of this expression and that of Eq. (5.3-3) is the 2 that precedes the $1 + j\omega/10$ term. This multiplication by 2 is due to the fact that this term is repeated; that is, it is a double zero. This means that the slope of the high-frequency asymptote is $+2$ rather than $+1$ so that beyond the break frequency $\omega = 10$ the plot rises by two decades of magnitude for every decade of frequency.

With this slight modification, the constituent terms of $|x_i(j\omega)/u(j\omega)|$ may now be placed on the log-log plot, as shown in Fig. 5.3-4a. The composite plot of the $|x_i(j\omega)/u(j\omega)|$ is obtained as before. Once again, the exact plot of the magnitude is also shown for comparison.

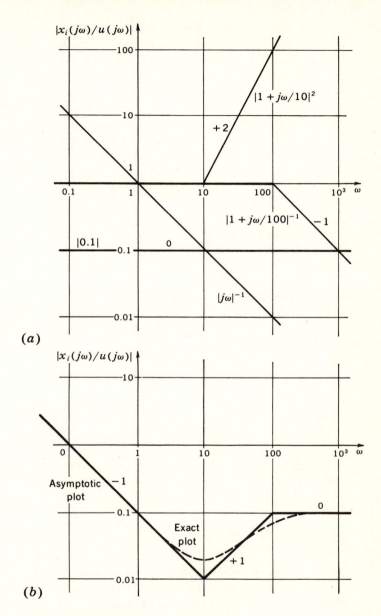

**Fig. 5.3-4**   Example 5.3-1.   (a) Constituent terms; (b) resultant plot.

With some practice, the reader should be able to draw the asymptotic plot for the $|W(j\omega)|$ directly from the time-constant form for $W(s)$ or $W(j\omega)$ without expanding the log $|W(j\omega)|$ or showing component terms on the diagram. It is suggested that the reader work several of the exercises at the end of this section until such a facility is gained, since it is valuable in our later work. Once such an ability is achieved, the reader should be able to determine the asymptotic plot for the $|W(j\omega)|$ with very little effort. Perhaps, even more important, he is able to sketch the general nature of the plot rapidly.

If the transfer function $W(s)$ contains complex conjugate poles or zeros, the approach must be somewhat altered, although the results are strikingly similar. To illustrate the treatment of such terms, let us suppose that $W(s)$ is actually a plant transfer function consisting of only a pair of complex conjugate poles so that

$$W(s) = G_p(s) = \frac{\omega_n{}^2}{s^2 + 2\zeta\omega_n s + \omega_n{}^2} = \frac{1}{1 + 2\zeta(s/\omega_n) + (s/\omega_n)^2}$$

Once again the use of the time-constant form is emphasized. The magnitude of $G_p(j\omega)$ is now

$$|G_p(j\omega)| = \frac{1}{|(1 - \omega^2/\omega_n{}^2) + 2j\zeta\omega/\omega_n|} \tag{5.3-7}$$

For $\omega \ll \omega_n$ the unity term in the real part predominates and the log $|G_p(j\omega)|$ is therefore

$$\log |G_p(j\omega)| \approx - \log 1 = 0 \qquad \text{for } \omega \ll \omega_n$$

On the other hand, for $\omega \gg \omega_n$, the quadratic term of the real part becomes predominant, and the log $|G_p(j\omega)|$ becomes

$$\log |G_p(j\omega)| \approx - \log \left|\left(\frac{\omega}{\omega_n}\right)^2\right| = -2 \log \omega + 2 \log |\omega_n|$$

and

$$\frac{d(-2 \log \omega + 2 \log |\omega_n|)}{d(\log \omega)} = -2$$

Hence the high-frequency asymptote has a slope of $-2$.

From the above discussion, we see that the pair of complex conjugate poles are identical to two real poles located at $s = \omega_n$ as far as the asymptotic plot is concerned. In other words, the asymptotic, straight-line approximation treats the situation as if $\zeta = 1$. A normalized asymptotic plot is shown in Fig. 5.3-5 where the double break occurs at $\omega/\omega_n = 1$.

Although the value of the damping ratio $\zeta$ does not influence the straight-line approximation, it does have an effect on the exact plot, as

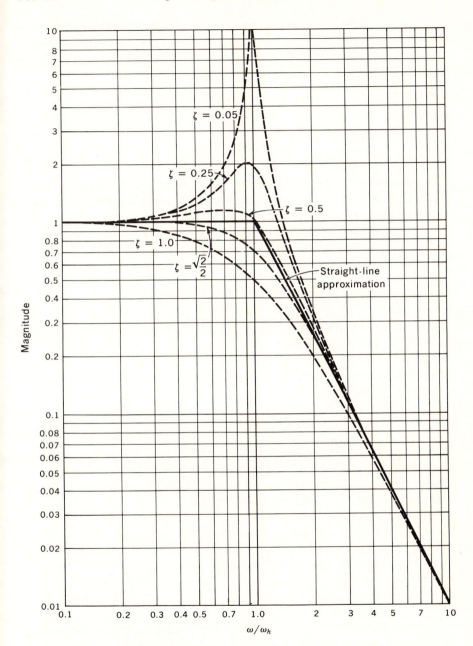

**Fig. 5.3-5** Asymptotic and exact plots of the magnitude of $1/[1 + 2j\zeta\omega/\omega_n + (\omega/\omega_n)^2]$.

shown in Fig. 5.3-5. As $\zeta$ approaches zero, the exact curve begins to rise above the asymptotic plot. In order to determine the exact value of $\zeta$ for which this effect first occurs, we solve the following expression for the frequency at which the maximum occurs:

$$\frac{d|G_p(j\omega)|}{d\omega}\bigg|_{\omega = \omega_{max}} = 0$$

From this equation we find

$$\omega_{max} = \omega_n \sqrt{1 - 2\zeta^2} \tag{5.3-8}$$

Therefore there is a maximum in the magnitude plot, that is, a real value for $\omega_{max}$, only if $\zeta < \sqrt{2}/2$. Otherwise, the exact plot always lies below the asymptotic plot and no maximum occurs.

If a maximum does occur, that is, if $\zeta < \sqrt{2}/2$, the maximum value of the $|G_p(j\omega)|$ may be obtained by substituting $\omega_{max}$ for $\omega$ in Eq. (5.3-7), with the result that

$$|G_p(j\omega_{max})| = \frac{1}{2\zeta \sqrt{1 - \zeta^2}} \quad \text{for } \zeta \leq 0.707 \tag{5.3-9}$$

Lest the reader become confused, it should be noted that the step response of the system has an overshoot whenever $\zeta < 1$. However, the magnitude portion of the frequency-response function does not display an overshoot until $\zeta < \sqrt{2}/2$. Note also that the value of the $|G_p(j\omega)|$ for $\omega = \omega_n$ is given by

$$|G_p(j\omega_n)| = \frac{1}{2\zeta} \tag{5.3-10}$$

From Fig. 5.3-5 we observe that for $0.5 < \zeta < 1$ the exact plot and the asymptotic approximation are actually closer than in the case of real roots. It is rather uncommon for control systems to have a damping ratio of less than 0.5. If $\zeta$ becomes larger than 1, the plant has two real poles and the previous procedures can be used.

In addition to the use of the asymptotic approximations for determining the magnitude plot, these approximations may also be used to form approximate analytic expressions for $|W(j\omega)|$. These analytic expressions can often be used to augment the graphical procedure and to determine more precise values than may be obtained graphically.

**Example 5.3-2**    Consider, for example, the problem of determining the frequency for which a loop transfer function

$$W(s) = KG_p(s)H_{eq}(s) = \frac{5}{s(1 + s)}$$

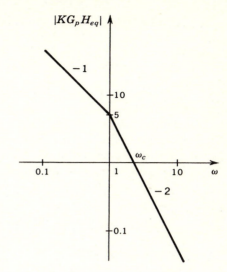

**Fig. 5.3-6** Asymptotic plot of the $|KG_p(j\omega)H_{eq}(j\omega)|$ for Example 5.3-2.

has a magnitude of unity. The asymptotic plot of the $|KG_p(j\omega)H_{eq}(j\omega)|$ is shown in Fig. 5.3-6. From this figure we see that the crossover frequency $\omega_c$ is between 1 and 10, but without a more exact graphical plot a precise value cannot be determined. If we use the asymptotic approximations, then, for $\omega > 1$, $|KG_p(j\omega)H_{eq}(j\omega)|$ becomes

$$|KG_p(j\omega)H_{eq}(j\omega)| = \frac{5}{\omega^2}$$

Therefore $\omega_c$, which is given by $|KG_p(j\omega)H_{eq}(j\omega)| = 1$, becomes

$$\omega_c = \sqrt{5} = 2.24$$

The exact value of $\omega_c$ is $[(\sqrt{101} - 1)/2]^{\frac{1}{2}}$.

Thus far the examples of this section have been concerned with the plotting of the magnitude of the frequency-response function for transfer functions that are concerned with an open-loop system. As a final example let us examine the case when $W(s)$ represents the closed-loop transfer function $y(s)/r(s)$.

**Example 5.3-3** Assume that the transfer function of a particular closed-loop system is given as

$$\frac{y(s)}{r(s)} = \frac{50(s + 5)}{[(s + 3)^2 + 4^2](s + 10)} = \frac{50(s + 5)}{(s^2 + 6s + 25)(s + 10)}$$

In time-constant form, this becomes

$$\frac{y(s)}{r(s)} = \frac{1 + s/5}{(1 + 6s/25 + s^2/25)(1 + s/10)} \tag{5.3-11}$$

The denominator term here contains a set of complex conjugate poles, with $\omega_n{}^2 = 25$, or $\omega_n = 5$. Thus a double break downward occurs at $\omega = 5$, owing to the denominator term. However, in the numerator there is a zero that occurs at $\omega = 5$, and hence a break upward also occurs at the same value of frequency. The net effect is that of a single break downward at $\omega$ equal to 5.

For small frequencies, as $\omega$ goes to zero, the magnitude of $y(j\omega)/r(j\omega)$ is 1, and the slope at that point is zero. With this knowledge and the knowledge of the break points, the final plot of the magnitude of the frequency-response function is made directly without bothering to write the frequency-response function. In other words, one works directly from the transfer function of Eq. (5.3-11). In addition, it seems like extra work to indicate the constituent elements that make up the final magnitude plot. Consequently the magnitude of the frequency-response function is drawn directly from Eq. (5.3-11) as shown in Fig. 5.3-7.

The facility with which this resulting magnitude curve is drawn

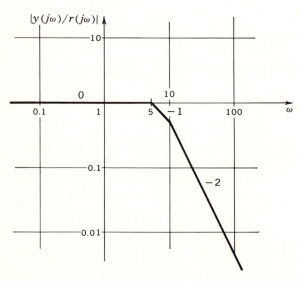

**Fig. 5.3-7**   Magnitude of the frequency-response function for Example 5.3-3.

is not due to the fact that in this example we are dealing with a
closed-loop transfer function, whereas in previous examples we have
dealt with open-loop transfer functions.   Rather, after one has established one point through which the magnitude curve passes at low
frequency, the remainder of the curve may be sketched rapidly from
the knowledge of the break points and the slopes.   The reason that
this example appears so simple is not because anything different has
been done but simply because the reader now understands the
procedure.

In this section, we have considered the problem of determining the
magnitude plot for the frequency-response function.   By making use of
asymptotic approximations and a log-log plot, we are able to obtain a
very satisfactory solution to this problem.   Unfortunately, the investigation of the phase portion of the frequency-response plot is not quite as
rewarding.   For this reason, we find it necessary to examine the phase
plot from two separate views.   These methods are the subject of the next
two sections.

**Exercises 5.3**   *5.3-1.*   Draw the asymptotic, straight-line magnitude
plots for each of the following transfer functions.   In addition,
determine a few points of the exact plot, especially near the break
points, and sketch the exact plot for comparison.

(a)   $\dfrac{s(s + 100)}{(s + 1)(s + 10)}$     (b)   $\dfrac{10^2(s + 10)^2}{s^2(s + 100)}$

(c)   $\dfrac{10^4 s(s + 10)}{(s^2 + 5s + 100)(s + 1{,}000)}$     (d)   $\dfrac{-10s(s + 1)}{(s + 10)(s + 100)^2}$

(e)   $\dfrac{10(s + 1)}{s(s + 10)^2(s^2 + 10s + 100)}$

*5.3-2.*   For each of the plant transfer functions given below, write
the analytic expression for the straight-line approximation of the
magnitude portion of the frequency response in the range of
frequencies indicated.

(a)   $\dfrac{100(s + 1)}{s(s + 100)}$     (b)   $\dfrac{10(s + 10)}{s(s + 1)^2(s + 100)}$

   $1 < \omega < 100$        $1 < \omega < 10$

(c)   $\dfrac{100}{s^2(s + 1)(s + 10)}$

   $1 < \omega < 10$

*answers:*

(a)    $|G_p(j\omega)| \sim 1$          for $1 < \omega < 100$

(b)    $|G_p(j\omega)| \sim \dfrac{1}{\omega^3}$          for $1 < \omega < 10$

(c)    $|G_p(j\omega)| \sim \dfrac{10}{\omega^3}$          for $1 < \omega < 10$

## 5.4   Minimum-phase condition

In this and the following section, attention is concentrated upon the second component of the frequency-response function, namely, the phase characteristic or plot.   At any one frequency the phase angle may be determined by finding the phase angle of $W(j\omega)$.   However, if we wish to know the plot of the phase angle for all frequencies, this calculation has to be repeated many times.   Means of avoiding this detailed calculation are discussed here and in Sec. 5.5.   More specifically, we examine a class of transfer functions that are classified as minimum phase.   For this class of systems it is possible to demonstrate that the magnitude and phase characteristics are not independent but are, in fact, closely related.   In particular, H. W. Bode[1] has shown that the phase shift, arg $W(j\omega)$, at a frequency $\omega_1$ is related to the $|W(j\omega)|$ by the integral expression

$$\text{arg } W(j\omega_1) = \frac{1}{\pi} \int_{-\infty}^{\infty} \frac{d \ln |W(j\omega)|}{du} \ln \coth \left| \frac{u}{2} \right| du \qquad (5.4\text{-}1)$$

where $u = \ln (\omega/\omega_1)$.   Note the appearance of the logarithms of the magnitude and frequency, once again pointing out the naturalness of the logarithmic representation.

We do not attempt to derive Eq. (5.4-1) since this would involve mathematics beyond the intended scope of this book.   In addition, no attempt is made to use Eq. (5.4-1) by making a substitution for the $|W(j\omega)|$ and carrying out the indicated integration.   In fact, the work involved in such a process exceeds by far the work involved in a direct calculation of the phase shift.

However, this does not mean that Eq. (5.4-1) is not of interest; on the contrary, the expression is useful in our study of the phase plot.

An examination of Eq. (5.4-1) reveals that phase shift is obtained by integrating the product of the slope of the logarithmic magnitude plot and a weighting factor ln coth $|u/2|$, much like the convolution procedure

---

[1] H. W. Bode, "Network Analysis and Feedback Amplifier Design," D. Van Nostrand Company, Inc., Princeton, N.J., 1945.

of Chap. 4.    Hence we obtain the valuable clue that *the phase shift is related to the slope of the logarithmic magnitude plot.*    In fact, we can show by the use of Eq. (5.4-1) that, if the slope of the magnitude plot is constant for all frequencies at a value of $\alpha$, the phase shift is given by

$$\arg W(j\omega) = \alpha \frac{\pi}{2} \tag{5.4-2}$$

Although theoretically Eq. (5.4-1) indicates that the phase shift at any frequency is dependent on the slope of the magnitude plot for all frequencies, an investigation of the function $\ln \coth |u/2|$, as shown in Fig. 5.4-1, indicates that practically we need only consider frequencies for *one decade on either side of the frequency of interest.*    In other words, if the slope of the magnitude plot has a constant value of $\alpha$ for the decade immediately preceding and following the frequency of interest, the phase shift is approximately given by Eq. (5.4-2).

If the slope is not constant over this two-decade range, we may still make use of Eq. (5.4-2) by calculating an average slope to use in place of $\alpha$.    In order to compute this average slope, we could make use of the $\ln \coth |u/2|$ function shown in Fig. 5.4-1.    This, however, is not the procedure recommended, since the work involved would defeat the purpose of using an approximation, namely, to obtain answers quickly and easily. Rather, it is recommended that the average slope be obtained by simple linear averaging, with a small bias given to the frequency range near the frequency of interest.    This should be done without making any calculations by simply "eyeballing" the magnitude plot.    Such a procedure provides good qualitative answers and often relatively accurate quantitative information.    If a more exact plot is needed, one can use the methods

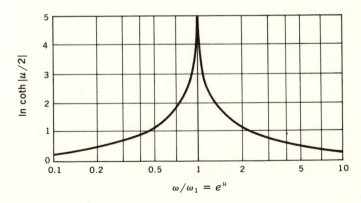

**Fig. 5.4-1**   The function $\ln \coth |u/2|$.

discussed in the next section or evaluate arg $W(j\omega)$ for particular values of $\omega$ that are of special interest.

*Example 5.4-1*    To illustrate the use of this procedure, let us consider the determination of the phase-shift plot of $W(s) = x_i(s)/u(s)$, where

$$W(s) = \frac{x_i(s)}{u(s)} = \frac{5(1 + s)(1 + s/10)}{(1 + s/0.1)(1 + s/100)}$$

This transfer function is similar to that of the lead-lag equalizer discussed in Example 5.3-1.   The magnitude plot for this transfer function is shown in Fig. 5.4-2a.

Let us determine the phase shift for $\omega = 0.01$.   If we examine the magnitude plot for one decade above and below this frequency, we see that the slope has a constant value of zero.   Hence

$$\arg W(j0.01) \approx 0\frac{\pi}{2} = 0$$

For $\omega = 0.1$ the average slope is $-\frac{1}{2}$, and therefore the phase shift becomes

$$\arg W(j0.1) \approx \left(-\frac{1}{2}\right)\frac{\pi}{2} = -\frac{\pi}{4}$$

The phase shift for $\omega = 1.0, 10, 100$, and $1,000$ is shown as Fig. 5.4-2b.   In addition, the exact phase plot is shown for comparison. For this example, the approximation is good and gives a good deal of qualitative information concerning the character of the phase plot.

The approximation procedure discussed here works best if the break points of the magnitude plot are separated by at least a decade.   The most accurate information is obtained for frequencies that are either at break points or a long distance from any break point.   This explains the success of the procedure for Example 5.4-1.   However, even if these features are not present, a qualitative picture of the phase plot can be obtained quickly and easily by using this procedure.

*Minimum-phase condition.*   As we have indicated, the relationship between the magnitude and phase plots given by Eq. (5.4-1) is valid only for minimum-phase systems, which are defined below.   To understand the necessity of this restriction, let us consider the frequency response of the following three systems:

$$W_1(s) = \frac{10(s+1)}{s+10} = \frac{1+s/1}{1+s/10}$$

$$W_2(s) = \frac{10(s-1)}{s+10} = \frac{-(1-s/1)}{1+s/10} \qquad (5.4\text{-}3)$$

$$W_3(s) = \frac{10(s+1)}{s-10} = \frac{-(1+s/1)}{1-s/10}$$

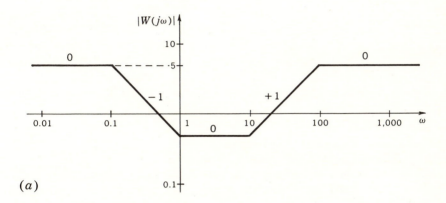

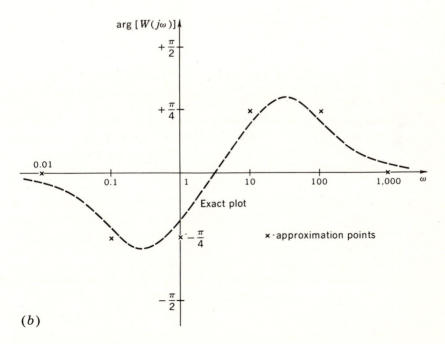

**Fig. 5.4-2** Example 5.4-1.  (a) Magnitude; (b) phase.

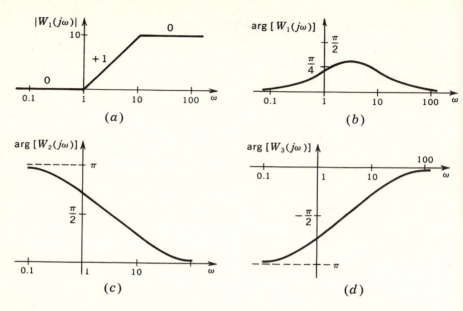

**Fig. 5.4-3**   Magnitude and phase plots for minimum-phase and nonminimum-phase systems.   (*a*) Magnitude; (*b*) arg $W_1(j\omega)$; (*c*) arg $W_2(j\omega)$; (*d*) arg $W_3(j\omega)$.

Here all three transfer functions have been expressed in both time-constant and pole-zero form.   The magnitude and phase plots for these three systems are shown in Fig. 5.4-3.   Since the magnitude plots of the three systems are identical, only one is shown.   Although the three systems have the same magnitude plot, they have radically different phase characteristics.   This difference between the phase plots for the three cases is due to the location of a pole or zero in the right half of the *s* plane.   Only for $W_1(s)$ is it possible to infer the phase-angle diagram from the amplitude diagram, because only $W_1(s)$ is minimum-phase.

*Systems,* such as $W_1(s)$, *that have no poles or zeros in the right half of the s plane are referred to as minimum-phase shift systems,* because they have the minimum amount of phase shift for a given magnitude plot. For the simple example we have considered, we have established that the Bode relation is valid only if the system is minimum-phase, that is, if it has no poles or zeros in the right-half plane.   In other words, the phase-shift plot is uniquely specified by the magnitude plot only if the transfer function is minimum-phase.   We have shown this condition to be true in only one specific example, but Bode[1] has shown that Eq. (5.4-1) is valid for any minimum-phase transfer function.   Thus, in the general

[1] *Ibid.*

case, the phase-angle diagram may be determined uniquely from the amplitude diagram if the corresponding transfer function has no poles or zeros in the right-half $s$ plane.

It is important to note that if poles are located on the $j\omega$ axis the transfer function is unstable, but it is still minimum-phase. Poles or zeros must be in the right-half $s$ plane, not simply on the boundary of the right-half $s$ plane, in order to violate the minimum-phase condition.

Special note should also be made of the fact that a system with a negative value for the gain $K$ is nonminimum-phase. Obviously the addition of a negative sign does not change the magnitude plot but does alter the phase characteristic by adding 180° of phase lag. This point can often be overlooked if one is not careful.

Fortunately, most of the transfer functions concerned with control systems are of the minimum-phase variety so that the approximations discussed above can be used to determine the phase plot. However, when nonminimum-phase transfer functions are encountered, it is convenient to be able to plot their phase characteristics with as little difficulty as possible. One approach is to return to the pole-zero form of the transfer function and the pole-zero plot on the $s$ plane. For this reason, the three transfer functions of Eqs. (5.4-3) are written in pole-zero form as well as the time-constant form associated with Bode diagrams.

In order to discuss the graphical approach on the $s$ plane, it is necessary to recall the graphical procedures used in Chap. 4. Consider Fig. 5.4-4. There a pole is located at $s = -\alpha$, and an arbitrary point $s$ on the plane has been chosen. A vector starting at $s = -\alpha$ and ending at the point $s$ represents the length and angle of the vector $s + \alpha$. In plotting the frequency response, values of $s$ along the $j\omega$ axis are of interest.

To illustrate the phase-angle calculation for a nonminimum-phase

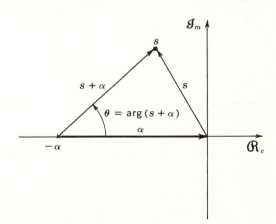

***Fig. 5.4-4*** Method of evaluating the amplitude and phase angle of the vector $s + \alpha$.

system, consider $W_2(s)$ as given in Eqs. (5.4-3).  The pole-zero plot for this transfer function is given in Fig. 5.4-5a.  Suppose we are interested in the phase angle for all positive frequencies, starting at $\omega = 0$.  For $\omega = 0$, the components of the transfer function

$$W_2(s) = \frac{10(s-1)}{s+10}$$

are shown in Fig. 5.4-5b.  The angle concerned with the vector originating on the zero at $s = +1$ is 180°.  The angle associated with the pole is 0°, so that at $s = 0$ the phase angle of $W_2(s)$ is just

$$\arg W_2(j0) = 180° - 0° = 180°$$

Figure 5.4-5c illustrates the calculation of the phase angle at $s = j1$.  Again the argument of $W_2(s)$ is just $\theta_1 - \theta_2$, where the angles are $\theta_1 = 135°$ and $\theta_2 \approx 6°$, or

$$\arg W_2(j1) = \theta_1 - \theta_2 = 135° - 6° = 129°$$

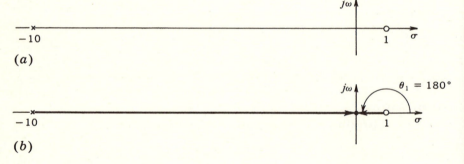

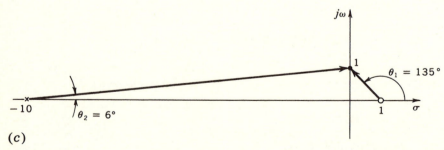

**Fig. 5.4-5**  Steps in calculating the phase angle of $W_2(j\omega)$ in Eqs. (5.4-3).

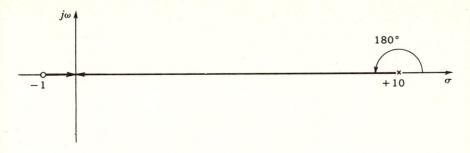

**Fig. 5.4-6**   Calculation of the phase angle of $W_3(j0)$.

At $s = j10$, $\theta_1$ is approximately 90°, and $\theta_2$ is 45°. As the frequency approaches infinity, both $\theta_1$ and $\theta_2$ approach 90° so that in the limit the argument of $W_2(j\omega)$ approaches zero. The phase plot has already been given as Fig. 5.4-3c.

If the transfer function $W_3(s)$ is considered, as in Fig. 5.4-6, note that, at zero frequency, the phase angle is $-180°$, since the phase shift is due to the angle associated with the pole. Again, as $\omega$ passes through 1, 10, and on to infinity, the phase-angle diagram of Fig. 5.4-3d is determined.

The approach described here is applicable not only to nonminimum-phase transfer functions but to any transfer function. Also, if the gain is taken into account and the vector distances measured, this approach is readily suited to the calculation of the amplitude diagram as well as the phase diagram. However, because the asymptotic straight-line Bode approximations are so easy to use, amplitude diagrams are usually drawn by that method rather than by the graphical approach.

Another method of determining the phase plot for a nonminimum-phase transfer function is to apply the Bode approximation to a *pseudo* minimum-phase magnitude plot. The idea is to determine a magnitude plot for a pseudo minimum-phase system having the phase plot equal to the phase of the given nonminimum-phase system. The formation of this pseudo minimum-phase transfer function may be easily accomplished. One must only replace every pole in the right-half plane at $s = +\alpha$ by a left-half-plane zero at $s = -\alpha$ and every right-half-plane zero by a left-half-plane pole. The resulting transfer function is then minimum-phase, and its phase characteristic is identical to the phase characteristic of the original system. The transfer function should be put into time-constant form *before* interchanging poles and zeros to form the pseudo minimum-phase transfer function.

***Example 5.4-2***   As an illustration of this procedure, consider the phase representation of the nonminimum-phase transfer function $W_2(s)$ discussed above.

$$W_2(s) = \frac{-(1 - s/1)}{1 + s/10}$$

The magnitude plot is given in Fig. 5.4-3a.   Applying the above procedure, we form the pseudo minimum-phase system by replacing the zero at $s = +1$ by a pole at $s = -1$, so that

$$W_{pseudo}(s) = \frac{-1}{(1 + s)(1 + s/10)}$$

The magnitude and phase plots for this system are shown in Fig. 5.4-7.   A comparison of Figs. 5.4-3c and 5.4-7b reveals that the phase plot of $W_2(s)$ is the same as the phase plot of $W_{pseudo}(s)$, as predicted.   Note that 180° has been added to the phase plot to account for the minus sign in $W_2(s)$.

Once familiarity with the use of this method has been gained, the reader may see that he no longer needs to find the pseudo transfer function or its magnitude plot.   Instead, he may simply mentally determine the slopes of the pseudo magnitude plot and write them on the actual magnitude plot as shown in Fig. 5.4-8.   The phase-shift plot is then obtained by using the pseudo slopes rather than the actual slopes.

By making use of this pseudo minimum-phase procedure, it is now possible to approximate the phase plot of any system by the use

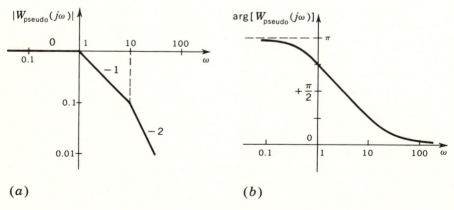

(a)                                      (b)

***Fig. 5.4-7***   Magnitude and phase plots for the pseudo minimum-phase system.

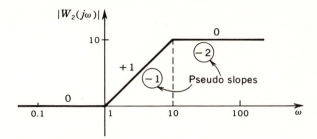

**Fig. 5.4-8**   Magnitude plot for $W_2(j\omega)$ with pseudo slopes shown.

of Eq. (5.4-2).   Although this procedure is adequate for an initial investigation of the qualitative nature of the phase characteristic, it often becomes necessary to refine the approximation in order to obtain more quantitative information.   This is the topic of the next section.

**Exercise 5.4**   *5.4-1.*   Find the approximate phase plot for the following transfer functions by drawing the asymptotic magnitude plot and applying the approximate Bode relation.   Check the answer by sketching the exact phase plot.

(a) $\dfrac{10^4}{s(s + 10)^2}$

(b) $\dfrac{10^4(s + 10)}{s^2(s + 100)}$

(c) $\dfrac{10^5(s - 10)}{s^2(s + 100)^2}$

(d) $\dfrac{-10s}{(s + 1)(s - 10)}$

(e) $\dfrac{100}{s(1 + s/10)}$

(f) $\dfrac{100(1 + s)}{s^2(1 + s/25)(1 + s/100)}$

(g) $\dfrac{100}{s(s^2 + 2s + 4)}$

(h) $\dfrac{10(s + 1)}{s(s - 10)}$

## 5.5   *Phase plot—arctangent approximation*

Although the approximation technique of the preceding section.provides a relatively simple means of plotting the phase characteristic of a transfer function $W(s)$, it is often convenient to have ᴀn easy-to-use analytic expression for the phase shift.   Consider, for example, the problem of finding the frequency at which the following plant has a phase shift of zero degrees:

$$G_p(s) = \frac{s}{(1 + s/10)(1 + s/100)^2} \tag{5.5-1}$$

The exact expression for the phase shift is

$$\arg G_p(j\omega) = \frac{\pi}{2} - \arctan \frac{\omega}{10} - 2 \arctan \frac{\omega}{100} \tag{5.5-2}$$

In order to find the value of $\omega$ for which $\arg G_p(j\omega) = 0$, it is necessary to solve a transcendental equation. This can be done only by trial and error, a tedious process at best. An alternative to this trial-and-error procedure that gives an analytic approach is the use of the arctangent approximation. To introduce this approach, let us consider once again a simple plant consisting of a single zero.

$$G_p(s) = 1 + \frac{s}{\omega_1}$$

The phase shift for this plant is given by

$$\arg G_p(j\omega) = \arctan \frac{\omega}{\omega_1}$$

We may obtain an approximate form for this equation by making a Taylor series expansion of the arctangent function. Two different expressions are obtained, depending on the relative magnitudes of $\omega$ and $\omega_1$:

$$\arctan \frac{\omega}{\omega_1} = \frac{\omega}{\omega_1} - \frac{1}{3}\left(\frac{\omega}{\omega_1}\right)^3 + \frac{1}{5}\left(\frac{\omega}{\omega_1}\right)^5 - \cdots \qquad \text{for } \omega < \omega_1 \tag{5.5-3}$$

$$\arctan \frac{\omega}{\omega_1} = \frac{\pi}{2} - \frac{\omega_1}{\omega} + \frac{1}{3}\left(\frac{\omega_1}{\omega}\right)^3 - \frac{1}{5}\left(\frac{\omega_1}{\omega}\right)^5 + \cdots \qquad \text{for } \omega > \omega_1 \tag{5.5-4}$$

These two series converge so rapidly that we normally use only the first term in Eq. (5.5-3) and the first two terms in Eq. (5.5-4), and so the approximate arctangent expressions become

$$\arctan \frac{\omega}{\omega_1} \approx \frac{\omega}{\omega_1} \qquad \text{for } \omega < \omega_1 \tag{5.5-5}$$

$$\arctan \frac{\omega}{\omega_1} \approx \frac{\pi}{2} - \frac{\omega_1}{\omega} \qquad \text{for } \omega > \omega_1 \tag{5.5-6}$$

These two approximations are plotted in Fig. 5.5-1 along with the exact expression for comparison.

The maximum error for both approximations occurs at $\omega = \omega_1$ and is equal to $1 - \pi/4$ rad, or approximately 12.3°. However, for $\omega < \omega_1/2$

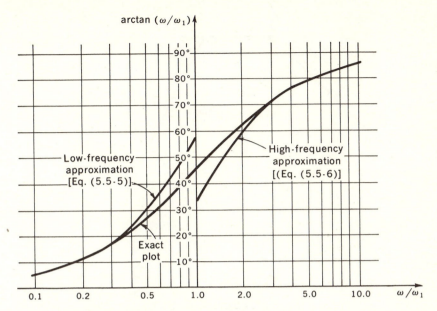

*Fig. 5.5-1*   Exact and approximate arctangent plots.

or $\omega > 2\omega_1$, the error is less than 0.04 rad, or about 2°.    Hence the approx-imation is fairly accurate as long as the frequency is not near a break point.    One way to overcome this difficulty partially is to assume a value of $\pi/4$ whenever the frequency is at or very close to a break point, that is, $\arctan \omega/\omega_1 \approx \pi/4$ for $\omega \approx \omega_1$.    In any case, one should be very cautious whenever working with near-break frequencies.

Now we return to the problem posed at the beginning of this section. Before the arctangent approximations may be applied to Eq. (5.5-2) to find the zero-phase-shift frequency, it is necessary to answer two qualita-tive questions.    First, Does the phase shift ever equal zero?    Second, In what range of frequencies does it occur or, more specifically, between which two break frequencies?    The second question must be answered so that we know which of the two arctangent approximations, Eq. (5.5-5) or Eq. (5.5-6), to apply to each term of Eq. (5.5-2).

Both questions can be quickly answered by making a rough sketch of the straight-line magnitude plot and then applying the approximate Bode method of Sec. 5.4 to sketch the qualitative nature of the phase characteristic.    The result of this process is shown in Fig. 5.5-2.    Although this plot is not accurate, it enables us to answer the two questions raised above.    Yes, there is a frequency for which the phase shift is zero, and

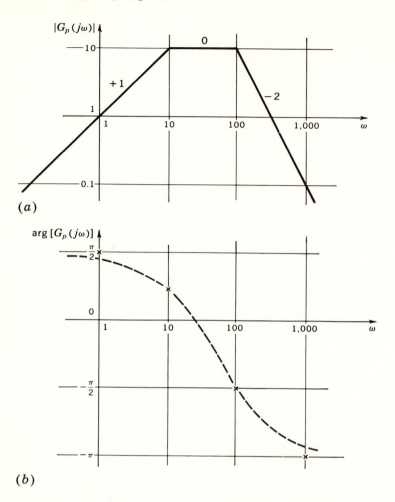

**Fig. 5.5-2**  Rough sketch of the frequency-response function for
$G_p(s) = s/[(1 + s/10)(1 + s/100)^2]$. *(a)*    Straight-line
magnitude plot; *(b)* approximate phase plot.

it appears to be between the break point at 10 rad/sec and the one at 100
rad/sec, that is, $10 < \omega < 100$.

For this frequency range, the second term of Eq. (5.5-2) becomes

$$\arctan \frac{\omega}{10} \approx \frac{\pi}{2} - \frac{10}{\omega}$$

since $\omega > 10$, and therefore the approximation of Eq. (5.5-6) is used.
On the other hand, the third term is approximated by using Eq. (5.5-5)

since $\omega < 100$ and is therefore

$$\arctan \frac{\omega}{100} \approx \frac{\omega}{100}$$

This term is multiplied by 2, since two poles exist at $s = -100$. The complete expression for the phase shift is therefore

$$\arg G_p(j\omega) \approx \frac{\pi}{2} - \left(\frac{\pi}{2} - \frac{10}{\omega}\right) - 2\frac{\omega}{100} \qquad \text{for } 10 < \omega < 100$$

or

$$\arg G_p(j\omega) \approx \frac{500 - \omega^2}{50\omega} \qquad\qquad \text{for } 10 < \omega < 100 \qquad (5.5\text{-}7)$$

In order to find the frequency $\omega_0$ for which the phase shift is zero, we set the right side of Eq. (5.5-7) equal to zero and solve for $\omega$. The result is

$$\omega_0 = \pm \sqrt{500} = \pm 22.4 \text{ rad/sec}$$

In this case, the negative answer is meaningless, and we conclude that

$$\omega_0 = 22.4 \text{ rad/sec}$$

Since $10 < \omega_0 < 100$, our initial assumption of the frequency range to investigate is verified. The exact phase shift for $\omega = \omega_0$ is $-1.2°$, and we observe that the accuracy of the answer is adequate.

The plus and minus values initially obtained for $\omega_0$ illustrate a feature of the arctangent-approximation approach. It is common to find that the method yields two solutions for a given problem. If only one solution falls within the frequency range assumed for the approximation, $10 < \omega < 100$ in the above example, that is the correct answer and the other answer must be discarded. If neither solution falls within the frequency range of interest, it means that either no solution exists or the wrong frequency range was selected.

Consider, for example, what would have happened if we had assumed that the zero-phase-shift frequency was greater than 100. In this case, the approximate phase-shift expression would have been

$$\arg G_p(j\omega) \approx \frac{\pi}{2} - \left(\frac{\pi}{2} - \frac{10}{\omega}\right) - 2\left(\frac{\pi}{2} - \frac{100}{\omega}\right) \qquad \text{for } \omega > 100$$

or

$$\arg G_p(j\omega) \approx -\pi + \frac{210}{\omega} \qquad\qquad \text{for } \omega > 100$$

If we set the right side of this equation equal to zero and solve for $\omega_0$, we obtain

$$\omega_0 = \frac{210}{\pi} < 100$$

Since $\omega_0$ is not in the frequency range assumed, the solution is not correct and must be discarded.

On the other hand, if both solutions fall within the range of interest, then either two solutions exist or the approximations are not accurate enough for the solution of the problem.    The latter situation occurs most often if the solutions obtained are close to break frequencies.    However, this problem is quite rare; the method works well for almost all problems.

An additional difficulty arises if the transfer function contains complex poles or zeros.    In Sec. 5.3 it was concluded that, for values of the damping ratio $\zeta$ in the range $0.5 < \zeta < 1$, the magnitude plot may be adequately approximated by a double break at $\omega = \omega_n$.    Let us use this approximation as the basis for the phase plot.    For example, we assume that

$$W(s) = \frac{x_i(s)}{x_j(s)} = \frac{K}{s^2 + 2\zeta\omega_n s + \omega_n^2} = \frac{K/\omega_n^2}{1 + 2\zeta s/\omega_n + s^2/\omega_n^2}$$
$$\approx \frac{K/\omega_n^2}{(1 + s/\omega_n)^2}$$

For the entire range of $\omega$ the phase shift associated with the approximate equation may be determined from Eq. (5.5-5) or (5.5-6) as

$$\arg \frac{x_i(j\omega)}{x_j(j\omega)} = -2\frac{\omega}{\omega_n} \qquad \omega < \omega_n$$

$$\arg \frac{x_i(j\omega)}{x_j(j\omega)} = -2\frac{\pi}{4} = -\frac{\pi}{2} \qquad \omega = \omega_n$$

$$\arg \frac{x_i(j\omega)}{x_j(j\omega)} = -2\left(\frac{\pi}{2} - \frac{\omega_n}{\omega}\right) \qquad \omega > \omega_n$$

Figure 5.5-3 illustrates the extent of the approximation involved.    Here the actual phase-shift curves for different damping ratios are compared with the approximate value.    As the damping ratio approaches 1, this approximation becomes a good one.

In the last two sections, we have considered two separate methods for treating the phase-shift portion of the frequency-response function. Ideally, the reader should not regard these methods as being competitive but rather as being complementary.    The reader should attempt to analyze each method for its strengths and its weaknesses so that, when a problem is to be handled, the appropriate combination of methods may be used.

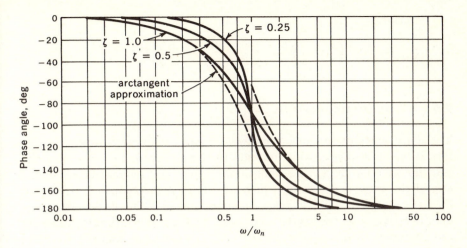

**Fig. 5.5-3**   Phase plots for a set of complex conjugate poles.

**Exercises 5.5**   *5.5-1.*   For the transfer function

$$G_p(s) = \frac{10s(s + 200)}{(s + 10)^2}$$

(a)   Use the arctangent approximations to write an expression for the phase shift in the region $10 < \omega < 200$.
(b)   Find the value of $\omega$ for which the phase shift is zero.
(c)   Find the value of $\omega$ for which the phase shift is minimum. What is the minimum phase shift?

*answers:*

(a)   $\arg G_p(j\omega) = \dfrac{\pi}{2} + \dfrac{\omega}{200} - 2\left(\dfrac{\pi}{2} - \dfrac{10}{\omega}\right)$
(b)   $\omega = 13.3$ rad/sec
(c)   $\omega = 63.2$ rad/sec, $-54°$

*5.5-2.*   For the transfer function

$$G_p(s) = \frac{10s}{(1 + s/3)(1 + s/50)^2(1 + s/100)}$$

write the arctangent approximation for the following:

(a)   The frequency range $\omega < 3$
(b)   The frequency range $3 < \omega < 50$
(c)   The frequency range $50 < \omega < 100$
(d)   The frequency range $\omega > 100$

*answer:*

$$(c) \quad +\frac{\pi}{2} - \left(\frac{\pi}{2} - \frac{3}{\omega}\right) - 2\left(\frac{\pi}{2} - \frac{50}{\omega}\right) - \frac{\omega}{100}$$

## 5.6   Plant identification

The early parts of this chapter developed the frequency-response function $W(j\omega)$ and related this function to the steady-state response of the transfer function $W(s)$ to a sinusoidal input. Subsequent sections discussed the various means by which this frequency-response function can be easily sketched, both in magnitude and phase. The latter are purely manipulative, and the reader may wonder what the ultimate utility of the frequency-response function is. The frequency-response function plays an important part in plant identification, closed-loop-system specification, and system synthesis. This section discusses plant identification.

Throughout this chapter we have treated the frequency-response function $W(j\omega)$ as though it were related to any transfer function $W(s)$; that is, $W(s)$ was not assumed to be related exclusively to either an open-loop plant or a closed-loop system. Since this section is concerned with plant identification, here we specifically restrict $W(s)$ to be associated with the plant to be controlled. Thus $W(s)$ may be the plant transfer function itself

$$W(s) = G_p(s) = \frac{y(s)}{u(s)}$$

or $W(s)$ may relate any one of the state variables to either the input or to another state variable. Therefore in this section $W(s)$ may also be

$$W(s) = \frac{x_i(s)}{u(s)}$$

or

$$W(s) = \frac{x_i(s)}{x_j(s)}$$

In the discussion no distinction is made between the three cases. In fact, the restriction that $W(s)$ be associated with the open-loop plant is in itself somewhat artificial. What is said in this section still applies to any general transfer function. The discussion here is directed toward the open-loop plant to emphasize the practical and important problem of plant identification that is readily solved through the use of frequency-response methods.

Plant identification is the determination of the plant transfer function from experimental measurements. This problem often arises in actual control situations. A given, unalterable plant is to be controlled by an automatic control system. The transfer function of this plant may be completely unknown or the form of the transfer function known but not the numerical values. The problem is to identify the important state variables and the transfer functions relating these state variables so that the plant may be controlled in an intelligent manner.

Three factors make the asymptotic amplitude and phase diagrams ideally suited to the solution of this problem. First, it is fairly easy to approximate an experimentally determined magnitude plot by means of a straight-line magnitude plot. Since the plot may have only integer values for its slope, the trial-and-error fitting of the straight-line approximation is considerably simplified.

Second, the form of the transfer function may be read, almost by inspection, from the straight-line approximation. This feature may be traced to the fact that the break points of the straight-line approximations correspond to the pole and zero locations of the transfer function. Only slight difficulty arises with regard to complex conjugate poles or zeros. This problem will be discussed in more detail later.

Third, joint consideration of the experimentally determined phase and magnitude plots uniquely specifies the transfer function. From the magnitude plot alone, the form of the transfer function is specified; that is, the break frequencies are all known. However, it is not known whether the system is minimum-phase or not, and thus some ambiguity remains unless the phase plot is considered. This is an important point because, although most plants are minimum-phase, such an assumption could prove disastrous if not true.

The use of the frequency-response function and the asymptotic approximation of this frequency-response function to a simple identification problem is illustrated in the following example.

***Example 5.6-1*** Assume that the state variables $x_3$ and $x_4$ are related by the transfer function

$$W(s) = \frac{x_3(s)}{x_4(s)} = \frac{K(s + \omega_1)}{s + \omega_2} = \frac{K(\omega_1/\omega_2)(1 + s/\omega_1)}{1 + s/\omega_2} \qquad (5.6\text{-}1)$$

In this simple example we have assumed that the form of the transfer function has been determined from a knowledge of the physical structure of the actual system in question. The problem is to find $K$, $\omega_1$, and $\omega_2$.

A typical set of experimental data points for this transfer func-

tion is shown in Fig. 5.6-1 where both amplitude and phase information are plotted.    From the phase plot Fig. 5.6-1*b* it is seen that the
phase angle initially starts at zero and that it also ends at zero.
This can be true only if the system is minimum-phase.    Once the
minimum-phase nature of the transfer function has been established,
the frequency-response function corresponding to Eq. (5.6-1) may

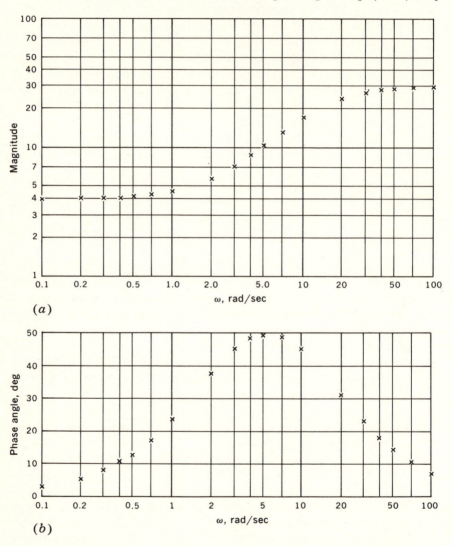

(*a*)

(*b*)

***Fig. 5.6-1***  Experimental data points for Example 5.6-1.    (*a*) Magnitude plot; (*b*)
phase plot.

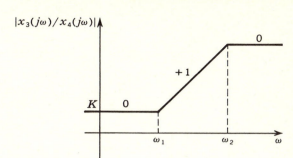

*(a)*

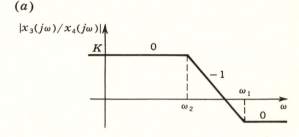

*(b)*

**Fig. 5.6-2**  Possible forms for the magnitude plot
for $x_3(s)/x_4(s) = K(1 + s/\omega_1)/(1 + s/\omega_2)$.
(a) $\omega_1 < \omega_2$; (b) $\omega_1 > \omega_2$.

still take two different forms, depending on the relative magnitudes
of $\omega_1$ and $\omega_2$.   This excludes the highly unlikely situation that $\omega_1$ is
equal to $\omega_2$, in which case the frequency response is flat at all frequencies.   If $\omega_1 < \omega_2$, the zero will "break" before the pole, and
the frequency-response function will have the general form of Fig.
5.6-2a.   On the other hand, if $\omega_1 > \omega_2$ the pole will "break" first,
and the general shape of the magnitude plot will be that of Fig.
5.6-2b.

A comparison of Figs. 5.6-1a and 5.6-2 easily reveals that the
frequency response must be of the form of Fig. 5.6-2a, or $\omega_1 < \omega_2$.
Once this decision has been reached, all that remains to be done is
to approximate the experimental data by horizontal (slope = 0)
high- and low-frequency asymptotes and an interconnecting line
segment with a slope equal to $+1$.   This has been done in Fig.
5.6-3.   From this figure we may easily read off the values of $\omega_1$,
$\omega_2$, and $K$ as $\omega_1 = 2.0$ rad/sec, $\omega_2 = 15.0$ rad/sec, and $K = 30.0$.
Therefore the transfer function $x_3(s)/x_4(s)$ becomes

$$\frac{x_3(s)}{x_4(s)} = \frac{4(1 + s/2)}{1 + s/15} = \frac{30(s + 2)}{s + 15} \qquad (5.6\text{-}2)$$

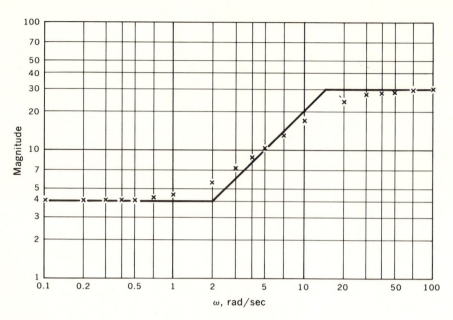

***Fig. 5.6-3***  Approximation of experimental data points by straight lines.

Although this problem is now completely solved, it is well to point out that the phase plot of Fig. 5.6-1*b* might also have been used to aid in writing Eq. (5.6-2). Note that the maximum value of the phase angle occurs at $\omega = 5$. The center of the line of slope plus 1 on Fig. 5.6-3 must also pass through this point, as it does in that figure.

Exactly this same procedure may be used in the general case when the plant is $n$th-order. Of course, in the general case the number of possible forms that the magnitude plot may take is much greater than the two found in Example 5.6-1. Because of this fact, it is generally not possible to examine an exhaustive list of all possible forms in order to select the proper one. Rather, it is necessary simply to fit the experimental data with possible forms for the magnitude plots by trial and error. Once again this procedure is considerably simplified by the fact that the plot may have only integer values for the slope.

If, in addition to the input-output data, one is able to make measurements of the internal state variables, it is possible to use this information to determine the state-variable representation of the plant in terms of the real physical variables. This is done by finding each of the internal transfer functions and then writing the corresponding differential equations. These equations may then be combined to form the state-variable representation. In addition, these internal transfer functions serve as consistency checks on each other as well as on the input-output transfer function. Note also that, since these internal transfer functions are of lower order than the input-output transfer function, their determination is simpler since there are fewer possible forms to consider. This is an important practical consideration.

If it is not possible to measure all the internal variables, it is necessary to determine some of the state-variable equations from a knowledge of the physical origin of the plant or by simply making mathematical definitions. This situation of inaccessible states often occurs if the plant transfer function contains one or more sets of complex conjugate poles.

In addition, complex conjugate poles present a more difficult identification problem since the straight-line approximation is not adequate for a complete identification. This inadequacy is due to the fact that the approximation is not affected by the value of the damping ratio $\zeta$. This difficulty is overcome through the use of Eq. (5.3-10), which relates the magnitude of the transfer function at $\omega = \omega_n$ to the damping ratio $\zeta$.

As in Sec. 5.3, let us assume that the plant is a typical second-order system so that

$$G_p(s) = \frac{\omega_n{}^2}{s^2 + 2\zeta\omega_n s + \omega_n{}^2} = \frac{1}{1 + 2\zeta s/\omega_n + s^2/\omega_n{}^2} \qquad (5.6\text{-}3)$$

The magnitude of $G_p(j\omega)$ is then

$$|G_p(j\omega)| = \frac{1}{|(1 - \omega^2/\omega_n{}^2) + 2j\zeta(\omega/\omega_n)|}$$

and when $\omega$ equals $\omega_n$, this magnitude becomes

$$|G_p(j\omega_n)| = \frac{1}{2\zeta} \qquad (5.6\text{-}4)$$

The frequency $\omega_n$ is readily determined by the intercept of the two asymptotes with slopes 0 and $-2$, as in Fig. 5.3-5. Or the frequency $\omega_n$ may be determined from the phase plot of Fig. 5.5-3. Here it is seen that, regardless of the damping ratio, the phase shift is $-90°$ when $\omega$ equals $\omega_n$. The damping ratio, and hence the location of the complex conjugate poles, may then be determined by the use of Fig. 5.5-3 and the measured value of the amplitude at the frequency $\omega_n$. The following example illustrates the determination of a plant transfer function involving complex conjugate poles.

*Example 5.6-2*    The transfer function of a completely unknown
plant is to be determined by frequency-response measurements.
The amplitude and phase data that result from frequency-response
experiments are given in Table 5.6-1.   As in Example 5.6-1, the
amplitude information is plotted on log-log paper with $\omega$ as the
abscissa and the amplitude as the ordinate.   The phase information
is plotted on semilog paper.   The amplitude and phase plots for
the data of Table 5.6-1 are shown in Fig. 5.6-4, and from this figure
we must determine the unknown plant transfer function.

*Table 5.6-1    Tabulation of Amplitude and
Phase Data for the Unknown
Plant of Example 5.6-2*

| $\omega$, rad/sec | $|G_p(j\omega)|$ | Phase Angle, deg |
|---|---|---|
| 1 | 5.02 | $-92.4$ |
| 2 | 2.07 | $-96.2$ |
| 4 | 1.36 | $-100.0$ |
| 5 | 1.17 | $-104.0$ |
| 6.3 | 1.03 | $-110.0$ |
| 8 | 0.965 | $-120.0$ |
| 10 | 0.967 | $-143.0$ |
| 12.5 | 0.743 | $-169.0$ |
| 20 | 0.128 | $-245.0$ |
| 31 | 0.0259 | $-258.0$ |

In the amplitude diagram, the initial slope is $-1$, and the
final slope is $-3$.   In the phase diagram, the initial phase lag is
$-90°$, whereas the final phase lag is $-270°$.   This can be possible
only if the system is minimum-phase.   (It might also be possible
if one pole and one zero were each in the right-half $s$ plane.   If this
were the case, it would not be possible to measure the transfer function of the plant in an open-loop configuration, since the plant would
be unstable.)   Because of the initial $-1$ slope, with the corresponding 90° phase shift, the plant must contain an integrator.   Because
of the rise in the amplitude diagram, the plant contains either one
or more zeros or complex conjugate poles.   Because of the final
$-3$ slope, and the corresponding 270° phase lag, the pole-zero excess
is 3.   Thus from a rather casual examination of the amplitude and
phase plots, we already have considerable information about the
plant transfer function.

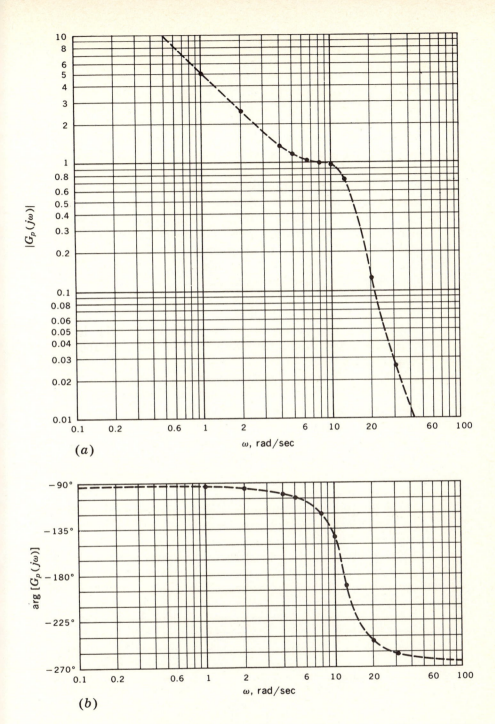

**Fig. 5.6-4** Plot of the data of Table 5.6-1 for the unknown plant of Example 5.6-2. (a) Magnitude; (b) phase.

Since it is clear that the plant contains an integrator, let us remove the frequency response of the integrator from the amplitude and phase plots. Removal of the 90° phase lag associated with the integrator is easily accomplished, and the resulting phase plot appears in Fig. 5.6-5. Removal of the effect of the integrator in the amplitude diagram is somewhat more difficult, as it must be done in a point-by-point fashion. What is required is that the transfer function $1/s$ be subtracted from the amplitude plot of Fig. 5.6-4$a$. This is just the reverse of the process used in early sections of this chapter to construct a composite transfer function from the straight-line approximations that made up its component parts. The subtraction is accomplished graphically in Fig. 5.6-6. In the resultant figure it is seen that the dc gain is 5. If this gain is also removed (division becomes subtraction when dealing with logarithms), the resulting amplitude diagram is as shown in Fig. 5.6-7. If this amplitude diagram is compared with those of Fig. 5.3-5, it is apparent that complex conjugate poles exist. We still must find the undamped natural frequency $\omega_n$ and the damping ratio $\zeta$ to know the transfer function completely.

From the phase plot of Fig. 5.6-5 it is seen that the $-90°$ phase shift occurs at $\omega = 12$. On the amplitude diagram the extension of the low- and high-frequency asymptotes also meet at $\omega = 12$. From either diagram we may thus conclude that $\omega_n$ is 12. In order to determine $\zeta$, we examine the value of $|G_p(j\omega_n)|$ and use either Eq. (5.6-4) or Fig. 5.3-5 to determine $\zeta$. From Fig. 5.6-7, it is seen

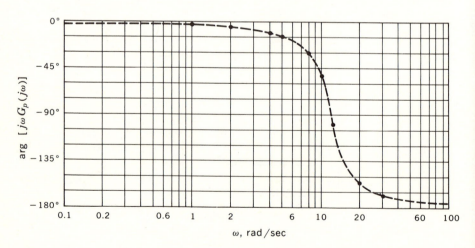

**Fig. 5.6-5**  Phase plot after the integrator phase shift has been removed.

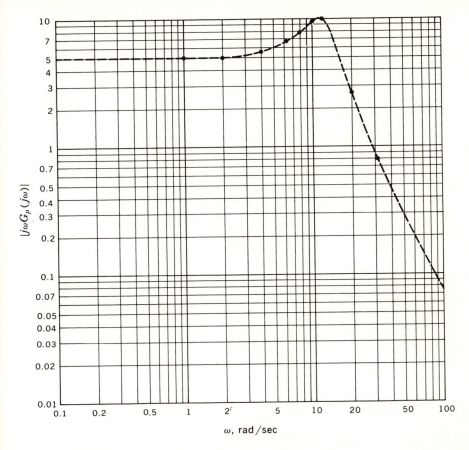

**Fig. 5.6-6**  Magnitude response after the effect of the integrator has been removed.

that $|G_p(j\omega_n)|$ is 2 so that $\zeta = 0.25$.   The plant transfer function is now known to be

$$G_p(s) = \frac{5}{s[1 + (2\zeta\omega/\omega_n)s + s^2/\omega_n^2]} = \frac{5}{s(1 + 5s/12 + s^2/144)}$$

or, in pole-zero form,

$$G_p(s) = \frac{720}{s(s^2 + 6s + 144)}$$

It is interesting to consider Example 5.6-2 from a state-variable point of view.   If measurements on the original system had been taken

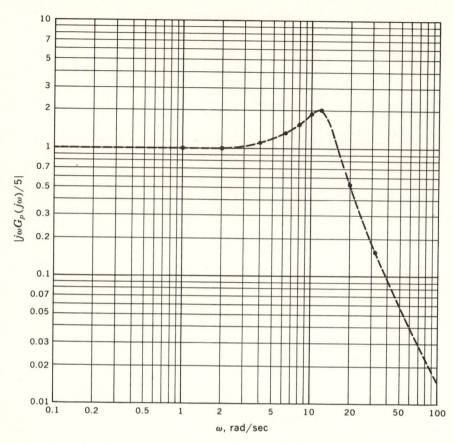

**Fig. 5.6-7**  Magnitude response after the effect of the integrator and the dc gain of 5 have been removed.

as indicated in Fig. 5.6-8, that is, if the frequency response due to the sinusoidal input had been recorded at $x_2$ as well as at the output, the frequency-response function $x_2(j\omega)/u(j\omega)$ would have appeared as in Fig. 5.6-6, with no subtraction required.  Of course, to determine the frequency response at $x_2$ would have required a transducer to measure that

**Fig. 5.6-8**  State-variable approach to frequency-response measurements.

state, just as a transducer is required to measure the behavior of the output. This brings up the subject of practical considerations in measuring frequency response.

The practical difficulties associated with frequency-response measurements are largely concerned with the generation of the sinusoidal input and the measurement of the frequency response at the various state-variable locations, including the output. If the component has an electrical input, an ordinary signal generator is sufficient to generate the required sinusoidal inputs. However, if the input is mechanical, as in a generator, for example, it is necessary to generate a sinusoidal motion. This is often difficult. The input may be almost any other physical quantity, such as temperature or flow rate of a given material, and then the plant identification is closely associated with the manner in which the plant component or components are to be utilized.

The measurement problem is concerned with the measurement of the output and the various state variables due to a sinusoidal input. Even in measuring a component transfer function as simple as that of an armature-controlled dc motor, certain difficulties are encountered. Here the output is an angle $\theta$, and the angle is usually measured with the aid of a potentiometer as an output transducer. At low-frequency inputs and even modest amplitudes, the amplitude of the output response is often sufficient to exceed the angular range of the output transducer. This is true because of the integration involved, and, as is clear from the frequency-response curve of an integrator, the amplitude goes to infinity at low frequency.

One method of at least partially avoiding these difficulties is to measure the frequency response in a closed-loop configuration. This requires $r(t)$ to be sinusoidal, and since $r(t)$ is the reference input, it is often electrical in nature. The output transducer must also be present to ensure that the loop may be closed. Measurements are made not from the reference input to the output but rather from $u$ to the output and to the intermediate state variables. At $x_i$, for instance, it is necessary to record the amplitude and phase of $x_i$ with respect to the amplitude and phase of $u$.

There is one inherent difficulty in closing the loop to aid in the measurement of the plant transfer function: The resulting closed-loop structure may be unstable. The topic of stability is not covered until the next chapter, and therefore about all that can be said here is that the engineer must know a good deal about the plant before he attempts to measure the plant transfer function in a closed-loop configuration. One approach is to close the loop with the gain $K$ set at zero and increase $K$ sufficiently to allow measurements to be made. This is often a safe

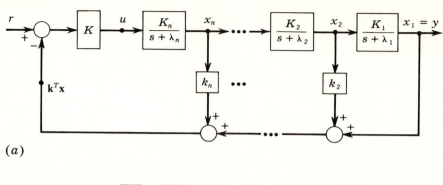

(a)

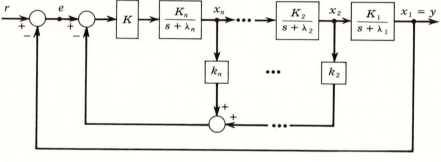

(b)

**Fig. 5.6-9**  Experimental measurement techniques.    (a) $G_p(s)$ and $H_{eq}(s)$; (b) $G_{eq}(s)$.

procedure if the system is minimum-phase.   The reader will not appreciate why this is true until the conclusion of the next chapter.

It is often of interest to determine not only the plant transfer function $G_p(s)$, but the feedback transfer function $H_{eq}(s)$, the loop transfer function $KG_p(s)H_{eq}(s)$, and the equivalent forward transfer function $G_{eq}(s)$. These may also be determined by frequency-response measurements made on either an open- or closed-loop configuration.   To see this, consider the two block diagrams of Fig. 5.6-9.   These two block diagrams are equivalent, but they are drawn in a slightly different way.   In both figures, $k_1$ has been set equal to 1.

All transfer functions but $G_{eq}(s)$ may be measured by using the configuration of Fig. 5.6-9a.   If the plant transfer function is to be determined by open-loop measurements, the physical system is actually opened at $u$ and frequency-response measurements made at the output and any other state variable of interest.   Since the loop is open, the effect of feedback is nonexistent.   Similarly $G_p(s)H_{eq}(s)$ may be measured by going from $u$ to the point $\mathbf{k}^T\mathbf{x}$.   With $G_p(s)$ and $G_p(s)H_{eq}(s)$ known, $H_{eq}(s)$ is also known.   If the loop is closed, then $r$, rather than $u$, is the sinusoidal input, but the measurements are made at exactly the same points.

The equivalent open-loop transfer function $G_{eq}(s)$ may be measured from $e(s)$ to $y(s)$, as indicated in Fig. 5.6-9$b$.   Again either open- or closed-loop methods may be employed.

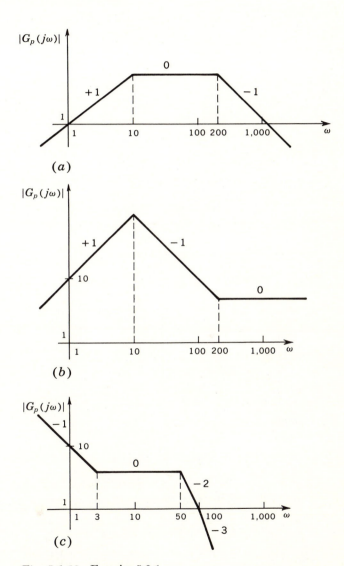

**Fig. 5.6-10**   Exercise 5.6-1.

**Exercises 5.6**   *5.6-1.*   Find the transfer function for each of the three
asymptotic magnitude plots shown in Fig. 5.6-10.   Assume that the
plants are minimum-phase, stable, and contain no complex conjugate poles or zeros.

*answers:*

(a)   $G_p(s) = \dfrac{s}{(1 + s/10)(1 + s/200)}$

(b)   $G_p(s) = \dfrac{10s(1 + s/200)}{(1 + s/10)^2}$

(c)   $G_p(s) = \dfrac{10(1 + s/3)}{s(1 + s/50)^2(1 + s/100)}$

*5.6-2.*   Find the values of $\omega_1$, $\omega_2$, $\omega_3$, and $K$ in the following transfer
function by using the experimental data given in Table 5.6-2:

$$G_p(s) = \frac{K(1 + s/\omega_1)}{s(1 + s/\omega_2)(1 + s/\omega_3)}$$

**Table 5.6-2   *Exercise 5.6-2***

*Experimental Data*

| $\omega$ | $|G_p(j\omega)|$ | $\omega$ | $|G_p(j\omega)|$ |
|---|---|---|---|
| 0.1 | 1,000 | 10 | 90 |
| 0.2 | 500 | 20 | 50 |
| 0.4 | 260 | 40 | 23 |
| 0.7 | 160 | 70 | 12 |
| 1.0 | 130 | 100 | 8 |
| 2.0 | 110 | 200 | 2.4 |
| 4.0 | 100 | 400 | 0.6 |
| 7.0 | 95 | 700 | 0.2 |
|  |  | 1,000 | 0.1 |

*answer:*

$$G_p(s) = \frac{100(1 + s)}{s(1 + s/10)(1 + s/100)}$$

*5.6-3.*   For the state-variable representation of a second-order plant
the following internal transfer functions were obtained:

$$\frac{x_2}{u} = \frac{10}{s + 1} \qquad\qquad \frac{x_1}{u} = \frac{10}{s(s + 1)}$$

$$\frac{x_1}{x_2} = \frac{1}{s} \qquad\qquad \frac{y}{u} = \frac{10(1 + 2s)}{s(s + 1)}$$

What is the state-variable representation?

*answer:*

$$\dot{\mathbf{x}} = \begin{bmatrix} 0 & 1 \\ 0 & -1 \end{bmatrix} \mathbf{x} + \begin{bmatrix} 0 \\ 10 \end{bmatrix} u \qquad y = \begin{bmatrix} 1 & 2 \end{bmatrix} \mathbf{x}$$

## 5.7   *Conclusions*

In this chapter the frequency-response function has been introduced, and several approximate and exact methods of representing the frequency response have been described. Initially the discussion considered the general transfer function $W(s)$, and no attempt was made to associate $W(s)$ with either the open- or the closed-loop system. In Sec. 5.6 $W(s)$ was restricted to the open-loop plant or transfer functions associated with the open-loop plant, and means were described by which the transfer function $W(s)$ could be identified. The subject of the frequency response of the closed-loop transfer function was not stressed, as this subject is closely associated with the idea of system specification as well as performance. These topics will be covered in detail in the following chapter and in Chap. 8 on specifications.

Another of the important uses of the frequency-response information is in answering the question of system stability, one of the basic questions in any control problem. Stability is the minimum requirement that any system design must satisfy. The frequency response provides a useful graphical tool for analyzing system stability. This subject will also be discussed in the next chapter.

Last but not least, the frequency-response information is valuable in the area of system synthesis. Many system specifications are entirely dependent on or closely related to frequency-response information. In addition, the frequency-response plot serves as a convenient tool for actually carrying out the design procedures. These concepts will be discussed in detail in Chaps. 8 and 10.

## 5.8   *Problems*

*5.8-1.* The closed-loop transfer function for a particular positioning servomechanism has been experimentally determined to be

$$\frac{y(s)}{r(s)} = \frac{125(s + 2)}{[(s + 3)^2 + 4^2](s + 10)}$$

From the frequency-response function, determine the analytic expression for $y(t)$ for the values of $\omega$ indicated if the input is

$$r(t) = 7 \sin \omega t \qquad \omega = 1, 5, 10, 100$$

Assume that the input was applied sufficiently long ago so that the transient solution is negligible.

5.8-2.   Plot the Bode diagrams, the approximate magnitude and phase plots, for the given transfer functions.

(a) $\dfrac{50(1 + s/0.5)}{s(1 + s/3)(1 + s/10)(1 + s/30)}$

(b) $\dfrac{25(1 + s/3)}{s(1 + s)(1 + s/16)(1 + s/30)}$

5.8-3.   Repeat Prob. 5.8-2 for the following two transfer functions involving complex poles and zeros.   Determine the actual phase shift at crossover, that is, when $|W(j\omega)| = 1$, and compare that value with that indicated by the approximate phase diagram.

(a)   $W(s) = \dfrac{100(s^2 + 2s + 3)}{s(1 + s/0.1)(1 + s/10)(1 + s/50)}$

(b)   $W(s) = \dfrac{125(s + 2)}{[(s + 3)^2 + 4^2](s + 10)}$

5.8-4.   The form of an unknown plant transfer function is indicated in Fig. 5.8-1, and the various parameter values are to be found.   This is

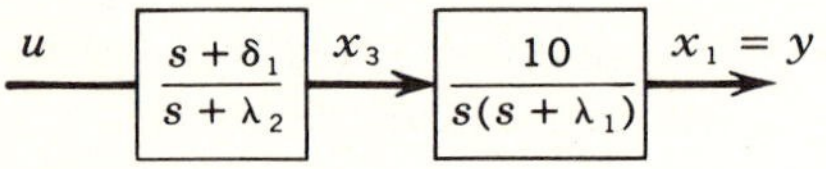

*Fig. 5.8-1*   Problem 5.8-4.

done by making frequency-response measurements, where $u$ is of the form

$$u = 7 \sin \omega t$$

Amplitude and phase data are taken between $u$ and the available states, and from one available state to the other, that is, between $u$ and $x_3$, $u$ and $x_1$, and $x_3$ and $x_1$, and these data are recorded in Table 5.8-1.   Use the data from $u$ to $x_3$ and $x_3$ to $x_1$ to determine the unknown plant parameters.   Then use only input-output data, $u$ to $x_1$, and repeat the problem.   Compare the difficulty involved with the two approaches.

**Table 5.8-1　Data for Prob. 5.8-4**

| $\omega$ | $\left|\dfrac{x_1(j\omega)}{u(j\omega)}\right|$ | $\arg\dfrac{x_1(j\omega)}{u(j\omega)},$ deg | $\left|\dfrac{x_3(j\omega)}{u(j\omega)}\right|$ | $\arg\dfrac{x_3(j\omega)}{u(j\omega)},$ deg | $\left|\dfrac{x_1(j\omega)}{x_3(j\omega)}\right|$ | $\arg\dfrac{x_1(j\omega)}{x_3(j\omega)},$ deg |
|---|---|---|---|---|---|---|
| 0.01 | 6.70 | $-84.6$ | 0.020 | 5.6 | 333 | $-90.2$ |
| 0.02 | 3.41 | $-79.3$ | 0.020 | 11.0 | 167 | $-90.4$ |
| 0.04 | 1.80 | $-69.5$ | 0.021 | 21.2 | 83.7 | $-90.8$ |
| 0.10 | 0.942 | $-48.0$ | 0.028 | 43.8 | 33.3 | $-91.9$ |
| 0.20 | 0.743 | $-32.7$ | 0.045 | 61.1 | 16.6 | $-93.8$ |
| 0.40 | 0.679 | $-26.2$ | 0.082 | 71.3 | 8.30 | $-97.6$ |
| 1.00 | 0.623 | $-35.6$ | 0.197 | 73.0 | 3.16 | $-108.4$ |
| 2.00 | 0.516 | $-58.3$ | 0.371 | 65.4 | 1.39 | $-123.6$ |
| 4.00 | 0.314 | $-93.0$ | 0.623 | 50.0 | 0.504 | $-143.0$ |
| 10.0 | 0.0857 | $-137.3$ | 0.894 | 26.0 | 0.0958 | $-163.3$ |
| 20.0 | 0.0241 | $-157.7$ | 0.970 | 13.8 | 0.0248 | $-171.4$ |
| 40.0 | 0.00624 | $-168.7$ | 0.992 | 7.0 | 0.00629 | $-175.7$ |
| 100.0 | 0.0010 | $-175.5$ | 0.999 | 2.8 | 0.00100 | $-178.3$ |

**5.8-5.**　This problem is an identification problem similar to that of Prob. 5.8-4, but here only the output state variable is available. However, it is known that the plant contains no zeros. Frequency-response data are given in Table 5.8-2. What is the plant transfer function?

**Table 5.8-2　Data for Prob. 5.8-5**

| $\omega$ | $|x_1(j\omega)/u(j\omega)|$ | $\arg\,[x_1(j\omega)/u(j\omega)],\ deg$ |
|---|---|---|
| 0.1 | 58.8 | $-90.7$ |
| 0.2 | 29.5 | $-91.3$ |
| 0.4 | 14.9 | $-92.7$ |
| 1.0 | 6.20 | $-97.1$ |
| 2.0 | 3.68 | $-107.0$ |
| 2.5 | 3.37 | $-115.2$ |
| 4.0 | 3.12 | $-171.8$ |
| 5.0 | 1.55 | $+141.0$ |
| 8.0 | 0.258 | $+109.0$ |
| 10.0 | 0.117 | $+103.5$ |
| 20.0 | 0.0131 | $+96.0$ |
| 40.0 | 0.0016 | $+92.9$ |
| 100.0 | 0.0001 | $+91.1$ |

*5.8-6.* Using the frequency-response data given in Table 5.8-3, find the unknown first-order transfer functions $G_1(s)$ and $G_2(s)$ for the plant shown in Fig. 5.8-2.

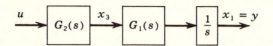

**Fig. 5.8-2**  Problem 5.8-6.

*Table 5.8-3    Data for Prob. 5.8-6*

| $\omega$ | $\lvert x_3(j\omega)/u(j\omega)\rvert$ | arg $[x_3(j\omega)/u(j\omega)]$, deg | $\lvert x_1(j\omega)/x_3(j\omega)\rvert$ | arg $[x_1(j\omega)/x_3(j\omega)]$, deg |
|---|---|---|---|---|
| 0.1 | 1.0 | 0 | 100.5 | +83.7 |
| 0.2 | 1.0 | −0.1 | 51.1 | 77.6 |
| 0.4 | 1.0 | −0.2 | 27.0 | 66.0 |
| 1.0 | 1.0 | −0.6 | 14.1 | 39.3 |
| 2.0 | 1.0 | −1.1 | 11.0 | 15.3 |
| 4.0 | 1.0 | −2.3 | 9.58 | −7.6 |
| 10.0 | 1.0 | −5.7 | 7.11 | −39.3 |
| 20.0 | 0.98 | −11.3 | 4.49 | −60.5 |
| 40.0 | 0.93 | −21.7 | 2.44 | −74.5 |
| 100.0 | 0.71 | −45.0 | 0.995 | −83.7 |
| 200.0 | 0.45 | −63.4 | 0.500 | −86.8 |
| 400.0 | 0.24 | −75.9 | 0.251 | −88.4 |
| 1,000.0 | 0.10 | −84.3 | 0.100 | −89.4 |

*Table 5.8-4    Data for Prob. 5.8-7*

| $\omega$, rad/sec | $\lvert G_p(j\omega)\rvert$ | arg $G_p(j\omega)$, deg |
|---|---|---|
| 0.1 | 100.5 | 83.6 |
| 0.2 | 51.1 | 77.4 |
| 0.4 | 27.0 | 65.8 |
| 1.0 | 14.1 | 38.7 |
| 2.0 | 11.0 | 14.2 |
| 4.0 | 9.57 | −9.9 |
| 10.0 | 7.07 | −45.0 |
| 20.0 | 4.40 | −71.8 |
| 40.0 | 2.26 | −96.2 |
| 100.0 | 0.704 | −129.0 |
| 200.0 | 0.224 | −150.2 |
| 400.00 | 0.0612 | −164.3 |
| 1,000.00 | 0.00995 | −173.6 |

*5.8-7.* The frequency-response data given in Table 5.8-4 were taken on a plant containing one zero which is known to be nonminimum-phase. Use the frequency-response data of Table 5.8-4 to find $G_p(s)$.

*5.8-8.* For the plant transfer function

$$G_p(s) = \frac{10}{s(1 + s)(1 + s/10)}$$

draw the magnitude and phase plots for the frequency range $0.1 < \omega < 10$. Now plot the frequency-response information on a single rectangular plot, using the logarithm of the magnitude for the vertical axis and the phase angle as the horizontal axis, with $\omega$ as a parameter along the plot. This type of frequency-response plot is known as a Nichols[1] chart and is an alternative method of representing the frequency-response information. Note that semilog paper may be used to plot magnitude vs. phase angle directly.

*5.8-9.* Plot the magnitude and phase information from Prob. 5.8-8 as a single polar plot. (Magnitude is not plotted on a logarithmic scale in this case.) This type of frequency-response plot, known as a Nyquist plot, will be used in the stability discussion of the next chapter.

*5.8-10.* For each of the following transfer functions

$$W(s) = \frac{10(s + 1)^2}{s(s + 10)(s + 100)^2}$$

$$W(s) = \frac{10^4 s(s + 10)}{(s + 1)(s^2 + 10s + 100)}$$

find:

(a) The straight-line magnitude plot
(b) The approximate phase plot
(c) The arctangent approximation in the range $1 < \omega < 10$

[1] J. J. D'Azzo and C. H. Houpis, "Feedback Control System Analysis and Synthesis," 2d ed., McGraw-Hill Book Company, New York, 1966.

# *six*      *stability*

## *6.1  Introduction*

This chapter is devoted to one of the most basic requirements of any control system: stability.   Independent of other performance specifications that the closed-loop system is to meet, it must always exhibit the characteristics that we shall define for stability. For this reason, the study of system stability forms an important part of both the analysis and design of control systems.

In a sense, the concept of this chapter is the opposite of that of Chap. 4.   There we sought the exact and complete behavior of a system due to a specific input for all time.   Here, on the other hand, our interest is in the asymptotic behavior of a system for all inputs as time approaches infinity.

The study of stability of linear systems deals with two basic questions. The first question, the absolute-stability problem, is qualitative in nature and seeks a simple "yes" or "no" statement concerning the system stability. The second question, the relative-stability problem, is quantitative in nature and is associated with the problem of determining how stable a system is.

Although the second question is more difficult to answer, its answer is by far the more valuable. In effect, relative-stability information provides a bridge between the minimal information of absolute stability and the complete information provided by the total time response, with labor also falling somewhere between the two extremes. That is, the use of the relative-stability concept is an attempt to characterize the behavior of a system by means of one or more relative-stability measures. Of course, relative stability is only one of the important performance measures that must be satisfied to ensure satisfactory system performance. Although performance measures will be discussed in Chap. 8, relative stability is of sufficient importance to warrant the special attention given in this chapter.

In Sec. 6.2 several definitions of stability are presented and discussed in order to make the concept of stability more precise. An algebraic method of determining whether the characteristic polynomial of a closed-loop system contains poles on the $j\omega$ axis or in the right-half plane is discussed in Sec. 6.3. This is the Routh-Hurwitz approach. A quite different approach to the stability problem, through the use of the frequency-response function and the Nyquist criterion, is discussed in Sec. 6.4. Different measures of relative stability are the subject of Sec. 6.5, and there the graphical appeal of the Nyquist diagram becomes evident.

The reader will notice that no apparent use is made of state-variable concepts in this chapter. The reason for this should be self-evident. Stability is a concept that involves the entire system. If one of the state variables is unstable, then all of them are unstable, since they have the same characteristic equation. Hence in this chapter our interest is in the characteristic polynomial and in the closed-loop transfer function. The effect of the individual state variables is felt through their involvement in the control $u$; in this manner they affect the characteristic polynomial and hence stability.

## 6.2   Definitions of stability

In the preceding chapter, the concept of stability was used in two places in a rather loose sense to describe two different properties. In Secs. 5.2

and 5.6, a stable system was described as a system in which the complementary solution decays to zero as time approaches infinity. In the discussion of the Bode relation in Sec. 5.4, on the other hand, we defined a stable system as one in which all the poles are in the left half of the $s$ plane.

This dual use of the concept of stability is acceptable for two reasons. First, these interpretations of stability should have some intuitive appeal, and second, as we shall see later, these two concepts of stability are equivalent. Hence there is no lack of consistency in our use of them.

It may appear that, if stability has an intuitive meaning, there is no need for this section. On the contrary; just the opposite is the case. Concepts having intuitive meaning usually escape analytic treatment until precise mathematical definitions are imposed. At the same time, care must be taken in making the definition to ensure that the intuitive appeal is not lost.

As the title of this section indicates, there is not one but several definitions of stability. In particular, there are three definitions that are most widely used. It is the normal procedure to select one of these definitions as *the* definition and then prove that the other two follow directly. Rather than follow this procedure and elevate one of them above the others, a slightly different approach is taken here. All three definitions are offered as definitions of equivalent importance, and we then undertake to establish that the three are equivalent. The three equivalent definitions are as follows:

1. A linear system is stable if its output remains bounded for every bounded input.
2. A linear system is stable if its weighting function is absolutely integrable over the infinite range, that is, if $\int_0^\infty |W(t)|\, dt$ is finite.
3. A linear system is stable if all the poles of the closed-loop transfer function $y(s)/r(s)$ lie in the left half of the $s$ plane.

The reader will note that the first two definitions deal with the time domain, whereas the third is a frequency-domain condition. Although the third definition proves to be the most useful in our work, the first two definitions are more closely related to reality. Hence, the three definitions tend to complement each other by providing alternative ways of viewing the same problem.

Let us begin by showing the equivalence of the first and second definitions; we then demonstrate that the second and third definitions are equivalent.

We consider first the problem of showing that, if the weighting function is absolutely integrable, every bounded input yields a bounded output. By using the convolution integral, we may write the output as

$$y(t) = \int_0^\infty W(\tau)r(t - \tau)\, d\tau \tag{6.2-1}$$

where $W(s) = y(s)/r(s)$. We choose this particular $W(s)$ because we are interested in the stability of closed-loop systems. If we take the absolute value of both sides of Eq. (6.2-1), we obtain

$$|y(t)| = \left| \int_0^\infty W(\tau)r(t - \tau)\, d\tau \right| \le \int_0^\infty |W(\tau)||r(t - \tau)|\, d\tau \tag{6.2-2}$$

Since the input is assumed to be bounded, there must be a positive number $M_1$ such that $|r(t)| \le M_1$ for all $t$, so that Eq. (6.2-2) becomes

$$|y(t)| \le M_1 \int_0^\infty |W(\tau)|\, d\tau$$

If the weighting function is absolutely integrable, there must exist a positive number $M_2$ such that

$$\int_0^\infty |W(\tau)|\, d\tau \le M_2$$

Therefore the magnitude of the output is bounded by $M_1 M_2$, or

$$|y(t)| \le M_1 M_2$$

which is the desired result since $|y(t)|$ is thus bounded.

So far, we have shown that, if the second definition of stability is satisfied, the first one is also satisfied. In order to establish the complete equivalence of these two definitions, we must now show that the second definition is satisfied if the first is. That is, we must show that, if every bounded input yields a bounded output, the weighting function is absolutely integrable.

An alternative but entirely equivalent approach is to show that, if the second definition is not satisfied, the first also fails. In other words, we must prove that if the weighting function is *not* absolutely integrable then there exists at least one bounded input that yields an unbounded output. This is the approach that is employed.

We consider once again Eq. (6.2-1); let us construct an input whose value at each instant of time is given by the relation

$$r(t - \tau) = +1 \quad \text{if } W(\tau) \ge 0$$
$$= -1 \quad \text{if } W(\tau) < 0$$

This input is bounded by $\pm 1$, and for this $r(t - \tau)$ the product $W(\tau)r(t - \tau)$ becomes

$$W(\tau)r(t - \tau) = |W(\tau)|\mu(t - \tau)$$

where $\mu(t - \tau)$ implies a unit (step) input for all $t > \tau$. If this change is made in the integrand of Eq. (6.2-1), that equation becomes

$$y(t) = \int_0^\infty |W(\tau)|\mu(t - \tau)\, d\tau$$

Since the input (a unit step) exists only for $t > \tau$, the range of $\tau$ in the upper limit of integration may be made $t$, or

$$y(t) = \int_0^\infty |W(\tau)|\mu(t - \tau)\, d\tau = \int_0^t |W(\tau)|\, d\tau \qquad (6.2\text{-}3)$$

Now let us make use of the assumption that the magnitude of $W(t)$ [or $W(\tau)$] is not absolutely integrable. If the weighting function is not absolutely integrable, it must be possible to make the right-hand side of Eq. (6.2-3) larger than any given number by selecting $t$ appropriately. Since $y(t)$ is equal to the right-hand side of Eq. (6.2-3), then of course $y(t)$ also becomes unbounded. Therefore, although the input is bounded, the assumption that the weighting function is not absolutely integrable allows the output to become unbounded, thereby establishing the complete dependence of the first two definitions.

It is interesting to compare the second definition with our previous requirement that the impulse response or weighting function of the system must decay to zero as time increases to infinity. Obviously the weighting function must decay to zero if it is to be absolutely integrable over the infinite time range. However, simple decay to zero is not sufficient to ensure that the weighting function is absolutely integrable. Consider, for example, the function $W(t) = 1/(t + 1)$. Although this function approaches zero as $t$ approaches infinity, it is not absolutely integrable. Hence it appears that the second definition is actually more restrictive than our previous requirement. However, for the class of linear systems in which we are interested, whose transfer functions are ratios of polynomials in $s$, this distinction becomes nonexistent. This is shown as part of the next development, which demonstrates the equivalence of the second and third definitions of stability.

Let us begin by showing that the condition of the third definition is met if the second definition is satisfied. In other words, we wish to demonstrate that all the poles of the system transfer function are in the left-half plane if the weighting function is absolutely integrable over the infinite time range. In order to do this, we make use of the Laplace transform definition to write $W(s)$ as

$$W(s) = \int_0^\infty W(t)e^{-st}\, dt \qquad (6.2\text{-}4)$$

Taking the magnitude of the complex functions on both sides of Eq. (6.2-4), we obtain

$$|W(s)| = \left| \int_0^\infty W(t)e^{-st}\, dt \right| \le \int_0^\infty |W(t)|\, |e^{-st}|\, dt$$

For all values of $s$ outside the left-half plane, that is, $\mathcal{R}_e(s) \ge 0$, the magnitude of the function $e^{-st}$ is less than or equal to unity so that

$$|W(s)| \le \int_0^\infty |W(t)|\, dt \qquad \text{for } \mathcal{R}_e(s) \ge 0 \tag{6.2-5}$$

Since the weighting function is absolutely integrable, there exists a positive number $M$ such that

$$\int_0^\infty |W(t)|\, dt \le M$$

Equation (6.2-5) therefore becomes

$$|W(s)| \le M \qquad \text{for } \mathcal{R}_e(s) \ge 0 \tag{6.2-6}$$

indicating that $W(s)$ is bounded for all values of $s$ outside the left-hand plane. If this is true, all poles of $W(s)$ must lie in the left-hand plane. If a pole of $W(s)$ were to lie on the $j\omega$ axis or in the right-half $s$ plane, then there is a value of $s$ with a real part greater than or equal to zero for which the denominator of $W(s)$ is zero. This would mean $W(s)$ would not be bounded, a contradiction of Eq. (6.2-6). Hence the poles of $W(s)$ must lie in the left-half $s$ plane, which is the desired result.

Next we must show that the conditions of the second definition are satisfied if the third definition holds; that is, the weighting function is absolutely integrable if all the poles of $W(s)$ are in the left-half plane. In order to establish this property, we need only to recall that the weighting function is composed of a finite sum of terms whose general form is

$$W_i(t) = \alpha t^r e^{\lambda_i t} \tag{6.2-7}$$

Here $\lambda_i$ is a pole of $W(s)$ which is repeated $r + 1$ times. It is easy to show that functions of the form described by Eq. (6.2-7) are always absolutely integrable over the infinite time range if the $\mathcal{R}_e(\lambda_i) < 0$, if the poles of $W(s)$ are in the left-half plane. This is true because the exponential decreases faster than $t^r$ for large values of $t$. Therefore we have established that the weighting function $W(t)$ is absolutely integrable if the poles of the transfer function $W(s)$ are all in the left-half plane.

In the light of the above discussion, the requirement that the weighting function decay to zero is seen to be equivalent to the requirement of absolute integrability. Because of the exponential nature of each term, the rate of decay to zero is sufficiently rapid so that the function is always absolutely integrable. A possible fourth definition of stability is therefore:

4. A linear system is stable if its weighting function decays to zero as time approaches infinity, that is,

$$\lim_{t \to \infty} W(t) = 0$$

The equivalence of this definition to the third definition assumes that $W(s)$ is a ratio of polynomials, and so the terms that make up $W(t)$ are of the form of Eq. (6.2-7).

Of the four definitions of stability given above, all but the third definition deal with the time-domain behavior of the system or its weighting function. The third definition, on the other hand, deals with the location of the closed-loop poles, a frequency-domain feature of the system. It is for exactly this reason that the third definition is so useful in our study of closed-loop-system stability. In fact, our entire study of stability is based on an examination of the closed-loop-pole locations.

Since the eigenvalues of the matrix $\mathbf{A}_k$ are identical to the closed-loop poles, the third definition can be equivalently phrased in terms of the state-variable representation:

3a. A linear system is stable if all the eigenvalues of the matrix $\mathbf{A}_k$ lie in the left half of the $s$ plane.

Although the transfer-function approach is, in general, more convenient, there are situations in which the state-variable approach is preferable. The latter situation occurs, for example, in cases where the plant is initially described in state-variable form and the plant transfer function is difficult to determine because of extensive interrelations of the state variables. In order to retain flexibility, both the transfer function and state-variable representations will be discussed in the following sections.

In discussing the topic of stability, the question of the meaning of instability cannot be avoided. A system is unstable if it is not stable; hence a system is unstable if any of the equivalent stability definitions are not satisfied. Specifically, a system is unstable if its output does not remain bounded for a bounded input, *or* if the weighting function is not absolutely integrable, *or* if the poles of the closed-loop transfer function lie on the $j\omega$ axis or in the right-half $s$ plane, *or* if the eigenvalues of $\mathbf{A}_k$ have a zero or positive real part.

Some authors choose to make a distinction between systems that have poles on the $j\omega$ axis but not in the right-half $s$ plane, as compared with systems that have poles with the real part of *s* greater than zero. If the poles of the system are in the right-half plane, the output becomes unbounded, regardless of the nature of the input.   On the other hand, if the poles are on the *jω* axis, the output becomes unbounded only for special inputs.   If the system in question contains an integration, a pole at the origin of the *s* plane, the special input that causes the output to become unbounded is just a constant input.   For example, a dc motor is a physical device which contains one integration.   For a constant-voltage input the output angle *θ* continues to increase without bound.   Of course, this is the type of behavior normally associated with a dc motor; hence one does not intuitively feel that such a device should be classified as unstable.   For this reason systems with poles on the *jω* axis are often classified as being *marginally stable*.   "Marginally unstable" may actually be a better description.

In this book we make no distinction between systems that have poles only on the *jω* axis and those that have poles in the right-half *s* plane.   All are classified as unstable, and all violate *any* of the equivalent stability definitions given at the beginning of this section.

## 6.3   The Routh-Hurwitz criterion

One obvious and direct method for determining the stability of a closed-loop system is to factor the denominator of the closed-loop transfer function $y(s)/r(s)$.   Once all the roots are known, it is immediately obvious whether all the poles lie in the left-half plane.   The actual factoring of the denominator of $y(s)/r(s)$, however, may be very difficult or even impossible if literal coefficients are present.[1]   In addition, this complete factoring procedure provides far more information than is actually needed to answer the stability question.

We do not need to know the exact location of each pole, which we obtain by factoring, but only whether each pole is in the left-half plane. In fact, it is only necessary to know if there are *any* poles not in the left-half plane.

Fortunately, there are procedures that enable one to obtain this information with much less work than complete factoring would require. As an added bonus, these procedures are also useful in terms of system design.   The remainder of this chapter presents two of the more common and more useful procedures of obtaining stability information.

In this section we consider an approach known as the *Routh-Hurwitz criterion*.   This approach is based on algebraic tests on the coefficients of

[1] See Appendix C for a discussion of procedures for factoring a polynomial.

the denominator polynomial of $y(s)/r(s)$. In the following sections the Nyquist criterion will be discussed. There a graphical procedure is used, based on the frequency-response functions of Chap. 5. Because of the sharp differences in their treatment of the stability problem, these two approaches complement each other well.

Before beginning the development of the Routh-Hurwitz criterion, let us review briefly the various possible procedures for finding the denominator polynomial of $y(s)/r(s)$. We consider first the transfer-function procedures for determining the closed-loop transfer function $y(s)/r(s)$. In terms of the $H_{eq}(s)$ or $G_{eq}(s)$ configurations, shown in Fig. 6.3-1, the closed-loop transfer function is given by

$$\frac{y(s)}{r(s)} = \frac{KK_pN_p(s)}{D_k(s)} = \frac{KG_p(s)}{1 + KG_p(s)H_{eq}(s)} = \frac{G_{eq}(s)}{1 + G_{eq}(s)} \tag{6.3-1}$$

where $G_p(s)$, $H_{eq}(s)$, and $G_{eq}(s)$ are in the form of a ratio of polynomials in $s$ as

$$G_p(s) = \frac{K_pN_p(s)}{D_p(s)} \tag{6.3-2}$$

$$H_{eq}(s) = \frac{K_hN_h(s)}{N_p(s)} \tag{6.3-3}$$

and

$$G_{eq}(s) = \frac{KK_pN_p(s)}{D_{eq}(s)} \tag{6.3-4}$$

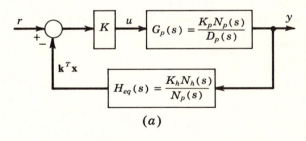

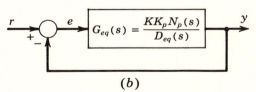

**Fig. 6.3-1**   The $H_{eq}(s)$ and $G_{eq}(s)$ configurations: (a) $H_{eq}(s)$; (b) $G_{eq}(s)$.

Here use has been made of the fact that the numerator polynomial of $y(s)/r(s)$, $G_p(s)$ and $G_{eq}(s)$, and the denominator polynomial of $H_{eq}(s)$ all contain $N_p(s)$, where $N_p(s)$ includes all the zeros, but not the gain, associated with the plant being controlled. Thus $N_p(s)$ is of the form

$$N_p(s) = (s + \delta_1)(s + \delta_2) \cdots (s + \delta_m)$$

Let us now find a number of expressions for $y(s)/r(s)$ in terms of the polynomial numerators and denominators defined in Eqs. (6.3-2) to (6.3-4). By writing $y(s)/r(s)$ in terms of $H_{eq}(s)$ we obtain

$$\frac{y(s)}{r(s)} = \frac{KK_pN_p(s)/D_p(s)}{1 + [KK_pN_p(s)/D_p(s)][K_hN_h(s)/N_p(s)]}$$
$$= \frac{KK_pN_p(s)}{D_p(s) + KK_pK_hN_h(s)} = \frac{KK_pN_p(s)}{D_k(s)}$$

Therefore the denominator of the closed-loop transfer function $y(s)/r(s)$ is equal to the denominator polynomial of the open-loop, plant transfer function $G_p(s)$, plus $KK_pK_h$ times the numerator polynomial of $H_{eq}(s)$, or

$$D_k(s) = D_p(s) + KK_pK_hN_h(s) \tag{6.3-5}$$

Using the $G_{eq}(s)$ representation, we find that the closed-loop transfer function is

$$\frac{y(s)}{r(s)} = \frac{KK_pN_p(s)/D_{eq}(s)}{1 + KK_pN_p(s)/D_{eq}(s)} = \frac{KK_pN_p(s)}{D_{eq}(s) + KK_pN_p(s)} = \frac{KK_pN_p(s)}{D_k(s)}$$

Here we find that $D_k(s)$ is equal to the sum of the denominator polynomial of $G_{eq}(s)$ and $KK_p$ times the numerator polynomial of $G_p(s)$, or

$$D_k(s) = D_{eq}(s) + KK_pN_p(s) \tag{6.3-6}$$

If we choose to make use of the state-variable representation, we have seen in Chap. 3 that $y(s)/r(s)$ may be expressed as

$$\frac{y(s)}{r(s)} = K\mathbf{c}^T(s\mathbf{I} - \mathbf{A}_k)^{-1}\mathbf{b} = K\mathbf{c}^T(s\mathbf{I} - \mathbf{A} + K\mathbf{b}\mathbf{k}^T)^{-1}\mathbf{b}$$

The denominator polynomial is therefore equal to the characteristic polynomial of the matrix $\mathbf{A}_k$, or

$$D_k(s) = \det(s\mathbf{I} - \mathbf{A}_k) = \det(s\mathbf{I} - \mathbf{A} + K\mathbf{b}\mathbf{k}^T) \tag{6.3-7}$$

The three procedures discussed above are only three of many possible ways in which the denominator of $y(s)/r(s)$ may be found. These are usually the most convenient means for finding $D_k(s)$. Either by using Eq. (6.3-5), (6.3-6), or (6.3-7) or some other method, one may find the denominator polynomial of the closed-loop transfer function in the unfactored form, as

$$D_k(s) = s^n + d_n s^{n-1} + \cdots + d_2 s + d_1 \qquad (6.3\text{-}8)$$

The problem is then to determine, without factoring $D_k(s)$, whether all its roots lie in the left-hand plane. The Routh-Hurwitz criterion provides a simple method for solving this problem. The procedure consists of an initial screening, which is referred to as the *Hurwitz test*, plus the application of the *Routh criterion*.

*Hurwitz test.* The Hurwitz test consists of a trivial examination of the coefficients of the polynomial to ensure the following:

1. All the $d_i$ coefficients are present.
2. All the $d_i$ coefficients are positive.

In order for a polynomial to have all its roots in the left-half plane, it is necessary but not sufficient that it pass the Hurwitz test. In other words, if the characteristic polynomial fails to meet either or both of these conditions, one may immediately conclude that the polynomial has at least one root that does not lie in the left-half plane, and hence the closed-loop system is unstable.

On the other hand, if the polynomial satisfies both conditions, no conclusion may be reached. Consider, for example, the following third-order polynomial

$$D_k(s) = s^3 + 0.5s^2 + 3.5s + 4$$

Although this polynomial satisfies both the above conditions, it is not difficult to factor the polynomial to find that two roots are located in the right-half plane.

The above conditions are a direct consequence of the algebraic property that

$$d_i = (-1)^{n-i+1} \Sigma(\text{product of the roots taken } n - i + 1 \text{ at a time})$$
$$i = 1, 2, \ldots, n \quad (6.3\text{-}9)$$

If all the roots are in the left-half plane, each of the coefficients must be nonzero and positive.

Because of the ease with which the above conditions may be checked, the Hurwitz test plays a valuable role in stability analysis. However, there is a large percentage of cases in which a polynomial satisfies both conditions but still has roots in the right-half plane. In these situations the Routh criterion discussed below may be utilized to determine the stability of the system.

*Routh test.* The first step in the application of the Routh test is the formation of the Routh array which takes the following form:

| | | | | |
|---|---|---|---|---|
| $s^n$ | $\delta_{01} = 1$ | $\delta_{02} = d_{n-1}$ | $\delta_{03} = d_{n-3}$ | $\cdots$ |
| $s^{n-1}$ | $\delta_{11} = d_n$ | $\delta_{12} = d_{n-2}$ | $\delta_{13} = d_{n-4}$ | $\cdots$ |
| $s^{n-2}$ | $\delta_{21}$ | $\delta_{22}$ | $\delta_{23}$ | $\cdots$ |
| $s^{n-3}$ | $\delta_{31}$ | $\delta_{32}$ | $\delta_{33}$ | $\cdots$ |
| $\cdots$ | | | | |
| $s^1$ | $\delta_{n-1,1}$ | 0 | 0 | $\cdots$ |
| $s^0$ | $\delta_{n1}$ | 0 | 0 | $\cdots$ |
| | Pivot column | | | |

where

$$\delta_{ij} = \frac{\delta_{i-1,1}\delta_{i-2,j+1} - \delta_{i-2,1}\delta_{i-1,j+1}}{\delta_{i-1,1}} \qquad i = 2, 3, \ldots, n$$

$$j = 1, 2, \ldots \qquad (6.3\text{-}10)$$

The first two rows of the array are formed by simply writing the coefficients of $D_k(s)$ in the order of decreasing powers of $s$ alternately in the first and second row. The remaining rows of the array are then consecutively formed by making use of Eq. (6.3-10). Note that the computation of these remaining elements always involves the two elements in the pivot column of the two preceding rows.

Although Eq. (6.3-10) looks somewhat complicated, once one has mastered the procedure, the elements of the array may be computed quite easily. The procedure used is similar to that employed in computing determinants, although the reader should note that the signs are reversed. Consider, for example, the following schematic form for the calculation of the $\delta_{22}$ element.

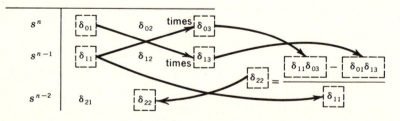

The first column of the Routh array consisting of the descending powers of $s$ is used only as a rough check on the formation of the array.

The number of nonzero elements in any row cannot be greater than one plus one-half the power of $s$ associated with that row. For example, in the $s^6$ row, there cannot be more than $1 + \frac{6}{2} = 4$ elements that are non-zero. This means that there may be only one nonzero element in the $s^1$ and $s^0$ rows.

In terms of the Routh array, the Routh criterion may be stated as follows:

> The characteristic polynomial $D_k(s)$ has roots on the $j\omega$ axis or in the right-half $s$ plane if there are any zeros or sign changes in the pivot column of the Routh array.

Of course, if the characteristic polynomial has roots on the $j\omega$ axis or in the right-half plane, then the system from which this polynomial originated is unstable.

**Example 6.3-1**   Consider, for example, the following polynomial:

$$D_k(s) = (s + 2)(s - 1)(s - 3) = s^3 - 2s^2 - 5s + 6$$

Since the polynomial does not pass the Hurwitz test, it is obvious that it must have at least one root that is not in the left-half plane. Hence there is no need to use the more elaborate Routh criterion.

**Example 6.3-2**   As a second example, let us consider the general third-order polynomial given by

$$D_k(s) = s^3 + d_3 s^2 + d_2 s + d_1$$

In order to satisfy the Hurwitz condition, it is necessary that $d_3$, $d_2$, and $d_1$ be greater than zero. On the basis of the given polynomial, the first two rows of the Routh array are written immediately, with the remaining elements indicated in terms of $\delta_{ij}$.

| | | |
|---|---|---|
| $s^3$ | $1$ | $d_2$ |
| $s^2$ | $d_3$ | $d_1$ |
| $s^1$ | $\delta_{21}$ | $0$ |
| $s^0$ | $\delta_{31}$ | $0$ |

The $\delta_{21}$ element of the array is found from Eq. (6.3-10) to be

$$\delta_{21} = \frac{(\delta_{11})(\delta_{02}) - (\delta_{01})(\delta_{12})}{\delta_{11}} = \frac{d_3 d_2 - 1(d_1)}{d_3}$$

and

$$\delta_{31} = \frac{\delta_{21}d_1}{\delta_{21}} = d_1$$

so that the complete array is finally

| $s^3$ | 1 | $d_2$ |
|---|---|---|
| $s^2$ | $d_3$ | $d_1$ |
| $s^1$ | $(d_3d_2 - d_1)/d_3$ | |
| $s^0$ | $d_1$ | |

There are no roots of the given polynomial $D_k(s)$ in the right-half $s$ plane or on the $j\omega$ axis as long as the elements of the pivot column do not change sign. This is true if

$$d_3 > 0$$
$$d_3d_2 - d_1 > 0 \qquad\qquad (6.3\text{-}11)$$
$$d_1 > 0$$

Note that from the Routh criterion alone it does not appear, at first glance, that the requirement $d_2 > 0$ is necessary. However, if $d_3$ and $d_1$ are both greater than zero, then $d_2$ must also be greater than zero if

$$d_3d_2 - d_1 > 0$$

Hence the Routh and Hurwitz criteria are in complete agreement.

For the general second-order polynomial, it is easy to show that the Hurwitz condition is both necessary and sufficient. Therefore a second-order polynomial has both its roots in the left-half $s$ plane if all the coefficients are nonzero and of the same sign.

**Example 6.3-3**  Let us consider the development of the Routh array for the polynomial

$$D_k(s) = s^4 + 10s^3 + s^2 + 15s + 3$$

The Routh array is

| $s^4$ | 1 | 1 | 3 |
|---|---|---|---|
| $s^3$ | 10 | 15 | |
| $s^2$ | $\delta_{21}$ | $\delta_{22}$ | |
| $s^1$ | $\delta_{31}$ | | |
| $s^0$ | $\delta_{41}$ | | |

Here $\delta_{21}$ is

$$\delta_{21} = \frac{(10)(1) - 1(15)}{10} = -0.5$$

Since $\delta_{21}$ has a negative sign, whereas $\delta_{11} = 10$ is positive, a sign change has occurred in the pivot column, and there is no need to proceed. The system for which $D_k(s)$ of this example is the characteristic polynomial is unstable.

In the formation of the Routh array, the following property often proves to be convenient:

Any row of the Routh array may be multiplied or divided by any positive constant without changing the sign of any element in the array.

The use of this property can simplify the development of the array by making the numbers more convenient. One approach is to reduce the numbers in every row by a common factor, if one exists. If this is done for the array of Example 6.3-3, the second row is divided by 5. That array then becomes

| $s^4$ | 1 | 1 | 3 |
|---|---|---|---|
| $s^3$ | 2 | 3 | |
| $s^2$ | $\delta_{21}$ | $\delta_{22}$ | |
| $s^1$ | $\delta_{31}$ | | |
| $s^0$ | $\delta_{41}$ | | |

Now the $\delta_{ij}$'s have values that are different from the corresponding elements in Example 6.3-3. This is unimportant since only the sign of the elements in the pivot column indicates stability or instability. The element $\delta_{21}$ is now

$$\delta_{21} = \frac{2 - 3}{2} = -0.5$$

and $\delta_{21}$ is negative, as before.

Another approach that makes use of the property that any row may be divided by a positive constant is to force all the numbers in the pivot column to have unit magnitude by dividing every element of the row by the magnitude of the number in the pivot column. If this approach is taken in this same example, the second row is divided by 10, with the result that

| $s^4$ | 1 | 1 | 3 |
|---|---|---|---|
| $s^3$ | 1 | $\frac{3}{2}$ | |
| $s^2$ | $\delta_{21}$ | $\delta_{22}$ | |
| $s^1$ | $\delta_{31}$ | | |
| $s^0$ | $\delta_{41}$ | | |

and $\delta_{21}$ is now just

$$\delta_{21} = 1 - \tfrac{3}{2} = -0.5$$

which is still negative.

**Example 6.3-4**   Consider, for example, the formation of the Routh array for the polynomial

$$D_k(s) = s^5 + s^4 + 2s^3 + 2s^2 + 3s + 15$$

The partial Routh array is

| $s^5$ | 1 | 2 | 3 |
|---|---|---|---|
| $s^4$ | 1 | 2 | 15 |
| $s^3$ | 0 | $\delta_{22}$ | |

Again there is no need to proceed further, as $\delta_{21}$ is zero, and, according to the Routh criterion, either a zero or a sign change in the pivot column is sufficient to indicate that all the roots are not in the left-half plane.

On the basis of Examples 6.3-3 and 6.3-4, it may appear that $\delta_{21}$ always changes sign if the poles are outside the left-half plane.  This is often, but not always, the case, as illustrated by the next example.

**Example 6.3-5**   The characteristic polynomial to be tested in this example is

$$D_k(s) = s^4 + 2s^3 + 11s^2 + 18s + 18$$

The Routh array for this polynomial is

| $s^4$ | 1 | 11 | 18 |
|---|---|---|---|
| $s^3$ | 2 | 18 | 0 |
| $s^2$ | 2 | 18 | |
| $s^1$ | 0 | | |
| $s^0$ | $\delta_{41}$ | | |

Since $\delta_{31}$ is zero, $D_k(s)$ has at least one root on the $j\omega$ axis or the right-half $s$ plane.

Our basic use of the Routh-Hurwitz criterion is not in those cases for which $D_k(s)$ is completely specified, but rather in the stability analysis of closed-loop systems. The following two examples illustrate various applications of the Routh-Hurwitz criterion to the study of system stability in which parameter adjustments are available.

***Example 6.3-6*** As our first example of this type, let us consider the closed-loop system shown in Fig. 6.3-2. The closed-loop transfer function for this system is

$$\frac{y(s)}{r(s)} = \frac{K}{s(s+2)(s+3)+K} = \frac{K}{s^3+5s^2+6s+K}$$

so that $D_k(s)$ is

$$D_k(s) = s^3 + 5s^2 + 6s + K$$

The Routh array for this polynomial is

| $s^3$ | 1 | 6 |
|---|---|---|
| $s^2$ | 5 | $K$ |
| $s^1$ | $(30-K)/5$ | |
| $s^0$ | $K$ | |

In order that there be no sign changes in the pivot column, it is necessary that

$$K > 0 \qquad \text{and} \qquad 30 - K > 0$$

or

$$0 < K < 30$$

Therefore the system shown in Fig. 6.3-2 is stable if the above condition on $K$ is satisfied. Note that the above result could have been obtained directly by making use of the Routh-Hurwitz conditions given in Example 6.3-2, namely, Eqs. (6.3-11).

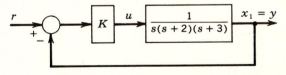

**Fig. 6.3-2** Closed-loop system for Example 6.3-6.

In the situation where a parameter, such as the gain $K$ above, has a maximum value beyond which instability occurs, it is possible to define a stability margin for that parameter. In effect, a stability margin describes the amount by which a given value of the parameter may be increased in order just to achieve marginal stability. In terms of a mathematical expression, this becomes

$$\text{Stability margin} = \frac{\text{maximum stable value}}{\text{actual value}} \qquad (6.3\text{-}12)$$

In Example 6.3-6, we may speak of the stability margin of the gain $K$, or more commonly just the *gain margin*, GM, which becomes

$$\text{GM} = \frac{\text{maximum stable gain}}{\text{actual gain}} = \frac{30}{K}$$

If the gain $K = 15$, then GM $= 2$, indicating that the gain may be increased by a factor of 2 before instability occurs.

The gain margin is one of the simplest relative-stability indices or figures of merit for a closed-loop system. Unfortunately, it is also one of the least useful. First, the concept of a gain margin does not apply to all systems, since in some systems there is no maximum stable gain, for example. In addition, even in the case of systems to which it applies, the gain margin does not give uniformly reliable information. For example, two systems that have the same gain margin may have entirely different behaviors.

> **Example 6.3-7**  As a second example of stability analysis using the Routh-Hurwitz tests, we consider the system shown in Fig. 6.3-3a. The basic plant is the same as that of Example 6.3-6 except that here state variables have been identified and all the state variables are fed back, rather than just $x_1$. The effect of the state-variable feedback on system stability is interesting. The $H_{eq}(s)$ form for this system is shown in Fig. 6.3-3b. Using Eq. (6.3-5), we find that $D_k(s)$ is given by
>
> $$D_k(s) = s(s+2)(s+3) + \frac{2K}{7}(s^2 + 3s + \tfrac{7}{2})$$
>
> $$= s^3 + s^2\left(5 + \frac{2K}{7}\right) + s\left(6 + \frac{6K}{7}\right) + K$$
>
> The Hurwitz conditions require that $K$, $6 + 6K/7$, and $5 + 2K/7$ all be greater than zero. The most restrictive of these is that $K$

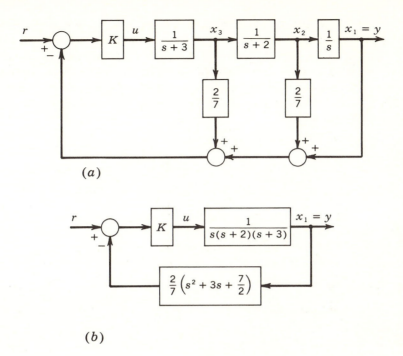

**Fig. 6.3-3** Example 6.3-7. (a) State-variable feedback; (b) $H_{eq}(s)$ form.

itself be greater than zero. The Routh array is

| $s^3$ | $1$ | $6 + 6K/7$ |
|---|---|---|
| $s^2$ | $5 + 2K/7$ | $K$ |
| $s^1$ | $\delta_{21}$ | |
| $s^0$ | $\delta_{31}$ | |

where $\delta_{21}$ is

$$\delta_{21} = \frac{(5 + 2K/7)(6 + 6K/7) - K}{5 + 2K/7} = \frac{30 + 5K + 12K^2/49}{5 + 2K/7}$$

and $\delta_{31}$ is $K$.

The last element of the pivot column, $\delta_{31}$, is positive if and only if $K$ is greater than 0. The $\delta_{11}$ term $5 + 2K/7$ is also positive if $K > 0$, as is $\delta_{21}$. Thus in this example the Hurwitz conditions proved to be both necessary and sufficient, although this must be checked in every case.

Here it has been shown that the system of Fig. 6.3-3 is stable for all positive gain $K$.   Hence the use of state-variable feedback has caused the system to be stable for all positive values of gain.

As an additional illustration of the advantages of state-variable feedback, let us suppose that the pole located at $s = -3$ in the above system is at some arbitrary point on the negative real axis $s = -\lambda$, $\lambda > 0$.   In this case, the characteristic polynomial becomes

$$D_k(s) = s^3 + (2 + \lambda + \tfrac{2}{7}K)s^2 + (2\lambda + \tfrac{6}{7}K)s + K$$

Once again the Routh-Hurwitz conditions are satisfied as long as $K > 0$ and $\lambda > 0$.

The preceding examples illustrate one of the advantages of feeding back all state variables, as compared with just the output.   The results obtained are typical even though only one specific example has been treated here.   The validity of this statement will be established in the next section by means of the Nyquist criterion.

Before proceeding to the Nyquist diagram, it should be noted that the form of the Routh criterion given in this section is not the most general form.   Means are available for determining how many poles are in the right-half plane and how many are on the $j\omega$ axis.   The number of roots of $D_k(s)$ with positive real parts is equal to the number of sign changes of the coefficients of the pivot column of the Routh array.   The number of poles on the $j\omega$ axis may be determined by a slightly more complicated procedure from the rows having all zero elements.   The point of view adopted here is that, from the standpoint of stability, it is unimportant how many poles are outside the left-half $s$ plane.   A much more important question than how "unstable" is a system is the question of relative stability.   Both questions of stability and relative stability are adequately handled by the Nyquist diagram.

**Exercises 6.3**   *6.3-1.*   Find the characteristic polynomial $D_k(s)$ for the system of Example 6.3-7 by the use of Eqs. (6.3-6) and (6.3-7).

*6.3-2.*   Each of the polynomials shown below represents the characteristic polynomial of some closed-loop system.   Determine as much of the Routh array as necessary to make a definite statement concerning the stability or instability of the system.

(a)    $s^4 + 5s^3 + 13s^2 + 19s + 10$
(b)    $s^4 + 2s^3 + 4s^2 - 2s - 5$
(c)    $s^5 + 4s^4 + 7s^3 + 8s^2 + 6s + 4$
(d)    $s^4 + 2s^3 + s + 2$
(e)    $s^4 + s^3 - s^2 + s - 2$
(f)    $s^5 - 9s^3 - 22s^2 - 22s - 8$

*answers:*

| | | |
|---|---|---|
| (a)  Stable | (b)  Unstable | (c)  Unstable |
| (d)  Unstable | (e)  Unstable | (f)  Unstable |

*6.3-3.*    Find the values of $K$ and $T$ for which the system shown in Fig. 6.3-4 is stable.

*answer:*

$$K > 0, \, 2K > T > 0$$

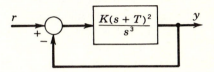

**Fig. 6.3-4**    Exercise 6.3-3.

## 6.4   *The Nyquist criterion*

The Routh-Hurwitz procedure discussed in the preceding section is ana-lytic in nature and deals directly with the closed-loop characteristic poly-nomial $D_k(s)$.    The Nyquist criterion presented here, on the other hand, is essentially a graphical method and deals with the open-loop transfer function $KG_p(s)H_{eq}(s)$.    The graphical character of the Nyquist criterion is probably one of its most appealing features.

If we write the closed-loop transfer function in the form

$$\frac{y(s)}{r(s)} = \frac{KG_p(s)}{1 + KG_p(s)H_{eq}(s)} = \frac{KK_pN_p(s)}{D_k(s)}$$

we see that the closed-loop poles are equal to the zeros of the function

$$1 + KG_p(s)H_{eq}(s) = 1 + \frac{KK_pN_p(s)K_hN_h(s)}{D_p(s)N_p(s)} = \frac{D_p(s) + KK_pK_hN_h(s)}{D_p(s)}$$

$$(6.4\text{-}1)$$

Of course, the numerator of Eq. (6.4-1) is just $D_k(s)$, from Eq. (6.3-5), so that

$$1 + KG_p(s)H_{eq}(s) = \frac{D_k(s)}{D_p(s)} \qquad\qquad (6.4\text{-}2)$$

In other words, we can determine the stability of the closed-loop system by locating the zeros of $1 + KG_p(s)H_{eq}(s)$.    This is not a new result, but the reader's attention is drawn to this fact, since it is of prime importance in the following development.

For the moment, let us assume that $1 + KG_p(s)H_{eq}(s)$ is known in factored form so that we have

$$1 + KG_p(s)H_{eq}(s) = \frac{(s + \lambda_{k1})(s + \lambda_{k2}) \cdots (s + \lambda_{kn})}{(s + \lambda_1)(s + \lambda_2) \cdots (s + \lambda_n)} \qquad (6.4\text{-}3)$$

Obviously, if $1 + KG_p(s)H_{eq}(s)$ were known in factored form, there would be no need for the use of the Nyquist criterion, since we could simply observe whether any of the zeros of $1 + KG_p(s)H_{eq}(s)$, poles of $y(s)/r(s)$, lie in the right half of the $s$ plane. In fact, the primary reason for using either the Routh-Hurwitz or Nyquist criterion is to avoid this factoring. Although it is convenient to think of $1 + KG_p(s)H_{eq}(s)$ in factored form at this time, no actual use is made of that form.

Let us suppose that the pole-zero plot of $1 + KG_p(s)H_{eq}(s)$ takes the form shown in Fig. 6.4-1a. Consider next an arbitrary closed contour, such as $\Gamma$ in Fig. 6.4-1a, which encloses one and only one zero of $1 + KG_p(s)H_{eq}(s)$ and none of the poles. Associated with each point on this contour is a value of the complex function $1 + KG_p(s)H_{eq}(s)$. The value of $1 + KG_p(s)H_{eq}(s)$ for any value of $s$ on $\Gamma$ may be found analytically by substituting the appropriate complex value of $s$ into the function. Alternatively the value may be found graphically by multiplying the distances from $s$ on $\Gamma$ to the zeros and dividing by the distances to the poles, as was done in Chap. 4.

If the complex value of $1 + KG_p(s)H_{eq}(s)$ associated with every point on the contour $\Gamma$ is plotted, another closed contour $\Gamma'$ is created in the complex $1 + KG_p(s)H_{eq}(s)$ plane, as shown in Fig. 6.4-1b. The function $1 + KG_p(s)H_{eq}(s)$ is said to map the contour $\Gamma$ in the $s$ plane into the $\Gamma'$ contour in the $1 + KG_p(s)H_{eq}(s)$ plane. What we wish to demonstrate is that, if a zero is enclosed by the contour $\Gamma$, as in Fig. 6.4-1a, the contour $\Gamma'$ encircles the origin of the $1 + KG_p(s)H_{eq}(s)$ plane in the same sense that the $\Gamma$ contour encircles the zero in the $s$ plane. In the $s$ plane the zero is encircled in the clockwise direction; hence we must show that the origin of the $1 + KG_p(s)H_{eq}(s)$ plane is also encircled in the clockwise direction.

The key to the argument rests in considering the value of the function $1 + KG_p(s)H_{eq}(s)$ at any point $s$ as simply a complex number. This complex number has a magnitude and a phase angle. Since the contour $\Gamma$ in the $s$ plane does not pass through a zero, the magnitude is never zero. Now we consider the phase angle by rewriting Eq. (6.4-3) in polar form:

$$1 + KG_p(s)H_{eq}(s) = \frac{|s + \lambda_{k1}|/\!\underline{\arg\ (s + \lambda_{k1})} \cdots |s + \lambda_{kn}|/\!\underline{\arg\ (s + \lambda_{kn})}}{|s + \lambda_1|/\!\underline{\arg\ (s + \lambda_1)} \cdots |s + \lambda_n|/\!\underline{\arg\ (s + \lambda_n)}}$$

$$= \frac{|s + \lambda_{k1}| \cdots |s + \lambda_{kn}|}{|s + \lambda_1| \cdots |s + \lambda_n|} /\!\underline{\arg\ (s + \lambda_{k1}) + \cdots}$$

$$\underline{+\ \arg\ (s + \lambda_{kn}) - \arg\ (s + \lambda_1) - \cdots - \arg\ (s + \lambda_n)} \qquad (6.4\text{-}4)$$

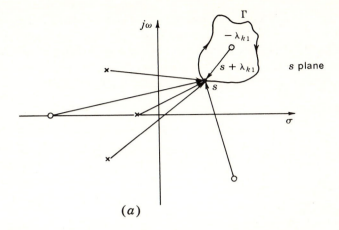

(a)

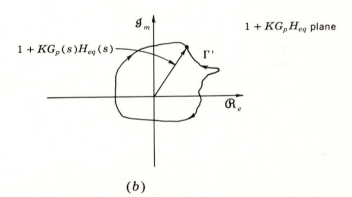

(b)

**Fig. 6.4-1**    (a) Pole-zero plot of $1 + KG_p(s)H_{eq}(s)$ in the $s$
plane; (b) plot of the $\Gamma'$ contour in the $1 +$
$KG_p(s)H_{eq}(s)$ plane.

We assume that the zero encircled by $\Gamma$ is at $s = -\lambda_{k1}$. Then the phase
angle associated with this zero changes by a full $-360°$ as the contour $\Gamma$
is traversed clockwise in the $s$ plane. Since the argument or angle
of $1 + KG_p(s)H_{eq}(s)$ includes the angle of this zero, the argument of
$1 + KG_p(s)H_{eq}(s)$ also changes by $-360°$. As seen from Fig. 6.4-1a, the
angles associated with the remaining poles and zeros make no net change
as the contour $\Gamma$ is traversed. For any fixed value of $s$, the vector associated with each of these other poles and zeros has a particular angle
associated with it. Once the contour has been traversed back to the
starting point, these angles return to their original value; they have not
been altered by plus or minus $360°$ simply because these poles and zeros
are not enclosed by $\Gamma$.

In a similar fashion, we could show that, if $\Gamma$ encircled two zeros of $1 + KG_p(s)H_{eq}(s)$ in the clockwise direction on the $s$ plane, the $\Gamma'$ contour would encircle the origin of the $1 + KG_p(s)H_{eq}(s)$ plane twice in the clockwise direction. On the other hand, if $\Gamma$ encircles only one pole and no zeros of $1 + KG_p(s)H_{eq}(s)$ in the *clockwise* direction, then the mapping of this contour $\Gamma'$ would encircle the origin of the $1 + KG_p(s)H_{eq}(s)$ plane in the *counterclockwise* direction. This change in direction comes about because angles associated with poles are accompanied by negative signs in evaluating $1 + KG_p(s)H_{eq}(s)$, as indicated by Eq. (6.4-4). In general, the following conclusion can be drawn. *The net number of clockwise encirclements by $\Gamma'$ of the origin in the $1 + KG_pH_{eq}$ plane is equal to the difference between the number of zeros $n_z$ and the number of poles $n_p$ of $1 + KG_p(s)H_{eq}(s)$ encircled in the clockwise direction by $\Gamma$.*

The above result means that the difference between the number of zeros and the number of poles enclosed by *any* contour $\Gamma$ may be determined simply by counting the net number of clockwise encirclements of the origin of the $1 + KG_p(s)H_{eq}(s)$ plane by $\Gamma'$. For example, if we find that $\Gamma'$ encircles the origin three times in the clockwise direction and once in the counterclockwise direction, then $n_z - n_p$ must be equal to $3 - 1 = 2$. Therefore in the $s$ plane $\Gamma$ must encircle two zeros and no poles, three zeros and one pole, or any other combination such that $n_z - n_p$ is equal to 2.

In terms of stability analysis, the problem is to determine the number of zeros of $1 + KG_p(s)H_{eq}(s)$, the poles of $y(s)/r(s)$, which lie in the right half of the $s$ plane. Therefore $\Gamma$ must be selected so that it encloses the entire right half of the $s$ plane. Accordingly, the contour $\Gamma$ is chosen as the entire $j\omega$ axis and an infinite semicircle enclosing the right-half plane as shown in Fig. 6.4-2a. This contour is known as the *Nyquist contour.*

In order to avoid any problems in plotting the values of $1 + KG_p(s)H_{eq}(s)$ along the infinite semicircle, let us assume that

$$\lim_{|s| \to \infty} KG_p(s)H_{eq}(s) = 0$$

This assumption is always justified since in general the open-loop transfer function $KG_p(s)H_{eq}(s)$ has $n - 1$ zeros and $n$ poles. With this assumption, the infinite semicircle portion of $\Gamma$ maps into the point $s = +1 + j0$ on the $1 + KG_pH_{eq}$ plane.

The mapping of $\Gamma$ therefore involves simply plotting the complex values of $1 + KG_p(s)H_{eq}(s)$ for $s = j\omega$ as $\omega$ varies from $-\infty$ to $+\infty$. For $\omega \geq 0$, $\Gamma'$ is nothing more than the polar plot of the frequency response of the function $1 + KG_p(s)H_{eq}(s)$ as discussed in Chap. 5. The

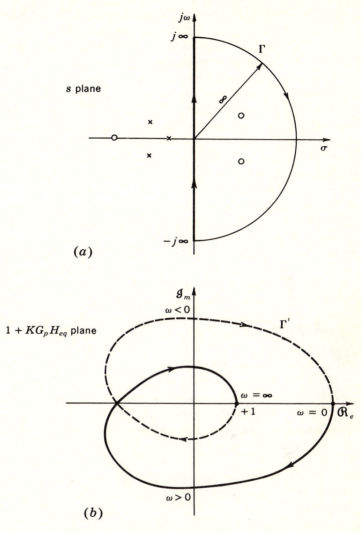

**Fig. 6.4-2**   (a) $\Gamma$ contour in the $s$ plane—the Nyquist contour; (b) $\Gamma'$ contour in the $1 + KG_p H_{eq}$ plane.

values of $1 + KG_p(j\omega)H_{eq}(j\omega)$ for negative values of $\omega$ are the mirror image of the values for $\omega > 0$ reflected about the real axis. The $\Gamma'$ contour may therefore be found by plotting the frequency response of $1 + KG_p(s)H_{eq}(s)$ for positive $\omega$ and then reflecting this plot about the real axis to find the plot for $\omega < 0$. The $\Gamma'$ plot is always symmetrical about the real axis of the $1 + KG_p H_{eq}$ plane.

From the $\Gamma'$ contour in the $1 + KG_p(s)H_{eq}(s)$ plane, as shown in Fig. 6.4-2$b$, the number of zeros of $1 + KG_p(s)H_{eq}(s)$ in the right half of the $s$ plane may be determined by the following procedure. The net number of clockwise encirclements of the origin by $\Gamma'$ is equal to the number of zeros minus the number of poles of $1 + KG_p(s)H_{eq}(s)$ in the right half of the $s$ plane. Note that we must know the number of poles of $1 + KG_p(s)H_{eq}(s)$ in the right-half plane if we are to be able to ascertain the exact number of zeros in the right-half plane and therefore whether the system is stable or not. This requirement usually poses no problem since the poles of $1 + KG_p(s)H_{eq}(s)$ correspond to the poles of the plant. In Eq. (6.4-2) the denominator of $1 + KG_p(s)H_{eq}(s)$ is just $D_p(s)$, which is usually described in factored form. Hence the number of zeros of $1 + KG_p(s)H_{eq}(s)$ or the number of poles of $y(s)/r(s)$ in the right-half plane may be found by determining the net number of clockwise encirclements of the origin of the $1 + KG_p(s)H_{eq}(s)$ plane and then adding the number of poles of the plant located in the right-half $s$ plane.

At this point the reader may revolt. Our plan for finding the number of poles of $y(s)/r(s)$ in the right-half $s$ plane involves counting encirclements in the $1 + KG_p(s)H_{eq}(s)$ plane and observing the number of plant poles in the right-half $s$ plane. Yet we were forced to start with the assumption that all the poles and zeros of $1 + KG_p(s)H_{eq}(s)$ are known, so that the Nyquist contour can be mapped by the function $1 + KG_p(s)H_{eq}(s)$. Admittedly we know the poles of this function, because they are the poles of the plant, but we do not know the zeros; in fact, we are simply trying to find how many of these zeros lie in the right-half $s$ plane.

What we do know is the poles and zeros of the open-loop transfer function $KG_p(s)H_{eq}(s)$. Of course, this function differs from $1 + KG_p(s)H_{eq}(s)$ only by unity. Any contour that is chosen in the $s$ plane and mapped through the function $KG_p(s)H_{eq}(s)$ has exactly the same shape as if the contour were mapped through the function $1 + KG_p(s)H_{eq}(s)$ except that it is displaced by one unit. Figure 6.4-3 is typical of such a situation. In this diagram the $-1$ point of the $KG_p(s)H_{eq}(s)$ plane is the origin of the $1 + KG_p(s)H_{eq}(s)$ plane. If we now map the boundary of the right-half $s$ plane through the mapping function $KG_p(s)H_{eq}(s)$, which we often know in pole-zero form, information concerning the zeros of $1 + KG_p(s)H_{eq}(s)$ may be obtained by counting the encirclements of the $-1$ point. The important point is that, by plotting the open-loop frequency-response information, we may reach stability conclusions regarding the closed-loop system.

A polar plot of the frequency-response function $KG_p(j\omega)H_{eq}(j\omega)$ for $-\infty \leq \omega \leq \infty$ is referred to as a *Nyquist diagram* of $KG_p(s)H_{eq}(s)$. Note that the remaining boundary of the right-half $s$ plane, the infinite

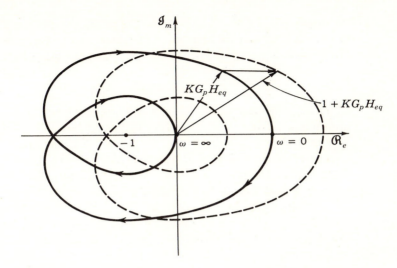

**Fig. 6.4-3**  Comparison of the $KG_p(s)H_{eq}(s)$ and $1 + KG_p(s)H_{eq}(s)$ plots.

semicircle, maps to the origin on the Nyquist diagram.   In terms of the Nyquist diagram of $KG_p(s)H_{eq}(s)$, the Nyquist stability criterion may be stated as: *The closed-loop system is stable if and only if the net number of clockwise encirclements of the point $s = -1 + j0$ by the Nyquist diagram of $KG_p(s)H_{eq}(s)$ plus the number of poles of $HG_p(s)H_{eq}(s)$ in the right-half plane is zero.*

Because the Nyquist diagram involves the open-loop transfer function $KG_p(s)H_{eq}(s)$, a good approximation of the magnitude and phase of the frequency-response plot can be obtained by using the straight-line approximation for the magnitude and the approximate Bode relation for the phase.   The Nyquist plot can then be obtained by transferring the magnitude and phase information to a polar plot.   If a more accurate plot is needed, the exact magnitude and phase may be determined for a few values of $\omega$ in the range of interest.   However, in most cases, the approximate plot is accurate enough for practical problems.

An alternative procedure for obtaining the Nyquist diagram is to plot accurately the poles and zeros of $KG_p(s)H_{eq}(s)$ and obtain the magnitude and phase by graphical means.   In either of these methods, the fact that $KG_p(s)H_{eq}(s)$ is known in factored form is important.   Even if $KG_p(s)H_{eq}(s)$ is not known in factored form, the frequency-response plot can still be obtained by simply substituting values of $s = j\omega$ into $KG_p(s)H_{eq}(s)$ or by frequency-response measurements on the actual system.

***Example 6.4-1***    To illustrate the use of the Nyquist criterion, let
us consider the simple first-order system shown in Fig. 6.4-4a.    For
this system the loop transfer function takes the following form:

$$KG_p(s)H_{eq}(s) = KG_p(s) = \frac{K}{s + 10} = \frac{50}{s + 10}$$

The magnitude and phase plots of the frequency response of $KG_p(s)$
are shown.    From these plots the Nyquist diagram for $KG_p(s)$ may
be easily plotted, as shown in Fig. 6.4-5.    For example, the point
associated with $\omega = 10$ rad/sec is found to have a magnitude of
$K/10\sqrt{2}$ and a phase angle of $-\pi/4$ rad.    The point at $\omega = -10$
rad/sec is just the mirror image of the values at $\omega = 10$ rad/sec.

From Fig. 6.4-5 we see that the Nyquist diagram can never
encircle the $s = -1 + j0$ point for positive values of $K$, and therefore the closed-loop system is stable for all positive values of $K$.
In this simple example, it is easy to see that this result is correct
since the closed-loop transfer function is given by

$$\frac{y(s)}{r(s)} = \frac{K}{s + 10 + K}$$

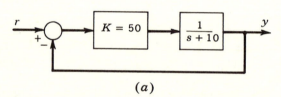

*(a)*

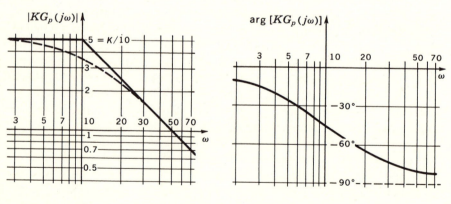

*(b)*

***Fig. 6.4-4***    Simple first-order example.    (*a*) Block diagram; (*b*) magnitude and phase
plots.

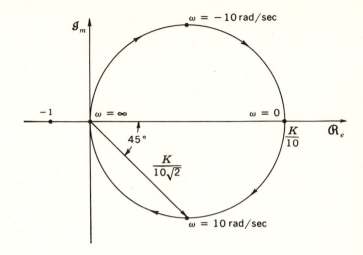

**Fig. 6.4-5**   Nyquist diagram for Example 6.4-1.

For all positive values of $K$, the pole of $y(s)/r(s)$ is in the left-half plane.

In the above example, $KG_p(s)H_{eq}(s)$ remains finite along the entire Nyquist contour. This is not always the case even though we have assumed that $KG_p(s)H_{eq}(s)$ approaches zero as $|s|$ approaches infinity. If a pole of $KG_p(s)H_{eq}(s)$ occurs on the $j\omega$ axis, as often happens at the origin because of an integration in the plant, a slight modification of the Nyquist contour is necessary. The method of handling the modification is illustrated in the following example.

**Example 6.4-2**   We consider again the state-variable-feedback system of Example 6.3-7. The loop transfer function for that system is given by

$$KG_p(s)H_{eq}(s) = \frac{(2K/7)[(s + \tfrac{3}{2})^2 + (\sqrt{5}/2)^2]}{s(s + 2)(s + 3)}$$

The pole-zero plot of $KG_p(s)H_{eq}(s)$ is shown in Fig. 6.4-6$a$. Since a pole occurs on the standard Nyquist contour at the origin, it is not clear how this problem should be handled. As a beginning, let us plot the Nyquist diagram for $\omega = +\epsilon$ to $\omega = -\epsilon$, including the infinite semicircle; when this is done, the small area around the origin is avoided. The resulting plot is shown as the solid line in Fig. 6.4-6$b$ with corresponding points labeled.

From Fig. 6.4-6$b$ we cannot determine whether the system is

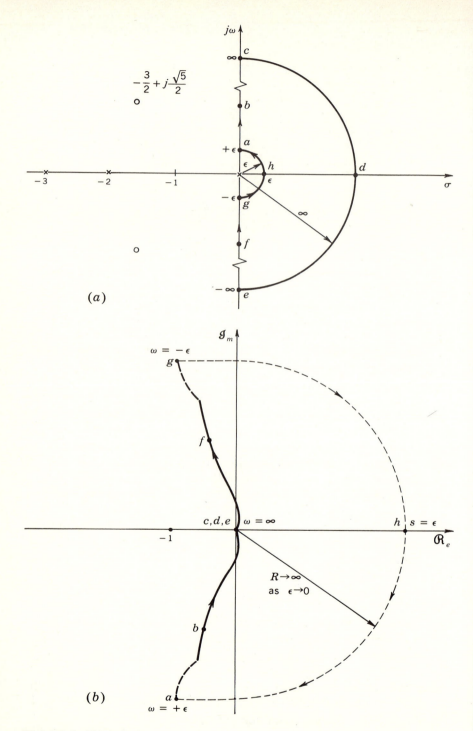

**Fig. 6.4-6** Example 6.4-2. (a) Pole-zero plot; (b) Nyquist diagram.
**274**

stable until the Nyquist diagram is completed by joining the points at $\omega = -\epsilon$ and $\omega = +\epsilon$. In order to join these points, let us use a semicircle of radius $\epsilon$ to the right of the origin, as shown in Fig. 6.4-6a. Now $KG_p(s)H_{eq}(s)$ is finite at all points on the contour in the $s$ plane, and the mapping to the $KG_pH_{eq}$ plane can be completed as shown by the dashed line in Fig. 6.4-6b. The small semicircle used to avoid the origin in the $s$ plane has mapped into a large semicircle in the $KG_pH_{eq}$ plane.

In order to ensure that no zeros of $1 + KG_p(s)H_{eq}(s)$ can escape discovery by lying in the $\epsilon$-radius semicircular indentation in the $s$ plane, $\epsilon$ is made arbitrarily small, with the result that the radius of the large semicircle in the $KG_pH_{eq}$ plane approaches infinity. As $\epsilon \to 0$, the shape of the Nyquist diagram remains unchanged, and we see that there are no encirclements of the $s = -1 + j0$ point. Since there are no poles of $KG_p(s)H_{eq}(s)$ in the right-half plane, the pole at the origin has been effectively moved to the left-half plane, and the system is stable. In addition, since changing the value of $K$ can never cause the Nyquist diagram to encircle the $-1$ point, the closed-loop system must be stable for all values of positive $K$. This result agrees with the result obtained for this system by the use of the Routh-Hurwitz approach in Example 6.3-7.

In each of the two preceding examples, the system was open-loop stable; that is, all the poles of $KG_p(s)H_{eq}(s)$ were in the left-half plane. The next example illustrates the use of the Nyquist criterion when the system is open-loop unstable.

***Example 6.4-3*** This example is based on the system shown in Fig. 6.4-7. The loop transfer function for this system is

$$KG_p(s)H_{eq}(s) = \frac{K(s + 1)}{(s - 1)(s + 2)}$$

Let us use Bode diagrams of magnitude and phase as an assistance in plotting the Nyquist diagram. The magnitude and phase plots are shown in Fig. 6.4-8a. If these plots look strange, the reader

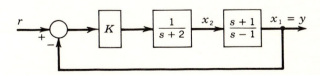

***Fig. 6.4-7*** Example 6.4-3.

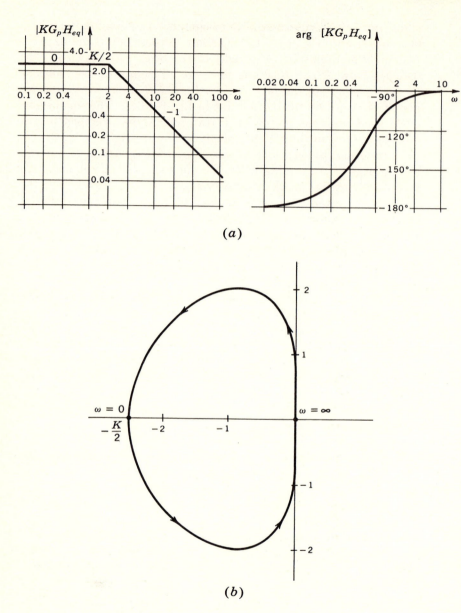

**Fig. 6.4-8** Example 6.4-3.    (a) Magnitude and phase plots; (b) Nyquist diagram.

should recall that we must put $KG_p(s)H_{eq}(s)$ into the time-constant form

$$KG_p(s)H_{eq}(s) = \frac{-K/2(1+s)}{(1-s)(1+s/2)}$$

before we use the straight-line approximation. In addition, it must be remembered that the unstable pole at $s = +1$ requires the use of the pseudo minimum-phase approach of Sec. 5.4 to obtain the phase plot.

The Nyquist diagram for this system is shown in Fig. 6.4-8*b*. Note that the exact shape of the plot is not very important since the only information we wish to obtain at this time is the number of encirclements of the $s = -1 + j0$ point. It is easy to see that the Nyquist diagram encircles the $-1$ point once in the *counterclockwise* direction if $K > 2$ and has no encirclements if $K < 2$. Since this system has one right-half-plane pole in $KG_p(s)H_{eq}(s)$, it is necessary that there be one counterclockwise encirclement if the system is to be stable. Therefore, this system is stable if and only if $K > 2$.

Since it is possible to determine stability of all closed-loop systems by means of the Routh-Hurwitz procedure, the reader may justifiably question why we introduce the Nyquist criterion. Assuredly, the Nyquist criterion is not easier to use than the Routh-Hurwitz procedure. The answer to this question is threefold.

First, because the Nyquist diagram is just a frequency-response plot of the loop transfer function, it may be determined experimentally. In this way, one can determine the stability properties of a closed-loop system without any analytical knowledge of the system transfer function. In fact, as we shall see later, it may even be possible to carry out some rudimentary forms of system design based on this information.

Second, the Nyquist procedure is graphical and therefore, by its very nature, appealing to engineers. The old cliché that "one picture is worth a thousand words" is very applicable to the engineering profession. An excellent example of the advantage gained by the use of a graphical representation is the sensitivity condition developed in Sec. 3.2. There it was shown that, in order to reduce the sensitivity of the closed-loop transfer function to variations in the plant and to reduce the effects of disturbances on the output,

$$|1 + KG_p(j\omega)H_{eq}(j\omega)| \geq 1 \qquad \text{for all } \omega \qquad (6.4\text{-}5)$$

This requirement takes a particularly appealing form of a circle criterion

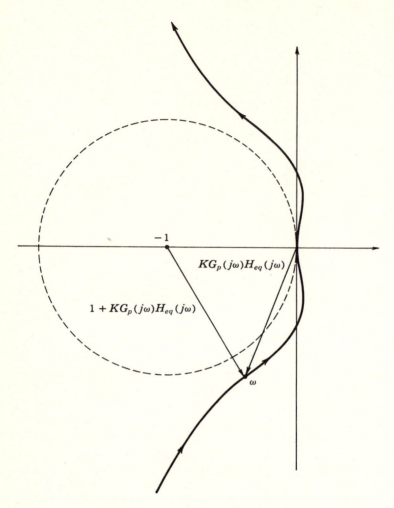

**Fig. 6.4-9**   Interpretation of the sensitivity condition
$|1 + KG_p(j\omega)H_{eq}(j\omega)| \geq 1.$

if interpreted graphically on a Nyquist diagram, as shown in Fig. 6.4-9. There the quantity $|1 + KG_p(j\omega)H_{eq}(j\omega)|$ is represented as the distance from the $s = -1 + j0$ point to the Nyquist diagram.   In order to satisfy the sensitivity condition of Eq. (6.4-5), it is necessary that the Nyquist diagram not penetrate the unit circle centered at $s = -1 + j0$.

The third and probably most significant reason for employing the Nyquist criterion is that simple and meaningful relative-stability measurements are based on the Nyquist diagram.   This is the topic of the next section.

**Exercises 6.4**   *6.4-1.*   Consider the Nyquist diagram for $KG_p(s)H_{eq}(s)$
shown in Fig. 6.4-10.   The data for this diagram were obtained
experimentally, according to the measurement technique outlined
in Fig. 5.6-9b.   Consider the five following possibilities regarding
the $KG_p(s)H_{eq}(s)$ transfer function:

(*a*)   No poles and zeros in the right-half *s* plane
(*b*)   No poles but one zero in the right-half *s* plane
(*c*)   One pole and no zeros in the right-half *s* plane
(*d*)   Two poles and no zeros in the right-half *s* plane
(*e*)   Two poles and two zeros in the right-half *s* plane

If the feedback loop is closed on each one of these transfer functions,
decide on stability for each case if the $-1$ point is located first in
region I and then in region II.

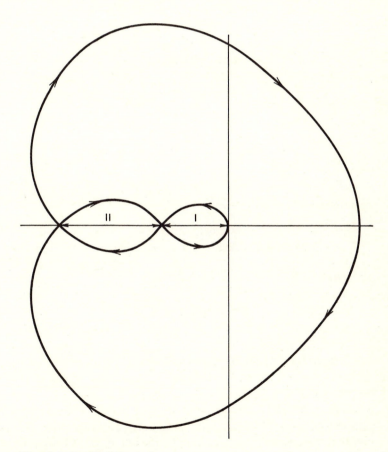

***Fig. 6.4-10***   Exercise 6.4-1.

*answers:*

(*a*) I, stable; II, unstable; (*c*) I, unstable; II, unstable; (*e*) I, unstable; II, unstable.

*6.4-2.* Make a rough sketch of the Nyquist diagram for the transfer functions of Exercise 5.4-1. Determine the stability of the resulting closed-loop systems, assuming that the transfer functions are open-loop transfer functions. (Do not bother to make a point-by-point plot unless accuracy is needed to determine stability.) Check your answers by using the Routh-Hurwitz criterion.

*answers:*

(*a*) Unstable; (*b*) stable; (*c*) unstable; (*d*) unstable; (*e*) stable; (*f*) stable; (*g*) unstable; (*h*) unstable.

*6.4-3.* Consider the system of Example 6.3-7 represented in state-variable form by

$$\dot{\mathbf{x}} = \begin{bmatrix} 0 & 1 & 0 \\ 0 & -2 & 1 \\ 0 & 0 & -3 \end{bmatrix} \mathbf{x} + \begin{bmatrix} 0 \\ 0 \\ 1 \end{bmatrix} u$$

$$y = x_1 \qquad u = K(r - [1 \quad \tfrac{2}{7} \quad \tfrac{2}{7}]\mathbf{x})$$

Assume that $K = 42$ and that the feedback path from the state variable $x_2$ has been opened. Use the Nyquist method to find if the resulting system is stable.

*answer:*

Stable.

*6.4-4.* Use the Nyquist criterion to show that the closed-loop system associated with the open-loop transfer function

$$KG_p(s)H_{eq}(s) = \frac{K(s + 1)}{s(s - 1)(s + 2)}$$

is unstable for all positive values of $K$.

## 6.5   *Relative stability and Nyquist diagrams*

This entire chapter is devoted to stability discussions. Until this section we have given only a "yes" or "no" answer with regard to the stability of a closed-loop system. By using the Routh-Hurwitz criterion or the Nyquist diagram, it is possible to decide if the system in question is

stable or not.   If it is not stable, it must be unstable.   By mentioning
the term marginally stable or marginally unstable, we have implied that
one may consider degrees of "instability"; that is, one may wish to distinguish between systems that are unstable because they have poles on
the $j\omega$ axis and systems that are unstable because they have poles in the
right-half plane.   But as yet only gain margin has been mentioned as a
measure of relative stability.   Other measures of relative stability are
the subject of this section.

The discussion of the Nyquist diagram in the preceding section is
based upon the representation of the system in the $H$-equivalent form.
If the poles and zeros of $KG_p(s)H_{eq}(s)$ are known, the plotting of the
Nyquist diagram is easy.   Although the poles of the plant may be known,
the zeros of $H_{eq}(s)$ must often be found by factoring, which is almost as
difficult as factoring $D_k(s)$.   Because of this difficulty, the Nyquist diagram would be of questionable value if it could be used only to answer
the stability question in a yes-no fashion.   The Routh-Hurwitz method
already does that.   The advantage of the Nyquist diagram is the variety
of relative-stability measures that can be read from it.

The one relative-stability measure that may be taken directly from
the Nyquist diagram of $KG_p(s)H_{eq}(s)$ is the gain margin.   Gain margin
was defined in Sec. 6.3 as the ratio of the maximum possible gain for
stability to the actual system gain.   The $KG_p(s)H_{eq}(s)$ form of the loop-
transfer-function representation is suited to the measurement of gain
margin because $KG_p(s)H_{eq}(s)$ can be written as

$$KG_p(s)H_{eq}(s) = \frac{KK_pN_p(s)}{D_p(s)} \frac{K_hN_h(s)}{N_p(s)} = K\frac{K_pK_hN_h(s)}{D_p(s)}$$

Here the adjustable gain $K$ is in no way involved with the pole-zero
locations, and so the shape of the Nyquist diagram is independent of the
value of $K$.   The adjustable gain $K$ acts only to adjust the scale on which
the Nyquist diagram is drawn.   If the plot of $KG_p(s)H_{eq}(s)$ for $s = j\omega$
intercepts the negative real axis at a point $-a$ between the origin and the
critical or $-1$ point, then the gain margin is simply

$$\text{Gain margin} = \text{GM} = \frac{1}{a} \qquad\qquad (6.5\text{-}1)$$

Examples 6.3-6 and 6.4-2 are repeated here as Example 6.5-1 in order to
illustrate the calculation of gain margin.

> ***Example 6.5-1***   This example is based upon Fig. 6.3-2 where
> $G_p(s)$ is $1/s(s+2)(s+3)$ and $H_{eq}(s)$ is unity.   Let us assume some

arbitrary value for $K$, say $K = 10$, and draw the Nyquist diagram for

$$KG_p(s)H_{eq}(s) = \frac{10}{s(s+2)(s+3)}$$

The portion of the Nyquist diagram for $\omega \geq 0$ for this loop transfer function is shown as Fig. 6.5-1. The frequency-response function crosses the negative real axis at approximately 0.35, and so the gain margin is approximately

$$GM \sim \frac{1}{0.35} \sim 2.86$$

The maximum gain for stability was found in Example 6.3-6 to be 30, and so the actual gain margin here should be 3. The accuracy with which the gain margin is found in this example is typical of accuracies that might be expected where the graphical Nyquist method is used.

The determination of gain margin for the system of Example 6.4-2 is actually trivial. The system is shown in Fig. 6.3-3, and the loop transfer function is

$$KG_p(s)H_{eq}(s) = \frac{(2K/7)[(s+\tfrac{3}{2})^2 + (\sqrt{5}/2)^2]}{s(s+2)(s+3)}$$

The Nyquist diagram is shown in Fig. 6.4-6$b$, and since the fre-

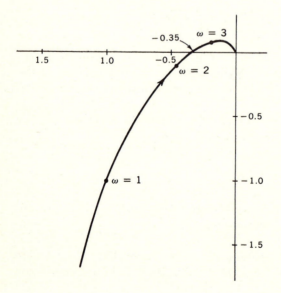

**Fig. 6.5-1**  Example 6.5-1.

quency-response function never crosses the negative real axis, the gain margin is infinity.    The system is stable for all gain.

As noted in the discussion of gain margin in Sec. 6.3, gain margin is only one measure of relative stability and not a particularly reliable one at that.    The other measures of relative stability discussed in this section are based on a $G_{eq}(s)$ representation, instead of the $H_{eq}(s)$ representation, used previously.

In terms of $G_{eq}(s)$, the closed-loop transfer function is given by

$$\frac{y(s)}{r(s)} = \frac{G_{eq}(s)}{1 + G_{eq}(s)}$$

Therefore the poles of $y(s)/r(s)$ are the zeros of $1 + G_{eq}(s)$.   We may determine the number of zeros of $1 + G_{eq}(s)$ in the right-half $s$ plane by plotting the Nyquist diagram for $G_{eq}(s)$.    As before, the Nyquist diagram is a polar plot of the complex values of $G_{eq}(s)$ for the contour of Fig. 6.4-2$a$. The net clockwise encirclements of the $-1$ point by this plot is equal to the number of poles of $y(s)/r(s)$ [zeros of $1 + G_{eq}(s)$] minus the number of poles of $G_{eq}(s)$ in the right half of the $s$ plane.    In other words, the absolute-stability information, namely, the number of poles of $y(s)/r(s)$ in the right-half $s$ plane, is obtained in the same manner from the Nyquist diagram of either $KG_p(s)H_{eq}(s)$ or $G_{eq}(s)$, even though the two plots are generally quite different in appearance.    To see that the Nyquist plots for the same system may be quite different, depending upon whether we plot $KG_p(s)H_{eq}(s)$ or $G_{eq}(s)$, let us return to the system of Example 6.4-2, as pictured in Fig. 6.3-3$a$.    For this system $G_{eq}(s)$ is

$$G_{eq}(s) = \frac{K}{s[s^2 + s(5 + 2K/7) + 6 + 4K/7]}$$

Note here that the poles of $G_{eq}(s)$ are a function of $K$, so that changing $K$ does not just change the scale factor to which the Nyquist diagram is drawn.    A completely different function must be plotted for every value of gain to be considered.    For this reason, the $G_{eq}(s)$ Nyquist plot is not suited to a determination of gain margin, although it is well suited to other relative-stability measures.

If $K$ is 42, the $G_{eq}(s)$ becomes

$$G_{eq}(s) = \frac{42}{s(s + 2)(s + 15)} \tag{6.5-2}$$

The corresponding Nyquist diagram is pictured in Fig. 6.5-2.    This is to be compared with the Nyquist diagram for the same system drawn for $KG_p(s)H_{eq}(s)$ as pictured in Fig. 6.4-6$b$.    Although the diagrams look different, they convey the same absolute-stability information.    In addition, the $KG_p(s)H_{eq}(s)$ diagram is suitable for the determination of gain margin.    We now examine the various relative-stability measures that may be determined from the $G_{eq}(s)$ diagram.

One important measure of relative stability is the frequency-response

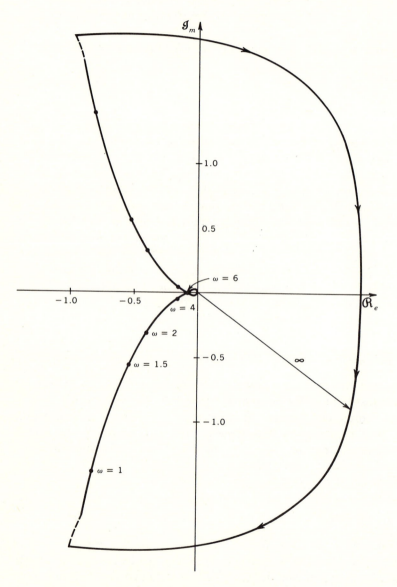

***Fig. 6.5-2***    Nyquist diagram of $G_{eq}(s)$ for Example 6.4-2.

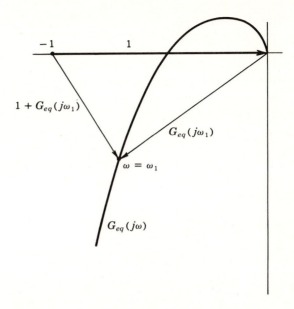

**Fig. 6.5-3**    Graphical determination of $M(j\omega)$.

function of the closed-loop system, often referred to as the $M$ curve. The $M$ curve is, of course, a function of frequency and may be determined from the analytic form of the closed-loop frequency-response function as

$$M(j\omega) = \frac{y(j\omega)}{r(j\omega)} = \frac{G_{eq}(j\omega)}{1 + G_{eq}(j\omega)} \tag{6.5-3}$$

Figure 6.5-3 illustrates how the value of $M$ may be determined directly from the Nyquist diagram of $G_{eq}(s)$ at one particular frequency $\omega_1$. In this figure the lengths 1 and $G_{eq}(j\omega_1)$ are indicated, as is the sum of these two vector quantities. The length of the vector $G_{eq}(j\omega_1)$ divided by the length of $1 + G_{eq}(j\omega_1)$ is thus the value of $M$ at the frequency $\omega_1$. The argument of $M(j\omega_1)$ is determined by subtracting the angle associated with the $1 + G_{eq}(j\omega_1)$ vector from that of $G_{eq}(j\omega_1)$. The complete $M$ curve may be found by repeating this procedure over the range of frequencies of interest. From the completed $M$ curve, the bandwidth of the closed-loop system may be read by inspection. The reader will recall that bandwidth (BW) was defined in Sec. 3.2 as that frequency at which the frequency-response function is 0.707 times its low-frequency value.

In terms of the magnitude portion of the $M(j\omega)$ plot, the point-by-point procedure illustrated above may be considerably simplified by plotting contours of constant $|M(j\omega)|$ on the Nyquist plot of $G_{eq}(s)$. The magnitude plot of $M(j\omega)$ can then be read directly from the Nyquist diagram of $G_{eq}(s)$. Fortunately, these contours of constant $|M(j\omega)|$ have a particularly simple form. For $|M(j\omega)| = M$, the contour is simply a circle centered at the point $-M^2/(M^2 - 1) + j0$ with a radius of $|M/(M^2 - 1)|$. This property may be easily demonstrated by the following development.

Let the complex function $G_{eq}(j\omega)$ be written as

$$G_{eq}(j\omega) = X(\omega) + jY(\omega) \tag{6.5-4}$$

where $X(\omega)$ and $Y(\omega)$ are real functions of $\omega$ which are, respectively, the real and imaginary parts of $G_{eq}(j\omega)$. In terms of this notation, $M(j\omega)$ becomes

$$M(j\omega) = \frac{X(\omega) + jY(\omega)}{[1 + X(\omega)] + jY(\omega)}$$

Therefore the magnitude of $M(j\omega)$ is given by

$$M(j\omega) = M = \frac{|X(\omega) + jY(\omega)|}{|[1 + X(\omega)] + jY(\omega)|} = \frac{\sqrt{X^2 + Y^2}}{\sqrt{(1 + X)^2 + Y^2}}$$

so that

$$M^2 = \frac{X^2 + Y^2}{(1 + X)^2 + Y^2} \tag{6.5-5}$$

After a few simple algebraic steps, this last result may be written as

$$\left(X + \frac{M^2}{M^2 - 1}\right)^2 + Y^2 = \left(\frac{M}{M^2 - 1}\right)^2 \tag{6.5-6}$$

This expression is the equation for a circle of radius $|M/(M^2 - 1)|$ centered at $X = -M^2/(M^2 - 1)$ and $Y = 0$ as predicted. These circles are referred to as constant-$M$ circles or simply $M$ circles. The center and radius specifications for a number of typical values of $M$ are given in Table 6.5-1.

If these constant-$M$ circles are plotted together with the Nyquist diagram of $G_{eq}(s)$, as shown in Fig. 6.5-4 for the $G_{eq}$ plot of Fig. 6.5-2, the values of $|M(j\omega)|$ may be read directly from the plot. Note that the $M = 1$ circle degenerates to the straight line $X = -0.5$. For $M < 1$ the constant-$M$ circles lie to the right of this line, whereas for $M > 1$ they lie to the left. In addition, the $M = 0$ circle is the point $0 + j0$, and $M = \infty$ corresponds to the point $-1.0 + j0$.

In an entirely similar fashion, the contours of constant arg $M(j\omega)$ can be found. Surprisingly, these contours turn out to be segments of circles. In this case, however, the circles are centered on the line

### Table 6.5-1   Constant-M-circle Specifications

| $M$ | Center <br> $-M^2/(M^2-1)+j0$ | Radius <br> $\lvert M/(M^2-1)\rvert$ |
|---|---|---|
| 0 | $0+j0$ | 0 |
| 0.7 | $0.96+j0$ | 1.37 |
| 0.8 | $1.77+j0$ | 2.21 |
| 0.9 | $4.25+j0$ | 4.72 |
| 1.0 | $\infty$ | $\infty$ |
| 1.1 | $-5.76+j0$ | 5.24 |
| 1.2 | $-3.27+j0$ | 2.73 |
| 1.3 | $-2.45+j0$ | 1.88 |
| 1.4 | $-2.04+j0$ | 1.46 |
| 1.5 | $-1.80+j0$ | 1.20 |
| 1.7 | $-1.53+j0$ | 0.90 |
| 2.0 | $-1.33+j0$ | 0.67 |
| 3.0 | $-1.13+j0$ | 0.38 |
| 4.0 | $-1.07+j0$ | 0.27 |
| 5.0 | $-1.04+j0$ | 0.21 |
| $\infty$ | $-1.00+j0$ | 0 |

$X=-\tfrac{1}{2}$.   The contour of the arg $M(j\omega)=\beta$ for $0<\beta<180°$ is the upper-half-plane portion of the circle centered at $-\tfrac{1}{2}+j1/(2\tan\beta)$ with a radius $\lvert 1/(2\sin\beta)\rvert$.   For $\beta$ in the range $-180°<\beta<0°$, the portion of the same circles in the lower half plane is used.   The specifications for the constant-phase contours for a few values of $\beta$ are given in Table 6.5-2. Figure 6.5-5 shows the plot of the constant-phase contours for the same values of $\beta$.

### Table 6.5-2   Constant-phase-contour Specifications

| $\beta$ <br> Phase, deg | Center <br> $-\tfrac{1}{2}+j1/(2\tan\beta)$ | Radius <br> $\lvert 1/(2\sin\beta)\rvert$ | Portion of Circle |
|---|---|---|---|
| $-180$ | $\infty$ | $\infty$ | From $-1$ to 0 |
| $-135$ | $-\tfrac{1}{2}+j\tfrac{1}{2}$ | $\sqrt{2}/2$ | Lower half plane |
| $-90$ | $-\tfrac{1}{2}+j0$ | $\tfrac{1}{2}$ | Lower half plane |
| $-45$ | $-\tfrac{1}{2}-j\tfrac{1}{2}$ | $\sqrt{2}/2$ | Lower half plane |
| 0 | $\infty$ | $\infty$ | All except $-1$ to 0 |
| 45 | $-\tfrac{1}{2}+j\tfrac{1}{2}$ | $\sqrt{2}/2$ | Upper half plane |
| 90 | $-\tfrac{1}{2}+j0$ | $\tfrac{1}{2}$ | Upper half plane |
| 135 | $-\tfrac{1}{2}-j\tfrac{1}{2}$ | $\sqrt{2}/2$ | Upper half plane |

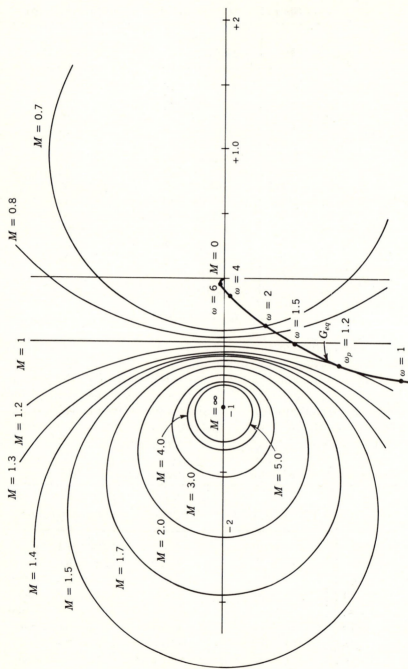

*Fig. 6.5-4* Constant-$M$ contours.

By using these constant-magnitude and -phase contours, it is possible to read directly the complete closed-loop frequency response from the Nyquist diagram of $G_{eq}(s)$. In practice it is common to dispense with the constant-phase contours, since almost all closed-loop systems are minimum-phase, and the phase shift may therefore be inferred from the magnitude plot, as discussed in Sec. 5.4. In fact, it is common to simplify the labor further by considering only one point on the magnitude plot, namely, the point at which $M$ is maximum. This point is referred to as $M$ peak or $M_p$, and the frequency at which the peak occurs is $\omega$ peak or $\omega_p$. The $M_p$ point may be easily found by considering the contours of larger and larger values of $M$ until the contour is found that is just tangent to the plot of $G_{eq}(s)$. The value associated with this contour is then $M_p$, and the frequency at which the $M_p$ contour and $G_{eq}(s)$ touch is

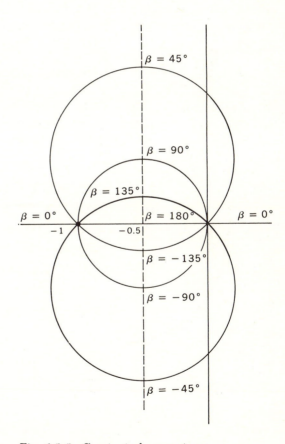

**Fig. 6.5-5**   Constant-phase contours.

$\omega_p$.  In the plot of $G_{eq}(s)$ shown in Fig. 6.5-4, for example, the value of $M_p$ is 1.2 at the radian frequency $\omega_p \approx 1.2$.

One of the primary reasons for determining $M_p$ and $\omega_p$, in addition to the obvious saving of labor as compared with the determination of the complete frequency response, is the close correlation of these quantities with the behavior of the closed-loop system.  In particular, for the simple second-order system

$$\frac{y(s)}{r(s)} = \frac{\omega_n{}^2}{s^2 + 2s\zeta\omega_n + \omega_n{}^2} \tag{6.5-7}$$

the values of $M_p$ and $\omega_p$ completely characterize the system.  In other words, for this second-order system, $M_p$ and $\omega_p$ specify $\zeta$ and $\omega_n$, the only parameters of the system.  In order to demonstrate this property, we need only recall Eqs. (5.3-8) and (5.3-9), which relate the maximum point of the frequency response of Eq. (6.5-7) to the values of $\zeta$ and $\omega_n$.  In our present terminology, these equations become

$$\omega_p = \omega_n \sqrt{1 - 2\zeta^2} \tag{6.5-8}$$

$$M_p = \frac{1}{2\zeta \sqrt{1 - \zeta^2}} \qquad \text{for } \zeta \leq 0.707 \tag{6.5-9}$$

From these equations one may determine $\zeta$ and $\omega_n$ if $M_p$ and $\omega_p$ are known, and vice versa.  Figure 6.5-6 graphically displays the relations between $M_p$ and $\omega_p$ and $\zeta$ and $\omega_n$.  Once $\zeta$ and $\omega_n$ are known, the results of Sec.

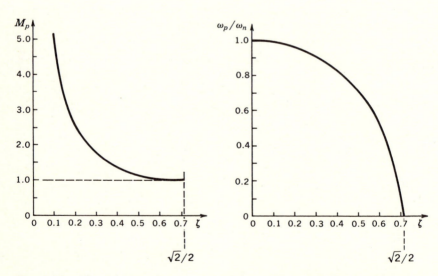

**Fig. 6.5-6**  Plots of $M_p$ and $\omega_p/\omega_n$ versus $\zeta$ for a simple second-order system.

4.3 may be used to determine the time behavior of this second-order system.

Not all systems are of the simple second-order form. However, it is common practice to assume that the behavior of many high-order systems is closely related to that of a second-order system with the same $M_p$ and $\omega_p$. This is, in fact, one of the primary reasons for the extensive consideration given to second-order systems in Sec. 4.3.

Two other measures of the qualitative nature of the closed-loop response which may be determined from the Nyquist diagram of $G_{eq}(s)$ are the phase margin and crossover frequency. The crossover frequency $\omega_c$ is the positive value of $\omega$ for which the magnitude of $G_{eq}(j\omega)$ is equal to unity, that is,

$$|G_{eq}(j\omega_c)| = 1 \tag{6.5-10}$$

The phase margin $\phi_m$ is defined as the difference between the argument of $G_{eq}(j\omega_c)$ and $-180°$. In other words, if we define $\beta_c$ as

$$\beta_c = \arg G_{eq}(j\omega_c) \tag{6.5-11}$$

the phase margin is given by

$$\phi_m = \beta_c - (-180°) = 180° + \beta_c \tag{6.5-12}$$

The phase margin takes on a particularly simple and graphic meaning in the Nyquist diagram of $G_{eq}(s)$. Consider, for example, the Nyquist diagram shown in Fig. 6.5-7. In that diagram, we see that the phase margin is simply the angle between the negative real axis and the vector $G_{eq}(j\omega_c)$. The vector $G_{eq}(j\omega_c)$ may be found by intersecting the $G_{eq}(s)$ locus with the unit circle, as shown in Fig. 6.5-7. The frequency associated with the point of intersection is $\omega_c$.

It is also possible to determine the $\phi_m$ and $\omega_c$ directly from the Bode plots of magnitude and phase of $G_{eq}(s)$ without plotting the Nyquist diagram. The value of $\omega$ for which the magnitude crosses unity is $\omega_c$. The phase margin is then determined by inspection from the phase plot by noting the difference between the phase shift at $\omega_c$ and $-180°$. Consider, for example, the straight-line magnitude and phase plots for the $G_{eq}(s)$ function of Fig. 6.5-2. For $K = 42$, the $G_{eq}(s)$ transfer function is given in Eq. (6.5-2). In time-constant form this transfer function becomes

$$G_{eq}(s) = \frac{1.4}{s(1 + s/2)(1 + s/15)}$$

The approximate Bode amplitude and phase diagrams are shown in Fig. 6.5-8. From this figure we see that $\omega_c \approx 1.4$ and $\phi_m = 60°$.

From the simple second-order system of Eq. (6.5-10), it is once

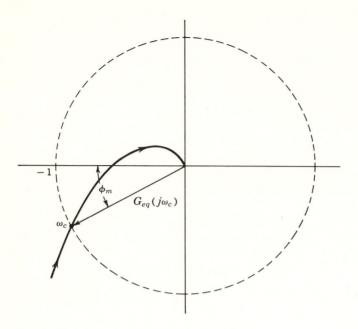

**Fig. 6.5-7**   Definition of phase margin.

again possible to show that $\phi_m$ and $\omega_c$ completely specify $\zeta$ and $\omega_n$. In particular, it is relatively easy to show that

$$\omega_c = \omega_n \sqrt{\sqrt{4\zeta^4 + 1} - 2\zeta^2} \qquad\qquad\qquad (6.5\text{-}13)$$

$$\phi_m = \arccos \left( \sqrt{4\zeta^4 + 1} - 2\zeta^2 \right) \qquad\qquad (6.5\text{-}14)$$

These relationships are graphically illustrated in Fig. 6.5-9. Since $\omega_n$ and $\zeta$ may be determined from $\phi_m$ and $\omega_c$, by the use of Eqs. (6.5-8) and (6.5-9) it is also possible to determine $M_p$ and $\omega_p$ from $\phi_m$ and $\omega_c$ for the system of Eq. (6.5-7).

As in the case of $M_p$, higher-order systems are handled by assuming that they are reasonably well approximated by second-order behavior. Unfortunately the correlation between response and phase margin is somewhat poorer than the correlation of response and $M$ peak. This lower reliability of the phase-margin measure is a direct consequence of the fact that $\phi_m$ is determined by considering only one point, $\omega_c$ on the $G_{eq}$ plot, whereas $M_p$ is found by examining the entire plot. Consider, for example, the two Nyquist diagrams shown in Fig. 6.5-10. The phase margin for these two diagrams is identical; however, it is obvious that the system of Fig. 6.5-10*b* has a far more oscillatory, underdamped behav-

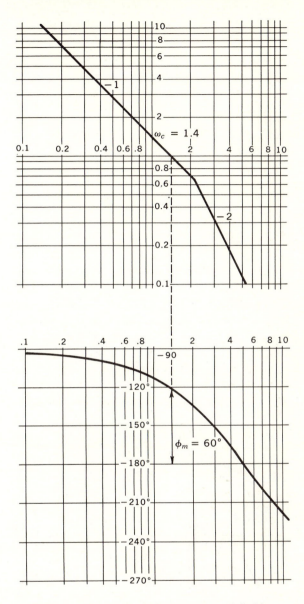

**Fig. 6.5-8**   Magnitude and phase plots of $G_{eq}(s)$.

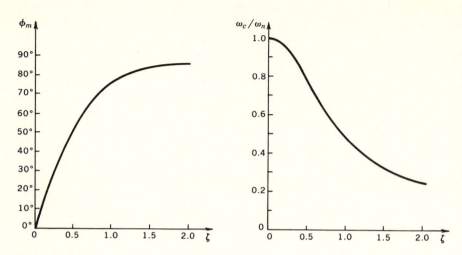

**Fig. 6.5-9** Plots of $\phi_m$ and $\omega_c/\omega_n$ versus $\zeta$ for a simple second-order system.

ior than that of Fig. 6.5-10a. Since the determination of $M_p$ requires that the entire $G_{eq}$ plot be examined, the more oscillatory behavior of the system of Fig. 6.5-10b is reflected in a much larger value of $M_p$ for that system than for that of Fig. 6.5-10a.

In other words, the relative ease of determining $\phi_m$ as compared with $M_p$ has been obtained only by sacrificing some of the reliability of $M_p$. Fortunately, systems such as that of Fig. 6.5-10b are rare, and

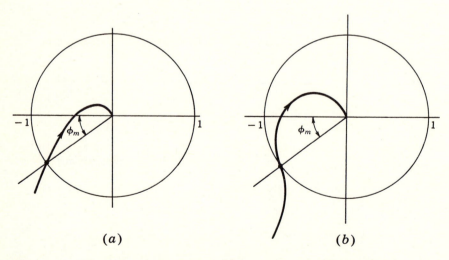

**Fig. 6.5-10** Two systems with the same phase margin but different $M_p$.

phase margin provides a simple and effective means of estimating the closed-loop response from the $G_{eq}$ plot. We shall make considerable use of phase margin in Chap. 10. However, one should never forget to check any results obtained by using phase margin as a measure of relative stability, lest a system such as the one of Fig. 6.5-10*b* slip by.

**Exercises 6.5**  *6.5-1*.  For the Nyquist plot of $KG_p(s)$ shown in Fig. 6.5-11, find the following:

(*a*)  $M_p$ and $\omega_p$
(*b*)  $\phi_m$ and $\omega_c$
(*c*)  Gain margin

Here assume that $H_{eq}(s) = 1$ so that

$$KG_p(s) = KG_p(s)H_{eq}(s) = G_{eq}(s)$$

*answers:*

(*a*) $M_p = 3$, $\omega_p \approx 50$; (*b*) $\omega_c = 10$, $\phi_m \approx 41°$; (*c*) gain margin $\approx 1/0.7$

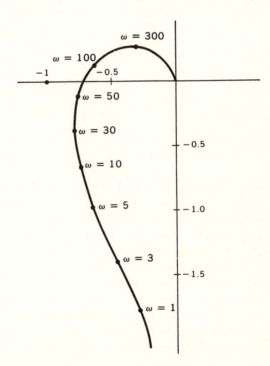

**Fig. 6.5-11**   Exercise 6.5-1.

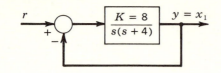

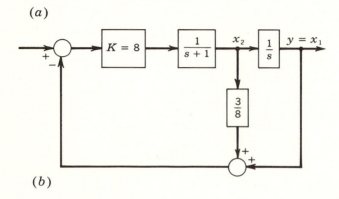

**Fig. 6.5-12**    Exercise 6.5-2.

*6.5-2.*    Consider the two feedback control systems of Fig. 6.5-12. For each case the closed-loop transfer function is

$$\frac{y(s)}{r(s)} = \frac{8}{(s + 2)^2 + 2^2}$$

Determine the value of $M_p$ and $\omega_p$ for both systems and compare these values for $K$ equal to 8, 16, and 32.

## 6.6    Conclusions

To summarize effectively the contents of this chapter, it is necessary to view the material in terms of all that has been said in the analysis portion of this book.    Chapter 4 discussed the means of obtaining the complete time solution for all the state variables for any input and/or initial conditions. As far as system response is concerned, this is the most complete information that one might seek.    In fact, it is often more information than can be effectively used, particularly in the early stages of system design or evaluation.

We now seek ways of obtaining a less complete answer to the question of system response, hopefully at the cost of less energy.    The most basic response question that can be asked is whether the closed-loop system is stable or not.   That question can be answered in a yes-no fashion by the use of the Routh-Hurwitz criterion.   If a system proves to be unstable, this often ends the analysis, as unstable operation is usually unacceptable from a physical as well as a mathematical point of view. The system must somehow be modified to make it stable before any more penetrating questions regarding system response can be answered.

If it is assumed that the answer to the stability question is "yes" and that the system in question is actually stable, the next level of questioning regards relative stability.   Relative-stability measures are easily discussed in the frequency domain, and for this reason the frequency-response function was introduced in Chap. 5.   Now in this chapter we have used the frequency-response function in terms of the Nyquist diagram as an alternative to the Routh-Hurwitz criterion to answer the absolute-stability question.   More important, a method is now available for discussing relative stability.

In using the frequency-response function, the complex variable $s = \sigma + j\omega$ is allowed to be a function only of $\omega$; that is, the frequency-response function is formed from the transfer function by setting $s = j\omega$, with $\sigma = 0$.   In terms of the $s$ plane, we make use only of the $j\omega$ axis and, in forming the Nyquist diagram, the infinite semicircle bounding the right-half $s$ plane.   One way in which more information regarding system behavior might be obtained is to make use of the entire $s$ plane. This is the procedure to be used in the next chapter on the root-locus method.

## 6.7   Problems

*6.7-1.*   Show that the closed-loop system given in Fig. 6.7-1 is stable by using (*a*) the Routh-Hurwitz criterion and (*b*) the Nyquist criterion.

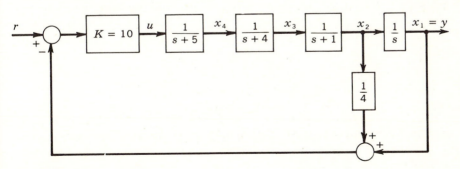

*Fig 6.7-1*   Problem 6.7-1.

Begin by sketching the magnitude and phase and proceed by calculating the phase shift at magnitude crossover.

*6.7-2.* Use the Routh-Hurwitz criterion to determine if the closed-loop systems corresponding to the following open-loop transfer functions are stable.

(a) $KG_p(s)H_{eq}(s) = \dfrac{4(s + 2)}{s(s^3 + 2s^2 + 3s + 4)}$

(b) $KG_p(s)H_{eq}(s) = \dfrac{2(s + 4)}{s^2(s + 1)}$

(c) $KG_p(s)H_{eq}(s) = \dfrac{4(s^3 + 2s^2 + s + 1)}{s^2(s^3 + 2s^2 - s - 1)}$

*6.7-3.* Use the Nyquist criterion to determine the stability of the closed-loop systems represented by the following open-loop transfer functions. Determine either the gain at phase crossover (180°) or phase at magnitude crossover in order to verify your answer.

(a) $KG_p(s)H_{eq}(s) = \dfrac{40}{s(1 + s/10)(1 + s/100)}$

(b) $KG_p(s)H_{eq}(s) = \dfrac{100(1 + s/5)}{s^2(1 + s/100)^2}$

*6.7-4.* Determine $M_p$, $\omega_p$, $\phi_m$, and $\omega_c$ for the Nyquist diagram of $G_{eq}(s)$ given in Fig. 6.7-2.

*6.7-5.* Find the $\phi_m$ and $\omega_c$ associated with the $G_{eq}(s)$ transfer functions given below. It is *not* necessary to plot the Nyquist diagrams.

(a) $G_{eq}(s) = \dfrac{50}{s(s + 1)(s + 10)}$

(b) $G_{eq}(s) = \dfrac{10(s + 1)}{s^2}$

*6.7-6.* Determine the stability of the closed-loop system given below by using (a) the Routh-Hurwitz criterion and (b) the Nyquist criterion.

$$\dot{x} = \begin{bmatrix} 0 & 1 & 0 \\ 0 & -1 & 1 \\ 0 & 0 & -10 \end{bmatrix} x + \begin{bmatrix} 0 \\ 0 \\ 100 \end{bmatrix} u \quad.$$

$y = x_1 \qquad u = (r - \begin{bmatrix} 1 & 0.5 & 0.5 \end{bmatrix} x)$

*6.7-7.* Find a relationship between $K$ and $\lambda$ such that the closed-loop system associated with the following open-loop transfer function is stable.

$$KG_p(s)H_{eq}(s) = \dfrac{K(s + 1)}{s^2(s + \lambda)}$$

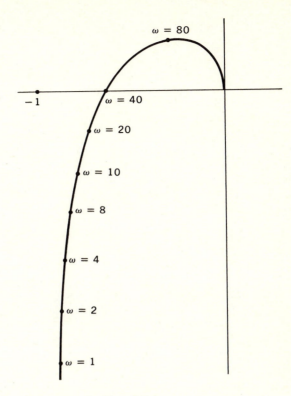

**Fig. 6.7-2**    Problem 6.7-4.

*6.7-8.*    For the system

$$\dot{\mathbf{x}} = \begin{bmatrix} 0 & 1 & 0 \\ 0 & -\alpha & 1 \\ 0 & 0 & -5 \end{bmatrix} \mathbf{x} + \begin{bmatrix} 0 \\ 0 \\ 1 \end{bmatrix} u$$

$y = x_1 \qquad u = K(r - [1 \quad \tfrac{3}{8} \quad \tfrac{1}{8}]\mathbf{x})$

Is the system stable for $\alpha \geq 0$ and $K \geq 0$?

# seven  *the root-locus method*

## 7.1  Introduction

In the discussion of the frequency-response methods in the two preceding chapters, it was assumed that the complex Laplace variable $s$ was replaced by an imaginary variable $j\omega$. In this chapter we consider the entire complex plane rather than restrict attention to that line in the complex plane where $s = j\omega$. Once again, however, we wish to obtain information about the closed-loop system by examining the open-loop transfer function. In this case the entire closed-loop transfer function is to be obtained in factored form, and from this information the closed-loop time and frequency responses to any input may be determined. The root-locus method studied here is also a useful synthesis tool.

The root-locus method is a graphical means of determining the poles of the closed-loop transfer function $y(s)/r(s)$. Conceptually, the procedure may be different from any the reader has previously encountered. A difficult problem, such as factoring the characteristic polynomial, is embedded in a more general problem. If one is able to solve the more general problem, the particular problem is automatically solved. The root locus is a picture of the behavior of the closed-loop poles as one of the system parameters, usually the gain $K$, is varied. If the closed-loop poles may be found for all values of the parameter, then the closed-loop poles for the specific value of interest are also known.

In Chap. 4 root-locus diagrams were drawn for a second-order system, though at that time they were not so called. The reader may recall that we considered the behavior of the closed-loop poles as $\omega_n$, $\zeta$, and $\zeta\omega_n$ were varied. Let us repeat that development here but in slightly different terms. We consider the closed-loop system of Fig. 7.1-1a, where $k_2$ has been set equal to 0, and the corresponding pole-zero plot of the open-loop transfer function $KG_p(s)H_{eq}(s)$ shown in Fig. 7.1-1b. The closed-loop

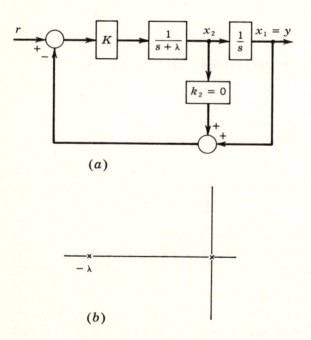

(a)

(b)

**Fig. 7.1-1**  Simple second-order system.   (a) Block
diagram; (b) pole-zero plot of $KG_p(s)H_{eq}(s)$.

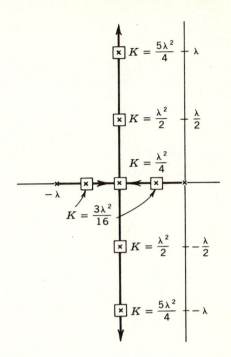

**Fig. 7.1-2** Variation of closed-loop poles as a function of $K$.

transfer function of this system is

$$\frac{y(s)}{r(s)} = \frac{K}{s^2 + \lambda s + K}$$

and two closed-loop poles are given by

$$s = -\frac{\lambda}{2} \pm \sqrt{\frac{\lambda^2}{4} - K}$$

As $K$ is varied from zero to infinity, the positions of the closed-loop poles change. If this change is plotted on the same $s$ plane as that on which the open-loop poles are displayed, Fig. 7.1-2 results.

In Chap. 4 we also allowed $\omega_n^2$ to be a constant and we varied $\zeta$. The result was Fig. 4.3-5a, repeated here as Fig. 7.1-3. In this case the parameter that is varying is not the gain $K$ but the feedback coefficient $k_2$. However, the resulting picture once again displays the behavior of the closed-loop poles as this parameter is allowed to vary. Note that, in both Figs. 7.1-2 and 7.1-3, many problems are solved, depending upon the value of $K$ or $k_2$ of interest. If this general solution for a variety of values of $K$ or $k_2$, as given pictorially by the diagrams, is known, it is

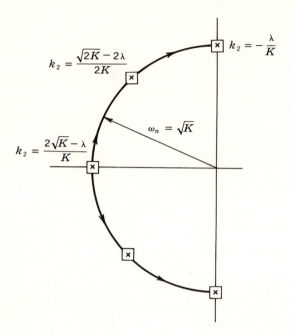

$$k_2 = \frac{\sqrt{2K} - 2\lambda}{2K}$$

$$k_2 = -\frac{\lambda}{K}$$

$$\omega_n = \sqrt{K}$$

$$k_2 = \frac{2\sqrt{K} - \lambda}{K}$$

***Fig. 7.1-3***    Variation of closed-loop poles as a function of $k_2$.

easy to find the particular closed-loop poles for any given value of $K$
or $k_2$.

In Figs. 7.1-2 and 7.1-3 the path of the closed-loop poles as a function of $K$ or $k_2$ was determined by actually solving for the closed-loop
poles. This is possible because the system is second-order, and we could
use the quadratic formula. If the system is of higher order, this convenience is lost. Hence the object of the first sections of this chapter is
to enable the drawing of the root-locus diagram without first having to
solve for the locations of the closed-loop poles. In fact, that is just what
we are trying to avoid: solving analytically for the closed-loop-pole
locations.

Sections 7.2 and 7.3 are devoted entirely to construction rules for
root-locus diagrams. Section 7.4 contains additional examples and continues the discussion of information that may be obtained from the completed root-locus diagram. These three sections assume that the gain $K$
is the parameter that is varying. In Sec. 7.6 methods for plotting the
root locus vs. any system parameter are discussed, particularly with
regard to sensitivity. Section 7.5 deals with the determination of the
closed-loop response from the root-locus diagram.

## 7.2   *The root-locus method*

As noted above, the root-locus method provides a graphical means of determining the poles of the system transfer function $y(s)/r(s)$, usually as a function of $K$. The approach may appear at first to be somewhat devious, like going from Chicago to New York by proceeding west around the world. This is partially true, but the reader will find that, once he has achieved facility with graphical construction procedures, the root-locus technique proves to be a powerful tool for analysis and synthesis.

Initially we consider only changes in the closed-loop poles as $K$ is varied and make use of the $H_{eq}(s)$ form shown in Fig. 7.2-1. The closed-loop transfer function is given by

$$\frac{y(s)}{r(s)} = \frac{KG_p(s)}{1 + KG_p(s)H_{eq}(s)} \tag{7.2-1}$$

It is a simple matter to express the closed-loop transfer function in terms of $KG_p(s)$ and $H_{eq}(s)$, but, as pointed out in the discussion of stability in Chap. 6, the poles of $y(s)/r(s)$ are not known. Not only are the exact closed-loop-pole locations unknown, but we may not even know if the closed-loop system is stable, that is, if the closed-loop poles lie in the left-half $s$ plane.

The necessity of factoring to determine the closed-loop poles is clearly illustrated by writing the closed-loop transfer function as a ratio of polynomials in $s$:

$$\frac{y(s)}{r(s)} = \frac{K[K_p N_p(s)/D_p(s)]}{1 + K[K_p K_h N_p(s) N_h(s)/D_p(s) N_p(s)]} \tag{7.2-2}$$

or

$$\frac{y(s)}{r(s)} = \frac{KK_p N_p(s)}{D_p(s) + KK_p K_h N_h(s)} = \frac{KK_p N_p(s)}{D_k(s)} \tag{7.2-3}$$

Even if $G_p(s)$ and $H_{eq}(s)$ are known in factored form, as they usually are,

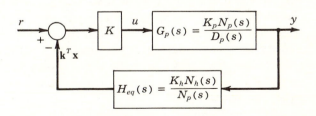

*Fig. 7.2-1*    Basic block diagram used in the root-locus method.

in order to find the closed-loop poles, the direct approach is to factor the denominator polynomial of $y(s)/r(s)$, $D_k(s) = D_p(s) + KK_pK_hN_h(s)$. This direct factoring[1] approach is not simple if the order of the system is higher than 3.

The root-locus method provides a simple graphical method of factoring that gives considerable insight into the nature of the closed-loop system. Let us consider, once again, the closed-loop transfer function of Eq. (7.2-1). The closed-loop poles occur at the values of $s$ for which the denominator of $y(s)/r(s)$ is zero. Hence let us set the denominator equal to zero

$$1 + KG_p(s)H_{eq}(s) = 0 \qquad\qquad (7.2\text{-}4)$$

and solve the resulting equation to obtain

$$KG_p(s)H_{eq}(s) = -1 = \underline{1/180°} \qquad\qquad (7.2\text{-}5)$$

The key idea of the root locus is to write Eq. (7.2-5) as two equations expressing the separate satisfaction of the magnitude and phase-angle requirements of Eq. (7.2-5) as

$$|KG_p(s)H_{eq}(s)| = 1 \qquad\qquad (7.2\text{-}6)$$

and

$$\arg KG_p(s)H_{eq}(s) = 180° \qquad\qquad (7.2\text{-}7)$$

The root-locus diagram is a plot of all the points, values of $s$, that satisfy the phase-angle criterion of Eq. (7.2-7), independent of the magnitude criterion of Eq. (7.2-6). Once the phase-angle condition is met, one may select $K$ to satisfy the magnitude condition.[2] In other words, the root locus is the locus of the closed-loop poles for all values of $K$ from 0 to $\infty$. This range for $K$ obviously includes any specific positive value of $K$. The closed-loop poles for a specific $K$ may be found by determining the points on the root locus that satisfy the magnitude criterion for the given $K$. Note that in this procedure we need only examine points on the root locus, not every point of the $s$ plane, because the closed-loop poles for all positive $K$ are by definition on the root locus. The advantage of this two-step procedure is that the root locus, that is, the points that satisfy the phase-angle criterion, may be determined with relative ease. In addition, the location of the closed-loop poles for all other values of gain have also been found, as well as the manner in which the poles move with gain variations.

The construction of the locus of points that satisfy the phase-angle

---

[1] Methods of factoring are discussed in Appendix C.
[2] See Rule 5 given below.

criterion is accomplished directly on the complex $s$ plane by the consideration of $KG_p(s)H_{eq}(s)$ as a complex number.   We begin by placing the
poles and zeros of $KG_p(s)H_{eq}(s)$ on the $s$ plane.   Since $KG_p(s)H_{eq}(s)$ may
be written as

$$KG_p(s)H_{eq}(s) \;=\; KK_p\frac{N_p(s)}{D_p(s)}\,K_h\frac{N_h(s)}{N_p(s)} \;=\; KK_pK_h\frac{N_h(s)}{D_p(s)} \qquad (7.2\text{-}8)$$

we see that we place the zeros of $H_{eq}(s)$ and the poles of $G_p(s)$ on the $s$
plane.   *If there are common factors in $N_h(s)$ and $D_p(s)$, these must not be
canceled.*   The reason for this restriction will become clear later in the
discussion.

Once the poles of $G_p(s)$ and the zeros of $H_{eq}(s)$ have been placed on
the $s$ plane, the determination of the root locus involves the determination
of all values of $s$ that satisfy the angle criterion of Eq. (7.2-7).   One can
find out if a given point satisfies the angle criterion by graphically determining the arg $KG_p(s)H_{eq}(s)$ from the $s$ plane on which the poles and
zeros of $KG_p(s)H_{eq}(s)$ have been placed.

Consider, for example, the pole-zero plot for $KG_p(s)H_{eq}(s)$ shown
in Fig. 7.2-2.   At the point $s = -1 + j1$, the arg $KG_p(s)H_{eq}(s)$ is

$$\text{arg } KG_p(s)H_{eq}(s)\,\Big|_{s=-1+j1} \;=\; -135° - 26.6° + 18.4° \;=\; -143.2°$$

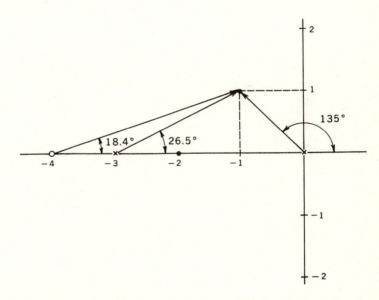

**Fig. 7.2-2**   Pole-zero plot of $KG_p(s)H_{eq}(s)$ used to determine the
root locus.

and this point does not lie on the root locus. On the other hand, the point $s = -2 + j0$ does lie on the root locus since

$$\arg KG_p(s)H_{eq}(s) \big|_{s=-2+j0} = -180° - 0° + 0° = 180°$$

It is clear that, if we proceeded in a random fashion to examine individual points on the $s$ plane, the determination of the root locus would be a lengthy process. Fortunately there are a number of simple rules that make the sketching of the root locus a simple task. In the remainder of this section we consider the simplest rules by discussing only situations in which the branches of the root locus remain on the real axis. These rules are applicable in the general case, however, and additional construction procedures are given in the following section.

> **Rule 1    Number of branches.**    There is one branch of the root locus for every closed-loop pole of $y(s)/r(s)$, and the total number of branches is equal to the number of poles of the open-loop transfer function $KG_p(s)H_{eq}(s)$, that is, the number of poles of $G_p(s)$.

A consideration of Eq. (7.2-3) easily establishes the veracity of the rule. Since $D_k(s) = D_p(s) + KK_pK_hN_h(s)$, $D_k(s)$ is of the same order as $D_p(s)$ since $N_h(s)$ is of $(n-1)$st order and cannot cancel the $s^n$ term in $D_p(s)$.

> **Rule 2    Starting points** $(K = 0)$. The branches of the root locus start at the poles of the open-loop transfer function $KG_p(s)H_{eq}(s)$, that is, at the poles of $G_p(s)$.

The truth of this statement is easily seen from Eq. (7.2-6). As $K$ approaches zero, the only way for the product to remain a constant is for $G_p(s)H_{eq}(s)$ to approach infinity. The quantity $G_p(s)H_{eq}(s)$ approaches infinity as $s$ approaches the poles of the open-loop transfer function. Thus the open- and closed-loop poles are identical as $K$ goes to zero.

An alternative method of establishing this rule is to consider Eq. (7.2-3) again. Since $D_k(s) = D_p(s) + KK_pK_hN_h(s)$, the zeros of $D_k(s)$, that is, the poles of $y(s)/r(s)$, approach the zeros of $D_p(s)$, that is, the poles of $G_p(s)$, as $K$ approaches zero. Obviously for $K = 0$, $D_k(s) = D_p(s)$ and the open- and closed-loop poles are identical.

> **Rule 3    End points** $(K = \infty)$. The branches of the root locus end at the zeros of the open-loop transfer function $KG_p(s)H_{eq}(s)$, that is, the zeros of $H_{eq}(s)$.

The argument here is similar to that used in the discussion of Rule 2. Again look at Eq. (7.2-6). If $K$ is now going to infinity, the product can remain constant only if $G_p(s)H_{eq}(s)$ is going to zero. Thus the closed-loop poles are at the open-loop zeros when $K$ is infinite. We shall see in the examples of this section that a zero might lie at infinity. In the following section we shall be concerned with the question "where at infinity." For the present it is sufficient to point out that the branches of the root locus terminate at the open-loop zeros.

> **Rule 4    Behavior along the real axis.** A point on the real axis is a point on the root locus if an odd number of poles and/or zeros lie to the right of the point.

This statement is easily verified from the phase-angle criterion [Eq. (7.2-7)]. Consider an arbitrary point on the real axis, such as the point $s = -a$ in Fig. 7.2-3. At this point the phase angle of the open-loop transfer function $KG_p(s)H_{eq}(s)$ is the sum of all the phase angles associated with the poles and zeros of $KG_p(s)H_{eq}(s)$. These phase angles are indicated in Fig. 7.2-3. The contribution to the phase angle from a set of complex conjugate poles or zeros is zero, since their angles are of opposite signs. The phase-angle contribution of each real pole or zero to the left of the point $s = -a$ is zero. The contribution of each pole or

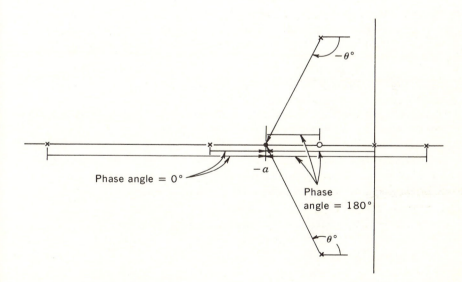

*Fig. 7.2-3*   Illustration of Rule 4.

zero to the right of the point is 180° plus some multiple of 360° for an effective phase angle of 180°. In Fig. 7.2-3 there are two poles and one zero to the right of $s = -a$, so that arg $KG_p(-a)H_{eq}(-a) = 2(-180°) + 180° = 180°$. The phase-angle criterion is satisfied; hence the point $s = -a$ must be a point on the root locus.

> ***Rule 5   Gain determination.*** The gain at an arbitrary point $s_1$ on the root locus is given as

$$K = \frac{1}{|G_p(s)H_{eq}(s)|}\bigg|_{s=s_1} = \frac{1}{K_pK_h}\frac{|D_p(s)|}{|N_h(s)|}\bigg|_{s=s_1} \tag{7.2-9}$$

This rule follows directly from the magnitude criterion [Eq. (7.2-6)]. We have assumed that $K$ is positive, and so Eq. (7.2-6) can be rewritten

$$K|G_p(s)H_{eq}(s)| = 1$$

Then, for $s = s_1$, $K$ is just

$$K = \frac{1}{|G_p(s)H_{eq}(s)|}\bigg|_{s=s_1}$$

The gain $K$ at any point may be determined by simple substitution into $1/|G_p(s)H_{eq}(s)|$. Usually this is done graphically, in much the same way that the residues were determined in Chap. 4. In this case, the gain associated with a point $s = s_1$ is proportional to the product of the vector distances from the poles of $KG_p(s)H_{eq}(s)$ to $s_1$ divided by the product of the vector distances from the zeros of $KG_p(s)H_{eq}(s)$ to $s_1$. The constant of proportionality is $1/K_pK_h$. In other words, $K$ is given by

$$K\bigg|_{s_1} = \frac{1}{K_pK_h}\frac{\Pi[\text{distance to poles of } KG_p(s)H_{eq}(s)]}{\Pi[\text{distance to zeros of } KG_p(s)H_{eq}(s)]}\bigg|_{s=s_1} \tag{7.2-10}$$

The application of these simple rules to the construction of a root-locus diagram is indicated in the following example.

> ***Example 7.2-1*** This example is based upon the system of Fig. 7.2-4. The problem is to determine the root-locus diagram for this closed-loop configuration. Of particular interest is the nature of $y(s)/r(s)$ for $K = 10$.
>
> For this particularly simple problem, $G_p(s) = 1/[s(s + 6)]$, and $H_{eq}(s)$ is found by inspection to be $\frac{1}{2}(s + 2)$ so that the open-loop transfer function is
>
> $$KG_p(s)H_{eq}(s) = \frac{K\frac{1}{2}(s + 2)}{s(s + 6)}$$

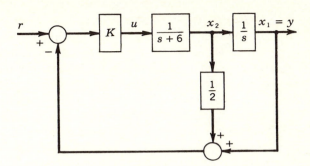

**Fig. 7.2-4**   Closed-loop system for Example 7.2-1.

where $K_p = 1$ and $K_h = \frac{1}{2}$. The first step is to mark on the $s$ plane
the open-loop pole and zero locations, as shown in Fig. 7.2-5a. The
next step is to examine the real axis to determine regions in which
the branches of the root locus might lie. Points on the positive real
axis have no poles or zeros to the right of them. Hence no point on
the positive real axis is a point on the root locus. Said in another
way, there is no value of $K > 0$ for which a closed-loop pole lies
on the positive real axis. However, for points between $s = -2$ and
$s = 0$, one pole lies to the right, namely, the pole at $s = 0$. Hence
every point on the line from 0 to $-2$ must lie on the root locus.

     For points on the real axis between $s = -6$ and $s = -2$, one
pole and one zero lie to the right. Since the sum of the number of
poles and zeros to the right of any point on a line between $-2$ and
$-6$ is 2, the root locus does not exist at any point on this line segment. For any value of $s$ less than $-6$, the sum of poles and zeros
to the right is 3; hence this is also a branch of the root locus. The
final root-locus diagram is given in Fig. 7.2-5b. The arrows on each
of the branches indicate the direction of increasing $K$. Note that
one branch of the root locus is going to infinity along the negative
real axis. It is easily verified that $s$ equal to minus infinity is a zero

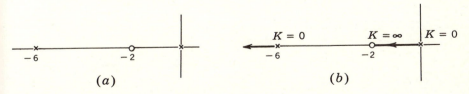

**Fig. 7.2-5**   Development of the root locus for Example 7.2-1.   (a) Pole-zero plot;
(b) root locus.

of $G_p(s)H_{eq}(s)$ since, at $s = -\infty$, $|G_p(s)H_{eq}(s)|$ is zero, so that the two branches of the root locus do terminate on zeros of $G_p(s)H_{eq}(s)$.

Now let us turn our attention to finding the closed-loop poles for $K = 10$. There is a point on *each* branch of the root locus corresponding to the gain of 10, and we must find these two points, as they determine the closed-loop-pole locations. Suppose that we start with the leftmost branch of the root locus, that from $-6$ to negative infinity. Our approach here is to guess an arbitrary point, to determine the gain at that point, and proceed to the correct answer. As an initial guess, we assume a closed-loop pole exists at $s = -8$. The gain at $s = -8$ is computed from the vector lengths indicated in Fig. 7.2-6a. This gain is

$$K = \frac{1}{\frac{1}{2}} \frac{(8)(2)}{6} = 5.34$$

This is below the desired gain of 10; hence we must move along the locus in the direction of higher gain. As a second try, let us assume the closed-loop pole is at $s = -12$. The vector lengths involved in the calculation of $K$ are indicated in Fig. 7.2-6b, with the result that

$$K = \frac{1}{\frac{1}{2}} \frac{(6)(12)}{10} = 14.4$$

This value is above the desired value of 10, and so we know on the basis of these two trials alone that the closed-loop pole lies between $s = -8$ and $s = -12$. The closed-loop pole is actually at $s = -10$, which might appear as a logical third choice.

One closed-loop pole also exists on the branch of the root locus

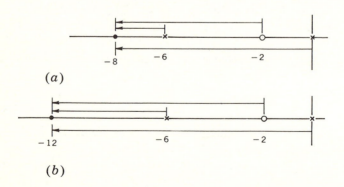

(a)

(b)

**Fig. 7.2-6**  Vector lengths involved in the determination of
$K$.   (a) Point at $s = -8$; (b) point at $s = -12$.

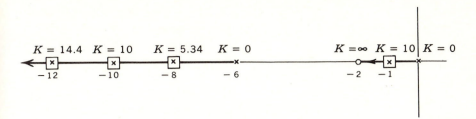

*Fig. 7.2-7*    Root locus for Example 7.2-1.

between $s = 0$ and $s = -2$. At the point $s = -1$, the vector lengths necessary to determine $K$ are 1, 1, and 5, so that at that point $K$ is the required value of 10. Now not only are the closed-loop poles known but also the whole closed-loop transfer function, since the numerator of $y(s)/r(s)$ is the numerator of $G_p(s)$. In this simple problem, $N_p(s)$ is just 1. By substituting into Eq. (7.2-3) it is seen that $y(s)/r(s)$ is

$$\frac{y(s)}{r(s)} = \frac{KN_p(s)}{D_k(s)} = \frac{K}{(s+1)(s+10)} = \frac{10}{(s+1)(s+10)} \qquad (7.2\text{-}11)$$

Because this problem is second-order, clearly it would have been easier simply to factor the denominator in order to find the closed-loop poles. The point here is to illustrate the use of the simple rules for drawing the root locus and to show in a graphical way the additional information regarding system performance that is readily gained from the root-locus method. Figure 7.2-7 is a redrawing of Fig. 7.2-5b with the addition of calculated gain values along the two branches of the root locus. From the very few values of gain indicated, it is clear that for low gain the closed-loop pole exists very near the origin or the system has a large time constant. As gain is increased to 10, the value of interest in this problem, the dominant time constant becomes $-1$. As gain is changed from 10 to infinity, this closed-loop pole moves only from $s = -1$ to $s = -2$. On the other hand, the closed-loop pole to the left of the point $s = -6$ moves toward infinity as the gain is increased.

Two other points are worth mentioning with respect to Example 7.2-1. In evaluating $K$ from the measurement of vector lengths, no attention is given to the phase angles of the vectors involved since it is already known that any point along the root locus satisfies the angle condition. This is, in fact, the way in which the root locus is drawn, to satisfy the angle condition.

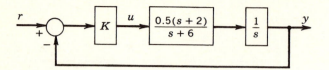

**Fig. 7.2-8**    Exercise 7.2-1.

The second point concerns conclusions regarding the closed-loop-system performance.   As discussed in Chap. 4, it is possible to determine graphically the contribution to the output made by each pole.   This essentially amounts to finding the residue, and we developed in Chap. 4 a method of finding the residues, based on measured lengths and angles of the system pole-zero plot.   The root locus is used to find the closed-loop poles but *not* the closed-loop zeros; these are already known.   In Example 7.2-1, for instance, one might be tempted to conclude that, as the gain is increased, the closed-loop pole that is approaching the zero at $s = -2$ also has a decreasing residue because of the proximity of the zero.   This is not true, because this zero does not appear in the closed-loop system.   More will be said on this point in Sec. 7.5.

**Exercises 7.2**    *7.2-1.*   Show that the root locus for the system in Fig. 7.2-8 is identical to the root locus of Example 7.2-1.   Find the closed-loop poles, graphically on the root locus, for $K = 10$; check by using the quadratic formula.

*answer:*

Closed-loop poles at $s = -1$ and $-10$.

*7.2-2.*   Find the root locus for the system in Fig. 7.2-9.   Find the locations of the closed-loop poles if $K = 10$.

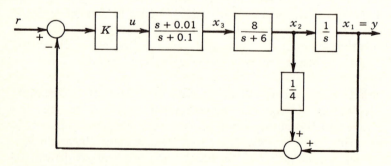

**Fig. 7.2-9**    Exercise 7.2-2.

*answer:*

Closed-loop poles at $s \sim -3.58$, $-22.5$, and $-0.01$.

## 7.3 *Additional root-locus construction rules*

The preceding section introduced a number of basic construction rules for drawing the root-locus diagram. These rules are aimed at providing a systematic means of determining just where on the $s$ plane it is possible for the poles of the closed-loop system to lie as $K$ is varied from zero to infinity. Although the construction rules cited in Sec. 7.2 are generally applicable, they were applied only to those cases in which the root locus is confined to the real axis. In this section further construction rules are given to handle those cases in which the closed-loop poles may be complex. Because complex roots always appear as complex conjugates, the root-locus diagram is always symmetric about the real axis and hence we have the following rule.

> *Rule 6 **Symmetry of locus.*** The root-locus diagram is always symmetric with respect to the real axis.

If the loop transfer function $KG_p(s)H_{eq}(s)$ has complex conjugate poles, then branches of the root locus must lie off the real axis, since they start at these complex conjugate poles. However, complex conjugate poles in the closed-loop system may also arise even if the open-loop transfer function has only real poles. If this occurs, it is then necessary for branches of the root locus to leave the real axis and enter the upper and lower portions of the $s$ plane. We begin the discussion by considering the case of breakaway from or reentry to the real axis.

> *Rule 7 **Breakaway or reentry points on the real axis.*** A point of breakaway from the real axis occurs at a relative maximum of the gain. Two branches of the root locus return to the real axis at a point of relative minimum gain.

Consider the situation in which every point on the real axis between two poles satisfies the angle condition. The simplest case in which this occurs is when the open-loop transfer function has only two real poles, as in Fig. 7.3-1*a*. In this simple case, it is known from Sec. 7.2 that there are two branches of the root locus and that each of these branches starts at one of the poles, as indicated in Fig. 7.3-1*b*. As the gain increases, these two separate branches of the root locus approach each other until

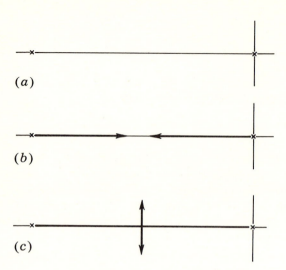

**Fig. 7.3-1** Illustration of breakaway. (*a*) Pole-
zero diagram; (*b*) branches move
together; (*c*) branches meet and break
away from the real axis.

they finally meet. Since the branches of the root locus must terminate on zeros, they cannot remain on the real axis, as no zeros exist there. Hence they must break away from the real axis to reach the zeros, wherever they exist, as indicated in Fig. 7.3-1*c*. Our interest here is in finding the point of breakaway.

From what was said immediately above, it is clear that the point of breakaway is the point of highest gain on the line segment joining the two poles. This point of highest gain can be found in a number of ways. The way that is advocated here is simply to determine the gain at a number of points along the line segment. The point of highest gain is then the point of breakaway.

Before doing an example, let us consider the analogous situation in which the line connecting two real zeros satisfies the angle condition. For infinite gain, a branch of the root locus terminates on each one of these zeros. Consequently, two branches of the root locus must meet somewhere between the two zeros, and the place of the meeting must be a point of minimum gain. The methods of finding this point of minimum gain are identical to those used to find the point of maximum gain associated with the breakaway point. Again, the easiest approach is to examine a number of points on the line segment until a point of minimum gain is found.

Once the point of breakaway or reentry has been established, the next question to be answered is: What angle do these branches of the root locus make with the real axis?

**Rule 8    *Breakaway or reentry angle.*** The branches of the root locus are separated by an angle of $180°/\alpha$ at a point of breakaway or reentry, where $\alpha$ is the number of branches intersecting at the point.

For $\alpha$ equals 2, which is really the only practical case, the angle separating branches of the root loci is 90°.

A simple example involving breakaway points is the case of two simple poles, illustrated in Fig. 7.3-1. This case was also discussed in the introductory section of this chapter, and the final root-locus diagram is as given in Fig. 7.1-2. A simple root locus involving both a breakaway point and a reentry point involves two poles and one zero. This is the case chosen for Example 7.3-1.

**Example 7.3-1** In this example we assume that the control system to be investigated is as shown in Fig. 7.3-2a so that

$$G_p(s) = \frac{4}{s(s+2)} \qquad H_{eq}(s) = \tfrac{1}{3}(s+3)$$

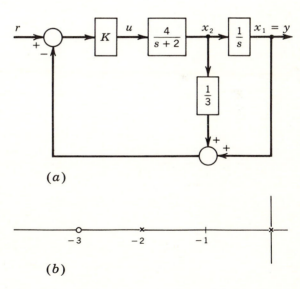

(a)

(b)

**Fig. 7.3-2** Example 7.3-1. (a) System; (b) pole-zero plot.

and thus

$$K_\rho = 4 \qquad K_h = \tfrac{1}{3}$$

and the loop transfer function is

$$KG_p(s)H_{eq}(s) = K\,\frac{4}{3}\,\frac{s+3}{s(s+2)}$$

The pole-zero plot of the loop transfer function is shown in Fig. 7.3-2b.  The complete root-locus diagram is given in Fig. 7.3-3, and a number of gain values are indicated.  The breakaway point is chosen as that point between $s = -2$ and $0$ where the gain is the highest.  The reentry point is chosen to the left of $s = -3$ at a point midway between two values of very nearly the same gain.  Clearly the minimum value of gain must exist between these two values, if gain is to increase as the branches of the root locus move from the reentry point to the real zero of $s = -3$ and to the other zero at minus infinity.

In order to complete the root locus it was necessary to examine a few points in the upper half plane.  (Recall that the root locus is symmetrical about the real axis so that the lower half plane need not be considered.)  This is done by finding the point on the line

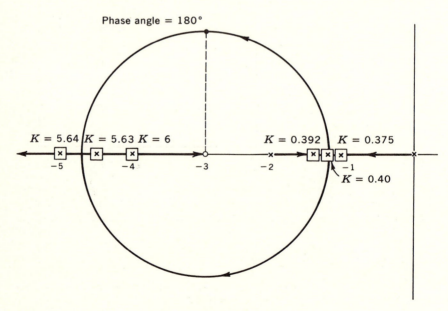

**Fig. 7.3-3**  Root locus for the system of Fig. 7.3-2.

$s = -3$ at which the phase angle equals 180°. This point is also indicated in Fig. 7.3-3. By Rule 8, the branches of the root locus must be 90° apart at the breakaway and reentry points. Hence the angles made by the root loci with the real axis must be $\pm 90°$, as shown in Fig. 7.3-3. The rest of the root locus is made up of the best smooth curve that passes through the point on the $s = -3$ line and is perpendicular to the real axis at the point of breakaway or reentry. In this case the curve is a semicircle, but in more complicated cases there may be no analytic expression for a given segment of the locus.

Before we leave this example, it is instructive to examine the effect of moving the zero location. If the zero is moved to minus infinity, the system has only two poles, and the root-locus diagram is that of Fig. 7.3-1. That is, the radius of the semicircle has gone from its present value to infinity, and the breakaway point is midway between the poles. If the zero is moved toward the pole at $s = -2$, the radius of the semicircle shrinks from its present value until, when the zero is at $s = -2$, the radius of the semicircle is zero. As the zero goes to $s = -2$ the breakaway point always moves toward the left. That is, the proximity of the zero to the pole influences the breakaway point. The closer the zero is to the leftmost pole, the closer the breakaway point is to that pole.

One more comment is in order with regard to Rule 7. Note that the statement of the rule used the terms "relative maximum" and "minimum." In Example 7.3-1, on each segment of the root locus there was only one maximum or minimum so that the absolute and relative maximum and minimum are the same. However, it is entirely possible that there may be more than one breakaway or reentry point on any line segment of the real axis that satisfies the angle condition. Cases in which this happens will be discussed in the following section, when more construction rules are available.

For large values of the gain $K$ the branches of the root locus approach the zeros at infinity along straight lines called asymptotes.

***Rule 9    Asymptotic behavior for large K.***  The asymptotes intersect the real axis at angles given by

$$\phi_{\text{asy}} = \frac{180° + k360°}{p - z} \qquad k = 0, 1, \ldots, p - z - 1 \qquad (7.3\text{-}1)$$

The single point of intersection on the real axis, called the origin of the asymptotes $OA$, is given by

$$OA = \frac{\Sigma(\text{pole locations}) - \Sigma(\text{zero locations})}{p - z} \qquad (7.3\text{-}2)$$

In Eqs. (7.3-1) and (7.3-2) the letter $p$ designates the number of poles of the loop transfer function $KG_p(s)H_{eq}(s)$ and the letter $z$ indicates the number of finite zeros of this same transfer function.[1]   The quantity $p - z$ is referred to as the *pole-zero excess* of $KG_p(s)H_{eq}(s)$.   If all the state variables are fed back, $p - z$ is normally equal to 1.

Here we are interested in cases where $p - z$ is $\geq 2$.   In all the examples of Sec. 7.2, $p - z$ was 1, and in that case the existence of a branch along the negative real axis out to infinity was apparent from Rule 4 concerning real-axis behavior.

It is relatively easy to verify Eq. (7.3-1) by the use of the magnitude condition

$$|KG_p(s)H_{eq}(s)| = 1$$

As $K$ becomes large, this equation can be satisfied only if $|G_p(s)H_{eq}(s)|$ becomes small.   Since this loop transfer function has more poles than zeros, $|G_p(s)H_{eq}(s)|$ becomes small as $|s|$ is increased so that, for large values of $K$, the quantity $KG_p(s)H_{eq}(s)$ behaves approximately as

$$KG_p(s)H_{eq}(s) \sim \frac{KK_pK_h}{s^{p-z}} = -1$$

or

$$s^{p-z} = -KK_pK_h = KK_pK_h\underline{/180°}$$

Clearly the magnitude of $s$ is going to infinity as $K$ goes to infinity. The angles of the values of $s$ that satisfy the above equation are

$$\arg s = \frac{180° + k360°}{p - z} \qquad k = 0, 1, \ldots, p - z - 1$$

For $p - z$ equal to 2, the angles of the asymptotes are just $\pm 90°$, and for $p - z$ equal to 3, the angles of the asymptotes are $\pm 60°$ and $180°$. In every case the asymptotes are always separated by equal angles, equal to $360°/(p - z)$.

It is more difficult to establish the truth of Eq. (7.3-2).   Rather than resort to an obscure theorem of linear equations, let us present a heuristic argument.   Recall that we are concerned with large values of $K$ and hence large values of $s$.   From a point near infinity along one of the asymptotes the original collection of poles and zeros is somewhat obscure, because the distance separating the individual poles and zeros is very small compared with the distance from any one of them to a point near infinity along one of the asymptotes.   As a consequence the individual character

---

[1] In the plant transfer function $G_p(s)$ we have consistently used $n$ to designate the number of poles and $m$ the number of finite zeros.

of each pole and zero is lost to an observer far away, so that the original pole-zero plot looks just like an isolated system of $p - z$ poles. The analogy often used is that of positive and negative charges, analogous to the poles and zeros of $G_p(s)H_{eq}(s)$. From far away a group of $p$-positive and $z$-negative charges appears as just a charge $p - z$. The location of the single charge of magnitude $p - z$ is the centroid of the individual charges that make up the composite charge $p - z$. Equation (7.3-2) is just the expression for the centroid of the poles and zeros of the open-loop transfer function, which, when viewed from a distance large compared with their separation, appears as $p - z$ poles located at their centroid.

The application of the rule for the asymptotic behavior of the branches of the root locus for large $K$ is illustrated in the following example of an output-feedback control system.

**Example 7.3-2**  This example is concerned with the positioning system of Fig. 7.3-4, where the open-loop transfer function is given by

$$KG_p(s)H_{eq}(s) = \frac{6K}{s(s + 1)(s + 2)}$$

Here only the output state variable is fed back, and the only parameter free for adjustment is $K$. A natural question concerns the behavior of this closed-loop system as $K$ is varied. Of particular interest is the question of stability for large $K$. We know from the Bode method of Chap. 6, for example, that the final slope of the straight-line amplitude diagram is $-3$, with an associated phase shift approaching $-270°$. Hence, for sufficiently high gain, the system can be unstable. A similar conclusion can be reached on the basis of the Nyquist criterion. The Nyquist plot approaches the origin from an angle of $-270°$ for large $\omega$, or from the second quadrant of the polar plane. Thus there is danger of encirclement of the $-1$ point. The root-locus diagram should give the same stability information, plus additional information on the location of the closed-loop poles for any value of gain.

In Fig. 7.3-5a the poles and zeros of the given loop transfer

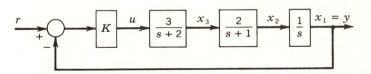

**Fig. 7.3-4**  System for Example 7.3-2.

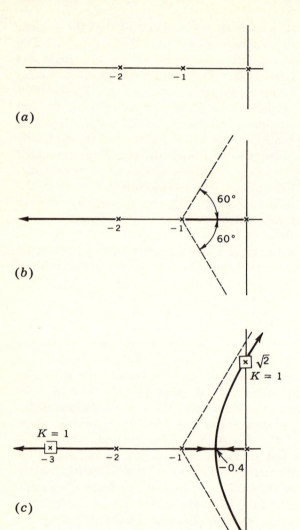

(a)

(b)

(c)

*Fig. 7.3-5*   Example 7.3-2.   (a) Pole-zero plot; (b)
asymptotes; (c) final root locus.

function are shown. Before finding the breakaway point, let us find the origin of the asymptotes and the angle of the asymptotes. From Eq. (7.3-2) the origin of the asymptotes is

$$OA = \frac{(-0 - 1 - 2) - 0}{3 - 0} = -1.0$$

and from Eq. (7.3-1) the angles of the asymptotes are

$$\phi_{asy} = 60°, 180°, 300°$$

These asymptotes are indicated in Fig. 7.3-5b, and the branches of the root locus along the real axis have been completed. The breakaway point from the real axis is found, as discussed under Rule 7, to be at approximately $s = -0.4$. In order to complete the locus, it would be helpful to know the point at which the branches of the root locus cross into the right-half plane. This is the topic of the next construction rule, and there it is shown that the crossover point is at $s = \pm j\sqrt{2}$. The measured gain at this point is $K = 1$. With this information, the final root-locus diagram is given in Fig. 7.3-5c.

The reader may wonder at the use of the word "final" in the preceding sentence. Clearly we are guessing at the exact root-locus diagram, since we have only approximately established the breakaway point and the point at which the root locus crosses into the right-half $s$ plane. The branches of the root locus that exist in the upper and lower half planes are drawn on the basis of these two points and the asymptotes and hence are not exact. However, the degree of accuracy of Fig. 7.3-5c, for example, is usually enough to serve as the basis for the selection of a desired gain $K$. Any more accuracy than this would be warranted only if one were sure that the pole positions were exactly as indicated, at $s = -1$ and $-2$. Because of the complex nature of many control systems, it is often difficult to establish the plant pole positions to within more than 5 or 10 percent.

Before leaving this example, note that, in this case, the breakaway point is to the right of the midpoint between the two poles. In Example 7.3-1 the breakaway point was to the left of the midpoint. The difference is that, in Example 7.3-1, the additional singular point on the $s$ plane was a zero rather than a pole, as it is in this case. It was pointed out with respect to the earlier example that, as the zero approached the left pole, the breakaway point also approached the same pole. In this case, if the pole at $s = -2$ is moved closer to the pole at $s = -1$, the breakaway point moves farther and farther to the right. If the pole at $s = -2$ is moved to the origin, the breakaway point is the origin itself. The point here is that the breakaway point between two poles on the real axis is influenced by the presence of other poles and zeros in the *s* plane. Often one can guess at which side of the midpoint the breakaway occurs. This is particularly true if there is another singularity on the real axis that is closer to either of the poles in question than other poles or zeros in the *s* plane. This is something of a rule of thumb but, with sufficient experience in drawing root-locus diagrams, the breakaway or reentry point can often be estimated with enough accuracy so that no calculation is needed.

> ***Rule 10  Imaginary-axis crossing.*** The branches of the root locus cross the imaginary axis at a point where the phase shift is 180°.

This rule is nothing more than a statement of the phase-angle criterion, here with respect to the imaginary axis. To find the point on the imaginary axis where the phase shift is 180°, one must start at an arbitrary point and measure the total phase-angle contribution from each of the poles and zeros of the open-loop control system. If the phase angle at the initially assumed point is not correct, that is, is not 180°, the phase angle must be measured at a point not too distant but still on the imaginary axis. This second point may also not have the necessary phase angle, but the two points are usually sufficient to establish a trend so that the third guess is often all that is needed to find the point to within the degree of accuracy required.

In a problem such as that of Example 7.3-2, the point at which the asymptotes cross the imaginary axis is often a good starting point. Consider once again Fig. 7.3-5*b*. There the asymptote is at 60° and crosses the imaginary axis at $s = j\sqrt{3}$. The calculation of the phase angle at that point is indicated in Fig. 7.3-6. The phase angles that are shown are all positive, but since these phase angles are associated with poles and hence are in the denominator of $G_p(s)H_{eq}(s)$, these angles are actually negative in determining the phase angle of $G_p(s)H_{eq}(s)$. Thus the phase angle at the point $s = j\sqrt{3}$ is $-191°$.

Consider the phase angle at the point $s = j\epsilon$. This is very nearly $-90°$, because the angle contributions from the poles at $s = -1$ and $-2$ are nearly zero, and the contribution from the pole at the origin is $-90°$. Thus the actual point of crossover into the right-half plane must lie somewhere between the crossing of the asymptote and the origin of the *s* plane. Clearly it must be nearer the crossover of the asymptotes, since the phase angle there is in error only by 11°. At $s = j\sqrt{2}$, the phase angle is

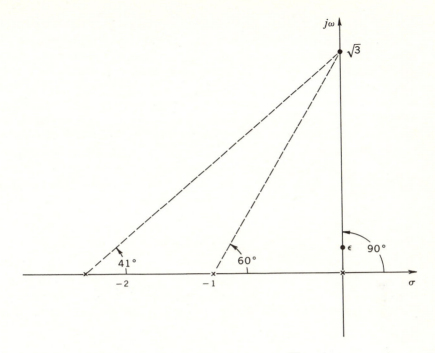

**Fig. 7.3-6**   Calculation of phase shift at $s = +j\sqrt{3}$.

measured to be

$$\phi = -(90° + 55° + 35°) = -180°$$

so that this establishes the crossover point.

>   ***Rule 11   The sum of the closed-loop poles.***   If the pole-zero excess of the open-loop transfer function $KG_p(s)H_{eq}(s)$ is $p - z \geq 2$, the sum of the closed-loop poles remains constant, independent of $K$, and is equal to the sum of the poles of the open-loop transfer function $KG_p(s)H_{eq}(s)$.

To demonstrate the truth of this statement, let us return to Eq. (7.2-3), repeated here as

$$\frac{y(s)}{r(s)} = \frac{KK_pN_p(s)}{D_p(s) + KK_pK_hN_h(s)} \tag{7.2-3}$$

The denominator of $y(s)/r(s)$ is the characteristic equation which is assumed to be of order $n$.    Equation (7.2-3) may be rewritten as a polynomial in $s$ with the highest power $n$.    This is done as

$$\frac{y(s)}{r(s)} = \frac{KK_pN_p(s)}{s^n + d_ns^{n-1} + \cdots + d_2s + d_1} \tag{7.3-3}$$

If the denominator of $G_p(s)H_{eq}(s)$ is of order 2 or higher than the numerator, from Eq. (7.2-3) it is seen that only terms in $D_p(s)$ are involved in $d_n$.    In particular, $d_n$ is not a function of $K$.    In factored form Eq. (7.3-3) becomes

$$\frac{y(s)}{r(s)} = \frac{KK_pN_p(s)}{(s + \lambda_{k1})(s + \lambda_{k2}) \cdots (s + \lambda_{kn})} \tag{7.3-4}$$

or by expanding this factored expression

$$\frac{y(s)}{r(s)} = \frac{KK_pN_p(s)}{s^n + (\lambda_{k1} + \lambda_{k2} + \cdots + \lambda_{kn})s^{n-1} + \cdots + \lambda_{k1}\lambda_{k2} \cdots \lambda_{kn}}$$

Therefore

$$d_n = \lambda_{k1} + \lambda_{k2} + \cdots + \lambda_{kn} \tag{7.3-5}$$

where the $\lambda_{ki}$'s are the closed-loop poles.    Since $d_n$ is a constant, the sum of the closed-loop poles is also a constant and independent of $K$.    When $K$ is zero, the closed-loop poles correspond to the open-loop poles so that this constant is just the sum of the poles of the open-loop transfer function $KG_p(s)H_{eq}(s)$.

Example 7.3-2 serves to illustrate the point.    In that example $n$ is 3 and there are no zeros, so that Rule 11 applies.    The sum of the open-loop poles is $-3$, so that the sum of the closed-loop poles is also $-3$. Thus, when a closed-loop pole exists at the point $s = -3$, this is the pole corresponding to the two pole locations on the $j\omega$ axis, since their sum is zero.    The gain $K$ at $s = -3$ is just

$$K = \tfrac{1}{6}(1)(2)(3) = 1$$

Hence this must be the gain at which the two complex conjugate roots cross into the right-half plane, the maximum gain for stability.    The gain corresponding to any other set of complex conjugate roots may be found in a similar way, as illustrated in the following example.

**Example 7.3-3**    In this example $KG_p(s)H_{eq}(s)$ is assumed to be

$$KG_p(s)H_{eq}(s) = \frac{K(s + 1)}{s(s - 1)(s + 6)}$$

This example is chosen to illustrate further the use of Rules 10 and 11 and to demonstrate that the location of a pole in the right-half plane has no effect upon the construction rules. The root locus for this system is shown in Fig. 7.3-7. A breakaway point exists in the right-half plane between $s = 0$ and $s = +1$. Because of the zero at $s = -1$, it is felt that the breakaway point should be to the left of the midpoint between the two poles, or to the left of 0.5. The gains at $s = 0.5$ and 0.4 are 1.08 and 1.09, respectively, and the gain at $s = 0.45$ is 1.11, or very nearly the maximum. Along the imaginary axis the phase shift is 180° at approximately $s = j1.2$. Rather than measure the value of $K$ at this point, let us apply Rule 11. Here the sum of the open-loop poles is $-5$, so that when the complex conjugate poles are on the $j\omega$ axis, the remaining closed-loop pole is at $s = -5$. The gain $K$ at this point is readily found

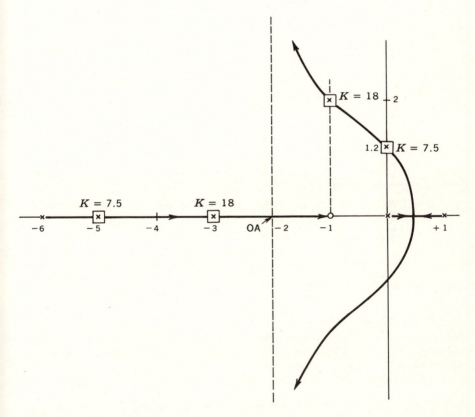

**Fig. 7.3-7**   Example 7.3-3.

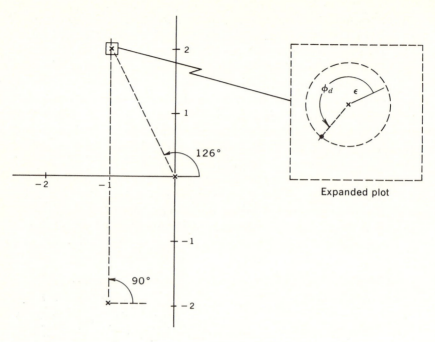

*Fig. 7.3-8*   Computation of departure angle.

to be $(6)(5)(1)/4$ or $7.5$.    Thus, in this system if the gain $K$ ever falls below $7.5$ the system is unstable.

In order to complete the locus in this example it was considered necessary to find one additional point.    This was done by finding the point along the line $s = -1$ at which the phase-angle criterion is satisfied.    The condition is met at $s \approx -1 + j2$, and the corresponding gain is 18, and the real pole is at $s = -3$.

The examples thus far in this chapter have dealt exclusively with those cases in which both the open-loop poles and zeros have all been real.    In cases where hydraulic actuators are used as power elements, complex conjugate poles are quite common, and state-variable feedback often results in complex conjugate zeros.    Hence the last construction rule deals with the angles made by the branches of the root locus as they either leave a complex pole or arrive at a complex zero.

***Rule 12   Angles of departure and arrival.***   The angle of departure from a complex conjugate pole and the angle of arrival at a complex conjugate zero are determined by satisfying the angle criterion at a point very close to the pole or zero in question.

To make the meaning of this rule clear, let us consider the pole-zero plot shown in Fig. 7.3-8. Let us determine the departure angle associated with the pole at $s = -1 + j2$. Suppose we consider a point on a circle of radius $\epsilon \sim 0$, as shown in the expanded plot. The angle at the point for the pole at $s = -1 + j2$ is indicated as $\phi_d$. The angles from this point to the other two poles are essentially the same as the angles to the pole at $s = -1 + j2$ since the radius of the circle is so small compared with the other lengths involved. If we assume now that the point on the circle is also on the root locus, the angle criterion must be satisfied

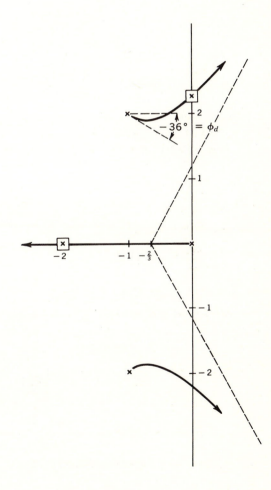

*Fig. 7.3-9*   Final root locus.

at this point so that

$$-(90° + 126° + \phi_d) + 0° = 180°$$

or

$$\phi_d = -36°$$

Near the pole at $s = -1 + j2$, the root locus must depart from the pole at an angle of $-36°$. When this information is used, the root locus takes the form shown in Fig. 7.3-9. Because the lower half of the root-locus diagram is just a mirror image about the real axis, the lower half of the diagram requires no further work.

The following example illustrates the use of Rule 12 in computing the arrival angle at a complex zero.

> **Example 7.3-4**  Complex conjugate zeros are often introduced through the use of state-variable feedback. The block diagram of Fig. 7.3-10 illustrates such a case. Here $KG_p(s)H_{eq}(s)$ is
>
> $$KG_p(s)H_{eq}(s) = \frac{(3K/4)[(s + 2.89)^2 + 2.23^2]}{s(s + 2)(s + 8)}$$
>
> The root locus for this system is shown in Fig. 7.3-11. Included also in this figure are the angles necessary for the calculation of the angle of arrival of the branch of the root locus at the zero at $s = -2.89 + j2.23$. If this angle is called $\phi_a$, then $\phi_a$ is calculated from the formula
>
> $$(90° + \phi_a) - (143° + 114° + 23°) = 180°$$
>
> so that $\phi_a$ is
>
> $$\phi_a = 370° = 10°$$

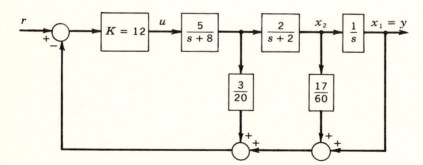

**Fig. 7.3-10**  Block diagram for Example 7.3-4.

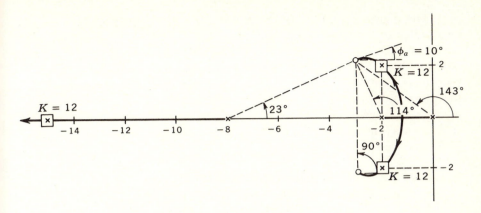

***Fig. 7.3-11***   Completed root locus for Example 7.3-4.

In this example $p - z$ is 1, so that Rule 11 does not apply, and the points corresponding to $K = 12$ must be calculated along each branch of the root locus. These gain points are indicated in the figure. Because the asymptote is at 180°, it is not shown.

Both this section and the preceding one have dealt with the construction rules that make the drawing of the root-locus diagram a relatively simple matter. Occasionally there is need for a few additional points, in order to ensure that the resulting root locus is an adequate approximation of the behavior of the closed-loop poles. In these cases one may always check the phase angle at any point on the $s$ plane to see if it is on the root locus. This was done in Example 7.3-3 in a systematic fashion by examining the phase shift along a vertical line. Any other systematic approach is satisfactory, although one usually chooses to check the phase angle along a straight line through which it is certain the root locus passes.

Before we leave this section it is worth noting that W. R. Evans,[1] the inventor of the root-locus method, has also developed an inexpensive drawing aid to facilitate the drawing of root-locus diagrams. The Spirule[2] is useful in summing angles and multiplying distances necessary in drawing root-locus diagrams. If one is required to draw root-locus diagrams more than just occasionally, it is certainly worthwhile to acquire a Spirule and to learn to use it.

[1] W. R. Evans, Graphical Analysis of Control Systems, *Trans. AIEE*, vol. 67, pt. II, pp. 547–551, 1948.

[2] Available from The Spirule Company, 9728 E. Venado, Whittier, Calif.

The digital computer is also a useful means of developing a root-locus diagram.[1]  If a computer factors the characteristic equation or, equivalently, finds the eigenvalues of the system matrix $\mathbf{A}_k$ for a range of values of $K$, this information may be displayed in graphical form as a root-locus diagram.   However, even if a digital computer is available to plot the root locus, it is often helpful to be able to sketch the general shape of the root locus quickly by hand and then use the digital computer for refinements if needed.   This man-machine procedure is especially helpful in design problems.

**Exercises 7.3**   *7.3-1.*   Rework Example 7.3-1 for $k_2 = \frac{1}{2}$ and $\frac{1}{4}$.   Compare the answers with that obtained in Example 7.3-1 for $k_2 = \frac{1}{3}$.
*7.3-2.*   Plot the root locus for the open-loop transfer function given below.   Determine the origin of the asymptotes, breakaway point, arrival angles, and $j\omega$-axis intersection.   Find the value of $K$ such that there is a closed-loop pole at $s = -4$.

$$KG_p(s)H_{eq}(s) = \frac{K[(s + 2)^2 + 2^2]}{s(s + 2)^2(s + 6)}$$

*answer:*

Breakaway points at $s \sim -0.75$ and $-4.25$; $OA = -3$; $\phi_a = +71°$; no $j\omega$-axis intersection for $\lambda_k = +4$, $K = 4$.

## 7.4   *Additional examples*

This section consists mainly of examples typical of situations that commonly occur in control work; they serve to illustrate further the construction rules of Secs. 7.2 and 7.3, which are summarized for easy reference in Table 7.4-1.

As noted in the introduction to this chapter, the root-locus diagram is a plot of the locus of the roots of the closed-loop system as the gain $K$ is allowed to vary from zero to infinity.   From viewing the entire diagram, we appreciate how gain changes affect the closed-loop-pole locations. For a specific value of gain we are able to state with reasonable accuracy the locations of the closed-loop poles.   The results derived from the root-locus diagram are, on one hand, very general, and on the other, very specific.   The general results concerning the overall behavior of the

---

[1] See J. L. Melsa, "Computer Programs for Computational Assistance in the Study of Linear Control Theory," McGraw-Hill Book Company, New York, 1970.

## Table 7.4-1   Summary of Root-locus Construction Rules

| | |
|---|---|
| Rule 1: *Number of branches* | There is one branch for each pole of the open-loop transfer function $KG_p(s)H_{eq}(s)$. |
| Rule 2: *Starting points* $(K = 0)$ | The branches of the root locus start at the poles of $KG_p(s)H_{eq}(s)$. |
| Rule 3: *End points* $(K = \infty)$ | The branches of the root locus end at the zeros of $KG_p(s)H_{eq}(s)$. |
| Rule 4: *Behavior along the real axis* | The root locus exists on the real axis at every point for which an odd number of poles and/or zeros lie to the right. |
| Rule 5: *Gain determination* | At a point $s_1$ on the root locus, the gain is given by $$K = \frac{1}{|G_p(s)H_{eq}(s)|}\Bigg|_{s=s_1}$$ |
| Rule 6: *Symmetry of locus* | The root locus is always symmetric with respect to the real axis. |
| Rule 7: *Breakaway or reentry points* | The root locus breaks away from the real axis at a point of relative maximum gain and returns to the real axis at a point of relative minimum gain. |
| Rule 8: *Breakaway or reentry angles* | At points of breakaway or reentry, the branches of the root locus are separated by an angle of $180°/\alpha$, where $\alpha$ is the number of branches that intersect. |
| Rule 9: *Asymptotic behavior for large K* | The asymptote angles are given by $$\phi_{\text{asy}} = \frac{180° + k360°}{p - z} \qquad k = 0, 1, 2, \ldots, p - z - 1$$ and the origin of the asymptotes is $$OA = \frac{\Sigma(\text{pole locations}) - \Sigma(\text{zero locations})}{p - z}$$ where $p - z$ is the pole-zero excess. |
| Rule 10: *Imaginary-axis crossing* | The branches of the root locus cross the imaginary axis at points where the phase shift is $180°$. |
| Rule 11: *Sum of the closed-loop poles* | If $p - z \geq 2$, the sum of the closed-loop poles is a constant. |
| Rule 12: *Angles of departure and arrival* | The angles of departure and arrival at complex conjugate poles and zeros are determined by satisfying the angle criterion near the pole or zero in question. |

closed-loop poles as the gain is varied are more important in terms of synthesis, when the pattern of system behavior is to be established. The specific results concerning the location of the closed-loop poles for one given value of gain are more important in the present discussion, since we are still concerned with analysis.   Once the specific closed-loop-pole locations are known, the analysis of the particular system may be carried out to any degree of completeness and accuracy that is desired. As one's facility with the root-locus method develops, so does his appreciation for both the general information and the specific facts that may be acquired from the root-locus diagram.

The first example of this section is meant to clarify the statement of Rule 7.   Note the use of the term "relative" maximum or minimum. In the examples considered thus far, there has been only one maximum or minimum along each real-axis segment of the root locus.   This is not always the case, however, as the following example illustrates.

> ***Example 7.4-1***    Here it is assumed that the loop transfer function $KG_p(s)H_{eq}(s)$ is
>
> $$KG_p(s)H_{eq}(s) = \frac{K(s + 2)}{s(s + 1)(s + 19)}$$
>
> and the complete root-locus diagram is desired.   The pole-zero plot is shown in Fig. 7.4-1.   Since the pole-zero excess is 2, the asymptotes are at $\pm 90°$, and the origin of the asymptotes is
>
> $$OA = \frac{(-19 - 1) - (-2)}{2} = -9$$
>
> A breakaway point must exist between the poles at $s = -1$ and 0. Gain calculations at $s = -0.5, -0.6$, and $-0.7$ yield the following values for $K$: 3.08, 3.15, and 2.96, respectively.   Thus the breakaway point is approximately $s = -0.6$.
>
> One might suspect at first glance that this is the only break-

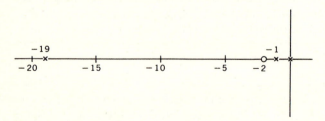

***Fig. 7.4-1***    Pole-zero plot for Example 7.4-1.

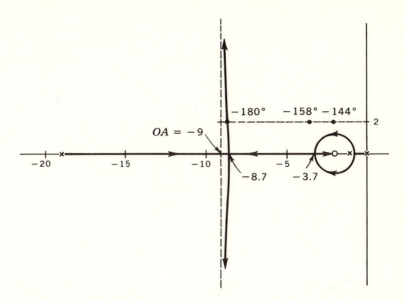

***Fig. 7.4-2*** Root locus for Example 7.4-1.

away point along the negative real axis.   A branch of the root locus exists between $-19$ and $-2$, but since this branch exists between a pole and a zero, no breakaway need occur.   In fact, there is one point of breakaway and one point of reentry, as shown in Fig. 7.4-2. One is led to this conclusion in two ways.

Once the breakaway point at $s = -0.6$ has been established, a first guess is that these two branches seek the asymptotes.   Because these asymptotes are fairly far removed from the breakaway point, it is probably desirable to establish another point on the branches of the root locus that are approaching the asymptote.   For example, one might select the line $s = j2$ and check the phase angle at various points along that line, looking for a point with a phase angle of 180°.   Several such points are indicated in Fig. 7.4-2, and in the region near the breakaway point, the phase angle is far from 180°.   Not until the asymptote is almost reached does the phase angle become the desired value.   This diversity of the two points might indicate that the locus does not proceed directly from the breakaway point to the asymptote.

The necessity of breakaway and reentry points becomes apparent when gain values are marked along the segment of the negative real axis from $s = -2$ to $s = -19$.   At $s = -10$ the gain is 101, whereas at $s = -6$ the gain is only 97.5. If the branch of the root locus were proceeding directly from the pole at $s = -19$ to the zero at $s = -2$, this decrease in gain could not occur. A bit of probing soon establishes that a point of relative maximum gain exists at about $s \approx -8.7$, and a point of relative minimum gain exists at $s \approx -3.7$. Since only two branches of the root locus are involved in each case, the breakaway and reentry angles must be $\pm 90°$. In this simple case the root locus was completed without checking any further points. Since the breakaway point at $s = -8.7$ is near the origin of the asymptotes, and since the point at $s = -8.9 + j2$ near the asymptote has a phase angle of 180°, the leftmost branch of the root locus is sketched by connecting the two points. Because the pole at $s = -19$ is so far removed from the other poles and zeros, its effect on the shape of the root locus near the other poles and zeros is only minor. This breakaway-reentry combination was encountered previously in Example 7.3-1, and the resulting locus was a semicircle. Hence a semicircle was assumed to be a good approximation to the locus, as indicated in Fig. 7.4-2, which is the required root-locus diagram.

An argument used in the above example was that, when a pole (or a zero) is far removed from other poles or zeros, it has only a minor effect on the shape of the root locus in the neighborhood of the other poles and zeros. Such a situation often occurs with pole-zero cancellation. Since we are dealing with physical quantities, perfect cancellation is possible only on paper. However, if cancellation is attempted, a pole-zero pair results with a separation that is small compared with the distances to other poles and zeros. The next example considers pole-zero cancellation on the real axis.

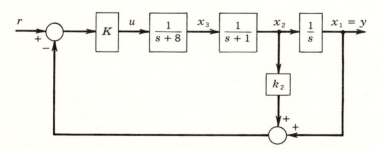

*Fig. 7.4-3*   Example 7.4-2.

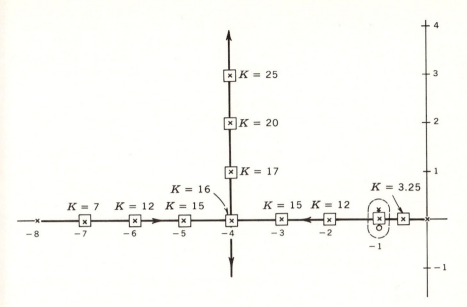

***Fig. 7.4-4***   Root locus for Example 7.4-2, with $k_2 = 1$.

**Example 7.4-2**   Let us consider the closed-loop system shown in Fig. 7.4-3 for which the open-loop transfer function is

$$KG_p(s)H_{eq}(s) = \frac{Kk_2(s + 1/k_2)}{s(s + 1)(s + 8)}$$

where $k_2$ and hence the zero location of $H_{eq}(s)$ are capable of adjustment. It is desired to cancel the pole at $s = -1$ with this zero so that $k_2$ is selected as $k_2 = 1$. Note that in this case a zero of $H_{eq}(s)$ is identical to a pole of $G_p(s)$, and it may appear that these terms should be canceled. The pole-zero combination should not be canceled, however, as indicated in the initial discussion of Sec. 7.2. Zeros of $H_{eq}(s)$ must never be canceled with poles of $G_p(s)$. In order to see why this is true, let us examine the root locus for $k_2 = 1$ as shown in Fig. 7.4-4.

Note that at $s = -1$ we have indicated the location of a pole, a zero, and a closed-loop pole. These should all be located at the point $s = -1$, but the open-loop pole and zero are located off the real axis to emphasize their distinct character. In every case of perfect cancellation, a closed-loop pole *always* exists at the point of cancellation. That is, a branch of the root locus is located between the pole and the zero that happen to be at the same point. In this case the branch takes the form of a single point.    If the original open-loop pole-zero pair had been canceled, this branch of the locus would have been suppressed and the closed-loop pole at $s = -1$ overlooked.

   The breakaway point is at $s = -4$, and the root-locus diagram is identical to that which would result if the pole-zero pair at $s = -1$ had been removed entirely from the drawing.    Angles and distances from the pole and zero are, of course, equal, and they cancel since one term appears in the numerator and the other in the denominator of the open-loop transfer function.    With respect to the rest of the root-locus diagram, the pole and zero that cancel might just as well not be present.    However, they are included so that the closed-loop pole between them is not overlooked.

   If $k_2$ is too small, say $k_2 = 1/1.2$, the zero is at $s = -1.2$ and the root locus of Fig. 7.4-5 results.    Note that the origin of asymptotes is now at $s = -3.9$ rather than at $s = -4$ and that the breakaway

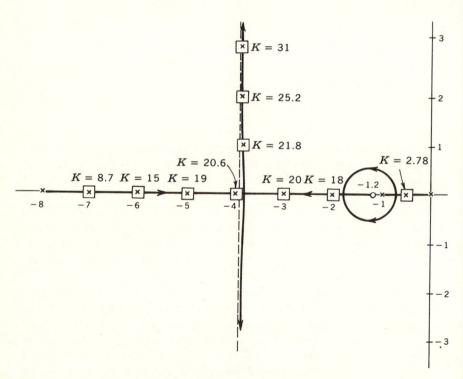

**Fig. 7.4-5**    Root locus for Example 7.4-2, with $k_2 = 1/1.2$.

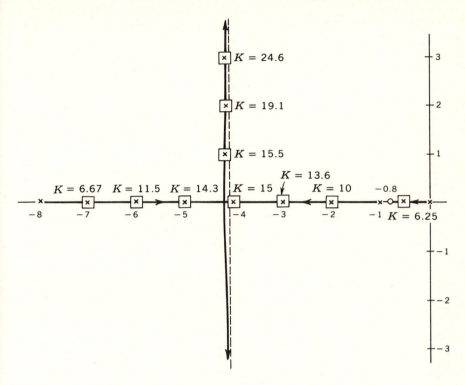

**Fig. 7.4-6**   Root locus for Example 7.4-2, with $k_2 = 1/0.8$.

point is slightly affected. On this diagram there are clearly three branches to the root locus and, of course, three closed-loop poles result, as in Fig. 7.4-4. The main difference here is that, near the point $s = -1$, it is possible for complex conjugate roots to exist. Note one striking similarity, however. For gains beyond approximately 20, the complex conjugate roots with real part $-3.9$ are very close to the same values that occurred when the cancellation was perfect except for the gain change due to the fact that $K_h$ is now equal to $1/1.2$. Also, the one remaining closed-loop pole is very close to the zero on the real axis, as before.

Figure 7.4-6 is the root-locus diagram for the same case with $k_2$ equal to $1/0.8$ so that the zero is at $s = -0.8$. Again the existence of three closed-loop poles is quite evident. This time the origin of the asymptotes is at $s = -4.1$, and there are no complex conjugate roots near $s = -1$. Once again, though, for gains beyond 20, the complex conjugate poles with real parts equal to $-4.1$ are very close to those shown in Fig. 7.4-4 for perfect cancellation, and the real pole is very close to the real zero.

Two conclusions can be drawn from the above example. Often in synthesis work, pole-zero cancellation is used, and as far as the paper-and-pencil analysis is concerned, the resulting root locus is simple to draw. That is, it is simpler to draw than if cancellation were not perfect. If in the actual system the cancellation is not perfect, this is not of great consequence, particularly if the gain is high. The closed-loop poles in the rest of the drawing are largely unaffected, and the closed-loop pole near the canceling zero seeks that zero.

The other conclusion to be drawn is that cancellation of a right-half-plane pole with a zero is utterly impractical. A closed-loop pole always exists in the right-half plane, causing the system to be unstable. A problem at the end of this chapter considers the case of pole-zero cancellation for complex poles.

In Examples 7.4-1 and 7.4-2 we encountered situations in which a portion of the root locus was essentially a semicircle. This occurs when breakaway and reentry points are in close proximity and other poles and zeros are relatively far removed. Essentially it was implied that this situation occurs frequently enough so that all one has to do is find the points of breakaway and reentry and sketch in the semicircle. In a sense this is a dangerous suggestion, as it might lead one to attempt to "remember" root-locus diagrams. The following example clarifies this point.

***Example 7.4-3*** This example deals with the situation in which the loop transfer function has only four poles and no zeros. Three similar pole configurations are considered, as indicated in Fig. 7.4-7. Notice that the only difference in these three cases is the spacing between the poles. In each case the origin of the asymptotes is the same, $s = -4$, and the angles of the asymptotes are also found to be $180°/4$, $(180° - 360°)/4$, etc. The root locus for case $a$ is given in Fig. 7.4-8$a$, and two breakaway points exist on the real axis at approximately $s = -0.8$ and $s = -7.2$. There is also a reentry point at $s = -4$ so that there are three places on the real axis at which branches of the root locus are separated from each other by 90°. Recall that breakaway or reentry always occurs at a 90° angle when only two branches of the root locus are involved.

The root locus for the pole-zero plot of Fig. 7.4-7$b$ is shown in Fig. 7.4-8$b$. Here the breakaway points have been moved to the same point as the point of reentry, and the branches of the root locus take off directly along the asymptotes. This is somewhat of a

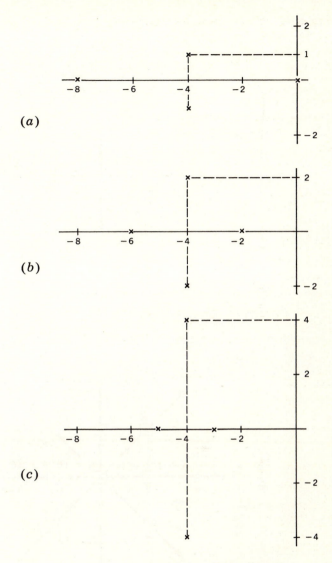

*Fig. 7.4-7*  Three pole configurations that are similar in
form: (*a*) case *a*; (*b*) case *b*; (*c*) case *c*.

pathological case, but the root-locus diagram is surely different
from that of Fig. 7.4-8*a*.

The root locus for the case of Fig. 7.4-7*c* is even more unusual,
as indicated by Fig. 7.4-8*c*.   There two branches of the root locus
meet in the upper and lower half planes rather than on the real

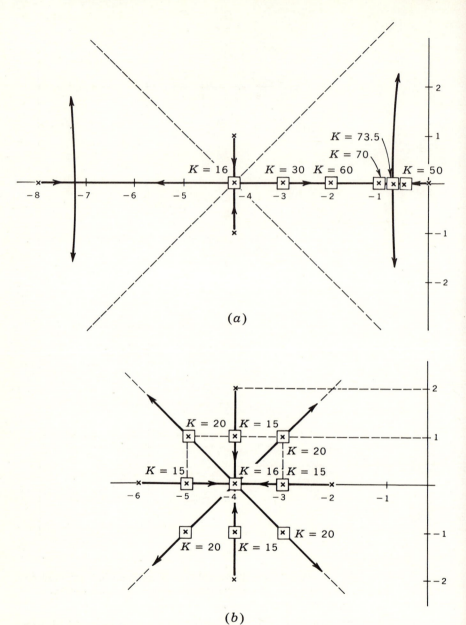

(a)

(b)

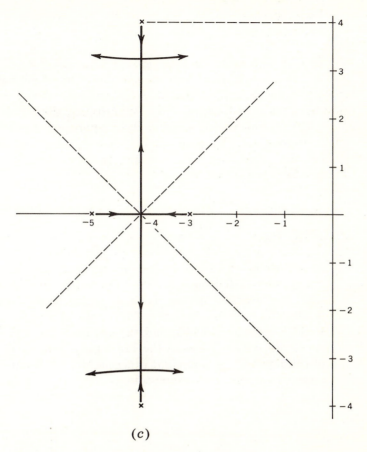

(*c*)

***Fig. 7.4-8***  Root-locus diagrams corresponding to Fig. 7.4-7: (*a*) case *a*; (*b*) case *b*; (*c*) case *c*.

axis.   This situation has not been encountered previously, and it is worthy of some additional comment.   The breakaway point on the real axis is easily located, as is the angle of departure from the uppermost pole.   This angle of departure is −90°.   Because the breakaway from the real axis is apparently on a "collision course" with the branch of the locus departing from the pole at $s = -4 + j4$, one is led to examine the phase angle along the line connecting the two.   Every point on that line satisfies the angle criterion.   Since the branches of the locus terminate only on zeros, the branches of the root locus must separate and approach the zeros at infinity along the asymptotes.   The point at which the branches of the root locus leave the vertical line to approach the asymptotes is a point of maximum gain.

It may appear to the reader that Example 7.4-3 is somewhat a special case and that therefore the caution of not "remembering" root-locus diagrams may not be particularly practical. Many other examples could be given. In Example 7.4-1, for instance, the zero location determines which branches of the root locus approach the asymptotes and which terminate on the zero. It is better to take a few moments and establish the actual root-locus diagram for each specific case than to try to remember typical examples.

**Example 7.4-4** Here the open-loop transfer function is given as

$$KG_p(s)H_{eq}(s) = \frac{K[(s + 1.5)^2 + 1^2]}{s^2(s + 0.5)(s + 8)(s + 9)}$$

A branch of the locus exists on the real axis for $s < -9$ and between $s = -0.5$ and $s = -8$. The maximum value of gain along the latter segment of the real axis occurs at $s = -2.4$, and this is the point of breakaway. The origin of the asymptotes is at $(-17.5 + 3)/3 = -4.83$, and the angles of the asymptotes are at $\pm 60°$ and $180°$. Because the breakaway point is near the origin of the asymptotes, one might expect that these branches of the root locus would follow the asymptotes to infinity. If this were the case,

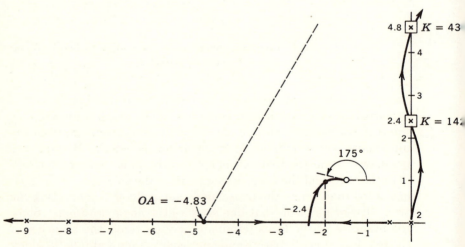

**Fig. 7.4-9** Completed root locus for Example 7.4-4.

the two branches starting at $s = 0$ would then terminate at the two finite complex conjugate zeros. However, the angle of arrival at the upper most complex zero is at 175°, and by determining one more point along the line $s = -2$ where the phase angle is satisfied it is evident that this portion of the locus is as indicated in Fig. 7.4-9.

The only remaining branch of the root locus to be determined is that which starts at the origin and approaches infinity along the asymptotes. Points of 180° phase shift are found to exist along the imaginary axis at $s = j2.4$ and $s = j4.8$. To complete this branch of the locus the phase angle was measured at several points along the lines $s = j1.5$ and $s = j3.5$. The completed root locus is shown in Fig. 7.4-9. Note that the system is stable only for a relatively narrow range of gain from $K = 142$ to $K = 433$.

Systems like the one in Example 7.4-4 that are unstable for both high and low values of gain are said to be *conditionally stable*. Such systems are undesirable in a practical sense, because during the initial turn-on the gain may be low enough to cause instability.

***Example 7.4-5*** The open-loop transfer function for a particular positioning system is

$$KG_p(s)H_{eq}(s) = \frac{0.0833K[(s + 3.25)^2 + 1.21^2]}{s(s + 1)(s + 2)}$$

The complete root-locus diagram is required, as well as the location of the closed-loop poles for a gain $K = 48$. The breakaway point is found to be at approximately $s = -0.5$, as the proximity of the pole at $s = -2$ is negated by the presence of the two complex zeros. The angle of arrival is 187°, and this is somewhat unexpected. The two branches of the root locus that break away from the real axis near $s = -0.5$ must negotiate a rather long path if they are to terminate on the complex zeros at an angle of 187°. The final root locus is shown in Fig. 7.4-10. Because the branch of the root locus in the upper half plane takes such an indirect path to the zero, it is somewhat more difficult to find than in the preceding examples. The construction rules give some help but not a great deal. It is necessary to find points along lines such as $s = j0.5, j1.0, j2.0, j3.0,$ and $j4.0$ that satisfy the angle criterion in order to determine the shape of the complex branches. By connecting these points the locus of Fig. 7.4-10 results. The closed-loop poles at $K = 48$ are indicated.

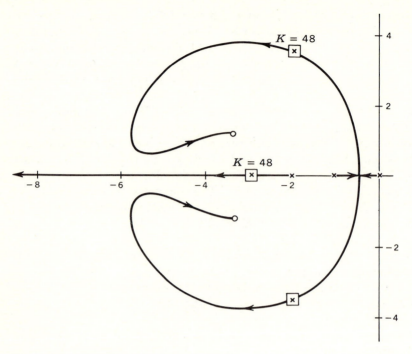

*Fig. 7.4-10*   Root locus for Example 7.4-5.

This concludes the discussion of drawing root-locus diagrams.    The following section is devoted to interpreting the diagram once it is completed.

**Exercises 7.4**   *7.4-1.*   Sketch the root locus for each of the open-loop transfer functions given below.    Determine and label, where applicable, breakaway and reentry points, asymptote origin, asymptote angles, arrival and departure angles, and $j\omega$-axis intersections.    Also indicate the value of gain at breakaway and reentry points and $j\omega$-axis intersection.

$(a)$   $KG_p(s)H_{eq}(s) = \dfrac{K(s+8)}{s(s+2)^2}$

$(b)$   $KG_p(s)H_{eq}(s) = \dfrac{K(s+1)}{s^2(s^2+2s+17)}$

$(c)$   $KG_p(s)H_{eq}(s) = \dfrac{Ks(s+2)}{(s+1)^2(s^2+2s+2)}$

$(d)$   $KG_p(s)H_{eq}(s) = \dfrac{K(s+1)}{s^2(s+2)(s+3)}$

*answers:*

(*a*)    Breakaway at $s \sim -0.7$, $K = 0.162$; $OA = +2$, $\phi_{asy} = \pm 90°$; $j\omega$-axis intersection $s = \pm j2\sqrt{2}$, $K = 4$.

(*b*)    Reentry at $s \sim -2$, $K = 68$; $OA = -\frac{1}{3}$, $\phi_{asy} = \pm 60°$, $180°$; $\phi_d = -30°$ at $s = -1 + j4$; $j\omega$-axis intersection $s = \pm j3.85$, $K = 29$.

(*c*)    $OA = -1$, $\phi_{asy} = \pm 90°$; $\phi_d = 90°$ at $s = -1 + j1$.

(*d*)    $OA = -\frac{4}{3}$, $\phi_{asy} = \pm 60°$, $180°$; $j\omega$-axis intersection $s = \pm j1$, $K = 5$.

*7.4-2.*    Draw the root locus for the open-loop transfer function given below if $\lambda = 3$, 4, and 5.    Superimpose these diagrams upon one another and note the significant trend in the diagrams as $\lambda$ is varied around $\lambda = 4$.

$$KG_p(s)H_{eq}(s) = \frac{K}{s(s + \lambda)(s^2 + 4s + 16)}$$

*7.4-3.*    The root locus for the open-loop transfer function

$$KG_p(s)H_{eq}(s) = \frac{K(s + 6)}{s(s + 2)(s + 4)}$$

is shown in Fig. 7.4-11.    Find the value for $K$ for which the dominant second-order poles have a damping ratio of 0.707.    For this value of gain, find the closed-loop transfer $y(s)/r(s)$ if

$$G_p(s) = \frac{2(s + 3)}{s(s + 2)(s + 4)}$$

*answer:*

(*a*)    $K = 1.05$,

$$\frac{y(s)}{r(s)} = \frac{2.10(s + 3)}{[(s + 0.8)^2 + (0.8)^2](s + 4.2)}$$

*7.4-4.*    Sketch the root locus for the open-loop transfer function given below and find the damping ratio of the dominant second-order poles if $K = 1.5$.

$$KG_p(s)H_{eq}(s) = \frac{Ks(s + 2)}{(s + 1)^2(s^2 + 2s + 2)}$$

*answer:*

$\zeta = 0.5$

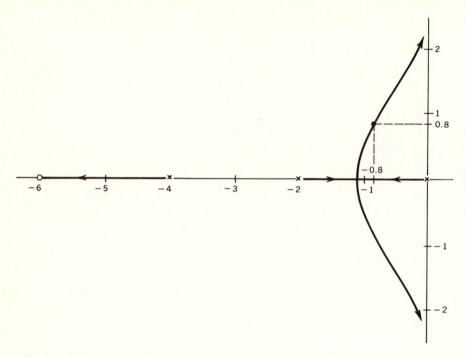

*Fig. 7.4-11*   Exercise 7.4-3.

## 7.5   The closed-loop-response plane

In Chap. 4 we discussed the determination of transient response from an $s$-plane diagram of poles and zeros. In particular, the graphical approach to the determination of both step response and impulse response was emphasized. In Chap. 5 means of determining frequency response from a pole-zero diagram were examined. In each case the pole-zero diagram in question is assumed to represent a transfer function. This transfer function may be the plant transfer function $G_p(s)$, the open-loop transfer function $KG_p(s)H_{eq}(s)$, or the closed-loop transfer function $y(s)/r(s)$.

Here we are concerned with any information about the behavior of the closed-loop system that may be gained from the pole-zero diagram of $y(s)/r(s)$. We have noted that, if $G_p(s)$ is given in factored form, the zeros of $y(s)/r(s)$ are already known. The three preceding sections of this chapter have been concerned with the root-locus method for determining the closed-loop-pole locations. Once the closed-loop-pole locations are known, the pole-zero information is complete.

   Although the root locus is drawn on the $s$ plane, in order to interpret results for a specific value of gain, it is necessary to transfer information from the $s$ plane on which the root locus is drawn to a *closed-loop-response plane* that contains only pole-zero information concerning the closed-loop system.   From this closed-loop-response plane we may find the transient and frequency responses of the closed-loop system.   In order to illustrate the need for the use of the closed-loop-response plane more clearly, let us consider the nature of the open-loop transfer function $KG_p(s)H_{eq}(s)$ and the closed-loop transfer function $y(s)/r(s)$.   Since the zeros of $G_p(s)$ and the poles of $H_{eq}(s)$ are the same, they cancel, and $KG_p(s)H_{eq}(s)$ becomes

$$K_pG(s)H_{eq}(s) = KK_pK_h\frac{N_h(s)}{D_p(s)}$$

In other words, the zeros of the open-loop transfer function are the zeros of $H_{eq}(s)$ and the poles are the poles of $G_p(s)$.   The zeros of the closed-loop transfer function, on the other hand, are the zeros of $G_p(s)$ so that $y(s)/r(s)$ is

$$\frac{y(s)}{r(s)} = \frac{KK_pN_p(s)}{D_k(s)}$$

   Since the root locus is based on the pole-zero plot of $KG_p(s)H_{eq}(s)$, the closed-loop response cannot be obtained directly from the root-locus diagram, since the zeros of $G_p(s)$ are not present on that diagram.   They have been canceled by the poles of $H_{eq}(s)$.   Therefore it is necessary to transfer the closed-loop-pole locations obtained from the root-locus diagram to another $s$ plane on which the zeros of $G_p(s)$ have been placed. Only then may the closed-loop behavior be determined.

   ***Example 7.5-1***   As an example of the utilization of the closed-loop-response plane, let us consider the system pictured in Fig. 7.5-1, where $G_p(s)$ and $H_{eq}(s)$ are readily seen to be

$$G_p(s) = \frac{1}{s(s + 1)} \qquad H_{eq}(s) = s + 1$$

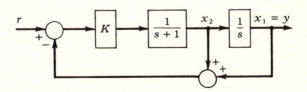

**Fig. 7.5-1**   System for Example 7.5-1.

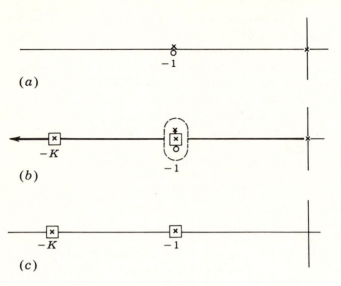

(a)

(b)

(c)

**Fig. 7.5-2** Example 7.5-1. (a) Pole-zero plot; (b) root locus; (c) closed-loop-response plane.

The open-loop transfer function $KG_p(s)H_{eq}(s)$ is therefore

$$K_pG(s)H_{eq}(s) = \frac{K(s+1)}{s(s+1)}$$

The zero in $H_{eq}(s)$ should not be canceled by the pole of $G_p(s)$, lest the closed-loop pole at this point be missed. The $s$ plane on which the root locus is to be drawn is shown in Fig. 7.5-2a. Because of the location of the open-loop pole and zero at the same point, the root locus may be drawn without regard to either of these two singularities. In measuring angles or distances, the pole and zero produce equal and opposite effects. Thus the completed root locus is given in Fig. 7.5-2b, with the location of a second closed-loop pole shown at the point $s = -K$. One is now ready to transfer the closed-loop poles determined from the root-locus diagram to the closed-loop-response plane to determine the response of the closed-loop system. The closed-loop-response plane showing the poles and zeros of the closed-loop system is given in Fig. 7.5-2c. [In this simple problem $G_p(s)$ has no zeros so that none occur in the closed-loop transfer function.]

It is a simple matter to verify analytically that the indicated closed-loop poles are indeed the correct ones. From the given

$G_p(s)$ and $H_{eq}(s)$, the closed-loop transfer function is found to be

$$\frac{y(s)}{r(s)} = \frac{K}{s^2 + s(K+1) + K} = \frac{K}{(s+1)(s+K)}$$

If the root locus had been plotted for $G_p(s)H_{eq}(s)$ and the forbidden cancellation made, then it would appear that the closed-loop system had only one pole at $s = -K$. This would, of course, have been a serious error.

As a second example, let us consider a more complex problem in which there are common factors in the poles of $H_{eq}(s)$ and the zeros of $G_p(s)$, as well as in the zeros of $H_{eq}(s)$ and the poles of $G_p(s)$.

**Example 7.5-2**  The closed-loop system discussed in this example is pictured in Fig. 7.5-3, and $G_p(s)$ and $H_{eq}(s)$ are readily found to be

$$G_p(s) = \frac{s+2}{s(s+3)(s+10)}$$

$$H_{eq}(s) = \frac{3}{125}\frac{s^2 + 18.3s + 83.3}{s+2} = \frac{3}{125}\frac{(s+10)(s+8.33)}{s+2}$$

Thus the open-loop transfer function $KG_p(s)H_{eq}(s)$ is

$$KG_p(s)H_{eq}(s) = \frac{(3K/125)(s+8.33)(s+10)}{s(s+3)(s+10)}$$

Note that in forming $KG_p(s)H_{eq}(s)$ we have canceled the zero in $G_p(s)$ by the pole in $H_{eq}(s)$ as always, but we have not canceled the pole of $G_p(s)$ at $s = -10$ with the zero in $H_{eq}(s)$ at $s = -10$. The latter cancellation is forbidden by the rule of never canceling poles

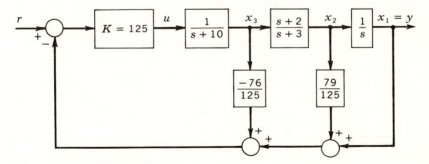

**Fig. 7.5-3**  Block diagram of the closed-loop system of Example 7.5-2.

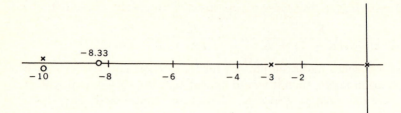

**Fig. 7.5-4** Pole-zero plot for the root locus of Example 7.5-2.

of $G_p(s)$ by zeros of $H_{eq}(s)$. Thus the $s$-plane plot of the poles and
zeros of $G_p(s)H_{eq}(s)$ from which the root locus is to be drawn is
given in Fig. 7.5-4.

As always, the object of drawing the root locus is to find the
closed-loop poles, this time at the given gain of $K = 125$. The completed root locus is shown in Fig. 7.5-5. This locus was rather
simple to draw, as only the two poles at $s = 0$ and $s = -3$ and the

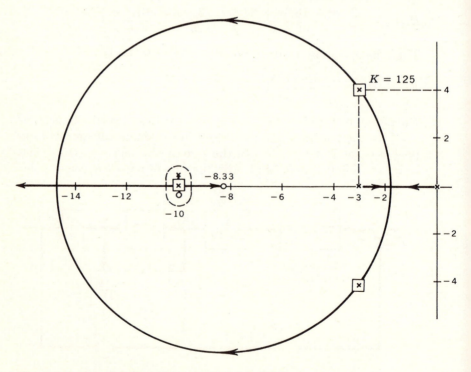

**Fig. 7.5-5** Completed root locus for Example 7.5-2.

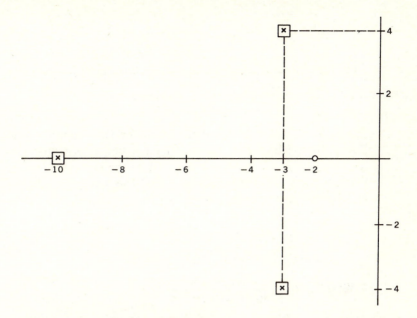

**Fig. 7.5-6** Closed-loop-response plane corresponding to the closed-loop system of Example 7.5-2.

zero at $s = -8.33$ affect the shape of the locus. The pole-zero pair at $s = -10$ locate a closed-loop pole at this location, regardless of gain.

In order to find the transient or frequency response of the closed-loop system, it is now necessary to move the closed-loop poles from the $s$ plane on which the root locus was drawn to the closed-loop-response plane and add the appropriate zeros. This is done in Fig. 7.5-6. Note that on the closed-loop-response plane the only zero that appears is the zero of $G_p(s)$. This zero did not appear on the $s$ plane from which the root locus was drawn. In fact, the only zeros that appear on the root-locus $s$ plane do not appear on the closed-loop-response plane.

Examples 7.5-1 and 7.5-2 have two features in common. In each case, all the state variables were fed back, and in each case the closed-loop-response plane was given but not used to determine a transient or frequency response. In the example to follow, only the output state variable is fed back, and the closed-loop-response plane is used to determine the step-function response for the closed-loop system.

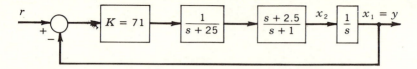

**Fig. 7.5-7**   Unity-ratio-feedback system of Example 7.5-3.

**Example 7.5-3**   The block diagram for this example is given in Fig. 7.5-7. Here $H_{eq}(s)$ is no longer a fictitious transfer function but is actually the feedback element of the system, namely, unity. However, since we have assumed that the denominator of $H_{eq}(s)$ is $N_p(s)$, we must write $H_{eq}(s)$ as

$$H_{eq}(s) = \frac{s + 2.5}{s + 2.5}$$

The open-loop transfer function is then

$$KG_p(s)H_{eq}(s) = \frac{K(s + 2.5)}{s(s + 1)(s + 25)}$$

The zero of the open-loop transfer function is still the zero of $H_{eq}(s)$. The $s$ plane on which the root locus is to be drawn is given in Fig. 7.5-8$a$, with the corresponding root locus in Fig. 7.5-8$b$. In

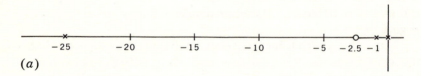

(*a*)

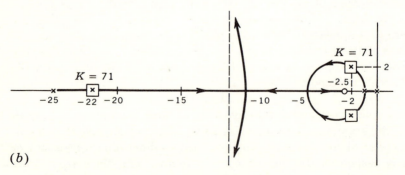

(*b*)

**Fig. 7.5-8**   Example 7.5-3.   (*a*) Pole-zero plot; (*b*) root locus with closed-loop poles for $K = 71$.

**Fig. 7.5-9**   Closed-loop-response plane for Example 7.5-3.

the latter figure the closed-loop poles are indicated for a gain of $K = 71$.

The closed-loop-response plane for this example is shown in Fig. 7.5-9. Note that, although a zero appears on the root locus and the closed-loop-response plane at $s = -2.5$, this is *not* the same zero. On the root locus, the zero is a zero of $H_{eq}(s)$, whereas on the closed-loop-response plane, it is a zero of $G_p(s)$. This point may appear rather academic, but it is helpful to keep the two zeros separate in order to avoid confusion.

Suppose it is desired to find the response of the closed-loop system to a step input. A step input involves a pole at the origin so that the correct expression for $y(s)$ is

$$y(s) = \frac{71(s + 2.5)}{s(s + 22)[(s + 2)^2 + 2^2]}$$

If, by incorrectly forming the closed-loop-response plane, the zero at $s = -2.5$ had been omitted, the expression for $y(s)$ for a step input would become

$$y_1(s) = \frac{176}{s(s + 22)[(s + 2)^2 + 2^2]}$$

Let us compare the time responses of these two systems as indicated in Fig. 7.5-10. Note that the two time trajectories are quite different, even though the correct and the incorrect systems have the same poles. These two time responses are included here to emphasize the need for transferring from the root-locus diagram not only the proper closed-loop poles but the proper closed-loop zeros as well. Any care exercised in drawing a careful root-locus diagram is wasted unless the correct information is transferred to the closed-loop-response plane.

In order to obtain the closed-loop frequency response, one may use

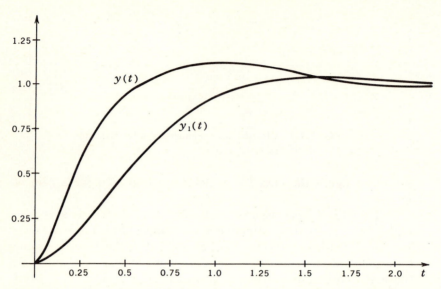

**Fig. 7.5-10**   Comparison of the time response for the correct and incorrect forms
for $y(s)$.

either graphical methods on the closed-loop-response plane or the approximate procedure of Secs. 5.3 to 5.5 on the factored closed-loop transfer
function.   The graphical method is probably best if the frequency
response is desired for only a few values of $\omega$, whereas the approximation
procedures are best if the entire closed-loop frequency response is needed.
In addition, the two approaches may be combined for greater flexibility.
The approximation methods can be used for a quick view of the complete
frequency response, with the graphical procedures used to evaluate more
accurately a few points of interest.

**Exercises 7.5**   *7.5-1.*   For each of the combinations of $G_p(s)$ and $H_{eq}(s)$
given below, sketch the general shape of the root locus.   (Do not
attempt to make the plot precise.)   Indicate on each root locus a
possible set of closed-loop poles and show the corresponding closed-
loop-response plane for $y(s)/r(s)$.

$(a)$   $G_p(s) = \dfrac{1}{s(s+2)(s+4)}$        $H_{eq}(s) = (s+1)^2 + 1^2$

$(b)$   $G_p(s) = \dfrac{2(s+3)}{s(s+2)(s+4)}$        $H_{eq}(s) = \dfrac{(s+1)(s+5)}{s+3}$

$(c)$   $G_p(s) = \dfrac{s+1}{s(s+2)(s+4)}$        $H_{eq}(s) = \dfrac{(s+2)(s+5)}{s+1}$

7.5-2.   Plot the root locus for the system shown in Fig. 7.5-11 and determine the closed-loop-response plane if $K = 40$.   Use this closed-loop information to plot the closed-loop frequency response by using the approximation methods of Chap. 5.   In addition, find the exact frequency response for $\omega = 2$ rad/sec and 4 rad/sec by using the graphical approach on the closed-loop-response plane.

*answers:*

$$\frac{y(s)}{r(s)}\bigg|_{s=j2} = 1.86\underline{/-21.8°}$$

$$\frac{y(s)}{r(s)}\bigg|_{s=j4} = 1.45\underline{/-79.4°}$$

Closed-loop poles are at $s = -5$, $-2 \pm j2$.

7.5-3.   The plant and equivalent feedback transfer functions for several systems are given below.   All these systems have the same root locus.   Draw the root locus and indicate a possible set of closed-loop poles.   Then draw the closed-loop-response plane for each system, assuming that these pole locations are correct.

(a)   $G_p(s) = \dfrac{1}{s(s+1)(s+2)}$     $H_{eq}(s) = s^2 + 2s + 2$

(b)   $G_p(s) = \dfrac{s+1}{s(s+1)(s+2)}$     $H_{eq}(s) = \dfrac{s^2 + 2s + 2}{s+1}$

(c)   $G_p(s) = \dfrac{s+3}{s(s+1)(s+2)}$     $H_{eq}(s) = \dfrac{s^2 + 2s + 2}{s+3}$

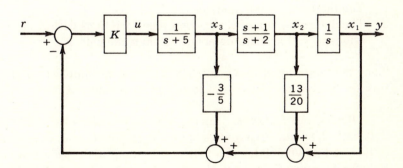

**Fig. 7.5-11**   Exercise 7.5-2.

## 7.6    Other uses of the root locus

As mentioned repeatedly throughout this chapter, the root locus is a graphical method of finding the locus of the roots of the characteristic equation of the closed-loop system as a function of the gain parameter $K$. The drawing of the root locus is accomplished by setting the denominator of $y(s)/r(s)$ equal to zero:

$$1 + KG_p(s)H_{eq}(s) = 0 \qquad\qquad (7.6\text{-}1)$$

or, in alternative form,

$$KG_p(s)H_{eq}(s) = -1 = 1\underline{/180°} \qquad\qquad (7.6\text{-}2)$$

This alternative form gives rise to the two criteria that must be satisfied to locate the actual closed-loop poles. The satisfaction of the phase-angle criterion gives the locus of roots for any finite value of gain from 0 to infinity. Satisfaction of the magnitude criterion locates the closed-loop poles for a specific value of gain.

In this section we wish to consider Eq. (7.6-2) in a more general context. The form of this equation is particularly interesting. Note that its left-hand side represents a system parameter times a transfer function. In Eq. (7.6-2) the system parameter is the controller gain $K$ and the transfer function is $G_p(s)H_{eq}(s)$. A more general form of Eq. (7.6-2) would be

$$\alpha W(s) = -1 = 1\underline{/180°} \qquad\qquad (7.6\text{-}3)$$

where $\alpha$ is any system parameter and $W(s)$ is any transfer function that is independent of $\alpha$.

In this section the root-locus method is applied to equations of the form of Eq. (7.6-3) in order to accomplish two separate ends:

1.  The factoring of polynomials
2.  The determination of closed-loop-pole variations with respect to variations of system parameters other than the gain $K$

Since the factoring of polynomials is perhaps the simplest application of this alternative use of the root locus, let us consider it first.

*Polynomial factoring.* In the previous sections, we have assumed that both $G_p(s)$ and $H_{eq}(s)$ were known in factored form, that is, as ratios of factored polynomials. Usually this assumption is valid for the plant transfer function $G_p(s)$, but this is not necessarily the case for $H_{eq}(s)$, since this is a fictitious transfer function that is formed through algebraic manipulations. The numerator of $H_{eq}(s)$ is a polynomial of order $n - 1$, which must be factored to find the specific zero locations. The poles of $H_{eq}(s)$ are assumed to be known since they are just zeros of the plant transfer function $G_p(s)$.

It was mentioned in Sec. 7.5 that often the factoring of the numerator of $H_{eq}(s)$ is not as formidable as it may seem in the most general case. Often the state variables are fed back to realize specific zero locations for $H_{eq}(s)$. In such cases, at least some of the factors of the numerator are known at the outset, and the required factoring involves a polynomial of order less than $n - 1$. From the point of view of analysis, however, we may not have this a priori information that is the basis for the design. Hence the numerator of $H_{eq}(s)$ must be factored. However, we are not concerned just with the factoring of the numerator polynomial of $H_{eq}(s)$ but rather with the general problem of polynomial factoring per se.

Before we give the general procedure, it is advantageous to consider a relatively simple example. For second-order polynomials the quadratic formula is convenient and easy to use, so that the first significant example involves a third-order polynomial. Consider the case when the cubic is

$$s^3 + 8s^2 + 38s + 56 = 0 \qquad (7.6\text{-}4)$$

It is known that at least one of the roots is real. Once the real root is found, the given cubic may be divided by the known factor with a quadratic remaining.

As a first step, Eq. (7.6-4) must be arranged into the form of Eq. (7.6-3). This is easily done by transferring the constant term to the right-hand side, as

$$s(s^2 + 8s + 38) = -56 \qquad (7.6\text{-}5)$$

If both sides of the above equation are now divided by $-s(s^2 + 8s + 38)$, the result is

$$\frac{56}{s(s^2 + 8s + 38)} = -1$$

or

$$\alpha \frac{1}{s[(s + 4)^2 + 4.76^2]} = 1/\underline{180°} \qquad (7.6\text{-}6)$$

with $\alpha = 56$. This is now of the required form, and we may draw the root locus, as required by Eq. (7.6-6). For this relatively simple example, the origin of the asymptotes is at $-\frac{8}{3}$, the asymptote angles are at $\pm 60°$ and $180°$, and the angle of departure from the upper complex conjugate pole is $-40°$.    The completed root locus is given in Fig. 7.6-1, where the desired factors for $\alpha = 56$ are shown.

In this example it was not necessary to draw the root locus; from the given pole configuration, it is known that a branch of the root locus lies along the negative real axis.    All that is necessary is to find the point on the negative real axis for which $\alpha$ is 56, which determines the real root. At $s = -2.2$, $\alpha$ is 55.5, so that a reasonable estimate of the real root is $s = -2.2$.    Long division of the polynomial in Eq. (7.6-4) yields the following result:

$$(s + 2.2)(s^2 + 5.8s + 25.2) = 0$$

The two roots of the resulting second-order polynomial are found from the quadratic formula to be

$$s = \frac{-5.8 \pm \sqrt{(5.8)^2 - 4(25.2)}}{2}$$

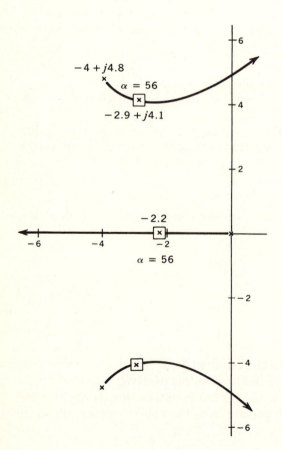

**Fig. 7.6-1**   Root locus for Eq. (7.6-6).

or

$$s = -2.9 \pm j4.1$$

Thus the desired result is

$$s^3 + 8s^2 + 38s + 56 \sim (s + 2.2)[(s + 2.9)^2 + 4.1^2]$$

The root locus of Fig. 7.6-1 indicates these same results.

The reader may have observed that the first step of the procedure illustrated above is somewhat arbitrary. As a first step, Eq. (7.6-4) is rearranged as Eq. (7.6-5). An alternative rearrangement that is equally satisfactory is the following:

$$s^3 + 8s^2 = -(38s + 56)$$

so that

$$\alpha \frac{s + 1.47}{s^2(s + 8)} = -1 = 1\underline{/180^\circ} \tag{7.6-7}$$

with $\alpha = 38$. Equation (7.6-7) requires the drawing of a root locus that looks quite different from that of Fig. 7.6-1, associated with Eq. (7.6-6). The root locus corresponding to Eq. (7.6-7) is given in Fig. 7.6-2 with roots shown for $\alpha = 38$. Again, the whole root locus need not be drawn. Once the closed-loop pole, or the root, lying on the negative real axis has been found, the remaining roots may be found through the use of the quadratic formula.

If the entire root locus is drawn in each case, Figs. 7.6-1 and 7.6-2 may be used together to determine the location of the complex conjugate poles. If Fig. 7.6-2 is laid over Fig. 7.6-1, the intersection of the various branches of the two root loci are points that satisfy the magnitude condition in each case. Thus these points of intersection are the desired closed-loop poles, which will give the two complex conjugate roots of the given polynomial. The two branches along the negative real axis intersect in a line, and this is not helpful in determining the location of the real pole.

On the basis of this one example, let us postulate a general procedure for finding the roots of an $n$th-order polynomial.

1. If the highest power of the polynomial to be factored is odd, first find the real root.
2. If the highest power of the polynomial to be factored is even, or if step 1 has already been accomplished, rearrange the given equation so that it is of the form of Eq. (7.6-3).
3. Plot the root locus for the result of step 2.

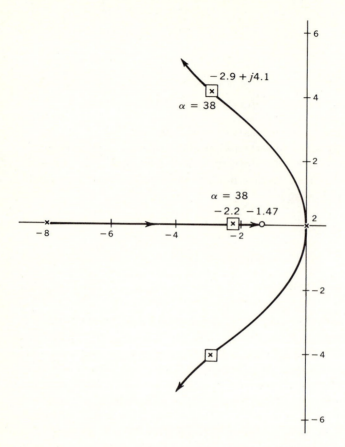

**Fig. 7.6-2**   Root locus for Eq. (7.6-7).

4.  Determine those points on the locus for which the magnitude criterion
    is satisfied.   These points are the remaining roots of the given
    polynomial.

    Step 1 reduces the order of the problem if the given polynomial is odd.
The second step is somewhat arbitrary, since we have seen that even in
the simplest case, that is, a third-order polynomial, at least two methods
of rearrangement proved satisfactory.   (Exercise 7.6-1 requires that this
same problem be worked in still another way.)   In general, there are at
least $n$ ways in which step 2 can be accomplished, although some of them
may be more advantageous than others.   Consider, for example, the

fourth-order problem

$$s^4 + 8s^3 + 39s^2 + 62s + 50 = 0$$

This fourth-order polynomial may be broken up in two different ways such that only second-order polynomials need be factored in order to find the pole-zero locations needed before the construction of the root locus can begin. These two ways are

$$s^3(s + 8) = -(39s^2 + 62s + 50)$$

and

$$s^2(s^2 + 8s + 39) = -(62s + 50)$$

If two distinct loci are plotted in step 3, the intersections determine all the complex conjugate poles, so that the magnitude determination mentioned in step 4 is not necessary for the complex conjugate roots. Usually it is difficult to realize the exact roots by the method discussed here, since the procedure is graphical, but the root-locus method does give a good approximation. The success of analytic methods for factoring, such as Lin's method, described in Appendix C, often depends on a good initial guess for the roots. The root-locus approach is capable of supplying this guess.

*Root locus vs. any parameter.* In the preceding discussion involving factoring, the key idea is to force the factored polynomial to appear in the form of Eq. (7.6-3). In the case of factoring, the multiplier $\alpha$ takes on a specific value, and although we draw the root locus for all values of $\alpha$, we are really interested only in the values of the roots of one particular value of $\alpha$, since this value of $\alpha$ is always a number in the given polynomial to be factored.

In the discussion to follow we wish to plot the root locus vs. a general system parameter. Again the key idea is to arrange the describing equation as Eq. (7.6-3). However, here our thinking is not necessarily concentrated on the closed-loop-pole locations for one specific value of the system parameter, although that parameter may have a given design value. Here we are interested in questions related to sensitivity: How do the closed-loop poles vary as the system parameter is varied about its design value?[1]

In the conventional root locus, the system parameter that is allowed to vary is the gain $K$. From an examination of the completed root locus with various values of $K$ indicated on the root-locus diagram, it is possible to obtain very specific information on the behavior of the closed-loop

[1] A comprehensive discussion of sensitivity is given in Chap. 8.

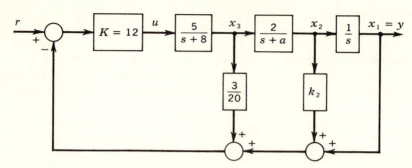

***Fig. 7.6-3***   Block diagram of the system.

poles as $K$ is allowed to vary.   Usually $K$ is selected primarily to obtain good closed-loop response, and sensitivity considerations are secondary. Here it is tacitly assumed that the plant parameter is fixed and not free to be set as $K$ is free to be set.   The primary concern here is sensitivity: How much do the closed-loop poles vary if one parameter of the actual plant deviates from the model?

The general procedure is fairly simple.   Rearrange the characteristic equation so that it is of the form of Eq. (7.6-3), with the parameter in question appearing as a pseudo gain term $\alpha$.   Then draw the root locus, marking on it the extremes expected in the system parameter.

The procedure is illustrated by the following example.   In this example, root loci are drawn for the case in which a plant pole is allowed to vary and then for the case in which a feedback coefficient is allowed to vary.   These two particular system parameters are chosen specifically to emphasize a point made in Chap. 2, namely, that feedback moves sensitivity problems from the plant, over which one may have no control, to the feedback paths, over which one may have complete control.

**Example 7.6-1**   This example is related to Example 7.3-4 and is described by the block diagram of Fig. 7.6-3.   In the earlier example, the pole at $s = -a$ was actually at $s = -2$, and the feedback coefficient $k_2$ was $\frac{17}{60}$.   Under these conditions the closed-loop transfer function is

$$\frac{y(s)}{r(s)} = \frac{120}{[(s + 2)^2 + 2^2](s + 15)} \tag{7.6-8}$$

The question we wish to ask now is how do the poles of the closed-loop system change if either the pole at $-2$ or the feedback coefficient $k_2$ is allowed to take on different values.

Let us consider first the case of the pole position. On the basis of information concerning the fixed plant, it was decided that the pole of the fixed plant was located at $s = -2$. Suppose that the pole location is not this value because of imprecise measurement or wrong initial assumptions concerning the fixed plant. We wish to find out how the closed-loop system is affected.

Let us proceed by finding the characteristic equation in terms of the system parameter $a$ and then by reshaping this equation to be of the form of Eq. (7.6-3). A simple procedure is to involve the variable parameter in the determination of the characteristic equation as few times as possible. Toward this end, the block diagram of Fig. 7.6-3 is reduced as indicated in Fig. 7.6-4. Note that this is neither a $G_{eq}(s)$ nor an $H_{eq}(s)$ reduction but just a convenient way of involving the parameter $a$ in the characteristic equation as little as possible. The characteristic equation is now

$$1 + \frac{120}{s(s + a)(s + 17)} \left[ \frac{17}{60} \left( s + \frac{60}{17} \right) \right] = 0$$

or

$$\frac{34(s + 3.53)}{s(s + a)(s + 17)} = -1 \tag{7.6-9}$$

If both sides of Eq. (7.6-9) are multiplied by the denominator of the left-hand side, the result is

$$-a(s^2 + 17s) = s^3 + 17s^2 + 34s + 120$$

or

$$a \frac{s^2 + 17s}{s^3 + 17s^2 + 34s + 120} = -1 \tag{7.6-10}$$

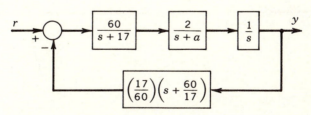

*Fig. 7.6-4*  Reduction of Fig. 7.6-3 to a more convenient form.

This is of the form of Eq. (7.6-3). Unfortunately the denominator is no longer in factored form. Hence before we can plot the root locus versus $a$, the denominator must be factored. This is rather easily done in this case because the denominator is only third-order. One root must be real, and this may be found from a root-locus diagram of the following auxiliary equation determined from Eq. (7.6-10):

$$\frac{120}{s(s^2 + 17s + 34)} = -1$$

If the root locus of this equation is drawn, the roots of the polynomial may be found, as discussed earlier in this section. These roots are

$$s^3 + 17s^2 + 34s + 120 \sim (s + 15.3)[(s + 0.85)^2 + 2.7^2]$$

and the root locus to be plotted versus $a$ is based upon the equation

$$a\,\frac{s(s + 17)}{(s + 15.3)[(s + 0.85)^2 + 2.7^2]} = -1 \qquad (7.6\text{-}11)$$

and is shown in Fig. 7.6-5. It is known in advance from Example 7.3-4 and Eq. (7.6-8) that when $a = +2$, the closed-loop poles are at $s = -2 \pm j2$ and $s = -15$. Note that a portion of the locus is of little consequence and hence no effort is made to determine that part of the locus with any degree of accuracy. The dotted portions of the locus are said to be of little consequence because in that region the values of the parameter $a$ are vastly different

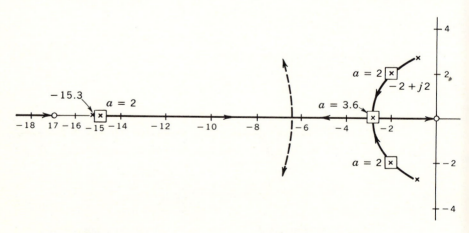

**Fig. 7.6-5**   Root locus for Eq. (7.6-11).

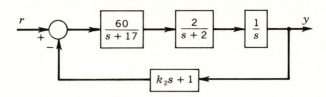

**Fig. 7.6-6**  Block diagram of Fig. 7.6-3 redrawn in a more convenient form.

from the design value.   It is seen that, regardless of how large $a$ is allowed to become, the system is always stable.   However, of more immediate importance is the location of the closed-loop poles if $a$ is different from the anticipated value by 50 or 100 percent.   It is seen from Fig. 7.6-5 that the pole farthest to the left on the $s$ plane is relatively insensitive to the value of $a$.   For values of $a$ less than the design value, the damping ratio is less, and for $a = 3.6$, both roots become real.   For very large values of $a$, the closed-loop response is dominated by the pole that is approaching the zero at the origin, and the exact location of the other closed-loop poles is unimportant.

To examine the effects in changes in $k_2$, it is convenient to redraw the block diagram of Fig. 7.6-3 as Fig. 7.6-6.   The purpose is exactly as before, to involve the parameter $k_2$ as few times as possible in the characteristic equation.   The characteristic equation for Fig. 7.6-6 is

$$\frac{120(k_2 s + 1)}{s(s + 2)(s + 17)} = -1 \tag{7.6-12}$$

This may be put into the form of Eq. (7.6-3), much as was done directly above.   The resulting equation is

$$k_2 \frac{120s}{s^3 + 19s^2 + 34s + 120} = -1$$

Again it is seen that the denominator must be factored, and again this is accomplished by the methods discussed earlier in this section.   The final form of the characteristic equation is

$$k_2 \frac{120s}{(s + 17.45)[(s + 0.775)^2 + 2.54^2]} = -1 \tag{7.6-13}$$

and the root locus is sketched in Fig. 7.6-7.   Perhaps the most important feature of system performance with regard to the feedback coefficient $k_2$ is the stability of the system when $k_2$ is zero.

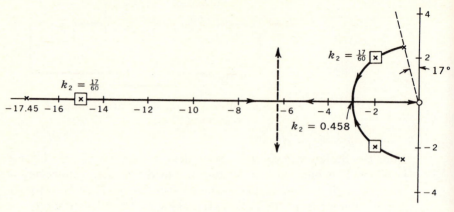

**Fig. 7.6-7**  Root locus corresponding to Eq. (7.6-13).

What happens to the system if the feedback path involving $k_2$ is opened?  In that case, the system is still stable, but the damping ratio decreases to the sin 17°, or 0.29.  The effects of increasing the velocity feedback are to increase damping and produce a more sluggish system response.  As the velocity feedback is increased by increasing $k_2$, the system once again becomes dominated by the closed-loop pole near the origin, and the remainder of the locus is unimportant.

**Exercises 7.6**  *7.6-1.*  Determine the roots of the polynomial

$$D(s) = s^3 + 8s^2 + 38s + 56$$

by plotting the root locus for

$$\frac{\alpha(s^2 + 7)}{s(s^2 + 38)} = -1$$

and evaluating the poles at $\alpha = 8$.

*answer:*

$$s \approx -2.9 \pm j4.1; \; -2.2.$$

*7.6-2.*  Factor each of the polynomials given below by the use of the root-locus method.  Plot two root loci for each polynomial and determine the complex roots from the intersections of the two diagrams.

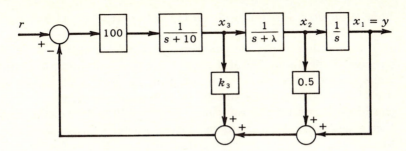

*Fig. 7.6-8* Exercise 7.6-3.

(a)  $D(s) = s^4 + 5s^3 + 10s^2 + 10s + 4$
(b)  $D(s) = s^4 + 6s^3 + 18s^2 + 30s + 25$
(c)  $D(s) = s^5 + 7s^4 + 24s^3 + 42s^2 + 40s + 16$

*answers:*

(a)  $s = -1;\ -2;\ -1 \pm j1$
(b)  $s = -2 \pm j1;\ -1 \pm j2$
(c)  $s = -1;\ -1 \pm j1;\ -2 \pm j2$

*7.6-3.*  For the closed-loop system shown in Fig. 7.6-8, determine:

(a)  The locations of the closed-loop poles as $\lambda$ is varied from 0.5 to 2 for $k_3 = 0.5$
(b)  The locations of the closed-loop poles as $k_3$ is varied from 0 to 1.0 for $\lambda = 1.0$
(c)  The locations of the closed-loop poles as $\lambda$ varies from 0.5 to 2.0 and $k_3$ from 0 to 1.0.

## 7.7  *Conclusions*

The root-locus method of this chapter serves as an ideal end to our dis-cussion of analysis methods.   Not only is the root-locus approach impor-tant as an analysis tool in its own right but, in addition, it may be used to replace or supplement many of the procedures discussed in the three pre-ceding chapters on analysis.   Time- or frequency-response determina-tions may be made through the use of the closed-loop-response plane. The stability of the closed-loop system may be investigated by examining the possibility of the locus crossing the $j\omega$ axis.   In addition, the graphical information dealing with the movement of the poles for parameter variations is of great assistance in sensing the general behavior of the closed-loop system.

In another sense, we could also have called this chapter the first chapter in the discussion of design or synthesis methods, since the root-locus method will serve as one of our primary design tools. Although the use of the root locus for analysis has been emphasized, we shall have considerable use for the root locus in Chaps. 9 and 10 dealing with synthesis methods. First, however, let us use our analysis ability to investigate the problem of specifying desired closed-loop behavior; this is the subject of Chap. 8.

## 7.8   Problems

*7.8-1.*   Sketch the root locus for each of the open-loop transfer functions given below as a function of $K$. Identify all critical points.

(a)   $KG_p(s)H_{eq}(s) = \dfrac{K(s+1)(s+2)}{s(s+3)(s+4)}$

(b)   $KG_p(s)H_{eq}(s) = \dfrac{K[(s+1)^2+1^2]}{s^2(s+2)(s+3)}$

(c)   $KG_p(s)H_{eq}(s) = \dfrac{K(s^2+1)(s+2)}{s(s+1)^2(s+2)}$

*7.8-2.*   Sketch the root locus for the open-loop transfer function given below and determine the value of $K$ such that the complex conjugate poles have a damping ratio of 0.5. Plot a few points of the closed-loop frequency response for this value of $K$; in particular, evaluate $M_p$.

$$KG_p(s)H_{eq}(s) = \frac{K(s+1)}{s(s+1)(s+2)}$$

The plant transfer function is

$$G_p(s) = \frac{1}{s(s+1)(s+2)}$$

*7.8-3.*   Plot the root locus for the closed-loop system given below (a) as a function of $K$ with $k_2 = 0.5$ and (b) as a function of $k_2$ with $K = 10$.

$$\dot{\mathbf{x}} = \begin{bmatrix} 0 & 1 & 0 \\ 0 & -1 & 2 \\ 0 & 1 & -2 \end{bmatrix}\mathbf{x} + \begin{bmatrix} 0 \\ 0 \\ 1 \end{bmatrix}u \qquad u = K(r - \begin{bmatrix} 1 & k_2 & 0.5 \end{bmatrix}\mathbf{x})$$
$$y = \begin{bmatrix} 1 & 1 & 0 \end{bmatrix}\mathbf{x}$$

*7.8-4.* Use the root-locus technique to find the range of values of $k_2$ for which the system given below is stable. Check your answer by using the Routh-Hurwitz criterion on the system.

$$\dot{\mathbf{x}} = \begin{bmatrix} 0 & 1 & 0 \\ 0 & -2 & 1 \\ 0 & 0 & -10 \end{bmatrix} \mathbf{x} + \begin{bmatrix} 0 \\ 0 \\ 1 \end{bmatrix} u \qquad u = r - 10[1 \quad k_2 \quad 0]\mathbf{x}$$

$$y = [1 \quad 0 \quad 0]\mathbf{x}$$

*7.8-5.* Use the root-locus technique to factor the following polynomials:

(a) $\quad s^4 + 2s^3 + 4s^2 + s + 5$
(b) $\quad s^5 + 30s^4 + 10s^3 + 15s^2 + 2s + 100$

*7.8-6.* Compare the root-locus diagrams for the two open-loop transfer functions

(a) $\quad KG_p(s)H_{eq}(s) = \dfrac{K[(s+2)^2 + 2.5^2]}{s[(s+2)^2 + 2^2]}$

(b) $\quad KG_p(s)H_{eq}(s) = \dfrac{K[(s+2)^2 + 1.5^2]}{s[(s+2)^2 + 2^2]}$

In each case the intent is cancellation of the complex conjugate poles by the complex conjugate zeros. For purposes of illustration, the imperfection of the cancellation has been exaggerated. Note that there may be a preferable way in which to miss cancellation.

*7.8-7.* Sketch the root locus for the open-loop transfer function given below as $K$ is varied from 0 to *minus* infinity. Note that this requires that the angle criterion become $\arg KG_p(s)H_{eq}(s) = 0°$. (See Prob. 7.8-12.)

$$KG_p(s)H_{eq}(s) = \dfrac{K(s+1)[(s+1)^2 + 1]}{s(s+2)(s+5)}$$

*7.8-8.* For the system shown in Fig. 7.8-1, plot the root locus as the

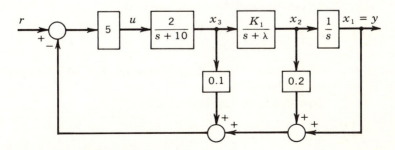

**Fig. 7.8-1** Problem 7.8-8.

parameter $K_1$ is varied from 0 to $\infty$ with $\lambda = 5$ and then as $\lambda$ is varied from 0 to 10 with $K_1 = 2$.

**7.8-9.**   Plot the root locus of the system shown in Fig. 7.8-2 as $K$ varies

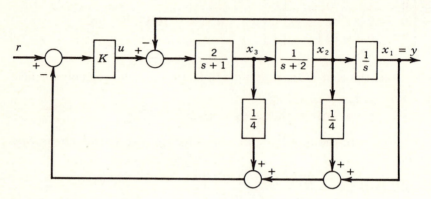

**Fig. 7.8-2**   Problem 7.8-9.

from 0 to $\infty$.   Note the insensitivity of the complex conjugate poles.

**7.8-10.**   For the system

$$\dot{x} = \begin{bmatrix} 0 & 1 & 0 \\ 0 & -2 & 1 \\ 0 & 0 & -4 \end{bmatrix} x + \begin{bmatrix} 0 \\ 0 \\ 2 \end{bmatrix} u$$

$$y = \begin{bmatrix} 1 & 0 & 0 \end{bmatrix} x \qquad u = K(r - \begin{bmatrix} 1 & 0 & k_3 \end{bmatrix} x)$$

(a)   Find the values of $k_3$ for which the closed-loop system is stable for all positive values of $K$.

(b)   Plot the root locus versus $K$ if $k_3 = 1$.

**7.8-11.**   Sketch the root locus for the following open-loop transfer function.   Determine and label, where applicable, breakaway, reentry, asymptotes, arrival angles, and $j\omega$-axis intersection.

$$KG_pH_{eq} = \frac{K(s^2 + 2s + 2)(s + 6)}{s(s + 1)(s + 2)(s + 4)}$$

**7.8-12.**   Make a summary of the construction rules for a zero-degree root locus similar to Table 7.4-1.

**7.8-13.**   Show that the closed-loop system

$$\dot{x} = \begin{bmatrix} 0 & 1 & 0 \\ 0 & -10 & 4 \\ 0 & 0 & -100 \end{bmatrix} x + \begin{bmatrix} 0 \\ 0 \\ 100 \end{bmatrix} u \qquad y = x_1 \qquad u = 10(r - x_1)$$

is stable by using the following methods:

(*a*)  Plotting a Nyquist diagram for the system.   This should include a calculation of the phase shift at crossover.

(*b*)  Plotting the root locus.   This should include an approximate determination of the closed-loop poles.

(*c*)  Applying the Routh-Hurwitz criterion.

# eight  *specifications*

## 8.1  Introduction

The four preceding chapters dealt with the analysis
of closed-loop control systems.  In retrospect this is
little more than busy work, much of which should be
programmed on a digital computer.  The job of the
analyst has been compared to that of the house painter,
necessary but not pleasant, whereas the job of the syn-
thesist, on the other hand, is likened to that of the
artist.  That is, the interesting and creative part of
control engineering is concerned with system synthesis.
This chapter on specifications is a transition between
analysis and synthesis.  Here the ideas and concepts,
discussed in previous chapters on analysis, are exam-

ined from the point of view of: What do we want the closed-loop system to do? Trying to decide what performance is desired is closely akin to analysis since the same methods and terminology are used. Thus, although we may now wish to relegate the pure analysis problem to a digital computer, the time spent in mastering the techniques of analysis surely is not wasted.

In the discussion of linear state-variable feedback in Sec. 3.4, we pointed out the following:

1. The poles of $y(s)/r(s)$ may be arbitrarily positioned by the proper selection of **k** and $K$.
2. The zeros of $y(s)/r(s)$ are the zeros of $G_p(s)$.

It will be shown in the following chapters on synthesis that with the addition of series compensation, the zeros as well as the poles of $y(s)/r(s)$ may be positioned arbitrarily. Thus, through the use of series compensation and linear state-variable feedback, one is able to realize any desired closed-loop transfer function. The only limitation on this statement is that the closed-loop system may not have a pole-zero excess smaller than the pole-zero excess of $G_p(s)$.

The outlook of this chapter is based upon this ability to realize *any* closed-loop transfer function. If *any* system transfer function can be realized, the important question then is what transfer function is desired. Surely no one transfer function is ideal in every case; rather, the selection of the desired system response depends upon the specifications that the particular system is expected to meet. If the specifications are given, the problem is to find the transfer function whose response satisfies these specifications. First, of course, the specifications must be generated, a job that often falls to the designer. In this chapter a variety of specifications which describe important performance measures are discussed.

It is important at the outset to distinguish between performance measures and specifications. By *performance measures* we mean those general properties or qualities of a closed-loop system that are important to its proper functioning. Performance measures thus are qualitative. The quantitative descriptions, that is, numbers, that are given to describe the performance measures are known as *specifications*. For example, one performance measure associated with closed-loop systems is the speed of response. A variety of specifications may be used to measure the speed of response, such as the rise time or the settling time or the bandwidth. Associated with each of these specifications is a number that must be known in order to describe the speed of response.

In the choice of a desirable closed-loop transfer function, the important performance measures are relative stability, speed of response, and accuracy. Relative stability and speed of response are discussed in Sec. 8.2 in terms of both time- and frequency-domain specifications. In addition, Sec. 8.2 serves to summarize and review much of the terminology and many of the specifications that have been mentioned earlier in the book.

Accuracy, discussed in Sec. 8.3, is most easily described in the time domain in terms of the final value of error for a particular input. On the other hand, error constants, a convenient means of measuring the final value of error, are most easily evaluated in the frequency domain. Although a variety of time and frequency specifications emerge from Sec. 8.2, the conclusion from Sec. 8.3 is that a critical specification is the velocity-error constant.

Sections 8.4, 8.5, and 8.8 develop graphical methods of choosing $y(s)/r(s)$ according to the specifications that must be satisfied. Section 8.4 treats the second-order case, and there it is seen that the addition of a zero considerably increases design freedom. The third-order case with one zero is considered in Sec. 8.5. This is about the most complicated case that the designer is able to handle without computational aids. The use of the third-order procedures of Sec. 8.5 for higher-order transfer functions is described in Sec. 8.8.

The principal concern of Secs. 8.6 and 8.7 is the relationship of performance measures and specifications to the plant. This includes such diverse things as power and weight requirements, maximum ranges of the state variables, and the sensitivity of the closed-loop system to disturbances that take place within the loop, due to either plant parameter changes or external disturbances. Whether the system is able to do what is desired is a function of the unalterable plant to be controlled. For example, in a horserace, the important performance measure is speed. One may well be able to specify a speed that is adequate to ensure victory, but whether or not this specification can be satisfied depends on the unalterable plant: the horse. Though important, the conclusions of these two sections are less general than those of the previous sections. This is because the discussion is centered on the plant, which is different for every problem.

## 8.2   *Relative stability and speed of response*

In this and the following section we define and discuss the specifications that are used to describe the important performance measures associated with closed-loop systems. Again the fact that we are considering only the input-output relationship for the closed-loop system is emphasized. First we attempt to describe an *ideal* response in terms of the specifications and then to find a transfer function that is compatible with the specifications. Since we are concerned with input-output relationships and transfer functions, the internal state variables that exist within the plant are temporarily ignored. Whether the realization of such a transfer function is meaningful for a given plant remains an unanswered question, at least temporarily. Consider, for example, the range of expected inputs. We know from previous discussions that any closed-loop transfer function may be realized, regardless of the given plant. However, this statement is true only if the assumption of linearity remains valid. If the range of expected inputs is sufficient to drive one or more of the state variables beyond its maximum allowable value, the system is no longer linear, and a linear analysis cannot predict the system behavior. The consideration of questions involving the plant is left to Sec. 8.7.

The performance measures discussed in this section are those of relative stability and speed of response. There are a multitude of specifications available to describe these two performance measures. Here we consider only the following specifications:

1. Those that convey an easily interpretable quality of the system's response
2. Those that are applicable to and valid for systems of any order or configuration
3. Those that provide a sensitive and discriminative measure

Two sets of specifications that meet these restrictions may be chosen, one set in the time domain and the other in the frequency domain. The time-domain specifications and their definitions are made with reference to a typical step-function response for an $n$th-order system, as indicated in Fig. 8.2-1. Note that here we have assumed that the desired input-output dynamics exhibits an underdamped response with zero steady-state error.

The time-domain specifications important in measuring relative stability and speed of response are percent overshoot, settling time, rise time, and delay time. These are defined and briefly discussed below. (See also Fig. 8.2-1.)

*Percent overshoot* (PO) is 100 times the maximum value of the response minus the final value of the response, divided by the final

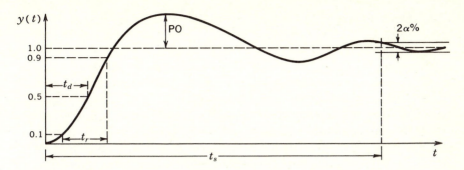

***Fig. 8.2-1***    Time-domain specifications.

value of the response, or PO is given as

$$PO = 100 \frac{(\text{maximum value of response}) - (\text{final value of response})}{\text{final value of response}}$$

Percent overshoot is a measure of relative stability, and the larger the overshoot, the smaller the degree of stability.

    *Settling time $t_s$* is the time required for the response to remain within $\pm \alpha$ percent of the final value.    Typical values for $\alpha$ are 2 and 5.

Settling time is a measure of both the speed of response and the relative stability.    The speed of response is clearly indicated, for if the settling time is small, the system must have responded rapidly.    The relation to relative stability is not quite as clear.    For a pair of complex conjugate poles near the $j\omega$ axis, a small change in the system parameters can force the system into a region of instability.    If the residues of these complex conjugate poles are small, their presence may go undetected during the initial transient portion of the step response.    However, because the poles are close to the $j\omega$ axis, their time constant is long, causing an increase in the settling time.    In this rather indirect manner the settling time is a measure of relative stability.

    *Rise time $t_r$* is the time required for the average system response to go from 10 to 90 percent of its final value.

Because rise time is concerned only with the early transient portion of the response, rise time is a measure of the speed of response.

*Delay time* $t_d$ is the time required for the average system to reach 50 percent of its final value.

Delay time, like rise time, is also a measure of the speed of response.

The word "average" in the definitions of rise time and delay time needs some explanation. For higher-order systems the step-function response may not always be as smooth as indicated in Fig. 8.2-1. For example, the response indicated in Fig. 8.2-2 is entirely possible for a linear system with more than one set of complex conjugate poles. With this type of response it can be seen that the literal interpretation of either of these definitions could give an uncertain or misleading result.[1]

This ambiguity can be removed and the definitions can be made fully meaningful if they are applied to an average response obtained by drawing the best smooth curve through the actual response. As there can be no fixed method for arriving at "the best smooth curve," these definitions remain somewhat imprecise.

The reader has no doubt observed the similarity between the terminology used here to describe the output of an $n$th-order system and that used in Sec. 4.3 to describe the response of a particular second-order system to a step input. The knowledge gained in the analysis of second-order systems proves to be valuable in specifying higher-order response.

[1] Gibson et al., A Set of Standard Specifications for Linear Automatic Control Systems, *Trans. AIEE*, vol. 80, pt. 2, pp. 65–77, 1961.

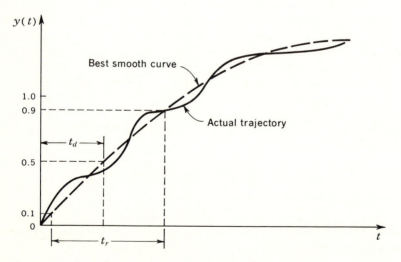

**Fig. 8.2-2**   Smooth approximation to an actual step response.

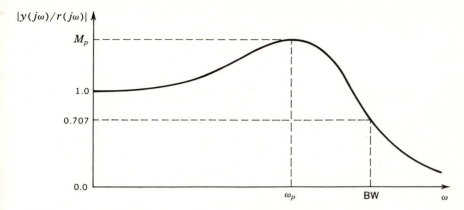

**Fig. 8.2-3**   Frequency-response curve for a typical $n$th-order system.

The discussion of the step-function response of the second-order system in Sec. 4.3 also made use of a number of other terms that are not even mentioned above. The most important of these are the damping ratio $\zeta$ and the undamped natural frequency $\omega_n$. These quantities are uniquely associated with second-order systems and, strictly speaking, should not be discussed for systems of higher order. Although these two quantities are not mentioned in the time-domain specifications cited above, it is impossible to avoid their use in attempting to determine model transfer functions. In particular, $\omega_n$ is used in Secs. 8.4 to 8.6 as at least a tentative measure of speed of response.

Consider now relative stability and speed of response based on frequency-response data. Figure 8.2-3 represents the frequency-response curve of a typical closed-loop system. The pertinent frequency-domain specifications and their definitions are briefly discussed below.[1]

*M peak* $M_p$ is the maximum value of the magnitude of the closed-loop frequency-response curve.

$M$ peak is a measure of relative stability. A high value of $M_p$, $M_p > 1.5$, indicates a pair of complex conjugate poles near the $j\omega$ axis.

*Peak frequency* $\omega_p$ is the frequency in radians per second at which $M_p$ occurs.

---

[1] Additional frequency-domain specifications include phase margin, crossover frequency, and gain margin. See Sec. 6.5.

Peak frequency is a measure of speed of response, and because of its intimate relationship with $M_p$ it is often mentioned along with $M_p$.

> *Bandwidth* BW is the frequency at which the closed-loop frequency-response curve reaches a magnitude of $\sqrt{2}/2 = 0.707$ times its low-frequency value.

Bandwidth is a measure of the speed of response. The connection between bandwidth and the speed of response is based upon the utilization of the superposition theorem and the Fourier-series representation of a square-wave input signal of frequency $\omega$. The Fourier-series representation of a square wave includes all the harmonics of the fundamental frequency $\omega$. In order for the system to reproduce the square wave at its output, the system would have to have infinite bandwidth. That is, associated with infinite bandwidth is zero rise time and zero delay time. In general, the larger the bandwidth, the shorter are the rise and the delay times. However, there is no analytic expression that even approximately relates the two for systems of arbitrary order.

The bandwidth specification is important also from the point of view of noise.[1] The goal of the control system is to follow the input signal. Often this input signal is corrupted with noise. If the bandwidth of the noise is identical with the signal spectrum of the input, it is impossible to discriminate between the two. Often, however, the noise is basically high-frequency in nature, calling for an upper limit on the bandwidth. This is somewhat of a paradox. In order for the system to follow extreme inputs, such as a step function, the bandwidth should be infinite. Yet in order to discriminate against unwanted noise signals, the bandwidth should be limited.

Another common situation requiring a bandwidth specification involves signals that are not necessarily present in the input but that enter the system elsewhere. Power-supply noise, or a signal at the line frequency, is such a case.

As in the case of the time-domain specifications cited earlier, the frequency-domain specifications mentioned above are familiar. They were discussed in Sec. 5.3 for a second-order system in terms of $\zeta$ and $\omega_n$. Thus it appears that a natural starting place in determining an ideal transfer-function response is a return to the second-order system. This is done after accuracy specifications are discussed in the next section.

---

[1] Noise rejection will be discussed further in Sec. 10.3.

## 8.3 *Accuracy*

Accuracy is a performance measure that needs some definition and discussion, particularly in terms of the assumptions and performance measures of Sec. 8.2. The accuracy referred to in this section is the final value of error due to inputs of form $t^p$, $p = 0, 1, 2, \ldots$. The most important of these inputs are $t^0 = 1$ and $t$.

In examples throughout this book we have often forced the final value of error due to a step function to be zero, perhaps without the reader's knowledge that this was being done. In systems with one integration in the output of $G_p(s)$ this is accomplished by setting $k_1$ to be unity, and the vast majority of examples discussed thus far involved plants with one integration in the output member of $G_p(s)$. In a sense, then, we have been assuming accuracy for step-function inputs. This was done because of the primary importance of this performance measure. In this section we generalize the idea of accuracy for inputs of the form $t^p$, $p = 0, 1, 2, \ldots$, and show how the final value of error for each input depends upon the number of equivalent integrations in the forward transfer function $G_{eq}(s)$.

The reader may question the discussion of accuracy as divorced from the speed of response. If a system has a large bandwidth, it can follow inputs with little error. The system may then be considered accurate, not only in its final value but for all time. Speed of response *is* a measure of accuracy for all time, and this is the reason for this preliminary discussion. We wish to make clear our meaning of accuracy as discussed in this section. If the steady-state errors for inputs that are powers of time are small, the system is accurate. The speed of response might be called the measure of dynamic accuracy. We have intentionally avoided that term in hope of also avoiding confusion.

Let us begin the discussion by a consideration of the error expression in terms of the $H_{eq}(s)$ and the $G_{eq}(s)$ methods of system representation. By definition, error is

$$e(t) = r(t) - y(t) \tag{8.3-1}$$

The transfer function relating the error and the input is found by taking the Laplace transform[1] of Eq. (8.3-1) and dividing by $r(s)$ to give

$$\frac{e(s)}{r(s)} = 1 - \frac{y(s)}{r(s)} \tag{8.3-2}$$

[1] Initial conditions are always assumed to be zero when transfer functions are involved.

In terms of the $H_{eq}(s)$ representation for $y(s)/r(s)$, the error-to-input transfer function becomes

$$\frac{e(s)}{r(s)} = 1 - \frac{KG_p(s)}{1 + KG_p(s)H_{eq}(s)} = \frac{1 + KG_p(s)[H_{eq}(s) - 1]}{1 + KG_p(s)H_{eq}(s)} \tag{8.3-3}$$

In terms of the $G_{eq}(s)$ representation of $y(s)/r(s)$, the error-to-input transfer function becomes

$$\frac{e(s)}{r(s)} = 1 - \frac{G_{eq}(s)}{1 + G_{eq}(s)} = \frac{1}{1 + G_{eq}(s)} \tag{8.3-4}$$

Since Eq. (8.3-4) is simpler than Eq. (8.3-3), this form will be used in the remainder of this section.

Equation (8.3-4) may also be written in another form by considering $G_{eq}(s)$ as a ratio of polynomials in $s$:

$$G_{eq}(s) = \frac{K(c_{m+1}s^m + c_m s^{m-1} + \cdots + c_1)}{s^n + b_n s^{n-1} + \cdots + b_2 s + b_1}$$

Then Eq. (8.3-4) also becomes a ratio of polynomials:

$$\begin{aligned}
\frac{e(s)}{r(s)} &= \frac{s^n + b_n s^{n-1} + \cdots + b_2 s + b_1}{s^n + b_n s^{n-1} + \cdots + b_2 s + b_1 + K(c_{m+1}s^m + c_m s^{m-1} + \cdots + c_1)} \\
&= \frac{b_1 + b_2 s + \cdots + b_n s^{n-1} + s^n}{b_1 + Kc_1 + (b_2 + Kc_2)s + \cdots + s^n}
\end{aligned}$$

Note that we have written the $s^0$ term in both the numerator and denominator to the left, contrary to previous practice. If a long division is carried out with the numerator and denominator in this form, the result is a power series in $s$:

$$\frac{e(s)}{r(s)} = \alpha_0 + \alpha_1 s + \alpha_2 s^2 + \cdots \tag{8.3-5}$$

where, for example, $\alpha_0$ is

$$\alpha_0 = \frac{b_1}{b_1 + Kc_1}$$

The idea is not that the long division *should* actually be carried out in any given problem but that it *can* be carried out. By combining Eqs. (8.3-4) and (8.3-5), $e(s)/r(s)$ may be written

$$\frac{e(s)}{r(s)} = \frac{1}{1 + G_{eq}(s)} = \alpha_0 + \alpha_1 s + \alpha_2 s^2 + \cdots \tag{8.3-6}$$

Now let us consider the steady-state error for inputs that are of the form $t^p$. As a first input, let the power of $t$ be zero, so that the input is a unit step function. In order to find the steady-state error for a step-function input, we may first find $e(s)$ and then use the final-value theorem.[1] Recall that the final-value theorem states that, if $se(s)$ has no poles on the $j\omega$ axis or in the right-half plane, then

$$\lim_{t \to \infty} e(t) = \lim_{s \to 0} se(s)$$

For a step-function input, $r(s) = 1/s$, so that from Eq. (8.3-6) $e(s)$ is

$$e(s) = \frac{1/s}{1 + G_{eq}(s)} = \frac{\alpha_0}{s} + \alpha_1 s^0 + \alpha_2 s^1 + \cdots$$

and $se(s)$ is just

$$se(s) = \frac{1}{1 + G_{eq}(s)} = \alpha_0 + \alpha_1 s + \alpha_2 s^2 + \cdots$$

Then the steady-state value of $e(t)$, $e(\infty)$, is

$$e(\infty)_{step} = \lim_{t \to \infty} e(t) = \lim_{s \to 0} \frac{1}{1 + G_{eq}(s)}$$

$$= \lim_{s \to 0} (\alpha_0 + \alpha_1 s + \alpha_2 s^2 + \cdots) = \frac{1}{1 + \lim\limits_{s \to 0} G_{eq}(s)} = \alpha_0$$

Now we define the *position-error constant* $K_{pos}$ as

$$K_{pos} = \lim_{s \to 0} G_{eq}(s)$$

so that $\alpha_0$ is

$$\alpha_0 = \frac{1}{1 + K_{pos}}$$

If the steady-state position error due to a step input is to be zero, that is, $\alpha_0 = 0$, then $K_{pos}$ must be infinite. The only way for $K_{pos}$ to be infinite is for

$$\lim_{s \to 0} G_{eq}(s) = \lim_{s \to 0} \frac{K(c_{m+1}s^m + c_m s^{m-1} + \cdots + c_1)}{s^n + b_n s^{n-1} + \cdots + b_2 s + b_1} = \infty$$

This can be true only if $G_{eq}(s)$ contains at least one integration or pole at the origin so that the denominator of $G_{eq}(s)$ goes to zero as $s$ goes to zero. The transfer function $G_{eq}(s)$ has one integration if $b_1$ is equal to zero, two integrations if $b_1$ and $b_2$ are both zero, etc. A *transfer function is said to be of type N if it has N integrations or poles at the origin.* If $G_{eq}(s)$ is type 1 (or higher) then the system follows step-function inputs with zero steady-state error.

---

[1] See Appendix A.

Let us examine the way in which $G_{eq}(s)$ can be type 1.   In order to do so let us return to the example problem considered in Sec. 3.3 in the initial discussion of the $G_{eq}(s)$ representation.   The block diagram is repeated here as Fig. 8.3-1a, and the $G_{eq}(s)$ form is indicated in Fig. 8.3-1b.   [Note that in this example $G_p(s)$ is type 1.]   From Fig. 8.3-1b it is clear that if $k_1$ is set equal to 1 the constant term in the denominator of $G_{eq}(s)$ vanishes, so that $G_{eq}(s)$ also becomes type 1 or has one integration.   If the plant does not have an integration, then $k_1$ must be chosen other than 1 to ensure zero steady-state error for a step-function input. (See Exercise 8.3-1.)

It is interesting to return to Eq. (8.3-2) and determine the condition required in the *closed-loop transfer function* so that the steady-state step-function error is zero.   From Eq. (8.3-2), $y(s)/r(s)$ is

$$\frac{y(s)}{r(s)} = 1 - \frac{e(s)}{r(s)} = 1 - \alpha_1 s - \alpha_2 s^2 - \cdots \tag{8.3-7}$$

In Eq. (8.3-7) $\alpha_0$ has been set equal to zero, the condition necessary for zero steady-state step-function error.   If $s$ is set equal to zero, then

$$\frac{y(0)}{r(0)} = 1 \tag{8.3-8}$$

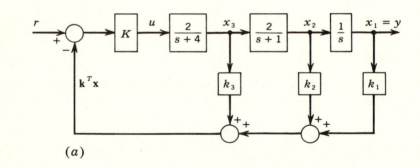

(a)

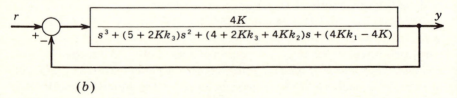

(b)

**Fig. 8.3-1**  System illustrating the requirements for zero steady-state position error. (a) Original system; (b) $G_{eq}(s)$ form.

If the closed-loop transfer function is written

$$\frac{y(s)}{r(s)} = \frac{KK_p(s + \delta_1)(s + \delta_2) \cdots (s + \delta_m)}{(s + \lambda_{k1})(s + \lambda_{k2}) \cdots (s + \lambda_{kn})}$$

the result of Eq. (8.3-8) may be expressed in terms of *the closed-loop poles and zeros:*

$$1 = \frac{KK_p\delta_1\delta_2 \cdots \delta_m}{\lambda_{k1}\lambda_{k2} \cdots \lambda_{kn}} \tag{8.3-9}$$

Now let us return to Eq. (8.3-6), with $\alpha_0$ equal to zero, and consider a ramp input. Here $r(s)$ is $1/s^2$ and $e(s)$ is

$$e(s) = \frac{1/s^2}{1 + G_{eq}(s)} = \frac{\alpha_1}{s} + \alpha_2 + \cdots$$

Again the final value of $e(t)$ as $t$ goes to $\infty$ may be found by using the final-value theorem so that

$$e(\infty)_{\text{ramp}} = \lim_{t \to \infty} e(t) = \lim_{s \to 0} se(s)$$

$$= \lim_{s \to 0} \frac{1/s}{1 + G_{eq}(s)} = \lim_{s \to 0} \alpha_1 + \alpha_2 s + \cdots$$

$$= \lim_{s \to 0} \frac{1}{s + sG_{eq}(s)} = \lim_{s \to 0} \frac{1}{sG_{eq}(s)} = \alpha_1 \tag{8.3-10}$$

Now we define the *velocity-error constant $K_v$* as

$$K_v = \lim_{s \to 0} sG_{eq}(s)$$

and, from Eq. (8.3-10), $\alpha_1$ is just

$$\alpha_1 = \frac{1}{K_v}$$

The condition necessary to make the steady-state error due to a ramp input zero is for $\alpha_1$ to equal zero or for $K_v$ to be infinite. As in the discussion of the position-error constant, let us examine the requirements on $G_{eq}(s)$ such that $K_v = \infty$. If $G_{eq}(s)$ is type 2 with $b_1 = b_2 = 0$, or if $G_{eq}(s)$ has two integrations, then $sG_{eq}(s)$ still has an $s$ in its denominator and

$$K_v = \lim_{s \to 0} sG_{eq}(s) = \lim_{s \to 0} \frac{K(c_{m+1}s^m + c_m s^{m-1} + \cdots + c_1)}{s(s^{n-2} + b_n s^{n-3} + \cdots + b_3)} = \infty$$

For $G_{eq}(s)$ to have two integrations is a much more stringent requirement than for $G_{eq}(s)$ to have only one integration.

In order to examine the conditions for $K_v = \infty$ in terms of the closed-loop transfer function, let us return to Eq. (8.3-7). With $\alpha_0$ equal to zero and $\alpha_1$ equal to $1/K_v$, this expression becomes

$$\frac{y(s)}{r(s)} = 1 - \frac{1}{K_v} s - \alpha_2 s^2 - \cdots \tag{8.3-11}$$

By taking the derivative of Eq. (8.3-11) with respect to $s$ and evaluating at $s = 0$, we see that

$$\frac{d}{ds} \frac{y(s)}{r(s)} \bigg|_{s=0} = -\frac{1}{K_v} \tag{8.3-12}$$

Since it has been assumed that the step-function steady-state error is zero, from Eq. (8.3-8) $y(0)/r(0)$ is 1. Equation (8.3-12) is not violated if both sides are divided by 1, and so Eq. (8.3-12) becomes

$$\frac{1/K_v}{y(0)/r(0)} = \frac{1}{K_v} = -\frac{(d/ds)[y(s)/r(s)]}{y(s)/r(s)} \bigg|_{s=0} \tag{8.3-13}$$

Since both numerator and denominator of the right-hand side are evaluated at $s = 0$, this can be written

$$\frac{1}{K_v} = -\frac{d}{ds} \ln \frac{y(s)}{r(s)} \bigg|_{s=0} \tag{8.3-14}$$

where we have used the relationship

$$\frac{d}{dx} \ln u = \frac{1}{u} \frac{du}{dx}$$

If $y(s)/r(s)$ is written in factored form, Eq. (8.3-14) becomes

$$\frac{1}{K_v} = -\frac{d}{ds} \ln \frac{KK_p(s + \delta_1)(s + \delta_2) \cdots (s + \delta_m)}{(s + \lambda_{k1})(s + \lambda_{k2}) \cdots (s + \lambda_{kn})} \bigg|_{s=0}$$

Now we can take advantage of the fact that we have the logarithm of a product and rewrite this expression as

$$\frac{1}{K_v} = -\frac{d}{ds} [\ln KK_p + \ln (s + \delta_1) + \ln (s + \delta_2) + \cdots + \ln (s + \delta_m)$$
$$- \ln (s + \lambda_{k1}) - \ln (s + \lambda_{k2}) - \cdots - \ln (s + \lambda_{kn})] \bigg|_{s=0}$$

Now the derivative of each term is easily taken so that

$$\frac{1}{K_v} = -\left( \frac{1}{s + \delta_1} + \frac{1}{s + \delta_2} + \cdots + \frac{1}{s + \delta_m} \right.$$
$$\left. - \frac{1}{s + \lambda_{k1}} - \frac{1}{s + \lambda_{k2}} - \cdots - \frac{1}{s + \lambda_{kn}} \right) \bigg|_{s=0}$$

If the entire expression is evaluated at $s = 0$, the final expression for $1/K_v$ is

$$\frac{1}{K_v} = \frac{1}{\lambda_{k1}} + \frac{1}{\lambda_{k2}} + \cdots + \frac{1}{\lambda_{kn}} - \frac{1}{\delta_1} - \frac{1}{\delta_2} - \cdots - \frac{1}{\delta_m}$$

or, in more compact notation,

$$\frac{1}{K_v} = \sum_{j=1}^{n} \frac{1}{\lambda_{kj}} - \sum_{j=1}^{m} \frac{1}{\delta_j} \tag{8.3-15}$$

This is the desired result.   We have succeeded in determining an expression for $K_v$ that involves only the *closed-loop* poles and zeros.   In order to make $K_v$ infinite so that zero steady-state error for a ramp input results, it is necessary that

$$\sum_{j=1}^{n} \frac{1}{\lambda_{kj}} = \sum_{j=1}^{m} \frac{1}{\delta_j} \tag{8.3-16}$$

This equation is analogous to Eq. (8.3-9), the requirement for zero step error in terms of the closed-loop-system parameters.   The reader should remember that, in order for the final-value theorem to be valid, the closed-loop system must be stable.

Let us summarize the results thus far with respect to the steady-state position error due to a ramp input.   The error can be made zero if $K_v$ is forced to be infinite, which can be accomplished either by requiring that $G_{eq}(s)$ have two integrations or by forcing the sum of the reciprocals of the closed-loop poles to equal the sum of the reciprocals of the closed-loop zeros, that is, requiring that Eq. (8.3-16) be satisfied.   Since we are presently interested in specifications involving a desired closed-loop transfer function, the satisfaction of Eq. (8.3-16) is of prime consideration. To illustrate how the satisfaction of Eq. (8.3-16) automatically forces $G_{eq}(s)$ to contain two integrations, we consider the following example.

***Example 8.3-1***   This example is based on the system pictured in Fig. 8.3-2.   A zero is created by feedforward, and feedback of the state variable $x_2$ is accomplished through the constant feedback element $k_2 = -0.25$.   The closed-loop transfer function is

$$\frac{y(s)}{r(s)} = \frac{4(s + 2)}{(s + 2)^2 + 2^2}$$

Here $y(0)/r(0)$ is 1, so that Eq. (8.3-8) is satisfied, and the closed-loop system has zero steady-state error due to a step input.   The

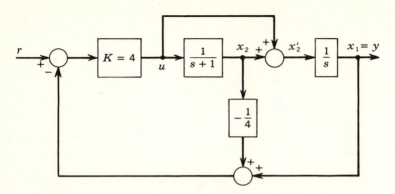

**Fig. 8.3-2** Block diagram for Example 8.3-1.

closed-loop poles are at $s = -2 \pm j2$, and the sum of the reciprocals is

$$\frac{1}{2 + j2} + \frac{1}{2 - j2} = \frac{4}{4 + 4} = \frac{1}{2}$$

Since the zero is at $s = -2$, the reciprocal of the zero position is also $\frac{1}{2}$, and Eq. (8.3-16) is also satisfied, ensuring an infinite velocity-error coefficient and zero steady-state error for ramp inputs.

Let us examine $G_{eq}(s)$ to see if it is actually type 2. If the inner loop in Fig. 8.3-2 is reduced, $G_{eq}(s)$ becomes

$$G_{eq}(s) = \frac{4(s + 2)}{s^2}$$

and $G_{eq}(s)$ has the necessary two integrations.

In the above example, $G_p(s)$ had an integration in the last block, and the position-error coefficient was forced to infinity by making $k_1$ equal to 1. Both $K$ and $k_2$ were chosen so that Eq. (8.3-16) was satisfied, making $K_v$ also equal to infinity. Thus, although we have not yet begun a study of synthesis procedures, this discussion of specifications is leading directly to such a study. The important accuracy specification that has emerged from this section is the specification of the velocity-error coefficient as a measure of the steady-state error for ramp inputs. The importance of being able to follow such inputs with small error cannot be overemphasized. Ramp inputs are common to all tracking situations, where common tracking devices might be a telescope, an antenna, an antiaircraft gun, or a machine tool. The ability of these devices to perform as required is often dependent to a large extent on their ability to follow constant-velocity inputs.

An infinite number of error constants may be defined and identified with the parameters of a closed-loop system. For example, the next logical input to consider is $t^2$, and the associated error constant is called the *acceleration-error constant $K_a$* which measures the steady-state error due to the parabolic input $t^2$. The acceleration-error constant is defined as

$$K_a = \lim_{s \to 0} s^2 G_{eq}(s)$$

By considering the second derivative of Eq. (8.3-11), the expression that results for $K_a$ is[1]

$$-\frac{2}{K_a} = \frac{1}{K_v^2} + \sum_{j=1}^{n} \frac{1}{\lambda_{kj}^2} - \sum_{j=1}^{m} \frac{1}{\delta_j^2}$$

It is seen that $K_a$ is also completely specified by the location of the closed-loop poles and zeros.

Control of the value of $K_a$ is less important than control of $K_v$, for the simple reason that a ramp input is much more common than a parabolic input. As we shall see in the following sections, the number of specifications that we are able to realize with a low-order model is limited; that is, there are more specifications than we are able to handle. For these reasons $K_v$ proves to be the most important accuracy measurement, provided that $K_{pos}$ has already been set equal to infinity.

Before leaving this section it is worthwhile considering the origins of the names of the error constants. First, the names of the error constants are associated with the system input, not the output. The position-, velocity-, and acceleration-error coefficients are all a measure of an error in the output variable due to a different input. The output variable in which the error is measured always stays the same. For example, if the input is $t$, or a multiple of $t$, as $\alpha t$, a typical response resulting from a finite velocity-error constant is shown in Fig. 8.3-3. If this were a position-servomechanism response, the error is in position, although it is associated with the velocity-error constant. Again, the name is associated with the input. For a positioning system, an input of $t$ requires the output to move at a constant velocity.

The names of the error constants arose from considerations of a positioning servomechanism. In such a system the inputs for which the

[1] John C. Truxal, "Control System Synthesis," p. 283, McGraw-Hill Book Company, New York, 1955.

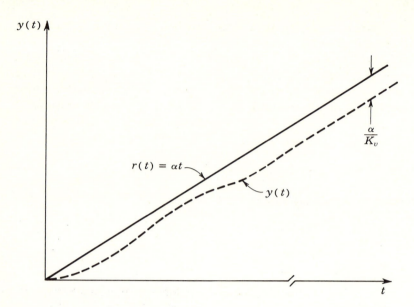

**Fig. 8.3-3**   Output (position) error due to a ramp input and a finite velocity-error constant.

error constants were named are actually position, velocity, and accelera-tion. In the general case, the output variable may not be position; it may be temperature or power or flow rate or almost any other variable. Then the corresponding inputs are not position, velocity, or acceleration. In a temperature regulator, for example, the position-error constant would correspond to a step in temperature demand, the velocity-error constant would correspond to a linear change in temperature, and the acceleration-error constant would correspond to a temperature changing as a squared function. Thus, although the error-constant names have intuitive appeal for a positioning servomechanism, they are in general misnomers.

**Exercises 8.3**   *8.3-1.*   Find the relationship between $k_1$ and $k_2$ such that the system shown in Fig. 8.3-4 has zero steady-state error for a step input. Note that, even though $G_p(s)$ has a pole at the origin, $k_1 = 1$ is not the correct solution.

*answer:*

$$k_1 + k_2 = 1$$

*8.3-2.*   Show that the system in Fig. 8.3-5 has zero steady-state error for step and ramp inputs by showing that $G_{eq}(s)$ has two integrations and that $y(s)/r(s)$ satisfies Eqs. (8.3-9) and (8.3-16).

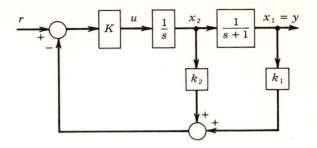

**Fig. 8.3-4**    Exercise 8.3-1.

**8.3-3.**    For the system

$$\dot{x} = \begin{bmatrix} 0 & 1 \\ 0 & -2 \end{bmatrix} x + \begin{bmatrix} 0 \\ 1 \end{bmatrix} u \qquad y = [1 \quad 1]x \qquad u = K(r - [k_1 \quad 0]x)$$

find the values of $k_1$ and $K$ such that the system has zero steady-state error for step and ramp inputs.

*answer:*

$k_1 = 1; K = 2.$

## 8.4   *Specification of second-order systems*

This and the following section are concerned with the selection of a model closed-loop transfer function to meet specifications describing relative stability, accuracy, and speed of response.[1]   We begin by considering the

[1] The material of Secs. 8.4 and 8.5 is based on Charles R. Hausenbauer, "Synthesis of Feedback Systems," University of Missouri, Columbia, Mo., 1957, and C. R. Hausenbauer and G. V. Lago, Synthesis of Control Systems Based on Approximation to a Third-order System, *Trans. AIEE*, pt. II, vol. 77, pp. 415–421, 1958.

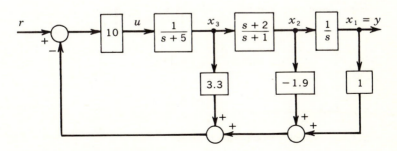

**Fig. 8.3-5**    Exercise 8.3-2.

second-order system with no zeros, which was discussed previously in Sec. 4.3. From the viewpoint of specifications the response of this simple system is found to be inadequate since there are not a sufficient number of parameters available to satisfy all the important performance measures. The addition of a zero brings added freedom of adjustment but increases the complexity of the problem.

The reader might well wonder: What happened to first-order systems? The answer is twofold. First, since there is only one parameter, the closed-loop-pole location $\lambda_k$, available to describe the response, there is simply not enough flexibility to adjust independently the speed, accuracy, and relative stability. For example, if $\lambda_k$ is chosen to satisfy a given rise-time specification, one must accept the relative-stability and accuracy measures that result from that particular choice of $\lambda_k$. Second, first-order models do not adequately describe the behavior of most control systems.

For the discussion of the second-order system without a zero, we wish to consider the transfer function

$$\frac{y(s)}{r(s)} = \frac{\omega_n{}^2}{s^2 + 2\zeta\omega_n s + \omega_n{}^2} \tag{8.4-1}$$

A typical system configuration having this closed-loop transfer function is pictured in Fig. 8.4-1. As the reader may recall from Sec. 4.3, it is convenient to discuss the behavior of the closed-loop system in terms of $\zeta$ and $\omega_n$ rather than in terms of the open-loop parameters of Fig. 8.4-1. For completeness, however, recall that for the system of Fig. 8.4-1

$$K = \omega_n{}^2 \quad \text{and} \quad \lambda + Kk_2 = 2\zeta\omega_n \tag{8.4-2}$$

In terms of $\zeta$ and $\omega_n$, the characteristic polynomial of this system is

$$D_k(s) = s^2 + 2\zeta s\omega_n + \omega_n{}^2$$

The roots of this polynomial are the system's closed-loop poles, and they

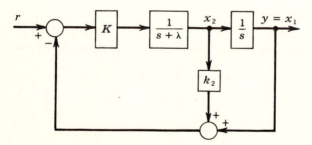

**Fig. 8.4-1**  Simple second-order system.

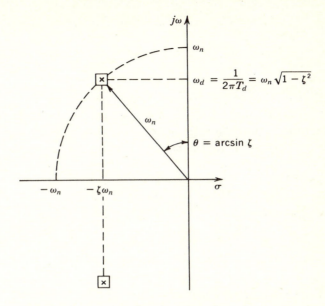

**Fig. 8.4-2**   Pole locations in terms of $\zeta$ and $\omega_n$.

may be real and unequal, real and equal, or a complex conjugate pair. Here, as in Sec. 4.3, we concentrate on the underdamped case when the roots are complex conjugates, since one is normally interested in achieving fast rise time.   The location of the closed-loop complex conjugate poles is shown in Fig. 8.4-2 in terms of the parameters $\zeta$ and $\omega_n$.

In Sec. 4.3 we also showed that the time response $y(t)$ for a step-function input is

$$y(t) = 1 - \frac{1}{\sqrt{1 - \zeta^2}} \, e^{-\zeta\omega_n t} \sin\left(\omega_n \sqrt{1 - \zeta^2}\, t + \psi\right) \tag{8.4-3}$$

where

$$\psi = \arctan \frac{\sqrt{1 - \zeta^2}}{\zeta} \tag{8.4-4}$$

To represent graphically the step response $y(t)$ for different values of $\zeta$ and $\omega_n$ would require a four-dimensional figure.   This is impossible to draw in a three-dimensional world.   Hence we are forced to some alternatives.   First we may normalize with respect to $\omega_n$.   This reduces the dimensionality to three, which is still undesirable.   The dimensionality may be reduced to two if we do not plot $y(t)$ versus $t$ but rather plot some characteristic of $y(t)$; for example, the rise time or percent overshoot.

In order to accomplish the normalization, we divide the numerator and denominator of Eq. (8.4-1) by $\omega_n^2$ and then let $s/\omega_n$ be equal to a new transform variable $s'$.   Equation (8.4-1) then becomes

$$\frac{y(s')}{r(s')} = \frac{1}{s'^2 + 2\zeta s' + 1} \qquad (8.4\text{-}5)$$

Since $s$ has the units of $1/\text{time}$, the substitution $s' = s/\omega_n$ is equivalent to changing the time scale in the time domain to $t' = \omega_n t$.   In terms of this new time scale, the response to a step-function input becomes

$$y(t') = 1 - \frac{1}{\sqrt{1 - \zeta^2}} \, e^{-\zeta t'} \sin \left( \sqrt{1 - \zeta^2} t' + \psi \right) \qquad (8.4\text{-}6)$$

where $\psi$ is still

$$\psi = \arctan \frac{\sqrt{1 - \zeta^2}}{\zeta}$$

We are now prepared to discuss the specifications that describe the important performance measures of relative stability and speed of response. We shall consider the time-domain specifications first.   The specification of relative stability in the time domain is percent overshoot, which is measured at the time $t$ equal to peak time $t_p$ or

$$t_p = \frac{\pi}{\omega_n \sqrt{1 - \zeta^2}}$$

In normalized time this peak time becomes $t'_p$ and is equal to

$$t'_p = \frac{\pi}{\sqrt{1 - \zeta^2}} \qquad (8.4\text{-}7)$$

If this value of time is substituted into the normalized step-function response, the result is

$$y(t')_{\max} = 1 - \frac{1}{\sqrt{1 - \zeta^2}} \exp \left( - \frac{\zeta \pi}{\sqrt{1 - \zeta^2}} \right) \sin (\pi + \psi)$$

$$= 1 + \exp - \frac{\zeta \pi}{\sqrt{1 - \zeta^2}}$$

The final value of $y(t')$ is 1, so that the percent overshoot is

$$PO = 100 \, \frac{1 + \exp (-\zeta\pi/\sqrt{1 - \zeta^2}) - 1}{1} = 100 \exp - \frac{\zeta \pi}{\sqrt{1 - \zeta^2}} \qquad (8.4\text{-}8)$$

This is the same value that resulted in the unnormalized or real-time case treated in Sec. 4.3.   This result might have been expected, since changing the time scale does not modify the maximum value of $y(t)$ or $y(t')$.

The measures of speed of response, however, are modified by the change of time scale. The time-domain specifications that indicate speed of response are the rise time $t_r$, the delay time $t_d$, and the settling time $t_s$. In Sec. 4.3 an analytic expression was found for the settling time for $\alpha = 5$; this expression was

$$t_s = \frac{3}{\zeta\omega_n}$$

The normalized settling time is therefore

$$t_s' = t_s\omega_n = \frac{3}{\zeta} \tag{8.4-9}$$

Analytic expressions cannot be obtained for the normalized delay times and rise time, but $t_r'$ and $t_d'$ may be measured from a normalized step-function-response curve for the different values of the damping ratio $\zeta$. These points may then be plotted and approximated by an analytic function of $\zeta$. Suitable approximations in the normalized time are

$$t_r' = t_r\omega_n = \frac{7.04\zeta^2 + 0.2}{2\zeta} \tag{8.4-10}$$

$$t_d' = t_d\omega_n = 1 + 0.7\zeta \tag{8.4-11}$$

Plots of the four time-domain specifications given in Eqs. (8.4-8) to (8.4-11) are shown in Fig. 8.4-3. Note that, in this normalized situation, selecting $\zeta$ uniquely specifies the relationships between all these performance specifications. Freedom to adjust speed of response still exists by selecting $\omega_n$, as illustrated by the following example.

> **Example 8.4-1** Suppose a 25 percent overshoot is specified, along with a settling time of 5 sec. From Fig. 8.4-3 we use the PO curve to determine that the required damping ratio for a 25 percent overshoot is $\zeta = 0.4$. In the same figure note that, for $\zeta = 0.4$, the normalized settling time is 7.4 sec or
>
> $$t_s' = \omega_n t_s = 7.4 = 5\omega_n$$
>
> Thus $\omega_n$ is just 1.48, and these values of $\zeta$ and $\omega_n$ may be substituted into the expression for the desired closed-loop transfer function Eq. (8.4-1) to yield
>
> $$\frac{y(s)}{r(s)} = \frac{(1.48)^2}{s^2 + 2(0.4)(1.48)s + 1.48^2} = \frac{2.2}{s^2 + 1.18s + 2.2}$$

Now let us turn to the frequency-domain specifications for relative stability and speed of response. The relative-stability specification is

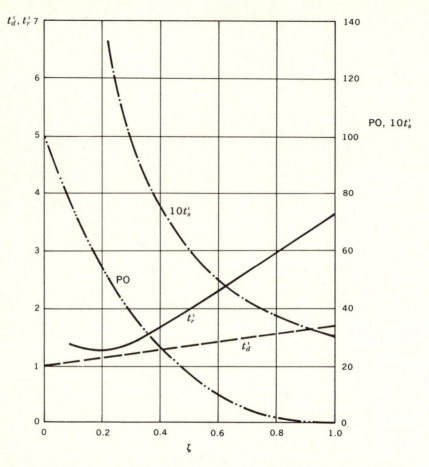

**Fig. 8.4-3**   Time-domain specifications versus $\zeta$ for the normalized second-order system of Eq. (8.4-5).

$M$ peak, and in terms of $\zeta$ and $\omega_n$, $M_p$ was shown in Sec. 6.5 to be equal to

$$M_p = \frac{1}{2\zeta\sqrt{1-\zeta^2}} \qquad \text{for } \zeta \le 0.707 \qquad (8.4\text{-}12)$$

Note that this equation is valid only if $\zeta \le 0.707$. Although the step-function response has overshoot for any value of the damping ratio less than 1, the frequency-response curve does not exhibit a peak unless $\zeta \le 0.707$. The value of $M_p$ does not change as the frequency or time scale is changed, since $M_p$ is not a function of $\omega_n$.

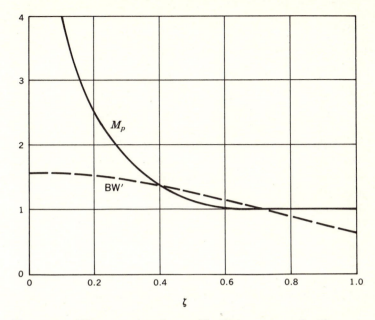

***Fig. 8.4-4***  Frequency-domain specifications versus $\zeta$ for the normalized second-order system of Eq. (8.4-5).

The frequency-domain specification of speed of response is the bandwidth. The expression for the normalized bandwidth in terms of $\zeta$ is

$$BW' = \frac{BW}{\omega_n} = [(1 - 2\zeta^2) + \sqrt{(1 - 2\zeta^2)^2 + 1}]^{\frac{1}{2}} \qquad (8.4\text{-}13)$$

A plot of Eqs. (8.4-12) and (8.4-13) is given in Fig. 8.4-4. Let us work a problem similar to that of Example 8.4-1 to illustrate the use of this chart.

***Example 8.4-2***  Suppose that the specifications require a second-order transfer function with a bandwidth of 10 rad/sec and an $M_p$ of 1.15. The requirement on $M_p$ indicates that a damping ratio of about 0.5 is needed. For this value of the damping ratio, the normalized bandwidth is approximately 1.25 [actually 1.275 if Eq. (8.4-13) is used]. Since the normalized bandwidth is

$$BW' = \frac{BW}{\omega_n} \qquad \text{or} \qquad \omega_n = \frac{BW}{BW'}$$

the necessary $\omega_n$ is just

$$\omega_n = \frac{10}{1.25} = 8$$

The required transfer function is

$$\frac{y(s)}{r(s)} = \frac{8^2}{s^2 + 2(0.5)(8)s + 8^2} = \frac{64}{s^2 + 8s + 64}$$

The inadequacy of the second-order response is indicated by an examination of the accuracy specification $K_v$. Implied, of course, is that the position-error constant is infinite so that the steady-state error due to a step input is zero. The velocity-error constant may be determined by using Eq. (8.3-15):

$$\frac{1}{K_v} = \sum_{j=1}^{n} \frac{1}{\lambda_{kj}} - \sum_{j=1}^{m} \frac{1}{\delta_j} \qquad (8.3\text{-}15)$$

In the normalized second-order system of Eq. (8.4-5) there are no zeros. The closed-loop poles are placed in evidence by rewriting Eq. (8.4-5) as

$$\frac{y(s')}{r(s')} = \frac{1}{(s' + \zeta)^2 + (\sqrt{1 - \zeta^2})^2}$$

$$= \frac{1}{(s' + \zeta + j\sqrt{1 - \zeta^2})(s' + \zeta - j\sqrt{1 - \zeta^2})}$$

so that $K'_v = K_v/\omega_n$, the normalized velocity constant, is

$$\frac{1}{K'_v} = \frac{1}{\zeta + j\sqrt{1 - \zeta^2}} + \frac{1}{\zeta - j\sqrt{1 - \zeta^2}} = 2\zeta$$

or

$$K'_v = \frac{1}{2\zeta} \qquad (8.4\text{-}14)$$

Equation (8.4-14) is plotted in Fig. 8.4-5. At first glance it might appear that any value of velocity-error constant may be chosen, since $K'_v$ ranges from $\frac{1}{2}$ to infinity. Since we are interested in small tracking errors, a large $K'_v$ is desired. But large values of $K'_v$ are associated with very small values of the damping ratio. For a value of $\zeta$ equal to 0.3, the magnitude of $K'_v$ is only 1.66. But from Fig. 8.4-4 the $M$ peak is 1.8. Both these measures of relative stability, $\zeta$ and $M_p$, indicate that it is dangerous to attempt to increase $K'_v$ any more by further decreasing the damping ratio. A practical lower limit on the damping ratio is usually around 0.5, with an associated overshoot of less than 20 percent and an $M_p$ near 1.2.

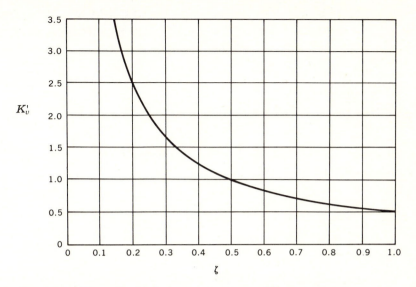

$K'_v$

**Fig. 8.4-5**   Normalized velocity-error constant $K'_v$ versus $\zeta$.

The reader may have anticipated this difficulty before we started to investigate the second-order system.    There are two parameters available for adjustment, $\zeta$ and $\omega_n$, and there are three performance measures to be satisfied: the relative stability, speed of response, and accuracy. Clearly, independent control of all three performance measures is not possible with only two adjustments.    Hence we must increase the number of degrees of freedom of adjustment in some manner.    Here we do it by adding a zero to the second-order system.    The closed-loop transfer function thus becomes

$$\frac{y(s)}{r(s)} = \frac{(\omega_n{}^2/\delta)(s + \delta)}{s^2 + 2\zeta\omega_n s + \omega_n{}^2} \tag{8.4-15}$$

and, as before, the gain of this closed-loop transfer function has been chosen so that $y(s)/r(s) \big|_{s=0} = 1$ and the steady-state position error is zero.

Once again we normalize this expression by letting $s = s'\omega_n$, and after both the numerator and denominator are divided by $\omega_n{}^2$, the result is

$$\frac{y(s')}{r(s')} = \frac{1}{\delta'} \frac{s' + \delta'}{s'^2 + 2\zeta s' + 1} \tag{8.4-16}$$

where $\delta' = \delta/\omega_n$.

The presence of the zero in Eq. (8.4-16) makes dealing with this equation more complicated than dealing with the unity-numerator system of Eq. (8.4-5). For a step-function input, Eq. (8.4-16) becomes

$$
\begin{aligned}
y(s') &= \frac{1}{\delta'} \frac{s' + \delta'}{s'(s'^2 + 2\zeta s' + 1)} \\
&= \frac{1}{\delta'} \frac{s' + \delta'}{s'[(s' + \zeta)^2 + (\sqrt{1 - \zeta^2})^2]}
\end{aligned}
\tag{8.4-17}
$$

By using entry 28 of the Laplace transform table in Appendix B, $y(t')$ may be written directly:

$$
y(t') = 1 + \frac{\sqrt{1 + \delta'(\delta' - 2\zeta)}}{\delta' \sqrt{1 - \zeta^2}}\, e^{-\zeta t'} \sin(\sqrt{1 - \zeta^2}\, t' + \psi)
\tag{8.4-18}
$$

where

$$
\psi = \arctan \frac{\sqrt{1 - \zeta^2}}{\delta' - \zeta} - \arctan \frac{\sqrt{1 - \zeta^2}}{-\zeta}
$$

We may now proceed as we did in the unity-numerator case. Once we had an expression for $y(t')$, we turned to the time-domain measures of relative stability and speed of response. In order to find PO, we must know the peak time which is then substituted into $y(t')$ to find $y(t')_{max}$ and eventually to find PO. After a number of lengthy algebraic manipulations, we find that

$$
\mathrm{PO} = 100 \left(1 + \frac{1}{\delta'^2} - \frac{2\zeta}{\delta'}\right)^{\frac{1}{2}} \exp \frac{-\zeta\phi}{\sqrt{1 - \zeta^2}}
$$

where

$$
\phi = \pi - \arctan \frac{\sqrt{1 - \zeta^2}}{\delta' - \zeta}
\tag{8.4-19}
$$

Since PO is a function of both $\delta'$ and $\zeta$, rather than use a three-dimensional figure to picture the dependence of PO on $\delta'$ and $\zeta$, let us plot PO versus $\delta'$ and use $\zeta$ as a parameter, as shown in Fig. 8.4-6. This is an interesting figure because it shows that the same percent overshoot may be realized with different values of $\zeta$, depending, of course, on the zero location $\delta'$. For example, if $\delta'$ is 0.6 and the damping ratio is $\zeta = 0.9$, 15 percent overshoot results. The same overshoot results if $\delta' = 2$ and $\zeta = 0.55$. In both cases the closed-loop poles lie on the unit circle in the $s'$ plane, although the relative positions of the poles and zero are quite different, as seen in Fig. 8.4-7.

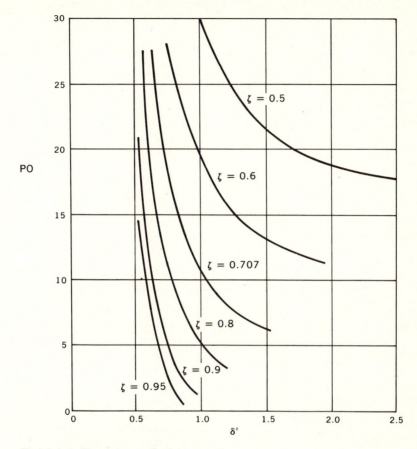

**Fig. 8.4-6**  Percent overshoot versus $\delta'$.

For the second-order case with no zero, selecting $\zeta$ and $\omega_n$ completely specifies the system response and hence all the properties of $y(t)$. After normalization, the selection of $\zeta$ completely describes the characteristics of $y(t')$. In the second-order case with a zero, knowledge of $\zeta$, $\omega_n$, and $\delta'$ completely specifies the system response and hence all the properties of $y(t)$. After normalization, the selection of $\zeta$ and $\delta'$ completely describes the characteristics of $y(t')$. Hence we should be able to describe all the other characteristics of $y(t')$ in a figure much the same as Fig. 8.4-6 with $\zeta$ as a parameter.

The expression for $K_v'$ is particularly easy to determine. Again we use Eq. (8.3-15). The location of the closed-loop poles remains as in the case with no zero, but now a zero appears at $\delta'$. Thus $K_v'$ is

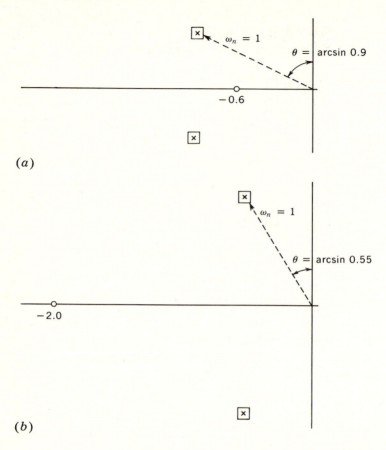

*Fig. 8.4-7*  Two closed-loop pole-zero configurations with the same percent overshoot: (*a*) $\zeta = 0.9$, $\delta' = 0.6$; (*b*) $\zeta = 0.55$, $\delta' = 2$.

found to be

$$\frac{1}{K_v'} = \frac{1}{\zeta + j\sqrt{1 - \zeta^2}} + \frac{1}{\zeta - j\sqrt{1 - \zeta^2}} - \frac{1}{\delta'} = 2\zeta - \frac{1}{\delta'}$$

or

$$K_v' = \frac{\delta'}{2\zeta\delta' - 1} \tag{8.4-20}$$

Equation (8.4-20) is plotted in Fig. 8.4-8 for various values of $\zeta$.  Here it is seen that the normalized velocity-error constant can have the same value for a number of different values of the damping ratio.   This is much the same situation that was described for Fig. 8.4-6 with respect to percent overshoot.

As previously noted, if one desires a 15 percent overshoot, either pole-zero configuration of Fig. 8.4-7 may be used.   However, for the pole-zero plot of Fig. 8.4-7a, with $\zeta = 0.9$ and $\delta' = 0.6$, it is seen from Fig. 8.4-8 that the velocity-error constant is approaching infinity.   For $\zeta = 0.55$ and $\delta' = 2$, as in Fig. 8.4-7b, Fig. 8.4-8 indicates that the velocity-error constant is less than 2.   Similarly, a specific vleocity-error constant may be realized for a number of different values of $\zeta$ and $\delta'$.   Of course, different values of overshoot result.   To be specific, assume that a normalized velocity-error constant of 2 is required.   With a damp-

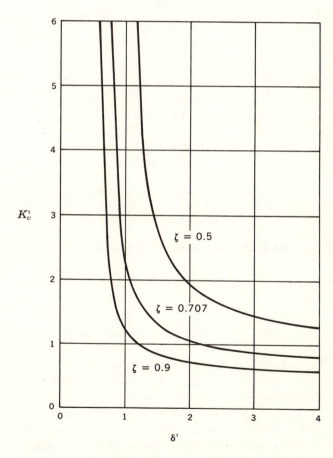

**Fig. 8.4-8**   Normalized velocity-error constant versus $\delta'$.

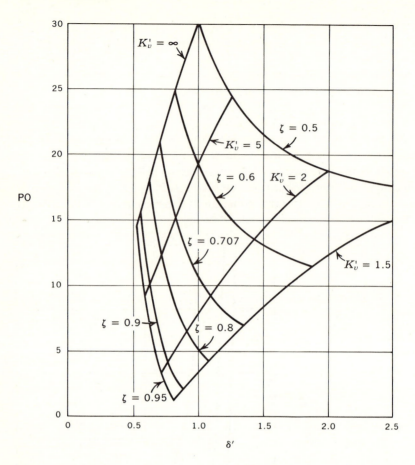

**Fig. 8.4-9**  Percent overshoot and normalized velocity-error constant versus $\delta'$.

ing ratio of 0.5, the necessary value of $\delta'$ is 2.    From Fig. 8.4-6 the percent overshoot for these values of $\zeta$ and $\delta'$ is about 18 percent.    If a damping ratio of 0.9 is chosen, a $\delta'$ of approximately 0.8 is needed, and the PO is less than 5 percent.

In order to display this interplay between percent overshoot and velocity-error constant more clearly, the data of Fig. 8.4-8 may be transferred in total to Fig. 8.4-6, as shown in Fig. 8.4-9.    This is accomplished in a point-by-point fashion, much as indicated in the discussion immediately above.    Because more data are presented in this one figure, it appears somewhat more confusing at first glance.    The reader should rework the simple examples discussed in the preceding paragraph to ensure that Fig. 8.4-9 is understood.

Thus far we have considered the specification of percent overshoot and velocity-error constant for the second-order case with one zero in the closed-loop transfer function. These two specifications are measures of relative stability and accuracy, respectively. Let us now consider the specification of bandwidth as a measure of speed of response.

The normalized bandwidth is defined as the point at which the magnitude of the normalized frequency-response function is $1/\sqrt{2} = 0.707$. The transfer function in question is given by Eq. (8.4-16), and the normalized frequency-response function is determined from this equation by replacing $s'$ by $j\omega'$, to yield

$$\frac{y(j\omega')}{r(j\omega')} = \frac{1}{\delta'} \frac{j\omega' + \delta'}{(j\omega')^2 + 2\zeta j\omega' + 1} \tag{8.4-21}$$

The magnitude of this normalized frequency-response function is just

$$\left| \frac{y(j\omega')}{r(j\omega')} \right| = \frac{1}{\delta'} \left[ \frac{\delta'^2 + \omega'^2}{(1 - \omega'^2)^2 + (2\zeta\omega')^2} \right]^{\frac{1}{2}} \tag{8.4-22}$$

After some algebraic manipulations the normalized bandwidth BW' is found to be

$$\text{BW}' = (a + \sqrt{1 + a^2})^{\frac{1}{2}} \tag{8.4-23}$$

where

$$a = 1 + \frac{1}{\delta'^2} - 2\zeta^2 \tag{8.4-24}$$

Figure 8.4-10 shows BW' versus $\delta'$ with $\zeta$ as a parameter.

To indicate the use of Fig. 8.4-10, let us return once again to the pole-zero configurations of Fig. 8.4-7. With $\zeta = 0.9$ and $\delta' = 0.6$ we have previously indicated an overshoot of 15 percent and a velocity-error constant approaching infinity. Here we see that the corresponding BW' is about 2.1. With $\zeta = 0.55$ and $\delta' = 2$, overshoot was again 15 percent with $K_v'$ less than 2. Figure 8.4-10 indicates that BW' is approximately 1.4.

Rather than use Fig. 8.4-10 in conjunction with Fig. 8.4-9, the composite figure made up of Figs. 8.4-6 and 8.4-8, it is possible to include all the data in one figure, as Fig. 8.4-11. In this figure the normalized bandwidth is indicated at the intersection of each of the constant-$\zeta$ and constant-$K_v'$ lines.

Figure 8.4-11 contains a wealth of information. The usable design range is the interior of the region bounded by the values of $K_v'$ between infinity and 1.5 and by the values of $\zeta$ between 0.95 and 0.5. The bounds on $K_v'$ were chosen between the best possible $K_v' = \infty$ and the value realizable by the second-order system with no zero, approximately

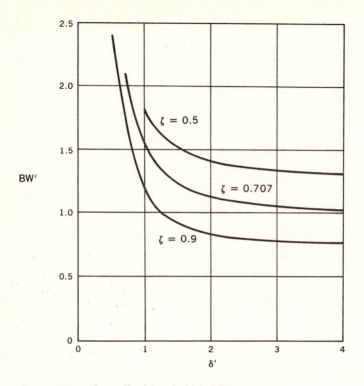

*Fig. 8.4-10* Normalized bandwidth BW′ versus δ′.

1.5. For values of ζ greater than 0.95, or as ζ approaches 1, the percent overshoot goes to zero. For values of ζ less than 0.5, the minimum allowable percent overshoot, associated with a reasonable $K'_v$, is too large. It is interesting to observe that, although the normalized velocity-error constant varies between 1.5 and ∞, the bandwidth variation indicated in Fig. 8.4-11 is restricted between the range 1.3 to 2.5. In general, as the velocity-error constant increases, so does bandwidth, but clearly not at the same rate.

Although Fig. 8.4-11 contains much valuable information, one seldom has a chance to make use of it. This is true because the pole-zero excess of the second-order system with one zero is only 1. A simple practical second-order system involves a power element with inertia and no inherent zero. If a zero is to be added to the closed-loop system, this is accomplished by the addition of a network in front of the plant. This network necessarily includes a pole as well as a zero, as will be discussed in Chaps. 9 and 10. The resulting plant and the closed-loop system

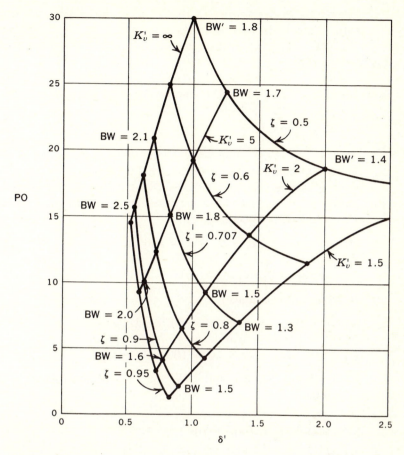

**Fig. 8.4-11** PO, $K'_v$, and BW' versus $\delta'$.

then both have three poles and one zero so that the pole-zero excess is 2 in this simple practical case.

An important feature of this section is the ability to control the velocity-error constant by positioning a zero of the closed-loop system. Equally important is the realization that, even with a system as simple as the ones considered here, the analysis can prove to be too difficult to accomplish by hand. For this reason we completely avoided such performance specifications as rise time or settling time. In the following section the data were acquired not by calculation but from an analog computer. Once curves of the response are available, all specifications can be discussed simultaneously.

It may be disheartening to realize that the study that went into this section has little practical use. It is the authors' opinion, however, that the following section would be quite unintelligible without this introduction.

**Exercises 8.4**   *8.4-1.*   Plot the step response of the system

$$\frac{y(s)}{r(s)} = \frac{1}{s^2 + 2\zeta s + 1}$$

for $\zeta = 1.0$, 0.707, and 0.5. From these plots determine PO, $t_r$, and $t_d$ and compare the results with Fig. 8.4-3.

*8.4-2.* Derive Eqs. (8.4-13) and (8.4-23).

*8.4-3.* Find the second-order transfer functions that satisfy each of the following specifications:

(a)   PO = 5%, $K'_v$ = 2, BW = 100 rad/sec
(b)   PO = 15%, $K'_v$ = ∞, BW = 100 rad/sec
(c)   PO = 20%, $K'_v$ = ∞, BW = 200 rad/sec

*answers:* (approximate)

(a)   $\zeta$ = 0.9, $\delta$ = 51.6, $\omega_n$ = 64.5 rad/sec
(b)   $\zeta$ = 0.9, $\delta$ = 28.6, $\omega_n$ = 47.7 rad/sec
(c)   $\zeta$ = 0.73, $\delta$ = 68, $\omega_n$ = 100 rad/sec

*8.4-4.* Find PO, $K'_v$, and BW' for each of the following transfer functions:

(a)   $\dfrac{y(s)}{r(s)} = \dfrac{10^4}{s^2 + 100s + 10^4}$

(b)   $\dfrac{y(s)}{r(s)} = \dfrac{50(s + 50)}{s^2 + 60s + 2{,}500}$

(c)   $\dfrac{y(s)}{r(s)} = \dfrac{6.67(s + 15)}{s^2 + 10s + 100}$

*answers:* (approximate)

(a)   PO = 16.5%, $K'_v$ = 1, BW' = 1.2
(b)   PO = 19.5%, $K'_v$ = 5, BW' = 1.75
(c)   PO = 23%, $K'_v$ = 3, BW' = 1.5

## 8.5   *Third-order systems*

The conclusions of Sec. 8.4 are disappointing. After much work, the practical results that followed are very small indeed. In this section just the opposite is true; the amount of analysis done is small but the results are the basis for the specification of all higher-order systems.

Let us begin by examining first a third-order system with no zero. Three adjustable parameters are associated with such a system: $\zeta$ and $\omega_n$ to locate the pair of complex conjugate poles, and the location of the real closed-loop pole $\lambda_k$. Although the number of adjustable parameters associated with second-order systems with one zero is also three, it is not possible to control the velocity-error constant by the addition of an extra pole, because no negative terms are available in Eq. (8.3-15). Thus the second-order system with one zero may be more accurate than the third-order system with no zero. The limited usefulness of a model transfer function having only three poles suggests the desirability of a third-order model with a zero.

A third-order-model transfer function with one zero is indicated as

$$\frac{y(s)}{r(s)} = \frac{K(s + \delta)}{(s^2 + 2\zeta\omega_n s + \omega_n{}^2)(s + \lambda_k)} \tag{8.5-1}$$

To ensure that the position-error constant $K_{\text{pos}}$ is infinite, $K$ must be $\omega_n{}^2\lambda_k/\delta$, so that Eq. (8.5-1) becomes

$$\frac{y(s)}{r(s)} = \frac{\omega_n{}^2\lambda_k}{\delta} \frac{s + \delta}{(s^2 + 2\zeta\omega_n s + \omega_n{}^2)(s + \lambda_k)} \tag{8.5-2}$$

Four parameters are available for adjustment: $\delta$, $\omega_n$, $\zeta$, and $\lambda_k$.

As in Sec. 8.4, we normalize by dividing Eq. (8.5-2) by $\omega_n{}^2$ in both numerator and denominator to obtain

$$\frac{y(s')}{r(s')} = \frac{\lambda_k'}{\delta'} \frac{s' + \delta'}{(s'^2 + 2\zeta s' + 1)(s' + \lambda_k')} \tag{8.5-3}$$

where $\lambda_k' = \lambda_k/\omega_n$, $\delta' = \delta/\omega_n$, and $s' = s/\omega_n$. The pole-zero configuration corresponding to Eq. (8.5-3) is shown in Fig. 8.5-1. The normalized time response $y(t')$ for a unit step-function input may be determined by partial-fraction expansion or by use of the table in Appendix B as

$$y(t') = 1 + \frac{\lambda_k' - \delta'}{\delta'[(\zeta - \lambda_k')^2 + \beta^2]} e^{-\lambda_k' t'}$$

$$+ \frac{\lambda_k'}{\delta'\beta} \left[\frac{(\delta' - \zeta)^2 + \beta^2}{(\lambda_k' - \zeta)^2 + \beta^2}\right]^{\frac{1}{2}} e^{-\zeta t'} \sin(\beta t' + \psi) \tag{8.5-4}$$

where $\beta = \sqrt{1 - \zeta^2}$, $t' = \omega_n t$, and

$$\psi = \arctan\frac{\beta}{\delta' - \zeta} - \arctan\frac{\beta}{\lambda_k' - \zeta} - \arctan\frac{\beta}{-\zeta}$$

Because of the complexity of this expression for $y(t')$, it is impossible to solve for the time-domain performance specifications explicitly. How-

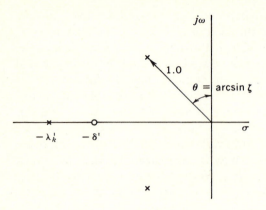

**Fig. 8.5-1**  Pole-zero configuration for Eq.
(8.5-3).

ever, by plotting $y(t')$ for various values of $\zeta$, $\delta'$, and $\lambda'_k$, it is possible to determine any of the time-domain characteristics previously discussed by direct measurement from the various time responses. From a frequency-response plot for the same range of values of $\zeta$, $\delta'$, and $\lambda'_k$, the frequency-domain characteristics may be determined in a similar fashion.

Obtaining plots of $y(t')$ for different combinations of the four parameters is a time-consuming job, but once it has been done, it need not be repeated. The problem that concerns us now is displaying the results.

Because of the four variables involved, the graphical presentation could be made in a variety of ways. As an initial step, let us include the normalized velocity-error constant $K'_v$ as one of the variables, in place of the location of the zero $\delta'$. This is possible because, if all pole positions are specified, $K'_v$ and $\delta'$ are related by Eq. (8.3-15). The method of presentation chosen is to select a value of damping ratio and then plot the various performance specifications versus $\lambda'_k$ for different values of the normalized velocity-error constant.

In order to avoid an excessive number of lines on the same graph, a separate graph is made for each value of $K'_v$. Figure 8.5-2$a$, $b$, and $c$ all involve a damping ratio of 0.5, with $K'_v$ ranging from 1.5 to 5. Compare the percent overshoot in Fig. 8.5-2$a$, $b$, and $c$ as $K'_v$ increases. For $K'_v$ equal to 5, the minimum value of overshoot, except for very small values of $\lambda'_k$, approaches 30 percent. For this reason, higher values of $K'_v$ are not considered for this relatively low value of damping ratio.

Included in Fig. 8.5-2 is also the required zero position $\delta'$ necessary to realize the indicated normalized velocity-error constant. As $\delta'$ and $\lambda'_k$ approach zero, the ratio $\delta'/\lambda'_k$ approaches unity, and the pole and zero

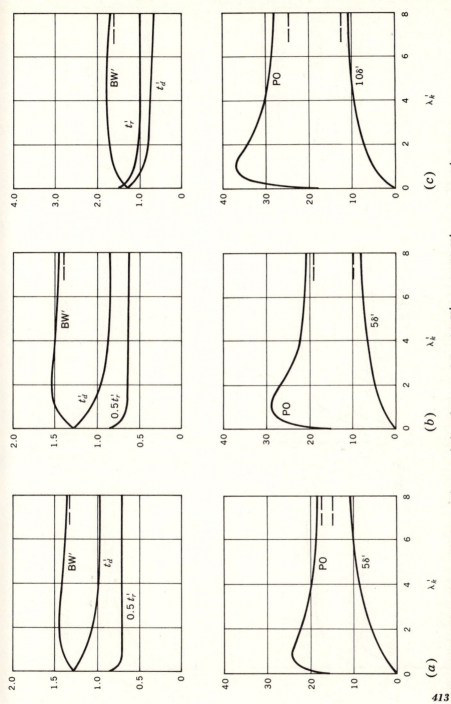

**Fig. 8.5-2** Closed-loop performance characteristics when $\zeta = 0.5$. (a) $K_v' = 1.5$; (b) $K_v' = 2.0$; (c) $K_v' = 5.0$.

413

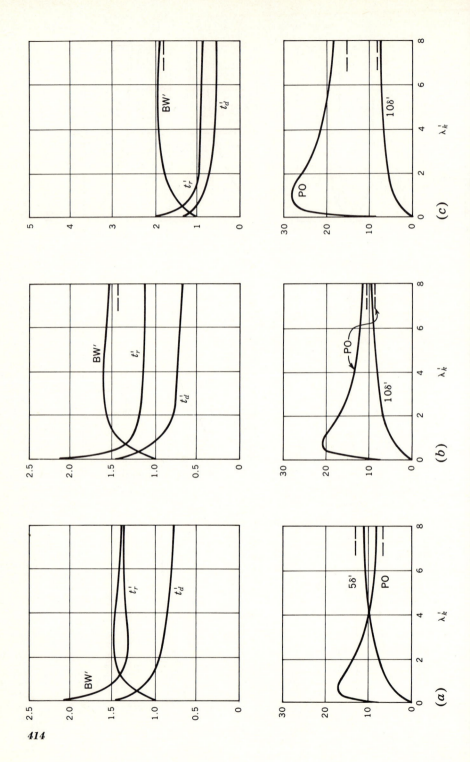

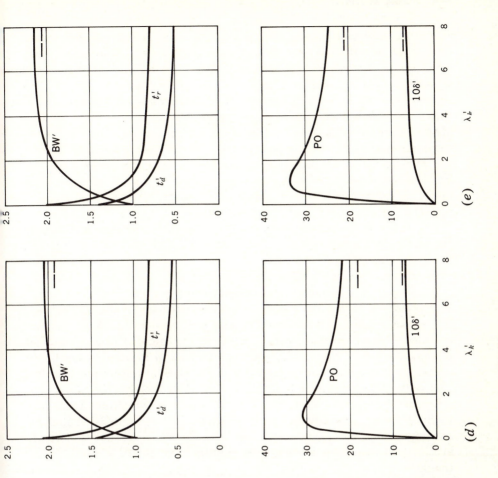

*Fig. 8.5-3* Closed-loop performance characteristics when $\zeta = 0.707$. (a) $K'_v = 1.5$; (b) $K'_v = 2.0$; (c) $K'_v = 5.0$; (d) $K'_v = 10$; (e) $K'_v = \infty$.

in effect cancel each other.    The system then degenerates to the second-order system with no zero, as discussed in Sec. 8.4.    To the extreme right of these figures, $\lambda'_k$ is equal to 8.    For this condition the real pole of $y(s)/r(s)$ has only a slight influence on the system performance, and the system behavior is very similar to that of the second-order system with a zero.    As $\lambda'_k$ approaches infinity the system performance is precisely that of the second-order system with one zero.    Thus the values of the specifications associated with the second-order system with one zero become the asymptotic values indicated in the figures by dashed lines at $\lambda'_k = 8$. For intermediate values of $\lambda'_k$ between 0 and 8, a variety of closed-loop performances is available.    Figure 8.5-3 corresponds to Fig. 8.5-2 except now the damping ratio is changed to 0.707.

In view of the usual tolerances on system-performance specifications, the curves prepared for $\zeta = 0.5$ in Fig. 8.5-2 and for $\zeta = 0.707$ in Fig. 8.5-3 are sufficient and representative of situations normally encountered with minimum-phase transfer functions, that is, overshoots between 5 and 35 percent and normalized velocity-error constants in the range of 1 to infinity.    If deemed necessary for a particular situation, additional curves for an appropriate $\zeta$ could be prepared by following the procedure outlined earlier.    In this regard it is interesting to note that a comparison of data for $\zeta = 0.707$ for $K'_v$ equal to 10 and to infinity, respectively, reveals that there are no appreciable differences in system performance. The only apparent major difference is in the velocity-error constant: a difference from 10 to infinity.    This difference is somewhat illusionary. Although one may mathematically design to satisfy Eq. (8.3-16) and thus realize an infinite velocity-error constant, the error constant that results in the actual system is less than infinity because of normal system tolerances.

The careful reader may have noticed that two specifications previously mentioned are not included in Figs. 8.5-2 and 8.5-3.    These specifications are the settling time and the $M$ peak.    As previously noted, settling time is a discontinuous curve, and although this was successfully approximated in the second-order case, it is not so easily done for third- and higher-order systems.    Since both $M$ peak and PO are specifications of relative stability, we have chosen to measure relative stability by percent overshoot rather than $M_p$.    Once the closed-loop transfer function is chosen, the entire $M$ curve may be drawn rapidly by use of the straight-line Bode approximations, and both $M_p$ and $\omega_p$ can be read directly from this $M$ curve.

Before considering specific examples of actually selecting the closed-loop transfer function on the basis of given specifications, let us examine several characteristic time responses.    In Fig. 8.5-4 the normalized time

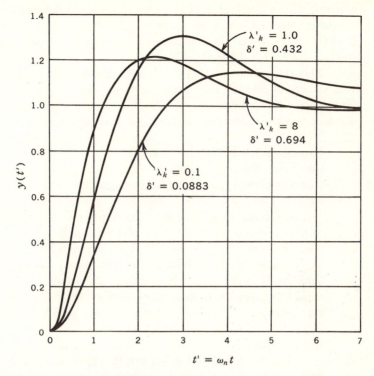

*Fig. 8.5-4*   Closed-loop step response for several values of $\lambda'_k$ with $K'_v = 10$ and $\zeta = 0.707$.

response for $K'_v = 10$ and $\zeta = 0.707$ is given.   The responses are shown for different values of the real-pole location and the corresponding zero locations.   The response curve for $\lambda'_k = 8$ is controlled to a very great extent by the conjugate poles and zero only.   From the composite figure for the second-order system with one zero (Fig. 8.4-9), it is seen that for a damping ratio of 0.707 and a $K'_v$ of 10 (interpolated) the percent over-shoot is between 15 and 20 percent.   From Fig. 8.5-3 the overshoot is about 22 percent so that presence of the real-axis pole far to the left on the $s$ plane does have negligible effect on the closed-loop response.

For the case when the $\lambda'_k = 1$, however, both the real pole and the zero have marked effects on the response.   If the real pole were not pres-ent, the overshoot due to the zero and the conjugate poles would be so large that it could not be read in Fig. 8.4-9 for the second-order system with one zero.   The overshoot is actually 53.1 percent.   With the pole present, as shown in Fig. 8.5-4, the overshoot is only 31 percent.   In general, the presence of the real pole has a prominent stabilizing influence on the system when $\lambda_k'$ is less than 4. If, however, $\lambda_k'$ becomes too small, then $t_r'$ and $t_d'$ rise sharply and the system becomes sluggish, as indicated by the response when $\lambda_k' = 0.1$.

The remainder of this section is devoted to examples using the various design charts that have been presented. Two examples have been chosen, in which the specifications have been given in the time domain and a combination of the time and frequency domains.

*Example 8.5-1* In this example all the specifications are given in the time domain. Consider a typical quick-acquisition, small-tracking-error-type problem, as might be associated with an anti-aircraft gun. In such a case, speed of response is important, and a rather large overshoot may be allowed in order to increase the response speed. Once target acquisition has been achieved, the tracking error must be small. If the expected tracking rate is given as $\omega_r$ rad/sec and an allowable steady-state position error is $\theta_p$ rad, then $K_v \geq \omega_r/\theta_p$. In a particular case assume that these general considerations have been translated to the following time-domain specifications: PO = 25 percent; $t_d = 0.1$ sec; $K_v = 50$. The problem is to select a closed-loop transfer function that meets these specifications.

Figures 8.5-2 and 8.5-3 are the basis for the design. Let us first decide what possibilities exist in terms of percent overshoot. In Fig. 8.5-2*b*, a 25 percent overshoot is available at approximately $\lambda_k' = 3$, with $\zeta$ equal to 0.5. The same PO is available for $\zeta = 0.707$ in Fig. 8.5-3*c*, *d*, and *e*. Consider Fig. 8.5-3*c*, for example. There PO is 25 percent for $\lambda_k'$ approximately equal to 3, with a normalized delay time $t_d'$ of about 0.7 and $K_v'$ of 5. Since $K_v' = 5 = K_v/\omega_n$ and we wish $K_v$ to be 50, we must have $\omega_n = 10$ rad/sec. For $\omega_n = 10$, the delay time becomes $t_d = t_d'/\omega_n = 0.07$. Since the normalized zero location is at $\delta'$, the actual zero is located at $s = -0.7\omega_n = -7$, and the real pole is at $s = -3\omega_n = -30$. The closed-loop transfer function is therefore

$$\frac{y(s)}{r(s)} = \frac{(10^2)(30)}{7} \frac{s+7}{(s^2 + 14.14s + 100)(s + 30)} \tag{8.5-5}$$

The step-function response for the transfer function has a PO = 25 percent, and the required $K_v$ of 50 has also been realized. The delay time of 0.07 sec is better than the required 0.1 sec; hence this closed-loop transfer function is more than satisfactory.

Many other choices could have been made for an ideal $y(s)/r(s)$

since we chose Fig. 8.5-3c in a rather arbitrary manner, as any of the four figures with PO = 25 percent could have been used.

In the previous discussion $K_v'$ was finite and governed the choice of $\omega_n$. If we use Fig. 8.5-3e, where $K_v'$ is infinite, then for $\lambda_k' = 8$ and $10\delta' \sim 7$, PO is once again 25 percent. Now $\omega_n$ may be chosen on the basis of delay time. From Fig. 8.5-3e, $t_d'$ is 0.5; since the required $t_d$ is 0.1, $\omega_n = t_d'/t_d = 5$ rad/sec. Since $\delta' = 0.7$, the actual zero location is at $s = -0.7\omega_n = -3.5$, and the real-axis pole is at $s = -40$. Use of Eq. (8.5-2) indicates that

$$\frac{y(s)}{r(s)} = \frac{(25)(40)}{3.5} \frac{s + 3.5}{(s^2 + 7.07s + 25)(s + 40)} \tag{8.5-6}$$

This transfer function exceeds the velocity-error-constant requirement, whereas the transfer function of Eq. (8.5-5) exceeded the delay-time requirement. Either is acceptable, in terms of the given specifications, as a desired closed-loop transfer function. Whether this ideal $y(s)/r(s)$ can be achieved depends upon the fixed plant that is to be controlled. The next two sections involve considerations associated with the unalterable fixed plant.

The time-domain specifications for Example 8.5-1 are not entirely adequate. For the given percent overshoot and velocity-error constant, a delay time as small as desired can be realized by just increasing $\omega_n$. This also increases the bandwidth, which is undesirable if any high-frequency noise is present. Another undesirable feature of increasing $\omega_n$ to an arbitrary size is the reduction of the range of allowable inputs for which the closed-loop system behaves in a linear fashion. More will be said on this topic in Sec. 8.7. In order to avoid this problem, a limitation on bandwidth is often included in the specifications. As a consequence of a bandwidth-limitation specification, mixed time- and frequency-domain specifications often occur, as in the following example.

***Example 8.5-2*** Assume the following specifications are given: PO = $20 \pm 1\%$; $K_v > 200$; $t_d < 0.02$ sec; and BW $< 100$ rad/sec.

The requirement for only 20 percent overshoot all but eliminates the use of the charts for a damping ratio of 0.5 given in Fig. 8.5-2. Hence we are restricted to the use of a damping ratio of 0.707. Since the ratio of bandwidth to $\omega_n$ is always greater than 1, $\omega_n$ must be less than 100 rad/sec. If $\omega_n$ is less than 100 rad/sec, then for $K_v$ to be greater than 200 requires $K_v/\omega_n$ to be larger than 2. This further restricts our attention to Fig. 8.5-3c, d, and e.

Let us consider the first of the possible design charts (Fig.

8.5-3$c$).   Here $K_v/\omega_n$ is 5, so that if $K_v$ is 200, then $\omega_n$ is 40.   The required 20 percent overshoot occurs with $\lambda_k'$ approximately 5.5 so that BW/$\omega_n$ is 1.9.   With $\omega_n$ already known to be 40, the actual bandwidth in the unnormalized system is 76, and the bandwidth specification is satisfied.   The normalized delay time is 0.6, so that the actual delay time is 0.6/40, or 0.015 sec, which is slower than allowed, and we must try a new design.

A percent overshoot of 20 percent can also be achieved in Fig. 8.5-3$c$ by the use of $\lambda_k' \sim 0.2$.   However, for this value of $\lambda_k'$, the system is so sluggish, as indicated by the high value of $t_d'$, that the required BW and $t_d$ specifications are incompatible.   The same problem arises with the use of Fig. 8.5-3$d$ and $e$, where a small $\lambda_k'$ is necessary to achieve PO = 20 percent.   It begins to appear as though our specifications cannot be met.

In retrospect, our first trial proved to be as close to satisfying specifications as any.   Let us return to that first trial which utilized design chart Fig. 8.5-3$c$.   In the first trial we picked $\omega_n$ on the basis of the minimum $K_v$, with the result that the bandwidth was actually smaller than the maximum allowable bandwidth.   Suppose we choose $\omega_n$ on the basis of maximum allowable BW.   For 20 percent overshoot we have the same value of $\lambda_k'$, and BW/$\omega_n$ is still 1.9, so that the maximum allowable $\omega_n$ is 100/1.9 or 52.6 rad/sec.   Of course, since we are reading at the same value of $\lambda_k'$, $t_d'$ is still 0.6 and $t_d$ becomes 0.6/52.6 or 0.0114 sec.   This meets the delay-time specification to ensure the necessary speed of response.   The velocity-error constant has increased to 5(52.6) = 263, which is greater than the minimum value of 200, and the given system specifications have been satisfied.   All that remains is to specify the actual transfer function to give the indicated response.

For $\lambda_k'$ equal to 5.5, 10$\delta'$ is about 7, so that $\delta'$ is 0.7.   Since $\lambda_k'$ and $\delta'$ are just scaled by the factor $\omega_n$, the actual $\lambda_k$ and $\delta$ are

$$\lambda_k = (5.5)(52.6) = 290$$
$$\delta = (0.7)(52.6) = 36.9$$

Substitution of these values, along with $\omega_n$ and the damping ratio of 0.707, into Eq. (8.5-2) yields

$$\frac{y(s)}{r(s)} = \frac{(52.6)^2(290)}{36.9} \frac{s + 36.9}{[s^2 + 2(0.707)(52.6)s + (52.6)^2](s + 290)}$$
$$= \frac{(2.17 \times 10^4)(s + 36.9)}{(s^2 + 74.4s + 2{,}770)(s + 290)}$$

This is the desired closed-loop transfer function.

A few comments are in order with regard to these two examples. It would be ideal if we could outline a procedure utilizing the design charts presented in this section. This would be possible if the specifications were always given in the same way. Unfortunately this is not always the case. In the second example, for instance, the speed-of-response specification could have been given in terms of a range of allowable bandwidths, rather than in terms of delay time and maximum allowable bandwidth. Hence about the only procedural outline that can be given is to state that each design chart must be tested against the specifications. Several design charts are usually eliminated by inspection, and trends become apparent as the others are checked. Finally, there may be none, one, or many system transfer functions that satisfy the specifications. If none satisfy the specifications, the system is probably over-specified. Overspecification is possible by giving more than one specification for one performance measure. In Example 8.5-2, for instance, the two speed-of-response specifications might have been in conflict. If the delay time had been set at 0.01 sec with the same maximum bandwidth, the specifications could not have been realized. If a large variety of transfer functions meet the specifications, other considerations must be used to decide which one to choose.

This last statement brings us to the introduction to the next section. Here we have discussed the selection of an ideal transfer function to meet a set of given specifications. The basis for these specifications was three performance measures: speed of response, relative stability, and accuracy. Sensitivity to parameter variations and external disturbances is a performance measure yet to be considered. It may be considered only by treating the plant and the model response jointly, as in the following section.

**Exercise 8.5**    *8.5-1.*  Use Figs. *8.5-2* and *8.5-3* to find PO, BW', $K_v'$, and $t_r'$ for each of the third-order transfer functions given below:

(a)  $\dfrac{y(s)}{r(s)} = \dfrac{2 \times 10^5}{7} \dfrac{s + 70}{(s^2 + 141.4s + 10^4)(s + 200)}$

(b)  $\dfrac{y(s)}{r(s)} = \dfrac{4{,}000}{9.5} \dfrac{s + 9.5}{(s^2 + 10s + 100)(s + 40)}$

*answers:*

(a)  PO = 17.2%, BW' = 1.6, $K_v' = 2$, and $t_r' = 1.2$
(b)  PO = 30.5%, BW' = 1.87, $K_v' = 5$, and $t_r' = 1$

## 8.6 Sensitivity and return difference

The last two sections have dealt with means of selecting an ideal closed-loop transfer function to satisfy a given set of specifications. No consideration was given to the plant that had to be controlled. This is, of course, an unrealistic approach; hence, before going on to the selection of a model transfer function for higher-order systems, let us examine how a given plant transfer function $G_p(s)$ affects the possible choices of the closed-loop transfer function $y(s)/r(s)$.

One of the important performance measures relating the plant and the system transfer functions is sensitivity. In this section, sensitivities to plant variations and external disturbances are considered in a more general sense than in the introductory material in Chap. 3. The specification of interest in regard to sensitivity is the magnitude of the *return difference* $1 + KG_p(s)H_{eq}(s)$. In order to ensure that feedback reduces the sensitivity of the closed-loop system to plant variations and external disturbances, the magnitude of the return difference must remain greater than 1 for the range of $s$ of interest, or

$$|1 + KG_p(s)H_{eq}(s)| > 1 \qquad (8.6\text{-}1)$$

Means of ensuring that Eq. (8.6-1) is satisfied are discussed in terms of the open-loop transfer function $KG_p(s)H_{eq}(s)$ and in terms of the transfer functions of the ideal model and of the unalterable plant.

The second important consideration involving the plant and the possible closed-loop-system configurations involves the state variables themselves. The subject of inherent state-variable limitations is discussed in the following section. Now we turn our attention directly to the subject of sensitivity.

Let us consider first the sensitivity of the system to variations in plant parameters. The sensitivity function $S_\alpha{}^T$, introduced in Chap. 3, is defined as the percent change in the quantity $T(s,\alpha)$ due to a percent change in the parameter $\alpha$.

$$S_\alpha{}^T = \frac{dT/T}{d\alpha/\alpha} = \frac{\alpha}{T}\frac{dT(s,\alpha)}{d\alpha} \qquad (8.6\text{-}2)$$

In general, $T(s,\alpha)$ is any transfer function, and $\alpha$ may be either a transfer function or an element of a transfer function as, for example, a pole or zero. For Fig. 8.6-1, for example, if $T$ is the closed-loop transfer function $y(s)/r(s)$ and $\alpha$ is the plant transfer function $G_p(s)$, we showed in Chap. 3 that the sensitivity of $y(s)/r(s)$ with respect to variations in $G_p(s)$ is

$$S_{G_p}{}^{y/r} = \frac{G_p}{y/r}\frac{d}{dG_p}\frac{y}{r} = \frac{G_p}{y/r}\frac{d}{dG_p}\frac{KG_p}{1 + KG_pH_{eq}}$$

so that

$$S_{G_p}{}^{y/r} = \frac{1}{1 + KG_p(s)H_{eq}(s)} \tag{8.6-3}$$

This result expresses the fact that increasing the loop gain of a system reduces the effects of variations of elements in the forward path.

The above result is based on the assumption that $H_{eq}(s)$ does not depend on $G_p(s)$. This assumption is *not* justified if state-variable feedback has been used; $H_{eq}(s)$ is generally a function of plant parameters. Even in this more general case, it is possible to show that the sensitivity of the closed-loop transfer function to plant variations is inversely proportional to the return difference $1 + KG_p(s)H_{eq}(s)$.[1] To illustrate this fact, let us consider the general closed-loop sensitivity function $S_\alpha{}^{y/r}$, where $\alpha$ may be either the entire plant or some part or parameter of the plant.

$$
\begin{aligned}
S_\alpha{}^{y/r} &= \frac{\alpha}{y/r}\frac{d}{d\alpha}\frac{y}{r} = \frac{\alpha}{y/r}\frac{d}{d\alpha}\frac{KG_p}{1 + KG_pH_{eq}} \\
&= \frac{\alpha(1 + KG_pH_{eq})}{KG_p}\frac{(1 + KG_pH_{eq})d(KG_p)/d\alpha - KG_pd(KG_pH_{eq})/d\alpha}{(1 + KG_pH_{eq})^2}
\end{aligned}
$$

or

$$S_\alpha{}^{y/r} = \frac{F(s)}{1 + KG_p(s)H_{eq}(s)} \tag{8.6-4}$$

where

$$F(s) = \frac{\alpha}{y(s)/r(s)}\frac{d}{d\alpha}KG_p(s) - \alpha\frac{d}{d\alpha}KG_p(s)H_{eq}(s) \tag{8.6-5}$$

Equation (8.6-4) shows that the sensitivity of $y(s)/r(s)$ to $\alpha$ is inversely proportional to $1 + KG_p(s)H_{eq}(s)$, but it is no longer clear that increasing $1 + KG_p(s)H_{eq}(s)$ necessarily reduces $S_\alpha{}^{y/r}$. Increasing $1 + KG_p(s)H_{eq}(s)$ is helpful only if the quantity $F(s)$ does not increase more rapidly.

---

[1] By the use of matrix methods, it is also possible to show that the sensitivity of $y(s)/r(s)$ to variations in the elements of $\mathbf{A}$, $\mathbf{b}$, or $\mathbf{c}$ is also inversely proportional to $1 + KG_p(s)H_{eq}(s)$.

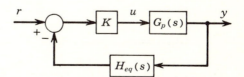

**Fig. 8.6-1**   $H_{eq}(s)$ representation.

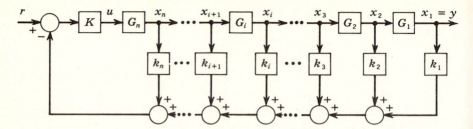

**Fig. 8.6-2** Series-cascade plant with state-variable feedback.

Consider, for example, the block diagram of Fig. 8.6-2 where $G_p(s)$ is composed of a series cascade of first-order blocks and state-variable feedback has been used. Let us determine the sensitivity of $y(s)/r(s)$ to variations in one of the blocks of $G_p(s)$, say $G_i(s)$. In order to find the effect of the variation of a parameter of $G_i(s)$, one is tempted to put the system of Fig. 8.6-2 into one of the standard forms, either the $H_{eq}(s)$ form or the $G_{eq}(s)$ form. If this is done, either $H_{eq}(s)$ or $G_{eq}(s)$ becomes a complicated function of the variable parameter. An easier approach is to use the method discussed in Sec. 7.6 and to cast the block diagram of Fig. 8.6-2 into that of Fig. 8.6-3.

In terms of the latter figure, $G_p(s)$ becomes

$$G_p(s) = G_a(s)G_i(s)G_b(s) \tag{8.6-6}$$

The two feedback transfer functions $H_a(s)$ and $H_b(s)$ are formed by referring all feedback paths to the points $x_1$ and $x_{i+1}$ in such a way that $G_i(s)$ is not involved in either $H_a(s)$ or $H_b(s)$. From the block diagram it is seen that the usual expression for $H_{eq}(s)$ is

$$H_{eq}(s) = H_a(s) + \frac{H_b(s)}{G_a(s)G_i(s)} \tag{8.6-7}$$

and *does* involve $G_i(s)$. The loop gain $KG_p(s)H_{eq}(s)$ from Eqs. (8.6-6)

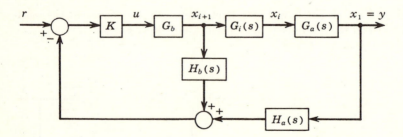

**Fig. 8.6-3** Reduction of Fig. 8.6-2 to a more convenient form.

and (8.6-7) is

$$KG_p(s)H_{eq}(s) = KG_aG_iG_bH_a + KG_bH_b \tag{8.6-8}$$

The closed-loop transfer is now

$$\frac{y(s)}{r(s)} = \frac{KG_p(s)}{1 + KG_p(s)H_{eq}(s)} = \frac{KG_a(s)G_i(s)G_b(s)}{1 + KG_a(s)G_i(s)G_b(s)H_a(s) + KG_b(s)H_b(s)} \tag{8.6-9}$$

In order to use Eqs. (8.6-4) and (8.6-5) to find $S_{G_i}{}^{y/r}$, we compute first

$$\frac{d}{dG_i} KG_p = \frac{d}{dG_i} KG_aG_iG_b = KG_aG_b \tag{8.6-10}$$

and

$$\frac{d}{dG_i} KG_pH_{eq} = \frac{d}{dG_i} (KG_aG_iG_bH_a + KG_bH_b) = KG_aG_bH_a \tag{8.6-11}$$

Substituting these expressions into Eq. (8.6-5), $F(s)$ becomes

$$F(s) = \frac{G_i(1 + KG_aG_iG_bH_a + KG_bH_b)}{KG_aG_iG_b} KG_aG_b - G_i(KG_aG_bH_a)$$
$$= 1 + KG_bH_b \tag{8.6-12}$$

Therefore the sensitivity function $S_{G_i}{}^{y/r}$ is

$$S_{G_i}{}^{y/r} = \frac{1 + KG_bH_b}{1 + KG_pH_{eq}} \tag{8.6-13}$$

Equation (8.6-13) is the desired result. We have shown that for the relatively general system of Fig. 8.6-2 or 8.6-3 the sensitivity of the closed-loop-system transfer function to variations in the transfer function $G_i(s)$ varies inversely as the return difference. Since $F(s)$ does not depend on the feedback coefficients $k_1$ to $k_i$, we could select these coefficients to maximize the return difference $1 + KG_pH_{eq}$ in order to minimize the sensitivity to variation in $G_i(s)$.

If we wish to determine the sensitivity of $y(s)/r(s)$ to variations in a parameter of $G_i(s)$, such as a pole or zero location, we may use the above result in the following fashion:

$$S_\alpha{}^{y/r} = \frac{\alpha}{y/r} \frac{d}{d\alpha} \frac{y}{r}$$

By using the chain of differentiation, $S_\alpha{}^{y/r}$ becomes

$$S_\alpha{}^{y/r} = \frac{\alpha}{y/r} \frac{d}{dG_i} \frac{y}{r} \frac{dG_i}{d\alpha} \frac{G_i}{G_i}$$

The last expression has been multiplied by $G_i/G_i$ so that the following result is more evident.   Thus $S_\alpha{}^{y/r}$ is

$$S_\alpha{}^{y/r} = \left(\frac{G_i}{y/r}\frac{d}{dG_i}\frac{y}{r}\right)\left(\frac{\alpha}{G_i}\frac{d}{d\alpha}G_i\right) = S_{G_i}{}^{y/r}S_\alpha{}^{G_i} \qquad (8.6\text{-}14)$$

If $G_i(s)$ is of the form

$$G_i(s) = \frac{K_i(s + \delta_i)}{s + \lambda_i}$$

then

$$S_{K_i}{}^{G_i} = 1 \qquad (8.6\text{-}15)$$

$$S_{\delta_i}{}^{G_i} = \frac{\delta_i}{s + \delta_i} \qquad (8.6\text{-}16)$$

$$S_{\lambda_i}{}^{G_i} = \frac{-\lambda_i}{s + \lambda_i} \qquad (8.6\text{-}17)$$

The sensitivity function $S_\alpha{}^{y/r}$ therefore is still minimized by maximizing $1 + KG_pH_{eq}$ while holding $F(s)$ constant.

By establishing that sensitivity for variations within an arbitrary $G_i(s)$ of Fig. 8.6-2 depends on the return difference, we have established a link between the plant transfer function and the closed-loop transfer function.   Once $y(s)/r(s)$ is selected, then, for a given $G_p(s)$, $H_{eq}(s)$ is also specified.[1]   Thus, once we have selected an ideal closed-loop response, in order to test whether or not the return difference is greater than 1 requires only algebra.   Unfortunately the algebra can be tedious; hence later in this section we shall establish an alternative method of examining whether or not the return difference satisfies Eq. (8.6-1).   Before doing that, however, let us show that the sensitivity of the system transfer function to external disturbances also depends inversely on the return difference.

Sensitivity to an output disturbance was discussed briefly in Sec. 3.2, where it was shown that the effect of disturbances could be minimized by maximizing the return difference.   Here we wish to show that, even in the more general case when the disturbance is applied to any part of the plant, the conclusion remains basically the same.   In order to treat this more general problem, let us consider the following state-variable representation:

$$\dot{\mathbf{x}} = \mathbf{A}\mathbf{x} + \mathbf{b}u + \mathbf{d}n \qquad (8.6\text{-}18)$$
$$y = \mathbf{c}^T\mathbf{x} \qquad (\mathbf{c})$$

and

$$u = K(r - \mathbf{k}^T\mathbf{x}) \qquad (\mathbf{k})$$

[1] The method of actually calculating $H_{eq}(s)$ is not important at present. This problem will be treated in detail in Chap. 9.

Here **d** is a *disturbance vector* which indicates how the disturbance $n$ affects the plant in much the same way that the control vector **b** relates $u$ to the plant equations. Since we are interested in the sensitivity of the output to the disturbance $n$, we set the reference input to zero so that Eqs. (8.6-18) and (**k**) become

$$\dot{\mathbf{x}} = (\mathbf{A} - K\mathbf{b}\mathbf{k}^T)\mathbf{x} + \mathbf{d}n \tag{8.6-19}$$

The transfer function $y(s)/n(s)$ is therefore

$$\frac{y(s)}{n(s)} = \mathbf{c}^T \Phi_k(s)\mathbf{d} = \frac{\mathbf{c}^T \text{ adj } (s\mathbf{I} - \mathbf{A}_k)\mathbf{d}}{\det (s\mathbf{I} - \mathbf{A}_k)} \tag{8.6-20}$$

Since the return difference may be written

$$1 + KG_p(s)H_{eq}(s) = \frac{D_k(s)}{D_p(s)} = \frac{\det (s\mathbf{I} - \mathbf{A}_k)}{\det (s\mathbf{I} - \mathbf{A})} \tag{8.6-21}$$

we may divide both the numerator and denominator of Eq. (8.6-20) by $\det(s\mathbf{I} - \mathbf{A})$ and obtain

$$\frac{y(s)}{n(s)} = \frac{\mathbf{c}^T \text{ adj } (s\mathbf{I} - \mathbf{A}_k)\mathbf{d}/\det (s\mathbf{I} - \mathbf{A})}{\det (s\mathbf{I} - \mathbf{A}_k)/\det (s\mathbf{I} - \mathbf{A})} = \frac{F_d(s)}{1 + KG_p(s)H_{eq}(s)} \tag{8.6-22}$$

where

$$F_d(s) = \frac{\mathbf{c}^T \text{ adj } (s\mathbf{I} - \mathbf{A}_k)\mathbf{d}}{\det (s\mathbf{I} - \mathbf{A})}$$

This equation is the desired result and establishes that the sensitivity to external disturbances can be reduced by maximizing $1 + KG_p(s)H_{eq}(s)$.

Before discussing means of ensuring that the return difference is greater than 1, let us briefly review the contents of this chapter. For satisfactory operation of a closed-loop system, it is necessary that the system satisfy requirements of stability, speed of response, accuracy, and sensitivity. Associated with each of these performance measures is a specification, and it is first necessary to find if there exists a model transfer function capable of realizing the desired results. This model transfer function is usually chosen on the basis of stability, speed, and accuracy, as discussed in the previous sections. Once the model has been chosen, it must then be tested to see if the model, in conjunction with the fixed plant, yields a sufficiently insensitive system; that is, Eq. (8.6-1) must be satisfied.[1] This is where we are now.

We have already mentioned one way in which the return difference may be calculated by solving for the required feedback coefficients **k**.

---

[1] We have seen that the sensitivity requirements are generally more involved than just Eq. (8.6-1), but we shall use Eq. (8.6-1) as a general requirement for simplicity.

These feedback coefficients must be used to find $H_{eq}(s)$. The newly found $H_{eq}(s)$ is then used with the given plant $G_p(s)$ to check if the magnitude of the return difference is greater than 1 for all frequencies. An easy way to check this is to employ the circle criterion on the Nyquist diagram as discussed in Sec. 6.4.

This procedure requires that the feedback coefficients and $H_{eq}(s)$ be calculated before applying the sensitivity criterion. A further difficulty is the requirement of plotting a frequency-response function. As we have noted, the asymptotic Bode diagram is much easier to draw. Actually it is quite easy to transform the circle criterion to one requiring the use of an asymptotic Bode diagram. This is done by recognizing that, since the return difference may be written

$$1 + KG_p(s)H_{eq}(s) = \frac{D_k(s)}{D_p(s)}$$

Eq. (8.6-1) becomes

$$\left|\frac{D_k(s)}{D_p(s)}\right| > 1 \tag{8.6-23}$$

Equation (8.6-23) states that the *denominator ratio*, that is, a transfer function whose numerator is the denominator of the model response and whose denominator is the denominator of the plant transfer function, must remain larger than 1, if Eq. (8.6-1) is to be satisfied or if the return difference is to be greater than 1. In this form the relationship between the open- and closed-loop systems is much more evident than in terms of $H_{eq}(s)$. The satisfaction of Eq. (8.6-23) may be easily checked by plotting an asymptotic Bode diagram of the amplitude of $D_k(j\omega)/D_p(j\omega)$. Because only the amplitude is required by Eq. (8.6-1), the phase-angle diagram need not be considered. Furthermore, the sensitivity condition may be tested as soon as the model transfer function is selected, without first finding the feedback coefficients and $H_{eq}(s)$. The use of Eq. (8.6-23) as the sensitivity condition is illustrated in the following example.

> **Example 8.6-1**  In this example we begin with the assumption that the desired closed-loop transfer function, chosen to satisfy requirements of relative stability, speed of response, and accuracy, is
>
> $$\frac{y(s)}{r(s)} = \frac{(382.4)(s + 5.23)}{(s^2 + 14.14s + 100)(s + 20)}$$
>
> The plant transfer function is
>
> $$G_p(s) = \frac{s + 5.23}{s(s + 10)(s + \lambda)}$$

and is completely specified with the exception of the pole location at $s = -\lambda$. Assume that possible choices for $\lambda$ are 1, 20, and 100 and that the problem is to examine each choice to see which best meets sensitivity requirements.

We may check to see if the plant is compatible with the desired closed-loop transfer function, in a sensitivity sense, by the use of the denominator ratio:

$$\frac{D_k(s)}{D_p(s)} = \frac{(s^2 + 14.14s + 100)(s + 20)}{s(s + 10)(s + \lambda)} \qquad (8.6\text{-}24)$$

In order to use the Bode diagram to ensure that the magnitude of the denominator ratio is greater than 1, we put Eq. (8.6-24) into time-constant form:

$$\frac{D_k(s)}{D_p(s)} = \frac{200}{\lambda} \frac{(s^2/100 + 0.1414s + 1)(s/20 + 1)}{s(s/10 + 1)(s/\lambda + 1)} \qquad (8.6\text{-}25)$$

Note the presence of the complex conjugate zeros in $D_k(s)/D_p(s)$. Here $\omega_n{}^2 = 100$, or $\omega_n$ is 10, and the damping ratio is 0.707. According to the straight-line Bode approximation methods, the complex conjugate zeros appear as two real zeros at $s = -10$, and because of the damping ratio, the error in making such an approximation is small.

The straight-line magnitude plots of the denominator ratio of Eq. (8.6-25) for $\lambda = 1$, 20, and 100 are shown in Fig. 8.6-4. For $\lambda = 1$ and 20, the magnitude plot is never below unity. From these two diagrams, one could conclude that $\lambda = 1$ is the better choice from the standpoint of sensitivity. For $\lambda = 100$, the denominator ratio is less than 1 for $2 \leq \omega \leq 100$ rad/sec, and the model and plant would appear to be incompatible in terms of sensitivity.

It is easy to show that if one or more of the poles of $G_p(s)$ is larger than all the poles of $y(s)/r(s)$, that is, an open-loop pole larger than a closed-loop pole, the sensitivity criterion of Eq. (8.6-1) or (8.6-23) is violated. Since both $D_k(s)$ and $D_p(s)$ are $n$th-order, the

$$\lim_{\omega \to \infty} |D_k(j\omega)/D_p(j\omega)| = 1$$

and the magnitude of the denominator ratio eventually becomes 1. If the last break point is a pole of $D_k(s)/D_p(s)$, that is, a pole of $G_p(s)$, then the slope change must be from some positive value to zero. However, if the magnitude plot arrives at the unity line with positive slope, it is necessary that the magnitude must have been less than unity for some previous range of frequency. This property provides an easy method of checking

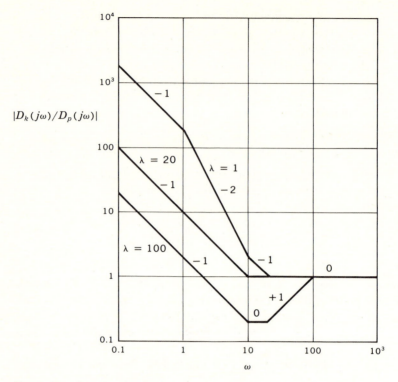

**Fig. 8.6-4**  Asymptotic plots of the denominator ratio for various values
of $\lambda$ for Example 8.6-1.

to see if the sensitivity criterion is not met but, like the Hurwitz criterion,
does not indicate that the sensitivity criterion is met if the last break
point is a zero of $D_k(s)/D_p(s)$.

   If we attempt to generalize the above example and comment, it
might appear that, if the closed-loop poles lie farther from the origin
than the open-loop poles, the system is insensitive.   This is true in a
sense.   From a root-locus point of view, having the closed-loop poles far
from the open-loop poles usually means a high gain is involved.   High
gain can make a system less sensitive to parameter changes and external
inputs.   However, high gain is not always sufficient; the judicious place-
ment of the closed-loop poles is also important.   In the example to follow,
a closed-loop pole is much farther removed from the origin than the open-
loop poles, and yet the system fails to satisfy the inequality of Eq.
(8.6-23).

***Example 8.6-2***  The model response is given as

$$\frac{y(s)}{r(s)} = \frac{K(s + \delta)}{(s^2 + 14.14s + 100)(s + 1,000)}$$

and $KG_p(s)$ is

$$KG_p(s) = \frac{K(s + \delta)}{s(s + 200)^2}$$

The numerators are not important, although presumably $K$ is chosen to ensure zero steady-state position error and $\delta$ is chosen to establish a desired velocity-error constant.  The denominator ratio is

$$\begin{aligned}
\frac{D_k(s)}{D_p(s)} &= \frac{(s^2 + 14.14s + 100)(s + 1,000)}{s(s + 200)^2} \\
&= \frac{(2.5)(s^2/100 + 0.1414s + 1)(s/1,000 + 1)}{s(s/200 + 1)^2}
\end{aligned} \tag{8.6-26}$$

A Bode plot of Eq. (8.6-26) appears in Fig. 8.6-5.  The inequality (8.6-23) is not satisfied, even though a closed-loop pole has been forced to be far removed from the open-loop poles.

In this section we have discussed the aspects of system performance associated with sensitivity to plant-parameter variations and external disturbances.   In both cases we have shown that sensitivity is inversely proportional to the return difference $1 + KG_p(s)H_{eq}(s)$.   Hence one may

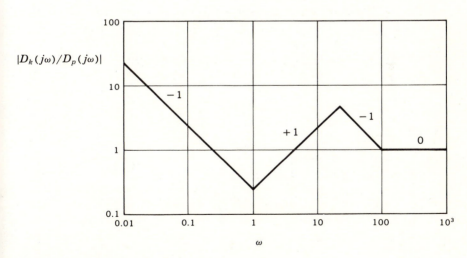

**Fig. 8.6-5**  Example 8.6-2.

reduce the sensitivity by increasing $1 + KG_p(s)H_{eq}(s)$ although this statement is tempered by the consideration of the numerator terms in the sensitivity functions and disturbance-to-output transfer functions.

Our treatment here has centered about the problem of analyzing the sensitivity of a given closed-loop model and plant combination rather than designing the model for minimum sensitivity. The latter topic is much more difficult, and its general treatment is beyond the scope of this book. We shall, nevertheless, discuss various specific techniques for reducing the sensitivity of the closed-loop system in Chap. 9.

**Exercises 8.6**  *8.6-1.*  Find the expressions, similar to Eqs. (8.6-4) and (8.6-5), for the sensitivity function $S_\alpha^{y/r}$, where $\alpha$ is a plant parameter in terms of the state-variable representation

$$\dot{x} = Ax + bu$$
$$y = c^T x$$
$$u = K(r - k^T x)$$

*answer:*

$$S_\alpha^{y/r} = \frac{F(s)}{1 + Kk^T\Phi(s)b}$$

where

$$F(s) = \frac{\alpha[1 + Kk^T\Phi(s)b]}{Kc^T\Phi(s)b} \frac{d}{d\alpha} Kc^T\Phi(s)b - \alpha\frac{d}{d\alpha} Kk^T\Phi(s)b$$

*8.6-2.*  Use the results of Exercise 8.6-1 to find $S_\lambda^{y/r}$ for the system given below. Does increasing $K$ decrease the sensitivity?

$$\dot{x} - \begin{bmatrix} 0 & 1 \\ 0 & -\lambda \end{bmatrix} x + \begin{bmatrix} 0 \\ 1 \end{bmatrix} u \qquad y = x_1 \qquad u = 10(r - [1 \quad 1]x)$$

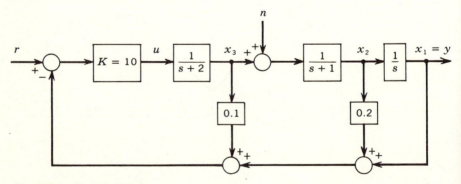

**Fig. 8.6-6**  Exercise 8.6-3.

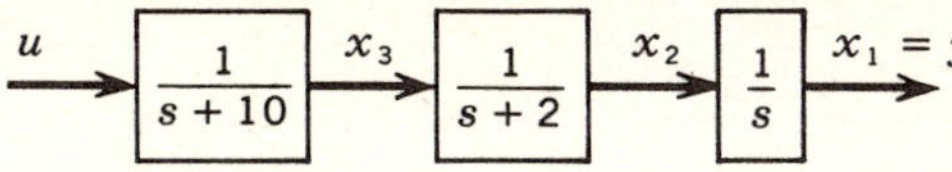

*Fig. 8.6-7*    Exercise 8.6-5.

*answer:*

$$S_\lambda{}^{y/r} = \frac{-\lambda s}{s^2 + (\lambda + 10)s + 10};   \text{yes.}$$

*8.6-3.*   Use Eq. (8.6-22) to find $y(s)/n(s)$ for the system shown in Fig. 8.6-6.   Can the sensitivity to the disturbance be reduced by increasing $K$ to 100?

*answer:*

$$\frac{y(s)}{n(s)} = \frac{s + 3}{s^3 + 4s^2 + 5s + 10};   \text{yes.}$$

*8.6-4.*   Repeat Exercise 8.6-3 but use block-diagram algebra to find the answer.

*8.6-5.*   For the plant of Fig. 8.6-7, which of the following closed-loop models leads to the lower sensitivity to variations in the controller gain $K$?

(a)   $\dfrac{y(s)}{r(s)} = \dfrac{10}{[(s + 1)^2 + 1^2](s + 5)}$

(b)   $\dfrac{y(s)}{r(s)} = \dfrac{40}{[(s + 1)^2 + 1^2](s + 20)}$

*answer:*

(b)

## 8.7   *State-variable constraints*

This section[1] is concerned with the effect of inherent physical constraints on the state variables on the selection of a model closed-loop transfer function.   Until now, in selecting a model closed-loop transfer function, we have been concerned only with the desired input-output behavior, that is, for a given step or sinusoidal input, how we wish the output to behave.   The behavior of the output has been given prime importance because that is the particular quantity of interest that must be controlled.

Although the remainder of the system exists for the sole purpose of controlling the output in a desired fashion, the behavior of the internal

[1] See also the last part of Sec. 4.4.

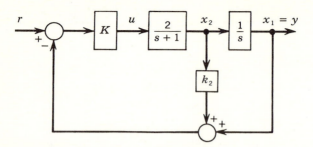

**Fig. 8.7-1**   Second-order control system.

state variables cannot be ignored, because the whole analysis and synthesis procedure has been based on an assumption of linearity. Perhaps the simplest example is a spring. The reaction force of a spring is proportional to the force acting on the spring until its elastic limit is exceeded. Then the spring ceases to act as a spring, and the linear equations are no longer valid.

To illustrate this point in a control-oriented example, we consider the simple positioning servomechanism of Fig. 8.7-1. By the proper selection of $K$ and $k_2$, any closed-loop transfer function of the form

$$\frac{y(s)}{r(s)} = \frac{\omega_n{}^2}{s^2 + 2\zeta\omega_n s + \omega_n{}^2} \tag{8.7-1}$$

may be realized. Assume for purposes of illustration that a damping ratio of 0.707 is chosen in order to give an overshoot of less than 5 percent for a step-function input of any size. This response satisfies a reasonable relative-stability performance measure. The speed of response, as measured by either delay time or rise time or bandwidth, depends directly on the choice of $\omega_n$. From Eq. (8.4-14) the velocity-error constant $K_v$ equals $0.707\omega_n$ and is also directly dependent on the choice of $\omega_n$. As long as $\omega_n > 1$ the denominator ratio remains greater than 1 for all $s = j\omega$. In this example the denominator ratio is

$$\frac{D_k(s)}{D_p(s)} = \frac{s^2 + 2\zeta\omega_n s + \omega_n{}^2}{s(s + 1)} = \frac{\omega_n{}^2(s^2/\omega_n{}^2 + 2\zeta s/\omega_n + 1)}{s(s + 1)} \tag{8.7-2}$$

The Bode plot of the magnitude of Eq. (8.7-2) is shown in Fig. 8.7-2. It is seen that by increasing $\omega_n$ this ratio becomes increasingly greater than 1 for larger and larger values of $\omega$. Thus, as far as speed of response, accuracy, and sensitivity are concerned, it is advantageous to select a large value for $\omega_n$.

Before deciding on the choice of $\omega_n$, let us examine the step-function response of the closed-loop system described by Eq. (8.7-1). The unit step-function response is discussed in Sec. 4.3 and is given by

$$y(t) = x_1(t) = 1 - \frac{1}{\sqrt{1 - \zeta^2}} e^{-\zeta\omega_n t} \sin\left(\omega_n \sqrt{1 - \zeta^2} t + \psi\right) \tag{8.7-3}$$

where

$$\psi = \arctan \frac{\sqrt{1 - \zeta^2}}{\zeta}$$

For a step size of magnitude $A$, $y(t)$ is just $A$ times greater, or

$$y(t) = x_1(t) = A \left[ 1 - \frac{1}{\sqrt{1 - \zeta^2}} e^{-\zeta\omega_n t} \sin\left(\omega_n \sqrt{1 - \zeta^2} t + \psi\right) \right] \tag{8.7-4}$$

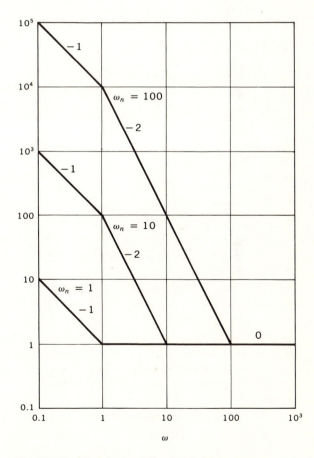

***Fig. 8.7-2***   Magnitude of the denominator ratio.

For a step-function input of magnitude $A$, $x_2(t)$ is just

$$x_2(t) = \frac{A\omega_n}{\sqrt{1 - \zeta^2}} e^{-\zeta\omega_n t} \sin (\omega_n \sqrt{1 - \zeta^2}t) \qquad (8.7\text{-}5)$$

Note here the presence of the multiplier $\omega_n$. The velocity at which the output must travel increases directly with $\omega_n$. Also the velocity at which the output must travel is directly dependent on the step size $A$.

Assume that the output member is a dc motor. The velocity of the dc motor is proportional to the field and armature current, if the motor is operated in its linear range. If the current becomes too large, the magnetic circuit saturates and the flux no longer increases, causing the motor to be velocity-limited. If the maximum velocity of the motor is $v_{max}$, to ensure that the motor does not saturate,

$$|x_2(t)|_{max} \leq v_{max}$$

The maximum value of $x_2(t)$ occurs at

$$t = \frac{1}{\omega_n \sqrt{1 - \zeta^2}} \arctan \frac{\sqrt{1 - \zeta^2}}{\zeta}$$

If this value of $t$ is substituted into Eq. (8.7-5), along with the value of $\zeta = 0.707$, $|x_2(t)|_{max}$ is found to be

$$|x_2(t)|_{max} = A\omega_n e^{-1}$$

so that

$$A\omega_n e^{-1} \leq v_{max} \qquad (8.7\text{-}6)$$

to ensure linear operation.

Equation (8.7-6) indicates that, if the closed-loop system is expected to experience a step-function input of amplitude $A$, then $\omega_n$ must be less than $v_{max}e/A$ if the system behavior is to remain linear. For $\omega_n$ larger than this value a step input of amplitude $A$ causes the system to exceed the linear range of operation. When this happens the form of the output cannot be predicted by linear techniques. That the system is operating in a nonlinear mode is not necessarily bad; at the level at which this book is written, there simply is no way of analyzing nonlinear behavior. The nonlinear operation may prove desirable, or it may make the system unstable. There is no indication at present of which to expect.

Although the closed-loop transfer function for a system such as that shown in Fig. 8.7-1 may be chosen on the basis of the response to a step-function input, the actual system may never experience an input that severe. If the system of Fig. 8.7-1 were never to experience a step input, then of course Eq. (8.7-6) has no meaning. But whatever the expected input, the magnitude of $x_2(t)$ is always a function of the undamped natural frequency. Unfortunately this function may not always be as simple as Eq. (8.7-5).

Two conclusions follow from this simple example. First, it is important to specify the magnitude and nature of the expected input signals in order to decide on the desired closed-loop transfer function. Often the exact nature of this influence is not known at the outset; hence we are forced to the second conclusion: that the last step in any synthesis procedure is *analysis*. Once the design is finished, the time response of all the state variables must be checked to ensure that none of them exceeds the inherent physical constraints on the state variables.

In a sense the contents of this section are somewhat out of place. First of all, the user usually specifies input-output requirements. The fact that internal state variables are constrained to specific maximum values does not interest the user at all. The designer's problem is to select components so that the user's desired results are realized. This section also assumes that the synthesis has been accomplished so that the complete time response may be obtained. Thus perhaps this section should have followed Chaps. 9 and 10. The section is included here as a sort of danger sign, like the waving of a red flag. Although *any* closed-loop transfer function can be realized for *any* given plant, this statement may be true only for very small inputs.

**Exercise 8.7**   *8.7-1.*   Find and plot the unit step response of $y(t)$, $x_1(t)$, and $x_2(t)$ for the system

$$\dot{\mathbf{x}} = \begin{bmatrix} -10 & 10 \\ 10 & -10 \end{bmatrix}\mathbf{x} + \begin{bmatrix} 1 \\ 0 \end{bmatrix}u \qquad y = [1 \quad 1]\mathbf{x} \qquad u = K(r - \mathbf{k}^T\mathbf{x})$$

for each of the following controller designs:

(a)   $K = 100$, $\mathbf{k} = \text{col}\,(1,1)$
(b)   $K = 75$, $\mathbf{k} = \text{col}\,(1.27, 0.73)$
(c)   $K = 50$, $\mathbf{k} = \text{col}\,(1.8, 0.2)$

Note the importance of considering the behavior of the internal state variables, especially $x_1(t)$, in addition to the output $y(t)$.

## 8.8   Specification of high-order systems

The specification of high-order systems is inherently a difficult problem simply because of the number of parameters involved. In specifying a third-order transfer function with one zero, four parameters, $\zeta$, $\omega_n$, $\lambda_k$, and $\delta$, were needed to characterize the response, and sets of design charts consisting of families of response characteristics had to be used. This amount of complexity is about the limit of our comprehension. To specify the response of higher-order systems, we must resort to a new approach.

The approach that is advocated here is to select the desired response in terms of a third-order system with one zero, designated as $[y(s)/r(s)]_m$, and then to approximate this response with a higher-order transfer function. Unfortunately the theoretical basis[1] for this approach is beyond the scope of this book. Thus in this section we simply outline the method for specifying higher-order systems and show that in the limit the procedure is correct.

The method of specifying high-order systems is based on the formation of a synthetic transfer function $\Gamma(s)$, where $\Gamma(s)$ is defined as

$$\Gamma(s) = \frac{KG_p(s)}{[y(s)/r(s)]_m} \tag{8.8-1}$$

Here we assume that the zeros of $[y(s)/r(s)]_m$ and $G_p(s)$ are identical. This restriction is a direct consequence of the inability of state-variable feedback to affect zero positions. After the discussion of an extension of the state-variable-feedback concept, which allows us to control the zeros of $y(s)/r(s)$ as well as the poles, this restriction can be removed.

The transfer function $\Gamma(s)$ is called a *synthetic transfer function* because it has only mathematical significance and has been created for our convenience. The five-step design procedure involving the synthetic transfer function is as follows:

1. Form the synthetic transfer function $\Gamma(s)$.
2. Form the associated synthetic transfer function $\Gamma(s)\Gamma(-s)$.
3. Plot the root locus of $\Gamma(s)\Gamma(-s)$ versus $K$.
4. On the resulting root-locus diagram, determine the location of the closed-loop poles that are associated with the maximum value of gain that is desirable for the particular system in terms of sensitivity or state-variable constraints discussed in the previous sections.[2]
5. The poles of the desired closed-loop transfer function are the poles

---

[1] D. G. Schultz and J. L. Melsa, "State Functions and Linear Control Systems," McGraw-Hill Book Company, New York, 1967.

[2] Once the general trend of the closed-loop-pole movement has been determined from the root locus and a value of $K$ selected, one may wish to use a digital computer to find the exact location of the closed-loop poles.

found in step 4 that appear in the *left-half s plane*.  The numerator of the closed-loop transfer function is the numerator of $KG_p(s)$.

Rather than comment on each step of the design procedure, let us illustrate the procedure by examples of increasing complexity.

*Example 8.8-1*  Consider as a first example the third-order plant of Fig. 8.8-1.  This is not a high-order plant; hence the procedure may not seem entirely applicable, since we have discussed in detail the specification of third-order systems.  However, here assume that we do not desire a third-order response but rather a first-order response; to be specific, assume that $[y(s)/r(s)]_m$ is

$$\left[\frac{y(s)}{r(s)}\right]_m = \frac{2}{s+2} \tag{8.8-2}$$

Note that we have chosen the gain of the closed-loop system so that $[y(0)/r(0)]_m$ is unity, to ensure zero steady-state position error for a step input.

Our problem now is to approximate this low-order transfer function with a third-order transfer function by the use of the five-step design procedure just outlined.  Here the forward transfer function is

$$KG_p(s) = \frac{3K}{s(s+1)(s+3)}$$

so that the synthetic transfer function $\Gamma(s)$ is

$$\Gamma(s) = \frac{KG_p(s)}{[y(s)/r(s)]_m} = \frac{\frac{3}{2}K(s+2)}{s(s+1)(s+3)} \tag{8.8-3}$$

The associated synthetic transfer function may be written immediately as

$$\Gamma(s)\Gamma(-s) = \frac{\frac{9}{4}K^2(s+2)(-s+2)}{s(s+1)(s+3)(-s)(-s+1)(-s+3)}$$

$$= \frac{\frac{9}{4}K^2(s+2)(s-2)}{s(s)(s+1)(s-1)(s+3)(s-3)}$$

with the pole-zero plot as indicated in Fig. 8.8-2.  This completes steps 1 and 2.

*Fig. 8.8-1*  Plant to be controlled for Example 8.8-1.

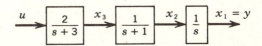

**Fig. 8.8-2**   Pole-zero plot of the associated synthetic transfer function for Example 8.8-1.

Since the pole-zero excess of the associated synthetic transfer function is 4, there must be four asymptotes on the root-locus diagram. Because of the complete symmetry of the pole-zero plot about the origin of the $s$ plane, the origin of the asymptotes is also the origin of the $s$ plane. The root locus for the associated synthetic transfer function is shown in Fig. 8.8-3. Only those branches of the root locus that lie in the second quadrant are shown, since the locus itself is also symmetrical about both the real axis and the imaginary axis. Several different gains are marked along the two branches of the root locus. Note that, for $K$ greater than 8.61,

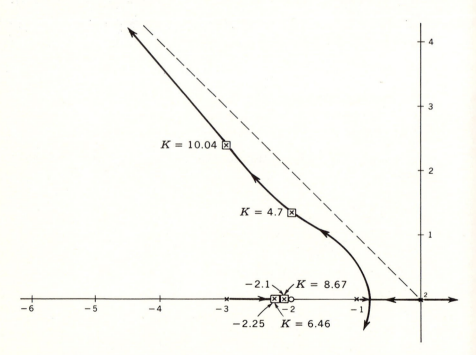

**Fig. 8.8-3**   Root locus for Example 8.8-1.

a closed-loop pole is between $s = -2.1$ and $-2$. Thus for values of gain around, say, 10, the real pole is essentially at $s = -2$. At $K$ very near to 10 the complex pole in the second quadrant is at $s = -3 + j2.5$.[1] If the maximum allowable system gain is 10, the required third-order system to approximate the given model of Eq. (8.8-2) is

$$\frac{y(s)}{r(s)} = \frac{3(10.0)}{(s + 2)[(s + 3.0)^2 + 2.5^2]} \tag{8.8-4}$$

In Eq. (8.8-4) the numerator is determined from the numerator of the forward transfer function, or $3K$. The poles are read directly from the root locus of Fig. 8.8-3, as required by step 5. In order to check if this closed-loop response approximates the model-closed-loop response of Eq. (8.8-2), we can check the steady-state error for a step-function input and the frequency-response curves of both Eqs. (8.8-2) and (8.8-4). Note that we are carefully avoiding a comparison of step-function responses, as there is no simple method of plotting $y(t)$ for a step input that corresponds to the asymptotic straight-line diagram method of plotting $y(s)/r(s)$. The evaluation of Eq. (8.8-4) for $s = 0$ yields

$$\frac{y(0)}{r(0)} = 0.988$$

This should be unity, and the error is due to the graphical construction of the root-locus diagram. Once this check has been made, one can alter the numerator of $y(s)/r(s)$ slightly to obtain zero steady-state error if desired. The frequency-response plots of the model and the third-order transfer functions [Eq. (8.8-4)] are compared in Fig. 8.8-4. The third-order system drops off at a $-3$ slope for frequencies beyond 3.9 rad/sec, but both the ideal model and the third-order approximation have the same bandwidth.

In the above example if the gain $K$ is allowed to go to infinity, the real closed-loop pole in the left-half plane is identically equal to the pole of the model. Also the other two closed-loop poles in the left-half plane go to infinity along the asymptotes. As $K$ goes to infinity, the time constant of the complex conjugate poles goes to zero, and so does the residue; that is, they do not contribute to the response. Thus in the limit, as $K$ goes to infinity, the response of the third-order system is identical to that of the first-order model. For all practical purposes,

---

[1] The exact values are $s = -2.07, -2.94 \pm j2.41$.

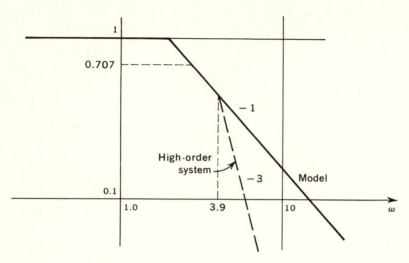

**Fig. 8.8-4**   Comparison of frequency responses for Example 8.8-1.

however, the low-order system may be considered to be closely approximated by the higher-order system having the same bandwidth.

The second example is similar to the first in that a first-order model is approximated by a low-order system.   The example is included to demonstrate that, when the pole-zero excess of the synthetic transfer function is an odd number, $\Gamma(-s)$ has a negative gain when written in the usual pole-zero form.   Thus the product $\Gamma(s)\Gamma(-s)$ has a negative gain, and it is necessary to draw a zero-degree locus rather than the usual 180° root-locus diagram.

***Example 8.8-2***   Here we assume that $KG_p(s)$ is given as

$$KG_p(s) = \frac{2K}{s(s+1)}$$

and the model transfer function is

$$\left[\frac{y(s)}{r(s)}\right]_m = \frac{3}{s+3}$$

Again we have chosen the gain of the low-order model in order to ensure a zero steady-state position error for a step input.   Since the given plant is second-order, the closed-loop transfer function is also second-order, and the specification problem is to choose a second-order transfer function to approximate closely the desired

**Fig. 8.8-5**  Pole-zero plot of the associated synthetic transfer function for Example 8.8-2.

first-order model.  For this example $\Gamma(s)$ is

$$\Gamma(s) = \frac{\frac{2}{3}K(s + 3)}{s(s + 1)}$$

and the associated synthetic transfer function is

$$\Gamma(s)\Gamma(-s) = \frac{\frac{2}{3}K(s + 3)}{s(s + 1)} \frac{\frac{2}{3}K(-s + 3)}{(-s)(-s + 1)}$$

$$= \frac{-(\frac{2}{3}K)^2(s + 3)(s - 3)}{s(s)(s + 1)(s - 1)}$$

Here it is seen that $\Gamma(s)\Gamma(-s)$ does indeed have a negative gain, which causes a 180° shift in the root-locus diagram that must be drawn.  The pole-zero plot of the associated synthetic transfer function is indicated in Fig. 8.8-5, and the root locus is shown in Fig. 8.8-6.  Because a zero-degree root locus must be drawn, the root locus exists on the real axis whenever an even number of poles and/or zeros is to the right.

On the basis of our experience with Example 8.8-1, we know that one way to ensure that the second-order system closely approx-

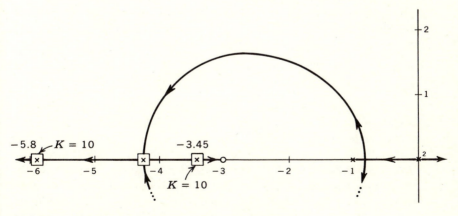

**Fig. 8.8-6**  Zero-degree root locus for Example 8.8-2.

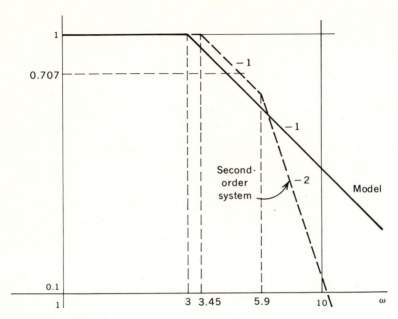

***Fig. 8.8-7***  Frequency-response curves for Example 8.8-2.

imates the first-order model is to let the gain $K$ approach infinity.
In most cases this is impossible, since the plant would saturate for
very small inputs or noise.  Thus let us assume, as we did in the
first example, that the gain $K$ is restricted to be less than 10.  For
$K = 10$, one closed-loop pole is at approximately $s = -5.8$ and
the other at approximately $s = -3.45$.  This determines the denominator of the required second-order transfer function to approximate
the model response.  Again the numerator is already known as $2K$,
or 20.  Thus the required $y(s)/r(s)$ is

$$\frac{y(s)}{r(s)} = \frac{20}{(s + 5.8)(s + 3.45)}$$

and here again $y(0)/r(0)$ is very nearly 1.  The two frequency-
response curves are compared in Fig. 8.8-7, and once again the
bandwidths are very close.

***Example 8.8-3***  The last example in this section is more physically
motivated than the previous two and involves the design of a tracking system.  Such a problem was considered briefly in Example
8.5-1, where it was decided that the prime requirements were fast
response and a relatively large velocity-error constant.  A set of

specifications discussed in that example included a percent overshoot of 25 percent, a delay time of 0.1 sec, and a velocity-error constant greater than 50. To meet these specifications, we determined that a satisfactory closed-loop transfer function would be as given in Eq. (8.5-6), or

$$\left[\frac{y(s)}{r(s)}\right]_m = \frac{(25)(40)}{3.5} \frac{s + 3.5}{(s^2 + 7.07s + 25)(s + 40)} \tag{8.5-6}$$

Assume that the plant to be controlled has a transfer function

$$G_p(s) = \frac{100(s + 3.5)}{s(s + 3)(s + 20)(s + 30)} \tag{8.8-5}$$

Since the denominator of the plant transfer function is fourth-order, the closed-loop system formed with feedback is also of fourth-order. Hence we are concerned with the approximation of the model of Eq. (8.5-6) by a fourth-order transfer function with one zero.

The synthetic transfer function is

$$\Gamma(s) = \frac{0.35K(s^2 + 7.07s + 25)(s + 40)}{s(s + 3)(s + 20)(s + 30)}$$

$$= \frac{0.35K[(s + 3.5)^2 + 3.55^2](s + 40)}{s(s + 3)(s + 20)(s + 30)}$$

The pole-zero excess is an odd number, so that a zero-degree locus for $\Gamma(s)\Gamma(-s)$ must be drawn. The associated synthetic transfer function is

$$\Gamma(s)\Gamma(-s) =$$
$$\frac{-(0.35K)^2[(s + 3.5)^2 + 3.55^2](s + 40)[(s - 3.5)^2 + 3.55^2](s - 40)}{s(s + 3)(s + 20)(s + 30)s(s - 3)(s - 20)(s - 30)} \tag{8.8-6}$$

The pole-zero plot of $\Gamma(s)\Gamma(-s)$ is shown in Fig. 8.8-8, and the

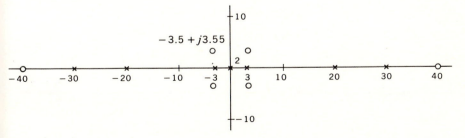

**Fig. 8.8-8** Pole-zero plot of the associated synthetic transfer function for Example 8.8-3.

root locus is sketched in Fig. 8.8-9. As in most realistic problems, the scaling presents somewhat of a problem. The only way to get an accurate root-locus diagram is to use a huge piece of paper or a computer.

In Examples 8.8-1 and 8.8-2, the gain $K$ was high enough so that the bandwidths of the resulting systems were very nearly the same. Assume here that a gain of 55 is allowed. Then, from the root-locus diagram of Fig. 8.8-9, the poles of the fourth-order system are at $s = -3.2 \pm j3$ and $s = -30 \pm j8.4$. The fourth-order transfer function is then

$$\frac{y(s)}{r(s)} = \frac{55(100)(s + 3.5)}{[(s + 3.2)^2 + 3^2][(s + 30)^2 + 8.4^2]} \tag{8.8-7}$$

and once again the graphical construction has resulted in a slight error, as $y(0)/r(0)$ is 1.02.

A comparison of the frequency-response curve of Eq. (8.8-7) with that of the model response of Eq. (8.5-6) is given in Fig. 8.8-10. Here the bandwidth is considerably smaller for the fourth-order approximation.

For gains substantially higher than 55, the closed-loop poles near the origin approach $s = -3.5 \pm j3.5$. Thus at $K = 128$, the closed-loop transfer function is approximately

$$\frac{y(s)}{r(s)} = \frac{128(100)(s + 3.5)}{[(s + 42)^2 + 11.25^2][(s + 3.5)^2 + 3.5^2]} \tag{8.8-8}$$

At this higher value of gain the frequency response of the model is closely approximated far beyond the closed-loop bandwidth, as shown in Fig. 8.8-10.

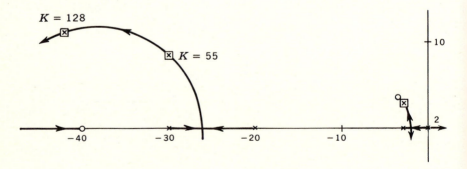

**Fig. 8.8-9**  Root locus for Example 8.8-3.

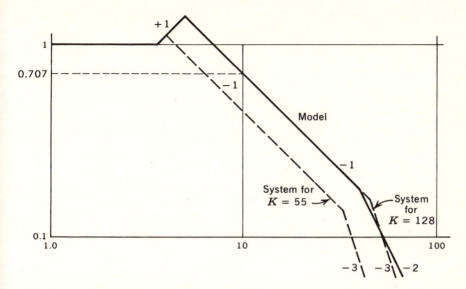

**Fig. 8.8-10**   Comparative frequency-response curves for Example 8.8-3.

In this section we have considered one method of extrapolating our procedures of selecting a model closed-loop transfer function to high-order systems. The approach is based on the approximation of a low-order model, usually third-order with one zero, by a high-order system.

**Exercises 8.8**   *8.8-1.*   Find the step responses of the model and second-order approximation for Example 8.8-2.

*8.8-2.*   Find the third-order closed-loop transfer function that approximates the second-order model

$$\left[\frac{y(s)}{r(s)}\right]_m = \frac{2(s+1)}{(s+1)^2 + 1}$$

if the plant transfer function is

$$G_p(s) = \frac{5(s+1)}{s(s+3)(s+2)}$$

Select $K$ such that the initial input to the plant does not exceed 10 for unit step inputs to the closed-loop system. Plot the step and frequency responses of the model and the third-order approximation.

*answer:*

$$\frac{y(s)}{r(s)} = \frac{50(s+1)}{[(s+0.995)^2 + (0.995)^2](s+25.25)}$$

*8.8-3.* Find the fourth-order closed-loop transfer function that approximates the third-order model

$$\left[\frac{y(s)}{r(s)}\right]_m = \frac{28(s + \frac{10}{7})}{[(s + 2)^2 + 2^2](s + 5)}$$

if $G_p(s)$ is

$$G_p(s) = \frac{10(s + \frac{10}{7})}{s(s + 1)(s + 3)(s + 10)}$$

Assume that $K = 10$. Plot the frequency response of the model and the fourth-order approximation.

*answer:*

$$\frac{y(s)}{r(s)} = \frac{100(s + \frac{10}{7})}{[(s + 1.4)^2 + (1.4)^2](s + 10.5)(s + 3.46)}$$

## 8.9   *Conclusions*

This chapter began with a discussion of three important performance measures: accuracy, relative stability, and speed of response. These qualitative performance measures were discussed quantitatively in terms of specifications; that is, the specifications assign numerical values to the general performance measures. Some specifications are given in the time domain, such as percent overshoot or rise time, whereas others, such as bandwidth, are given in the frequency domain. But regardless of how the specifications are given, the ultimate goal of this chapter is the selection of a system transfer function that satisfies the given specifications.

Sections 8.4 and 8.5 discussed methods of selecting second- and third-order transfer functions in order to satisfy given sets of performance specifications. In the selection of a desired closed-loop transfer function, one must consider not only the desired performance of the closed-loop system but also the limitations and capabilities of the given plant. The relationship between the plant and desired closed-loop behavior was emphasized by a consideration of sensitivity and state-variable constraints in Secs. 8.6 and 8.7.

The method of selecting a closed-loop transfer function for high-order systems, discussed in Sec. 8.8, is based on the initial selection of a third-order model. The concepts of sensitivity and state-space constraints are then used to assist in the eventual selection of a desired closed-loop transfer function. Thus Sec. 8.8 serves not only as a discussion of specifications of high-order systems but also as a review of the earlier material of this chapter.

Hence our present position with respect to the complete control problem is as follows: We have thoroughly discussed representation and analysis and in this chapter have attempted to bridge the gap between analysis and synthesis through the subject of specification. The only remaining aspects to be considered are the actual methods of synthesizing the desired closed-loop transfer function. This is the subject of the next two chapters.

## 8.10   Problems

*8.10-1.*   Determine the sensitivity of the closed-loop transfer function to changes in the parameter $\alpha$ for the systems given below:

(a)   $\dot{x} = \begin{bmatrix} 0 & 1 & 0 \\ 0 & -2 & 2 \\ \alpha & 0 & -4 \end{bmatrix} x + \begin{bmatrix} 0 \\ 0 \\ 1 \end{bmatrix} u \quad y = x_1 \quad u = 10(r - \begin{bmatrix} 1 & 1 & 0 \end{bmatrix}x)$

(b)   $\dot{x} = \begin{bmatrix} -1 & 1 & 0 \\ 0 & -5 & 1 \\ 0 & 0 & -2 \end{bmatrix} x + \begin{bmatrix} 0 \\ 1 \\ 2 \end{bmatrix} u \quad y = x_1 \quad u = 10(r - \begin{bmatrix} 1 & \alpha & 1 \end{bmatrix}x)$

*8.10-2.*   For the system given below, determine $K$ such that the initial value of $x_3(t)$ due to a unit impulse input is equal to 10.

$$\dot{x} = \begin{bmatrix} 0 & 1 & 0 \\ 0 & -1 & 1 \\ 0 & 0 & -5 \end{bmatrix} x + \begin{bmatrix} 0 \\ 0 \\ 1 \end{bmatrix} u \quad y = x_1 \quad u = K(r - \begin{bmatrix} 1 & 0.5 & 0.5 \end{bmatrix}x)$$

*8.10-3.*   A closed-loop transfer function is to be selected for the third-order plant given below such that the response approximates the first-order model:

$$\left[\frac{y(s)}{r(s)}\right]_m = \frac{2}{s+2}$$

Select the value of $K = 50$.

$$\dot{x} = \begin{bmatrix} 0 & 1 & 0 \\ 0 & 0 & 1 \\ 0 & 0 & -5 \end{bmatrix} x + \begin{bmatrix} 0 \\ 0 \\ 1 \end{bmatrix} u \quad y = x_1$$

*8.10-4.*   If the closed-loop transfer function is given by

$$\frac{y(s)}{r(s)} = \frac{KK_p(s^m + c_m s^{m-1} + \cdots + c_2 s + c_1)}{s^n + d_n s^{n-1} + \cdots + d_2 s + d_1}$$

show that the position- and velocity-error constants are given by

$$K_{pos} = \frac{KK_pc_1}{d_1 - KK_pc_1} \qquad \text{and} \qquad \frac{1}{K_v} = \frac{d_2}{d_1} - \frac{c_2}{c_1}$$

*8.10-5.*   Use the conditions on *both* $G_{eq}(s)$ and $y(s)/r(s)$ to show that the system given below has zero steady-state error for *both* step and ramp inputs.

$$\dot{\mathbf{x}} = \begin{bmatrix} 0 & 1 \\ 0 & -4 \end{bmatrix} \mathbf{x} + \begin{bmatrix} 0 \\ 1 \end{bmatrix} u \qquad y = [1 \quad 1]\mathbf{x} \qquad u = 4(r - x_1)$$

*8.10-6.*   For the system shown in Fig. 8.10-1, find the values of $K$, $\delta$, and $\lambda$ such that $K_v = 1$ and the closed-loop system has complex conjugate poles at $s = -1 \pm j1$.   Where is the real closed-loop pole located?

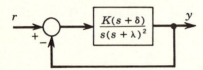

**Fig. 8.10-1**   Problem 8.10-6.

# nine  *system design with state-variable feedback*

## 9.1  Introduction

This chapter and the next form the design portion of this book and are its most important part, since no control engineer can be content with just formulating or analyzing a control system. He must ultimately become involved with the design or synthesis.

The material contained in the preceding chapters serves, to a large extent, to provide the reader with tools for attacking the synthesis problem. Although these tools are applicable to both analysis and synthesis, there is one important difference. The answer to an analysis problem is unique. However, in the realm of system design, problems often do not have

**451**

unique solutions.   This fact, which is one of the most frustrating and, at the same time, most exciting aspects of the subject, makes system design difficult to discuss in a textbook.   A given set of specifications, for example, may be satisfied by a number of system transfer functions, as we have seen in Chap. 8.   In addition, we see that a given system transfer function may be realized by several means.   It is often possible to judge between competing systems only on the basis of criteria partic- ular to the problem, such as power requirements, weight, size, and cost. The material in Chaps. 9 and 10 establishes guidelines for the synthesis of closed-loop systems and leaves the particular criteria for the reader's further study.

In these two chapters it is assumed that the desired closed-loop transfer function has been selected by making use of the concepts dis- cussed in Chap. 8.   In effect, a major portion of the synthesis problem has been solved once the specifications have been translated into a desired $y(s)/r(s)$.   The problem that remains to be treated is that of achieving or realizing the selected $y(s)/r(s)$ by finding the necessary system con- figuration.   This might involve the selection of feedback coefficients and forward-path gain, for example.

In a practical design problem, one might select several closed-loop transfer functions that meet the specifications and find the system con- figuration for each.   The actual system to be used would then be selected on the basis of simulation studies and/or the other engineering considera- tions discussed above.

Throughout this chapter it is assumed that all the state variables are directly available for measurement and control.   For this reason, the use of physical-system state variables is advocated.   In the next chapter the modification and extension of these design procedures to the case where all the state variables are not available will be discussed.   In particular, emphasis there will center on the situation when only the output is available.

In order to introduce the design procedures with a minimum amount of complication, the next section considers the simplest case in which neither the desired $y(s)/r(s)$ nor $G_p(s)$ has zeros.   By way of introduction, the properties of linear state-variable feedback discussed in Chap. 3 and elsewhere are reviewed.

Section 9.3 contains the extensions of the methods of Sec. 9.2 to the general design case.   In addition, this section introduces the concept of series compensation for further generality in the design procedure. The methods of Secs. 9.2 and 9.3 are illustrated in Sec. 9.4 by additional examples.

In addition to the standard procedures, there are a number of special procedures that may be applied in some but not all problems. When these procedures are applicable, they are, in general, highly successful. These procedures are discussed in Sec. 9.5.

## 9.2 Simplest case

Before beginning the discussion of the design procedure, let us review the conclusions made at the end of Sec. 3.4 concerning linear state-variable feedback. In terms of the equivalent feedback transfer function $H_{eq}(s)$, the following properties were established:

1. The zeros of $H_{eq}(s)$ may be located anywhere on the $s$ plane by suitable selection of the feedback coefficients **k**.
2. $H_{eq}(s)$ is not a function of $K$.
3. The poles of $H_{eq}(s)$ are zeros of $G_p(s)$.

It is important to emphasize once again the conditions under which statement 3 is valid, namely, that the closed-loop system must be described by Eqs. (**Ab**) and (**c**). Even if only the output is fed back, so that it appears that $H_{eq}(s)$ is just a constant $k_1$, the general expression for $H_{eq}(s)$

$$H_{eq}(s) = \frac{\mathbf{k}^T \mathbf{\Phi}(s)\mathbf{b}}{\mathbf{c}^T \mathbf{\Phi}(s)\mathbf{b}}$$

and statement 3 are still valid. The value of this ratio is, of course, $k_1$, because the poles and zeros of $H_{eq}(s)$ in this special case happen to lie at the same points. However, the poles of $H_{eq}(s)$ are still the zeros of $G_p(s)$, so that statement 3 is always true. Thus the loop-gain function $KG_p(s)H_{eq}(s)$ has $n - 1$ zeros, some of which may be at infinity, which are the zeros of $H_{eq}(s)$ and $n$ poles which are the poles of $G_p(s)$.

In addition to the above properties concerning $H_{eq}(s)$, the following properties of the closed-loop transfer function $y(s)/r(s)$ were also established:

1. The numerator of $y(s)/r(s)$ is the numerator of $KG_p(s)$ and is completely unaffected by state-variable feedback.
2. The poles of $y(s)/r(s)$ may be located anywhere on the $s$ plane by suitable selection of the feedback coefficients **k**.
3. If one auxiliary condition involving the feedback coefficients must be satisfied, the ability to position the $n$ poles of $y(s)/r(s)$ is retained as long as the forward-path gain $K$ is adjustable.

The auxiliary condition mentioned in property 3 most often arises from a requirement that the system have zero steady-state error for a step input.

From the above items, we see that there are only two restrictions on the $y(s)/r(s)$ that may be selected.

1.  The zeros of $y(s)/r(s)$ must be the same as the zeros of $G_p(s)$.
2.  The pole-zero excess of $y(s)/r(s)$ must be equal to the pole-zero excess of $G_p(s)$.

Although the first restriction may be eliminated completely, it is not possible to reduce the pole-zero excess, even though we are able to increase it. In this section, however, we consider only the situation in which neither $G_p(s)$ nor $y(s)/r(s)$ has any zeros. This situation is referred to as the *simplest case*. The more general situation in which both $G_p(s)$ and the desired $y(s)/r(s)$ may have zeros forms the subject of the *general case*, discussed in the next section.

In the light of the above discussion, it is obvious that it should be possible in the simplest case to find values for **k** and $K$ such that any desired $y(s)/r(s)$ may be achieved. Rather than state a general procedure for computing **k** and $K$ at the onset, let us consider a specific example of the problem. On the basis of the procedures that we use for this example, a general design procedure is formulated.

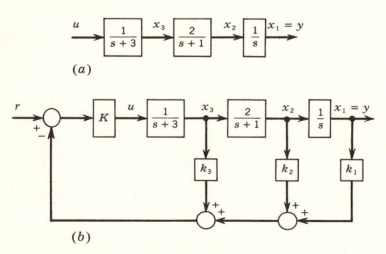

*(a)*

*(b)*

*Fig. 9.2-1*   Example of the simplest case.   (*a*) Plant to be controlled; (*b*) system configuration.

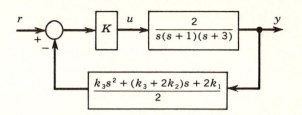

**Fig. 9.2-2** $H_{eq}(s)$ configuration.

For this example, let the plant be described by the following state-variable representation:

$$\dot{\mathbf{x}} = \begin{bmatrix} 0 & 1 & 0 \\ 0 & -1 & 2 \\ 0 & 0 & -3 \end{bmatrix} \mathbf{x} + \begin{bmatrix} 0 \\ 0 \\ 1 \end{bmatrix} u$$

$$y = \begin{bmatrix} 1 & 0 & 0 \end{bmatrix} \mathbf{x}$$

The block-diagram representation for this plant is shown in Fig. 9.2-1a. The state variables used here are assumed to be real physical variables so that they may be directly employed for control. The resulting closed-loop configuration is shown in Fig. 9.2-1b.

Let us assume that the desired closed-loop transfer function is given by

$$\frac{y(s)}{r(s)} = \frac{20}{[(s+1)^2 + (\sqrt{3})^2](s+5)} = \frac{20}{s^3 + 7s^2 + 14s + 20} \tag{9.2-1}$$

Note that the desired $y(s)/r(s)$ has been selected to give zero-steady-state error for a step input. The problem is then to find the values of $\mathbf{k}$ and $K$ such that the closed-loop transfer function of the system shown in Fig. 9.2-1b is equal to the desired closed-loop transfer function given by Eq. (9.2-1). The procedure used for computing the values for $\mathbf{k}$ and $K$ is to equate the closed-loop transfer function for Fig. 9.2-1b to Eq. (9.2-1).

In order to determine the closed-loop transfer function for the system shown in Fig. 9.2-1b, let us make use of the $H_{eq}(s)$ reduction. The resulting configuration is shown in Fig. 9.2-2. The expression for $H_{eq}(s)$ may be found by block-diagram manipulations or by either of the matrix expressions given in Table 3.4-1. In terms of $H_{eq}(s)$, the closed-loop transfer function becomes

$$\frac{y(s)}{r(s)} = \frac{KG_p(s)}{1 + KG_p(s)H_{eq}(s)}$$

$$= \frac{2K}{s^3 + (4 + Kk_3)s^2 + (3 + Kk_3 + 2Kk_2)s + 2Kk_1} \tag{9.2-2}$$

If we equate equal powers of $s$ in Eqs. (9.2-1) and (9.2-2), the following four equations result:

$$2K = 20 \qquad 4 + Kk_3 = 7$$
$$3 + Kk_3 + 2Kk_2 = 14 \qquad 2Kk_1 = 20 \qquad (9.2\text{-}3)$$

Although the last three expressions are nonlinear in $K$ and the elements of $\mathbf{k}$, they are linear in the quantities $Kk_1$, $Kk_2$, and $Kk_3$. In addition, the first expression involves $K$ only and in a simple linear fashion; these features are typical of the form of the expressions obtained. The four expressions given in Eqs. (9.2-3) may be solved as a simple equivalent set of linear simultaneous algebraic equations. The resulting values for $\mathbf{k}$ and $K$ are

$$\mathbf{k} = \text{col } (1.0, 0.4, 0.3) \qquad \text{and} \qquad K = 10 \qquad (9.2\text{-}4)$$

The reader is urged to carry out the simple algebra involved in solving these four equations.

Although the design problem has been solved in that $\mathbf{k}$ and $K$ have been determined, let us examine the root locus of $1 + KG_p(s)H_{eq}(s)$ as a function of $K$ to see if the advantages associated with state-variable feedback in the previous chapters have been realized in this case. In terms of the value for $\mathbf{k}$ given in Eq. (9.2-4), $H_{eq}(s)$ becomes

$$H_{eq}(s) = \frac{0.3s^2 + 1.1s + 2}{2} = \frac{0.3[(s + 1.83)^2 + (1.82)^2]}{2}$$

so that $KG_p(s)H_{eq}(s)$ is

$$KG_p(s)H_{eq}(s) = \frac{0.3K[(s + 1.83)^2 + (1.82)^2]}{s(s + 1)(s + 3)} \qquad (9.2\text{-}5)$$

The root locus for $1 + KG_p(s)H_{eq}(s)$ as a function of $K$ is shown in Fig. 9.2-3.

From Fig. 9.2-3 we observe that the system is stable for all positive values of $K$. In addition the damping ratio of the pair of complex conjugate poles of $y(s)/r(s)$ remains relatively constant for all values of $K$ greater than the design value of 10. The only other effect of increasing the gain is that the real pole moves farther out along the negative real axis, thereby making the transient response faster. Therefore the design appears to be not only acceptable but highly satisfactory from stability and sensitivity considerations.

Although the above example has been worked in terms of block diagrams and transfer functions, we could also have carried out the complete example in terms of matrix algebra. If we had taken the matrix approach, either Eq. (3.4-8) or Eq. (3.4-14) could have been used to find

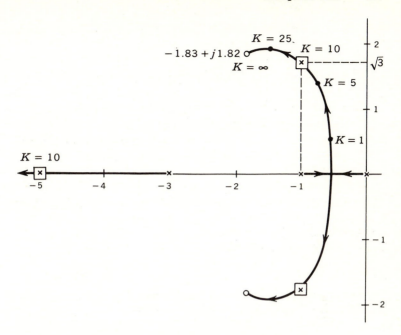

**Fig. 9.2-3**   Root locus for $KG_p(s)H_{eq}(s)$ given Eq. (9.2-5).

$y(s)/r(s)$.   This matrix approach is usually more desirable if there is a large amount of interrelation among the state variables.

On the other hand, the use of the $H_{eq}(s)$ approach allows one to check directly the stability and sensitivity features of the design.   Because of this fact, it is recommended that the $H_{eq}(s)$ approach be taken even in the case when $H_{eq}(s)$ is computed by the use of matrix algebra rather than block-diagram manipulations.

In terms of the above example, the following general design procedure can be formulated for the solution of the simplest case in which neither $G_p(s)$ nor $y(s)/r(s)$ has zeros.

1.   Begin with a state-variable representation of the plant (in terms of matrix equations and/or block diagrams) and assume the usual linear state-variable-feedback configuration for the closed-loop system.
2.   Select the desired closed-loop transfer function by making use of the material in Chap. 8.
3.   Find the actual $y(s)/r(s)$ in terms of **k** and $K$, preferably by making use of the $H_{eq}(s)$ configuration.
4.   Equate the expressions for $y(s)/r(s)$ from steps 2 and 3 and solve for **k** and $K$.

Once the design has been completed by using the above procedure, it is suggested that the following evaluation process be carried out:

1. Use the value of **k** found above to express $H_{eq}(s)$ in numerical form.
2. Factor the numerator of $H_{eq}(s)$ and plot the root locus of $KG_p(s)H_{eq}(s)$. It is convenient to place the closed-loop poles of $y(s)/r(s)$ on the plot to assist in drawing the root locus.
3. Examine the root locus to see if the desired stability and sensitivity benefits that can be achieved by state-variable feedback have been realized.

The $H_{eq}(s)$ approach is recommended for the design procedure because it is so well adapted to the above evaluation procedure. Although this evaluation procedure is recommended for the simplest case, it becomes a necessity in the general case treated in the next section.

**Exercises 9.2**   *9.2-1.*   Find the feedback coefficients **k** and the gain $K$ such that each of the systems given has the following closed-loop transfer function:

$$\frac{y(s)}{r(s)} = \frac{80}{[(s+2)^2 + 2^2](s+10)}$$

Plot the root locus for $KG_p(s)H_{eq}(s)$ for both cases.   Use the $H_{eq}(s)$ approach.

$(a)$   $\dot{\mathbf{x}} = \begin{bmatrix} 0 & 1 & 0 \\ 0 & -3 & 3 \\ 0 & 0 & -1 \end{bmatrix} \mathbf{x} + \begin{bmatrix} 0 \\ 0 \\ 1 \end{bmatrix} u$   $y = [1 \quad 0 \quad 0]\mathbf{x}$

$(b)$   $\dot{\mathbf{x}} = \begin{bmatrix} -1 & 1 & 0 \\ 0 & -2 & 2 \\ 0 & 0 & -3 \end{bmatrix} \mathbf{x} + \begin{bmatrix} 0 \\ 0 \\ 1 \end{bmatrix} u$   $y = [1 \quad 0 \quad 0]\mathbf{x}$

*answers:*

$(a)$ $K = 26.7,$   **k** $=$ col $(1.0, 0.1875, 0.375)$
$(b)$ $K = 40,$   **k** $=$ col $(0.5625, 0.1625, 0.2)$

*9.2-2.*   Repeat Exercise 9.2-1 but make direct use of Eq. (3.4-8) to find the values of **k** and $K$.

*9.2-3.*   Find the feedback coefficients **k** and $K$ for the unstable plant

$$\dot{\mathbf{x}} = \begin{bmatrix} 0 & 1 & 0 \\ 0 & 2 & 1 \\ 0 & 0 & -1 \end{bmatrix} \mathbf{x} + \begin{bmatrix} 0 \\ 0 \\ 1 \end{bmatrix} u \qquad y = x_1$$

such that $y(s)/r(s)$ is

$$\frac{y(s)}{r(s)} = \frac{80}{[(s+2)^2 + 2^2](s+10)}$$

Draw the root locus for the final system.

*answer:*

$K = 80, \mathbf{k} = \text{col}\ (1,1,\frac{3}{16})$

*9.2-4.*   Find the values of $K$ and $\mathbf{k}$ for the system

$$\dot{\mathbf{x}} = \begin{bmatrix} -1 & 1 & 0 \\ 0 & -4 & 2 \\ 0 & 0 & -5 \end{bmatrix} \mathbf{x} + \begin{bmatrix} 0 \\ 0 \\ 1 \end{bmatrix} u \qquad y = x_1$$

such that

$$\frac{y(s)}{r(s)} = \frac{180}{[(s+3)^2 + 3^2](s+10)}$$

*answer:*

$K = 90; \mathbf{k} = \text{col}\ (\frac{117}{180}, \frac{19}{180}, \frac{1}{150})$

## 9.3   *General case*

In the preceding section the state-variable-feedback method was applied to the simplest case.   There both $G_p(s)$ and $y(s)/r(s)$ were assumed to have no zeros.   Although there is a large number of practical situations that belong to the simplest case, there is also a large number that do not. The latter group is referred to as the *general case*, which is the subject of this section.

In the general case both $y(s)/r(s)$ and $G_p(s)$ may have zeros, and any zeros that appear in $G_p(s)$ may not be the same as those in the desired closed-loop system.   Since zeros of the plant are unaltered by state-variable feedback and appear also as closed-loop zeros, and since state-variable feedback does not generate any new zeros, a more general approach must be used to realize the desired $y(s)/r(s)$ in the general case.

The form of this new controller structure is shown in Fig. 9.3-1 along with the previous form, for comparison.   The basic difference between the two controller structures is the addition of an $n_c$th-order dynamical compensator $G_c(s)$ in series with the control input to the plant. In addition, the $n_c$ new states $\mathbf{x}_c$ generated by the series compensator have also been involved in the linear feedback.   With this more general controller structure, it is possible to achieve the desired generality for $y(s)/r(s)$.

In order to simplify the treatment of the general controller system, it is convenient to absorb the series compensator into the plant to form the compensated plant, as shown in Fig. 9.3-2. This compensated plant is then treated as an $(n + n_c)$th-order plant, and the distinction between the states of the original plant and the states of the compensator is dropped. In addition, the input is once again considered to be just the control input $u$. By making these changes, the general controller-structure system of Fig. 9.3-1$b$ takes the form of the simple system shown in Fig. 9.3-1$a$, as can be seen in Fig. 9.3-2. The only change is that the plant

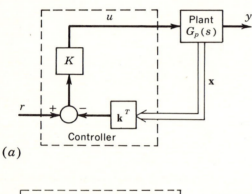

(a)

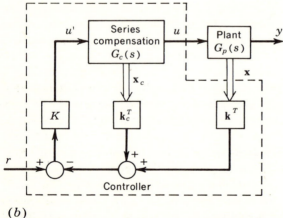

(b)

**Fig. 9.3-1**   Two controller structures. (a) Simple form; (b) general form including series compensation.

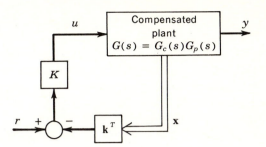

*Fig. 9.3-2* Series compensation absorbed in
the plant.

has been modified by the addition of the series compensator. The forward transfer function is now $KG(s) = KG_c(s)G_p(s)$.

The approach to be used to obtain the more general form for $y(s)/r(s)$ is now obvious. Since the desired generality of $y(s)/r(s)$ cannot be achieved with the original plant, the plant must be modified by appropriate selection of the series compensation so that the desired $y(s)/r(s)$ may be realized. Once a series compensation that makes $G(s)$ compatible with the desired $y(s)/r(s)$ has been chosen, the design proceeds in a manner very similar to that of the simplest case.

The only new problem, therefore, is the proper selection of the series compensator $G_c(s)$. This is the main subject of the discussion to follow. This discussion has been divided into four separate phases: (1) elimination of zeros, (2) increase of pole-zero excess, (3) movement of zeros, and (4) addition of zeros. This division was made not because the topics are unrelated but rather so that the reader might grasp the basic concepts of each without confusing complications. In a practical problem one often finds that more than one of the above procedures must be used.

*Elimination of zeros.* The problem that we consider here is that of eliminating from $y(s)/r(s)$ zeros of $G_p(s)$ that are not desired in $y(s)/r(s)$. This problem occurs whenever $G_p(s)$ has zeros and the desired $y(s)/r(s)$ has none. According to the properties of state-variable feedback, any zero of $G_p(s)$ must appear as a zero of $y(s)/r(s)$. Hence we have the problem of removing the unwanted zero(s) from $y(s)/r(s)$.

In a sense the discussion of the elimination of zeros belongs more logically in Sec. 9.2, since the process does not require the use of series compensation. On the other hand, the elimination of zeros forms such an integral part of the following procedures that it seems appropriate to introduce the concept at this time.

The procedure used to remove unwanted zeros from $y(s)/r(s)$ is very simple. We merely cancel these unwanted zeros by placing poles of $y(s)/r(s)$ at the zero locations.    Note that zeros of $G_p(s)$ that lie in the right-half plane may not be eliminated by using this cancellation since, as we have seen in Chap. 7, pole-zero cancellation in the right-half plane is impractical.    In general, it is impossible to eliminate right-half-plane zeros of $G_p(s)$ from $y(s)/r(s)$, and they therefore are a definite performance limitation.[1]

   One unfortunate side effect of this cancellation procedure is that freedom is lost in the selection of the poles of $y(s)/r(s)$.    In general, only $n - n_z$ poles of $y(s)/r(s)$ may be chosen arbitrarily, where $n_z$ is the number of unwanted zeros that must be eliminated.    This problem is a direct consequence of our inability at this time to increase the pole-zero excess. Later, after the method of increasing the pole-zero excess has been discussed, the elimination of unwanted zeros imposes no limitation on $y(s)/r(s)$.

   ***Example 9.3-1***    Although the procedure for eliminating an unwanted zero from $y(s)/r(s)$ is relatively straightforward, it may be of value to consider a simple illustration of the process.    For this example, let us use the second-order system shown in Fig. 9.3-3. The plant transfer function associated with this system is

$$G_p(s) = \frac{y(s)}{u(s)} = \frac{2(s + 1)}{s(s + 2)} \tag{9.3-1}$$

   Let us suppose that the desired closed-loop transfer function is

$$\frac{y(s)}{r(s)} = \frac{6}{s + 6} \tag{9.3-2}$$

[1] In Sec. 9.5 a special procedure for eliminating right-half-plane zeros will be introduced.    In the cases where this special procedure is applicable, this performance limitation may be removed.

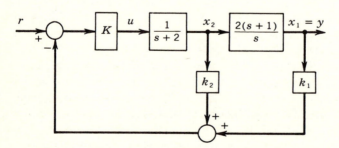

**Fig. 9.3-3**    System used to illustrate zero elimination.

Here $y(s)/r(s)$ has been selected with the knowledge that the pole-zero excess is not to be changed, and therefore since the pole-zero excess of $G_p(s)$ is 1, the pole-zero excess of $y(s)/r(s)$ must also be 1. In other words, if we wish to remove a zero from $y(s)/r(s)$, a pole must also be removed in order to maintain the same pole-zero excess.

Although the $y(s)/r(s)$ given by Eq. (9.3-2) is the desired closed-loop transfer function, the properties of state-variable feedback dictate that the zeros of $G_p(s)$ must appear as zeros of $y(s)/r(s)$. In order to remove the unwanted zero at $s = -1$, we must place a pole of $y(s)/r(s)$ at this same location. The actual $y(s)/r(s)$ to be realized is therefore

$$\frac{y(s)}{r(s)} = \frac{6(s + 1)}{(s + 1)(s + 6)} = \frac{6(s + 1)}{s^2 + 7s + 6} \tag{9.3-3}$$

This $y(s)/r(s)$ may be achieved by direct use of state-variable feedback.

In order to compute the values of $\mathbf{k}$ and $K$, let us determine the closed-loop transfer function for the system shown in Fig. 9.3-3.

$$\frac{y(s)}{r(s)} = \frac{2K(s + 1)}{s^2 + (2 + Kk_2 + 2Kk_1)s + 2Kk_1} \tag{9.3-4}$$

If the two forms for $y(s)/r(s)$ as given by Eqs. (9.3-3) and (9.3-4) are equated, the following three equations result:

$$2K = 6 \qquad 2 + Kk_2 + 2Kk_1 = 7 \qquad \text{and} \qquad 2Kk_1 = 6$$

Once again note that the equations are linear in $K$, $Kk_1$, and $Kk_2$. The solution of these equations yields

$$\mathbf{k} = \text{col} \ (1, -\tfrac{1}{3}) \qquad \text{and} \qquad K = 3$$

Using these values for $\mathbf{k}$, we find that $H_{eq}(s)$ is given by

$$H_{eq}(s) = \frac{5}{6} \frac{s + \tfrac{6}{5}}{s + 1}$$

so that the open-loop transfer function is

$$KG_p(s)H_{eq}(s) = \frac{5K(s + \tfrac{6}{5})}{3s(s + 2)}$$

The root locus for $1 + KG_p(s)H_{eq}(s)$ is shown in Fig. 9.3-4.

*Increase of pole-zero excess.* As we have seen in the preceding discussion, the elimination of zeros from $y(s)/r(s)$ imposes a limitation on

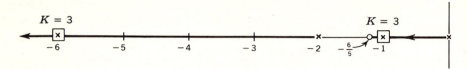

**Fig. 9.3-4**   Root locus of $KG_p(s)H_{eq}(s)$ for Example 9.3-1.

the possible form of $y(s)/r(s)$. In Example 9.3-1, for instance, it was impossible to realize a second-order closed-loop response, even though the given plant was second-order. One of the closed-loop poles had to be used to cancel the unwanted zero and hence was not available as part of the specified $y(s)/r(s)$. If $G_p(s)$ had had one additional pole, a second-order response could have been specified and the unwanted zero of $G_p(s)$ still removed from the closed-loop transfer function. Thus a means of restoring design flexibility to $y(s)/r(s)$ is to provide a means of increasing the pole-zero excess of $G_p(s)$.

Although one of the primary uses of increase of pole-zero excess is in conjunction with zero removal, the discussion here considers the increase of pole-zero excess as a distinct subject. In other words, here we need not be concerned with *why* the pole-zero excess is to be increased but just *how* it may be accomplished. The combination and interrelation of the four aspects of the general method will be discussed in detail in the next section.

Since the general properties of the state-variable-feedback procedure do not allow the pole-zero excess of $y(s)/r(s)$ to be different from the pole-zero excess of $G_p(s)$, it is clear that some new approach is needed. As mentioned in the introductory portion of this section, the needed flexibility may be gained by the introduction of series compensation. The series compensator is used to increase the pole-zero excess of $G(s)$, which causes the pole-zero excess of $y(s)/r(s)$ to increase by the same amount.

Assume, for example, that it is desired to make the pole-zero excess of $y(s)/r(s)$ one greater than the pole-zero excess of $G_p(s)$. In this case a series compensation of the form

$$G_c(s) = \frac{1}{s + \lambda}$$

is used so that the compensated or modified plant transfer function is

$$G(s) = G_c(s)G_p(s) = \frac{1}{s + \lambda} G_p(s)$$

Since the pole-zero excess of the plant has been increased by 1, the pole-zero excess of $y(s)/r(s)$ also increases by 1.

The complete closed-loop system including the series compensator is shown in Fig. 9.3-5. There the series compensation has been absorbed into the plant in the manner shown in Fig. 9.3-2. The state associated with the compensator is labeled $x_{n+1}$, and its associated feedback coefficient is $k_{n+1}$. The plant is now treated as being of order $n + 1$; $\mathbf{x}$ is therefore

$$\mathbf{x} = \text{col}\ (x_1, x_2, \ \ldots \ , x_{n+1})$$

and $\mathbf{k}$ is

$$\mathbf{k} = \text{col}\ (k_1, k_2, \ \ldots \ , k_{n+1})$$

It is important to note that the state $x_{n+1}$ of the compensator has been included in the state-variable feedback. If this is not done, some of the stability characteristics we have attached to state-variable feedback are lost.

For each pole of $G_c(s)$, two unknowns are introduced into the design problem. These unknowns are the location of the pole $\lambda$ and the value of the feedback coefficient associated with the new state, $k_{n+1}$ in the example. Since there is only one additional pole to be selected in $y(s)/r(s)$, some arbitrariness is now introduced into the problem. In other words, in terms of the previous example, the value of $\lambda$ can be selected *arbitrarily* with no loss in the freedom of selection of $y(s)/r(s)$. Alternatively the feedback coefficient $k_{n+1}$ may be selected arbitrarily and then the desired $y(s)/r(s)$ realized by proper selection of the remaining feedback coefficients, $K$ and $\lambda$.

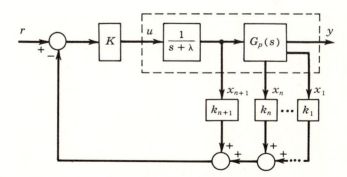

**Fig. 9.3-5** System structure for increasing the pole-zero excess of $y(s)/r(s)$ by 1.

Although the compensator-pole location *or* feedback coefficient may be selected arbitrarily, this is not the procedure that is recommended. On the contrary, it is recommended that this flexibility be used to achieve some desired feature in the resulting design. For example, it may be possible to select the pole location so that certain desired sensitivity or stability characteristics are achieved. Because the character of the resulting design may vary widely depending on how the compensator-pole locations or feedback coefficients are selected, it is imperative that the design-evaluation procedure outlined at the end of the preceding section be carried out for each design.

Because of the large number of possible ways that the compensator-pole location or feedback coefficients may be selected, it is impossible to discuss all possible situations even for a simple example. At the same time, it is impossible to establish hard-and-fast rules for making the selections. This is truly a case when the only limitation is one's imagination.

To illustrate the method for selecting the compensator-pole location that the authors have found to be most successful, let us consider a simple example. Several more complex examples will appear in the next section. The plant to be controlled is shown in Fig. 9.3-6a. The desired closed-loop transfer function is

$$\frac{y(s)}{r(s)} = \frac{8}{[(s+1)^2 + 1^2](s+4)} = \frac{8}{s^3 + 6s^2 + 10s + 8}$$

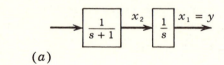

(*a*)

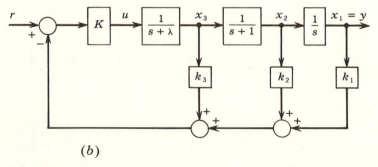

(*b*)

**Fig. 9.3-6**  Example of increasing pole-zero excess.   (*a*) Plant; (*b*) complete system.

Here the pole-zero excess of $y(s)/r(s)$ is greater than that of $G_p(s)$ by 1. A series compensator of the form

$$G_c(s) = \frac{1}{s + \lambda}$$

is therefore necessary to achieve the desired $y(s)/r(s)$. The complete closed-loop system is shown in Fig. 9.3-6b. The problem is now to select the values of **k**, $K$, and $\lambda$ so that the closed-loop transfer function of the system shown in Fig. 9.3-6b is equal to the $y(s)/r(s)$ given above.

We could make an arbitrary selection for $\lambda$; then the problem would reduce to the simplest case discussed in Sec. 9.2. Rather than substitute a numerical value for $\lambda$, let us assume that $\lambda$ is known, but let us continue to manipulate it as a literal term. The $H_{eq}(s)$ transfer function for the system shown in Fig. 9.3-6b is easily found to be

$$H_{eq}(s) = k_1 + (k_2 + k_3)s + k_3 s^2 \tag{9.3-5}$$

The closed-loop transfer function is therefore

$$\frac{y(s)}{r(s)} = \frac{K}{s^3 + (1 + \lambda + Kk_3)s^2 + (\lambda + Kk_2 + Kk_3)s + Kk_1}$$

By equating the two forms for $y(s)/r(s)$, we obtain the following equations:

$$K = 8 \quad 1 + \lambda + Kk_3 = 6 \quad \lambda + Kk_2 + Kk_3 = 10 \quad \text{and} \quad Kk_1 = 8$$

Note that the quantity $\lambda + Kk_3$ appears as an entity in both the middle equations. Because of this fact, the solutions for $K$, $k_1$, and $k_2$ do not depend on $\lambda$. The resulting solutions are therefore

$$k_1 = 1 \qquad k_2 = \tfrac{5}{8} \qquad k_3 = \frac{5 - \lambda}{8} \qquad \text{and} \qquad K = 8 \tag{9.3-6}$$

The fact that $k_1$, $k_2$, and $K$ are independent of $k_3$ or $\lambda$ is important. It means that any arbitrary selection of $\lambda$ or $k_3$ may be made without affecting $K$, $k_1$, or $k_2$. This selection may be made only on the criterion of computational ease. Two convenient choices are to let $\lambda = 0$ or $k_3 = 0$. Neither of these choices is a good one from the standpoint of the resulting system, as will be shown later. But that is of no concern at present.

If Fig. 9.3-6b is redrawn so that the series compensator and its feedback coefficient are isolated, as shown in Fig. 9.3-7, then, once $\lambda$ and $k_3$ are known, the transfer function $G_c^*(s) = x_3(s)/u^*(s)$ may be calculated. Any other combination of $k_3$ and $\lambda$ that yields the same $G_c^*(s)$ also yields a correct solution of the design problem, since the closed-loop transfer function is unchanged as long as $G_c^*(s)$ is constant. In effect,

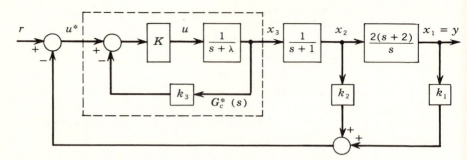

**Fig. 9.3-7**  Isolation of the series compensator and its feedback coefficient to form $G_c^*(s)$.

a simplified state-variable-feedback problem has been generated as an adjunct to the original problem. In this simplified problem the object is to select $\lambda$ and $k_3$ so that the desired $G_c^*(s)$ is realized.

As an illustration of this procedure, suppose that we had let $\lambda = 0$ in the previous calculation. Then from Eq. (9.3-6) we find that $k_3 = \frac{5}{8}$. Using these values for $\lambda$ and $k_3$ and the fact that $K = 8$, we find that $G_c^*(s)$ is

$$G_c^*(s) = \frac{8}{s + \lambda + 8k_3} = \frac{8}{s + 5} \tag{9.3-7}$$

From Eq. (9.3-7) it is obvious that any values of $\lambda$ and $k_3$ that satisfy the following equation also produce a correct solution.

$$\lambda + 8k_3 = 5 \qquad \text{or} \qquad k_3 = \frac{5 - \lambda}{8} \tag{9.3-8}$$

This result, which is identical with the expression contained in Eq. (9.3-6), has been obtained by realizing the correct $G_c^*(s)$ rather than $y(s)/r(s)$. Although the computational saving is not great in this simple example, the procedure is very helpful in more complex problems.

So far we have not discussed the problem of how to select either $\lambda$ or $k_3$. All that we have done is show first that $k_1$, $k_2$, and $K$ are independent of the selection of $\lambda$ or $k_3$ and, second, that the relation between the possible choices for $\lambda$ and $k_3$ may be obtained by solving the simplified state-variable-feedback problem of realizing $G_c^*(s)$. Although any values of $\lambda$ and $k_3$ that satisfy Eq. (9.3-8) and the values for $k_1$, $k_2$, and $K$ given by Eq. (9.3-6) solve the design problem, in the sense that the correct $y(s)/r(s)$ results, not all the choices are equally satisfactory. To illustrate this point, let us examine some of the possible choices for $\lambda$ by making use of the evaluation procedure outlined in Sec. 9.2.

Any negative values for λ may be ruled out since they would make the plant open loop unstable, an undesirable situation. In addition, we may rule out any values of λ greater than 5,[1] since according to Eq. (9.3-8) they would make $k_3$ negative. Although negative values for $k_3$ do not pose a great instrumentation problem, they cause a zero of $H_{eq}(s)$ to move into the right-half plane, so that the system becomes unstable for high gain. Therefore, it appears that the values for λ must be selected from the range $0 \leq \lambda \leq 5$. Let us begin by examining the resulting closed-loop system for the two possible extremes of this range.

For λ = 0 we have previously found the feedback coefficients and gain to be

$$\mathbf{k} = \text{col}\ (1,\tfrac{5}{8},\tfrac{5}{8}) \qquad \text{and} \qquad K = 8$$

The resulting $H_{eq}(s)$ as given by Eq. (9.3-5) is

$$H_{eq}(s) = \tfrac{5}{8}(s^2 + 2s + \tfrac{8}{5}) = \tfrac{5}{8}[(s + 1)^2 + (0.775)^2]$$

and $G(s) = G_c(s)G_p(s)$ is

$$G(s) = \frac{1}{s^2(s + 1)}$$

The root locus for $KG(s)H_{eq}(s)$ is shown in Fig. 9.3-8. Although the

[1] The sensitivity condition $|D_k(i\omega)/D(j\omega)| \geq 1$ is violated if λ > 4.

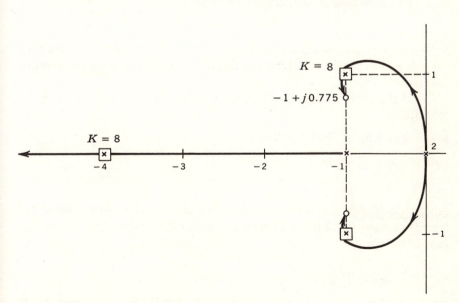

**Fig. 9.3-8**   Root locus of $KG(s)H_{eq}(s)$ for λ = 0.

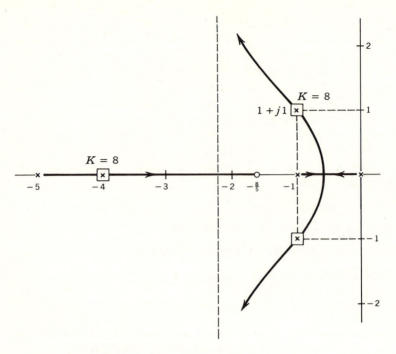

**Fig. 9.3-9** Root locus of $KG(s)H_{eq}(s)$ for $\lambda = 5$.

closed-loop system is stable for all positive values of gain, the damping ratio becomes rather small for low gain. This might be an undesirable feature.

At the other extreme, $\lambda = 5$, the feedback coefficients and gain are found to be

$$\mathbf{k} = \text{col } (1,\tfrac{5}{8},0) \qquad \text{and} \qquad K = 8$$

and $H_{eq}(s)$ is given by

$$H_{eq}(s) = \tfrac{5}{8}(s + \tfrac{8}{5})$$

Note that $H_{eq}(s)$ has only one zero. This is a direct consequence of the fact that $k_3 = 0$. The compensated forward transfer function in this case becomes

$$KG(s) = \frac{K}{s(s + 1)(s + 5)}$$

The resulting root locus for $1 + KG(s)H_{eq}(s)$ is shown in Fig. 9.3-9. This root locus has better low-gain properties than the case when $\lambda = 0$, but its high-gain situation is not as good.    As the gain increases above 8, the closed-loop poles move considerably from their design locations, and the damping ratio decreases.    Both these features are due to the fact that not all the state variables are fed back, since $k_3 = 0$.

It would appear likely that a value of $\lambda$ between 0 and 5 would produce a system that is a compromise between the behavior at the two extremes.    To show that this is the case, let us consider the value in the middle of this range, namely, $\lambda = 2.5$.    In this case, $k_3 = \frac{5}{16}$, $H_{eq}(s)$ becomes

$$H_{eq}(s) = \tfrac{5}{16}(s^2 + 3s + \tfrac{16}{5}) = \tfrac{5}{16}[(s + 1.5)^2 + (0.978)^2]$$

and the compensated forward transfer function is

$$KG(s) = \frac{K}{s(s + 1)(s + 2.5)}$$

The root locus for this situation is shown in Fig. 9.3-10.    As expected,

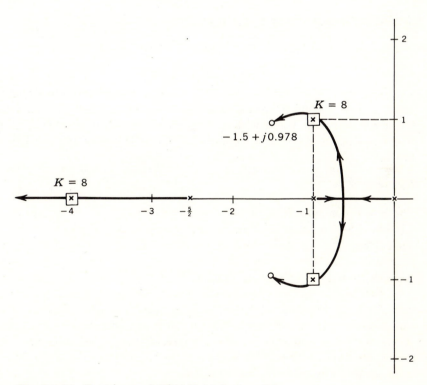

**Fig. 9.3-10**    Root locus of $KG(s)H_{eq}(s)$ for $\lambda = 2.5$.

this root locus is a reasonably good compromise of the good and bad features of Figs. 9.3-8 and 9.3-9.   It would appear that $\lambda \sim 2.5$ is a good design choice.

The choice of $\lambda = 2.5$ may be justified further on the basis of sensitivity considerations.   Let us consider the sensitivity of the closed-loop transfer function to parameter changes in $G_c^*(s)$.   The chain rule may be used to find the sensitivity of $y(s)/r(s)$ to changes in parameters of $G_c^*(s)$, designated as $p_i$, as

$$S_{p_i}^{y(s)/r(s)} = S_{G_c^*(s)}^{y(s)/r(s)} S_{p_i}^{G_c^*(s)}$$

The parameters of interest in this example are $\lambda$ and $k_3$.   In order to minimize sensitivity to changes in $p_i$, only the term $S_{p_i}^{G_c^*(s)}$ need be considered, since no $p_i$ are involved in $S_{G_c^*(s)}^{y(s)/r(s)}$.   The sensitivity of $G_c^*(s)$ with respect to the parameter $\lambda$ is

$$S_\lambda^{G_c^*(s)} = \frac{\lambda}{G_c^*(s)} \frac{\partial G_c^*(s)}{\partial \lambda} = \frac{-\lambda}{s + \lambda + Kk_3}$$

Similarly, the sensitivity of $G_c^*(s)$ to the parameter $k_3$ is

$$S_{k_3}^{G_c^*(s)} = \frac{-Kk_3}{s + \lambda + Kk_3}$$

Recall that all the elements of $G_c^*(s)$ are under the complete control of the designer.   Thus it seems likely that the component elements that are associated with $\lambda$ and with $k_3$ might be of the same quality.   Under this assumption it seems reasonable to equate the resulting sensitivities so that

$$Kk_3 = \lambda$$

The controller gain $K$ is already known to be 8, and from Eq. (9.3-8) $\lambda$ is just $5 - 8k_3$, and so the resulting values of $k_3$ and $\lambda$ are $\frac{5}{16}$ and 2.5, respectively.   Thus the value of $\lambda = 2.5$ is not only an intuitive one because it is midway between the extremes of 0 and 5, but that value also is desirable from a sensitivity point of view.

Before leaving this problem it is interesting to examine one more value of $\lambda$ in the range of $0 \leq \lambda \leq 5$, namely, $\lambda = 4.0$.   Note that this choice places the open-loop pole of the compensator at the location of one of the desired closed-loop poles.   Because of this, $H_{eq}(s)$ must also have a zero at that location.   For $\lambda = 4$, $k_3$ is found from Eq. (9.3-8) to be equal to $\frac{1}{8}$.   Therefore $H_{eq}(s)$ becomes

$$H_{eq}(s) = \tfrac{1}{8}(s^2 + 6s + 8) = \tfrac{1}{8}(s + 2)(s + 4)$$

and $KG(s)$ is

$$KG(s) = \frac{K}{s(s+1)(s+4)}$$

The root locus for this case is shown in Fig. 9.3-11. Although the overall character of this design does not appear to be as good as that shown in Fig. 9.3-10, this system has one interesting feature. Because of the pole-zero combination at $s = -4$, a closed-loop pole exists at that location for *all* gain. This fact is not necessarily of much importance at this time, but it will prove to be useful later. In addition, since we know that a zero of $H_{eq}(s)$ is at $s = -4$, the problem of factoring $H_{eq}(s)$ is simplified.

Although the preceding discussion of the procedure for increasing the pole-zero excess has centered on a simple, specific example, the concepts are applicable to the general case. The following is a step-by-step summary of the design procedure:

1. Select the desired closed-loop transfer function.
2. Add a sufficient number of series compensators of the form

$$G_c(s) = \frac{1}{s + \lambda} \tag{9.3-9}$$

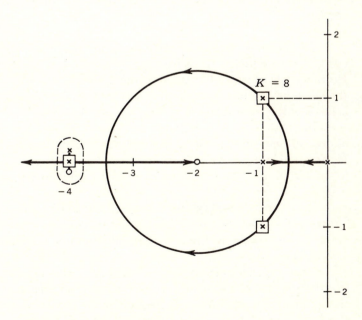

**Fig. 9.3-11**  Root locus of $KG(s)H_{eq}(s)$ for $\lambda = 4$.

to make the pole-zero excess of $G(s)$ compatible with $y(s)/r(s)$.   Let $n_c$ be the number of added poles.

3.   Represent the compensated system in the usual state-variable-feed-back form.    All the states associated with $G_c(s)$ are also fed back.

4.   Make any convenient selection of either the compensator poles or their associated feedback coefficients.   One convenient way is to let either all the compensator-pole locations or their feedback coefficients be zero.

5.   Find the closed-loop transfer function of the system and equate it with the desired $y(s)/r(s)$ to solve for the remaining $n + n_c + 1$ unknowns.

6.   Use the known values of pole locations and feedback coefficients to find $G_c^*(s)$.

7.   Use $G_c^*(s)$ to find the relations that exist between the possible choices of pole locations and feedback coefficients.

8.   Use the relations found in step 7 to select the proper value for the pole locations by evaluating the root locus of the resulting systems. The values of the feedback coefficients and the forward-path gain associated with the original system remain the same as the values found in step 5.

Additional examples illustrating this procedure will be given in the next section.

*Movement of zeros.*   In some cases, in order to achieve zero velocity error, for example, it is desired to have one or more zeros in $y(s)/r(s)$. However, the desired locations of the zeros in $y(s)/r(s)$ are usually not the same as the locations of the zeros of $G_p(s)$, if any exist.   Then one is faced with the problem of moving the zeros of $G_p(s)$ to the desired locations in $y(s)/r(s)$ or creating new zeros if not enough exist.   The process of creating new zeros will be discussed at the conclusion of this section. If there are more zeros in $G_p(s)$ than are desired in $y(s)/r(s)$, the process for eliminating unwanted zeros discussed earlier in this section may be used.

In order to simplify the following discussion, we assume that the desired number of zeros of $y(s)/r(s)$ is the same as the number of zeros of $G_p(s)$.   Examples of the more general situation will be taken up in the next section.

Since the rules of simple linear state-variable feedback dictate that the numerators of $G_p(s)$ and $y(s)/r(s)$ must be identical, it is obvious that a modified controller is necessary if we are to achieve the desired effect of having the zeros of $y(s)/r(s)$ at different locations from the zeros of $G_p(s)$. A series compensator once again is used to realize the desired flexibility. The basic form of the series compensator to be used in this case is

$$G_c(s) = \frac{s + \delta}{s + \lambda} = 1 + \frac{\delta - \lambda}{s + \lambda} \qquad (9.3\text{-}10)$$

Here $\delta$ is chosen such that the zero of $G_c(s)$ is at the location of one of the desired zeros of $y(s)/r(s)$ and the pole location $\lambda$ is arbitrary. One element of the above form is included in the series compensator for each zero that must be moved. All the states associated with $G_c(s)$ are fed back as before.

It is fairly easy to analyze the effect of the use of a series compensator of the form given by Eq. (9.3-10). Since the modified plant transfer function $G(s)$ contains a zero at the location of every desired zero in $y(s)/r(s)$, the properties of state-variable feedback cause these zeros to appear at the same locations in $y(s)/r(s)$, as required. In addition, zeros appear in $y(s)/r(s)$ at the locations of the original zeros of $G_p(s)$, since these zeros still appear in $G(s)$. At the same time, $y(s)/r(s)$ contains an extra pole for each zero to be moved, because $G_c(s)$ is made up of pole-zero combinations as shown in Eq. (9.3-10). These extra poles of $y(s)/r(s)$ may be used to cancel the unwanted zeros in $y(s)/r(s)$ by placing poles at the original locations of the zeros in $G_p(s)$ to be moved. Because these poles appear in addition to the usual $n$ poles of $y(s)/r(s)$, the ability to select the location of the $n$ closed-loop poles is retained. Of course, the distinction between the extra and the usual poles of $y(s)/r(s)$ is completely artificial since there is really nothing that distinguishes one from the other. The only point is that there are enough additional poles to cancel the original zeros without affecting the selection of poles for $y(s)/r(s)$.

The movement of a zero is therefore accomplished by creating a new zero at the desired location and canceling the unwanted zero by a closed-loop pole. Note that the pole-zero excess is never affected in this process since the pole-zero excess of the compensator $G_c(s)$ is always zero.

Since the pole locations of $G_c(s)$ are arbitrary, the previous discussion dealing with an increase of pole-zero excess is applicable here. However, a more definite guide can be given to the selection of the pole locations in the present situation than before. The suggested procedure is to place the poles of $G_c(s)$ at the locations of the zeros to be moved. Since these poles also appear in $y(s)/r(s)$, there must be zeros of $H_{eq}(s)$ at the same locations. As we have seen before, the effect of having a pole of $G(s)$ and a zero of $H_{eq}(s)$ at the same location is to force a pole of $y(s)/r(s)$ to exist at the same location for all values of gain. This result is particularly desirable in the present situation since it assures cancellation of the unwanted zero for all values of gain.

Although this is a suggested procedure for selecting the poles of $G_c(s)$, it is obviously not the only choice that may be made. In special situations it may well be desirable to make some other choice.

One word of caution is necessary with regard to the suggested procedure for selecting the poles of $G_c(s)$. On the surface it might appear that pole-zero cancellation is taking place in the open loop. This is not the interpretation to be given, as the cancellation actually takes place in the closed loop. One may be tempted to cancel the poles of $G_c(s)$ with the zeros of $G(s)$. This cancellation should be avoided lest closed-loop poles be missed. In addition, one must continue to use these poles and zeros in order to calculate the correct feedback coefficients.[1]

**Example 9.3-2** To illustrate the procedure of moving zeros, let us consider the simple second-order system shown in Fig. 9.3-12a. In this case, however, rather than remove the zero at $s = -2$, let us move it to $s = -3$. Since in the present situation we are free to select both the poles of $y(s)/r(s)$, let us choose a set of complex conjugate poles at $s = -1 \pm j1$ so that the desired $y(s)/r(s)$ becomes

$$\frac{y(s)}{r(s)} = \frac{\frac{2}{3}(s + 3)}{(s + 1)^2 + 1^2} \tag{9.3-11}$$

[1] This problem was discussed in detail in Chap. 7.

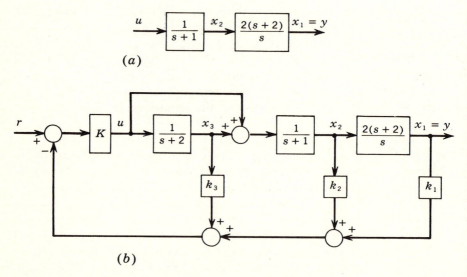

*(a)*

*(b)*

**Fig. 9.3-12**   Example 9.3-2.   (a) Plant; (b) compensated system.

Although this is the desired closed-loop transfer function, we know that $y(s)/r(s)$ must also contain the zero of $G_p(s)$ at $s = -2$ and that this zero is canceled by a pole of $y(s)/r(s)$ at the same location. The actual $y(s)/r(s)$ to be realized is therefore

$$\frac{y(s)}{r(s)} = \frac{\frac{2}{3}[(s+2)(s+3)]}{[(s+1)^2 + 1^2](s+2)} = \frac{\frac{2}{3}[(s+2)(s+3)]}{s^3 + 4s^2 + 6s + 4} \tag{9.3-12}$$

The series compensation to be used in this case is

$$G_c(s) = \frac{s+3}{s+2} = 1 + \frac{1}{s+2}$$

The complete compensated system is shown in Fig. 9.3-12b, where a feedforward realization of the required pole-zero compensator is indicated. This feedforward form for $G_c(s)$ has been chosen to maintain the describing differential equations in the $\dot{x} = \mathbf{A}x + \mathbf{b}u$ form, as discussed in Sec. 2.6. The use of the feedforward representation of $G_c(s)$ therefore causes the properties of $H_{eq}(s)$ and $y(s)/r(s)$ discussed in Sec. 9.2 to remain valid. The pole of $G_c(s)$ has been placed at the location of the unwanted original zero of $G_p(s)$ as suggested.

    The calculation of $H_{eq}(s)$ is not as simple as in previous cases, because of the summer between the $x_3$ state variable and the output. A simple block-diagram procedure for finding $H_{eq}(s)$ is shown in Fig. 9.3-13. First the connection to $x_3$ is shifted to the point $u$, as in Fig. 9.3-13a. The feedforward network is reduced to a simple pole-zero pair, as in Fig. 9.3-13b, and the feedback block is connected to the $x_3'$ position, as in Fig. 9.3-13c. Now $H_{eq}(s)$ may be found as before:

$$H_{eq}(s) = \frac{k_3(s+1)(s)}{2(s+3)(s+2)} + \frac{k_2 s}{2(s+2)} + k_1$$
$$= \frac{(k_3 + k_2 + 2k_1)s^2 + (k_3 + 3k_2 + 10k_1)s + 12k_1}{2(s+2)(s+3)} \tag{9.3-13}$$

and $KG(s)$ is just

$$KG(s) = \frac{2K(s+2)(s+3)}{s(s+1)(s+2)}$$

As always, the poles of $H_{eq}(s)$ are the zeros of $G(s)$ so that the open-loop transfer function is

$$KG(s)H_{eq}(s) = \frac{K[s^2(k_3 + k_2 + 2k_1) + s(k_3 + 3k_2 + 10k_1) + 12k_1]}{s(s+1)(s+2)}$$
$$\tag{9.3-14}$$

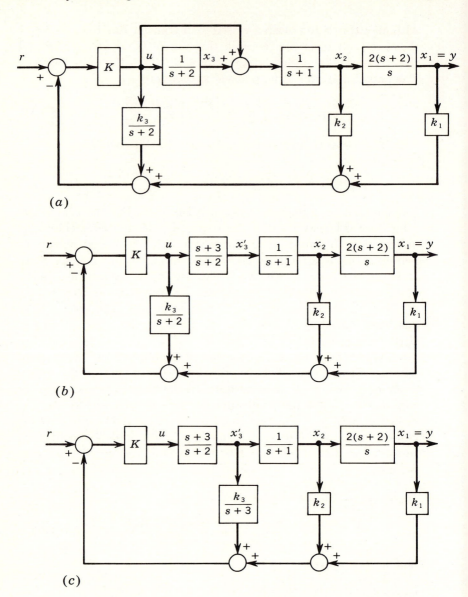

**Fig. 9.3-13**  Three steps in reducing Fig. 9.3-12*b* to the form from which $H_{eq}(s)$ may
be determined.

and $y(s)/r(s)$ in terms of $K$ and $\mathbf{k}$ is

$$\frac{y(s)}{r(s)} = \frac{2K(s + 2)(s + 3)}{s^3 + s^2(3 + Kk_3 + Kk_2 + 2Kk_1)}$$
$$+ s(2 + Kk_3 + 3Kk_2 + 10Kk_1) + 12Kk_1$$

$$(9.3\text{-}15)$$

If the two equations for $y(s)/r(s)$, Eqs. (9.3-12) and (9.3-15), are equated, four equations result:

$$\tfrac{2}{3} = 2K \qquad 3 + Kk_3 + Kk_2 + 2Kk_1 = 4$$
$$2 + Kk_3 + 3Kk_2 + 10Kk_1 = 6 \qquad 12Kk_1 = 4$$

These equations are readily solved for $K$ and $\mathbf{k}$ to yield

$$\mathbf{k} = \text{col } (1.0, 0.5, 0.5) \qquad \text{and} \qquad K = \tfrac{1}{3}$$

For these values $H_{eq}(s)$ becomes

$$H_{eq}(s) = \frac{3(s + 2)^2}{2(s + 2)(s + 3)}$$

and $KG(s)H_{eq}(s)$ is then

$$KG(s)H_{eq}(s) = \frac{3K(s + 2)^2}{s(s + 1)(s + 2)}$$

The corresponding root locus is shown in Fig. 9.3-14.   Note that,

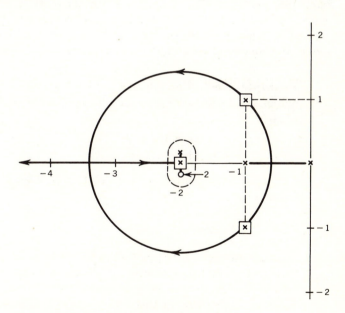

***Fig. 9.3-14***   Root locus for Example 9.3-2.

throughout the problem, no cancellation of common poles and zeros was made. Had this not been done, the location of a closed-loop pole at $s = -2$ might have been overlooked. Recall that the root locus is used only to find the system's closed-loop poles. The zeros of $y(s)/r(s)$ are already known to be the open-loop zeros. When the information regarding closed-loop poles and zeros is transferred to the closed-loop-response plane, no closed-loop pole should be overlooked. In this example the closed-loop pole at $s = -2$ might easily have been overlooked had cancellation of common poles and zeros in $G_p(s)$ or in $H_{eq}(s)$ taken place.

*Addition of zeros.* The procedure involved in adding zeros is closely allied to both increasing the pole-zero excess and shifting a zero. If a particular zero location is required in $y(s)/r(s)$, and this zero is not already present in $G_p(s)$, the zero must be added with the use of a compensating network of the form

$$G_c(s) = \frac{s + \delta}{s + \lambda} = 1 + \frac{\delta - \lambda}{s + \lambda}$$

This transfer function is the same as previously used in the discussion of shifting zeros. The value of $\delta$ is known in advance, since that is the particular zero position desired. However, there is no zero in $G_p(s)$ to cancel, and thus there is no simple procedure for choosing $\lambda$. As in the case of increasing pole-zero excess, both $\lambda$ and $k_{n+1}$ are free for adjustment, and one must be given an arbitrary value.

**Example 9.3-3**  For this example, let us consider the synthesis of one of the closed-loop transfer functions selected in Example 8.5-1, namely,

$$\frac{y(s)}{r(s)} = \frac{(3{,}000/7)(s + 7)}{(s^2 + 14.14s + 100)(s + 30)}$$

The plant to be controlled is given in Fig. 9.3-15a, and we see that $G_p(s)$ is

$$G_p(s) = \frac{10}{s(s + 5)}$$

In order to make this plant transfer function compatible with the desired $y(s)/r(s)$, it is necessary to add a series compensation consisting of a single pole-zero pair as shown in Fig. 9.3-15b. We are able to add the zero in this case because the pole-zero excess of the plant is not greater than the pole-zero excess of $y(s)/r(s)$.

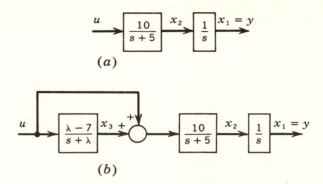

$$(a)$$

$$(b)$$

**Fig. 9.3-15**   Example 9.3-3.   (*a*) Original plant; (*b*) compensated plant.

It should be recalled that a basic performance limitation is that the pole-zero excess may not be decreased, which would require the addition of a free zero, that is, a zero without an associated pole, a physical impossibility.   If the original plant had been third-order with no zero so that the pole-zero excess was 3, it would be possible approximately to realize $y(s)/r(s)$ by adding a pole sufficiently far out on the negative real axis to be of little consequence to the response.

We are once again faced with the problem of selecting the value for the added pole, that is, $\lambda$.   By comparison with Example 8.6-1, the system is least sensitive if $\lambda = 0$.   However, for this value, the system has serious stability problems for low gain, since there would be two poles at the origin.   If $\lambda$ is chosen larger than 30, the frequency response of the denominator ratio is less than 1 for some values of $\omega$ below $\lambda$ and the system does not meet sensitivity conditions.   As a compromise, let us select $\lambda$ as 15 so that $G(s)$ becomes

$$G(s) = \frac{10(s + 7)}{s(s + 5)(s + 15)}$$

The $H_{eq}(s)$ for this system is

$$H_{eq}(s) = \frac{7k_1 + (k_1 + 7k_2 + 4k_3)s + (k_2 + 0.8k_3)s^2}{s + 7}$$

The closed-loop transfer function is therefore

$$\frac{y(s)}{r(s)} = \frac{10K(s + 7)}{s^3 + (20 + 10Kk_2 + 8Kk_3)s^2 \\ + (75 + 10Kk_1 + 70Kk_2 + 40Kk_3)s + 70Kk_1}$$

By equating coefficients with the desired $y(s)/r(s)$, we find that

$$K = 42.86 \quad \text{and} \quad \mathbf{k} = \text{col}\ (1.0, -0.117, 0.216)$$

For this value of $\mathbf{k}$ the open-loop transfer function becomes

$$KG(s)H_{eq}(s)$$
$$= \frac{K(0.563s^2 + 10.5s + 70)}{s^3 + 20s^2 + 75s} = \frac{0.563K[(s + 9.3)^2 + (6.2)^2]}{s(s + 5)(s + 15)}$$

The root locus for the system, shown in Fig. 9.3-16, indicates an acceptable solution.

In this section, the four phases of the general method have been treated as separate entities. Attention has been restricted to the simplest situations in each case. This simplification and separation were made in order that the basic concepts might be introduced without undue complication. In the next section these restrictions are removed and all four phases of the general method are combined in order to illustrate the complete flexibility of the general method.

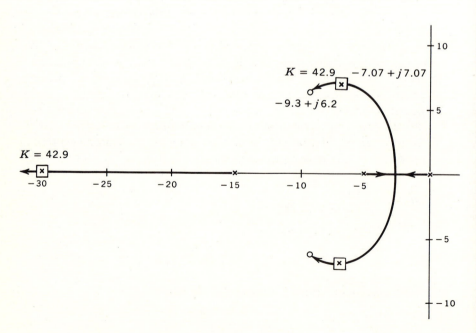

**Fig. 9.3-16**  Root locus for Example 9.3-3.

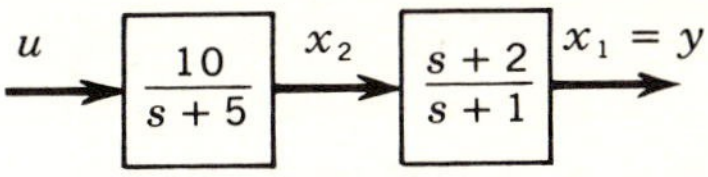

$u \longrightarrow \boxed{\dfrac{10}{s+5}} \xrightarrow{\,x_2\,} \boxed{\dfrac{s+2}{s+1}} \xrightarrow{\,x_1 = y\,}$

**Fig. 9.3-17**   Exercise 9.3-1.

**Exercises 9.3**  *9.3-1.*   Find the feedback coefficients **k**, the gain $K$, and the series compensation, if needed, such that the closed-loop transfer function for the plant shown in Fig. 9.3-17 is

(a)  $\dfrac{y(s)}{r(s)} = \dfrac{10}{s+10}$

(b)  $\dfrac{y(s)}{r(s)} = \dfrac{s+3}{(s+1)^2+2}$

Plot the root locus for the resulting system.

*answers:*

(a)  $K = 1$    $\mathbf{k} = \text{col } (0.9, -0.3)$

(b)  If the pole of the compensator is at $\lambda$, then $K = \frac{1}{10}$,
     $\mathbf{k} = \text{col } (1, -7.5, (45 - 10\lambda)/(3 - \lambda))$,
     and $G_c(s) = (s+3)/(s+\lambda)$.

*9.3-2.*   Find the series compensation, $K$, and **k** such that the plant

$$\dot{x}_1 = u \qquad y = x_1$$

has a closed-loop transfer function

$$\frac{y(s)}{r(s)} = \frac{2}{(s+1)^2 + 1^2}$$

Plot the root locus for the resulting system. If the compensator pole is at $s = -\lambda$, for what values of $\lambda$ is the closed-loop system stable for all positive values of $K$?

*answers:*

$G_c(s) = 1/(s+\lambda)$; $K = 2$; $\mathbf{k} = \text{col } (1, (2-\lambda)/2)$; system is stable for all $K > 0$, if $0 < \lambda < 2$.

## 9.4  *Applications*

In the two preceding sections the state-variable-feedback design method was presented. The discussion was initiated by the treatment of the simplest case where both $y(s)/r(s)$ and $G_p(s)$ were assumed to have no zeros, and the general case in which both $y(s)/r(s)$ and $G_p(s)$ are allowed to have zeros was discussed in Sec. 9.3. There the procedure for the general case was presented as four distinct concepts of (1) elimination of zeros, (2) increase of pole-zero excess, (3) movement of zeros, and (4) addition of zeros. In actual practice these four concepts are used together to obtain the solution for a given design problem. The purpose of this section is to illustrate, by means of a series of example problems, the interplay among the concepts presented in the preceding section.

Some of the examples treated here were chosen from problems discussed in Chap. 8. The intent of doing this is to illustrate further some of the implications of the design specifications developed in that chapter. The reader should not attach any particular physical significance to the problems discussed here. Emphasis has been placed on illustrating principles of the state-variable-feedback methods and the related specification problem.

> **Example 9.4-1** In this first example, we discuss some further features of and approaches to the problem of selecting the pole locations of the series compensation. Let us assume that the closed-loop transfer function to be achieved is given by

$$\frac{y(s)}{r(s)} = \frac{382.4(s + 5.23)}{(s^2 + 14.14s + 100)(s + 20)}$$

> and the plant transfer function is

$$G_p(s) = \frac{s + 5.23}{s(s + 10)}$$

> with the state-variable block diagram of Fig. 9.4-1a.

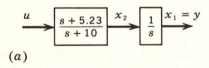

(a)

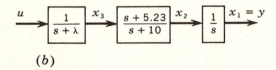

(b)

**Fig. 9.4-1**   Example 9.4-1.  (a) Original plant;
              (b) compensated plant.

A comparison of the desired $y(s)/r(s)$ and $G_p(s)$ reveals two problems. First, the desired closed-loop transfer function is third-order whereas the plant is only second-order; second, the pole-zero excess of $y(s)/r(s)$ is 2 and that of $G_p(s)$ is only 1. These two problems are related, and both may be solved by adding a series compensation consisting of one pole, as shown in Fig. 9.4-1b.

Let us suppose that $\lambda$ may be selected from three possible values: $\lambda = 1$, 20, and 100. This problem was considered in Sec. 8.6 on sensitivity as Example 8.6-1. There it was shown that from the standpoint of sensitivity $\lambda = 1$ is the best selection. Here we examine the design results, in particular the root locus, for each of the possible values of $\lambda$ and illustrate in a different manner the consequences of sensitivity reduction.

The equivalent feedback transfer function $H_{eq}(s)$ is independent of $\lambda$ and is given by

$$H_{eq}(s) = k_1 + k_2 s + k_3 \frac{s(s + 10)}{s + 5.23}$$

$$= \frac{5.23k_1 + s(k_1 + 5.23k_2 + 10k_3) + s^2(k_2 + k_3)}{s + 5.23}$$

Therefore the closed-loop transfer function $y(s)/r(s)$ becomes

$$\frac{y(s)}{r(s)} = \frac{KG(s)}{1 + KG(s)H_{eq}(s)}$$

$$= \frac{K(s + 5.23)}{s^3 + s^2(10 + \lambda + Kk_2 + Kk_3) + s(10\lambda + Kk_1 + 5.23Kk_2 + 10Kk_3) + 5.23Kk_1}$$

In order to achieve the desired $y(s)/r(s)$, the following four equations must be satisfied:

$$K = 382.4 \qquad 10 + \lambda + Kk_2 + Kk_3 = 34.14$$
$$10\lambda + Kk_1 + 5.23Kk_2 + 10Kk_3 = 382.8 \qquad 5.23Kk_1 = 2,000$$

For $\lambda = 1$, we find that

$$K = 382.4 \qquad \text{and} \qquad \mathbf{k} = \text{col}\ (1.0, 0.132, -0.0716)$$

If $\lambda = 20$, the results are

$$K = 382.4 \qquad \text{and} \qquad \mathbf{k} = \text{col}\ (1.0, 0.132, -0.121)$$

and for $\lambda = 100$, we find that

$$K = 382.4 \qquad \text{and} \qquad \mathbf{k} = \text{col}\ (1.0, 0.132, -0.330)$$

The root-locus diagrams for the resulting system for each of the three values of $\lambda$ are shown in Fig. 9.4-2.

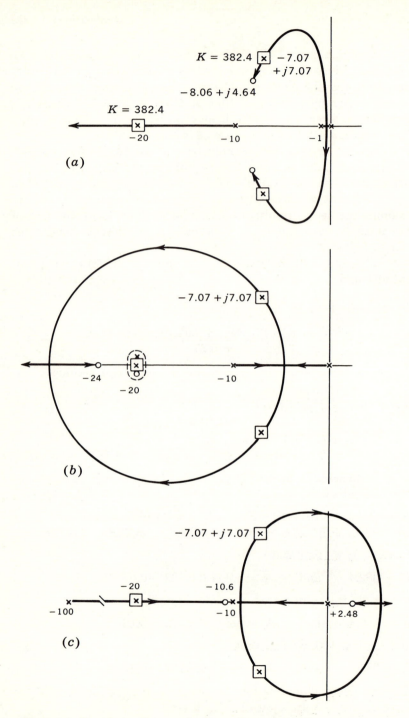

**Fig. 9.4-2** Root-locus diagrams for Example 9.4-1. (a) $\lambda = 1$; (b) $\lambda = 20$; (c) $\lambda = 100$.

486

These three root-locus diagrams are quite different and indicate that the minimum-sensitivity design associated with $\lambda = 1$ is probably the best solution. When $\lambda = 20$ the pole at $s = -20$ is completely insensitive to gain variations, although the other two poles tend to vary considerably from their design values. For $\lambda = 100$ one of the zeros of $H_{eq}(s)$ is in the right-half plane and the system becomes unstable for gains that are only slightly higher than the design value. In Example 8.6-1, it was shown that for $\lambda = 100$ the sensitivity criterion

$$|1 + KG_p(j\omega)H_{eq}(j\omega)| > 1$$

is violated for $2 < \omega < 100$ rad/sec. Hence the effort here confirms the undesirability of the $\lambda = 100$ solution.

The above example is quite similar to the original problem in Sec. 9.3 dealing with increasing the pole-zero excess. Here, however, we make use of sensitivity considerations to assist in the determination of $\lambda$. In the next example we consider a combination of two of the procedures discussed in Sec. 9.3, namely, zero removal and increase of pole-zero excess.

***Example 9.4-2*** The desired closed-loop transfer function for this example is given by

$$\frac{y(s)}{r(s)} = \frac{450}{(s + 25)[(s + 3)^2 + 3^2]}$$

with the plant given by the block diagram of Fig. 9.4-3 so that

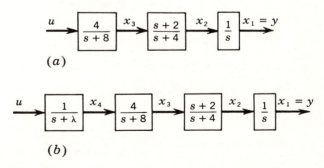

(a)

(b)

**Fig. 9.4-3**   Example 9.4-2.   (a) Original plant; (b) compensated plant.

$G_p(s)$ is

$$G_p(s) = \frac{4(s+2)}{s(s+4)(s+8)}$$

Here the desired $y(s)/r(s)$ is third-order with no zeros and hence a pole-zero excess of 3, whereas $G_p(s)$ is third-order with one zero and a pole-zero excess of only 2. The addition of a series compensation consisting of a single pole makes the plant compatible with $y(s)/r(s)$. In this case, we are combining the zero-removal and the pole-zero-excess-increase procedures.

Because of the rules of state-variable feedback, we know that the zeros of $G_p(s)$ appear as zeros in $y(s)/r(s)$. Therefore, in order to remove the undesired zero at $s = -2$, we must place a closed-loop pole at the same location so that the $y(s)/r(s)$ to be synthesized is actually

$$\frac{y(s)}{r(s)} = \frac{450(s+2)}{(s+2)(s+25)[(s+3)^2 + 3^2]}$$

The addition of the series compensation makes $G(s)$ become a fourth-order plant with a pole-zero excess of 3 so that it is compatible with the required $y(s)/r(s)$.

Once again we are faced with the problem of selecting the value of $\lambda$. It is not difficult to show that if $\lambda < 25$ the sensitivity criterion is met. For convenience, let us select $\lambda$ to be equal to 2.0. If a compensator pole is placed at the location of one of the closed-loop poles, a zero of $H_{eq}(s)$ must also lie at that point, and we may factor that term from $H_{eq}(s)$ and thereby be left with only a second-order polynomial to factor. In addition, $\lambda = 2$ provides a reasonable selection in terms of sensitivity and stability of the design. If we use $\lambda = 2$, then $G(s)$ becomes

$$G(s) = \frac{4(s+2)}{s(s+2)(s+4)(s+8)}$$

The resulting values of $K$ and $\mathbf{k}$ are

$$K = 112.5 \quad \text{and} \quad \mathbf{k} = \text{col}\ (1.0, 0.133, -0.253, 0.169)$$

and the open-loop transfer function is given by

$$KG(s)H_{eq}(s) = \frac{0.169K(s+2)[(s+3.58)^2 + (3.31)^2]}{s(s+2)(s+4)(s+8)}$$

The root locus for this system, shown in Fig. 9.4-4, indicates that the design is acceptable, and, in fact, the damping ratio associated

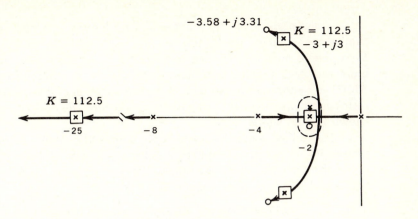

**Fig. 9.4-4**   Root locus for Example 9.4-3.

with the dominant pair of complex conjugate poles remains relatively constant for a large range of gain.

In the next example, we consider a case in which three of the procedures developed in Sec. 9.3 must be combined.

***Example 9.4-3***   Here let us consider the realization of the third-order closed-loop transfer function developed in Example 8.5-2.

$$\frac{y(s)}{r(s)} = \frac{(2.17 \times 10^4)(s + 36.9)}{[(s + 37.2)^2 + (37.2)^2](s + 290)}$$

The plant that we must control is represented in Fig. 9.4-5a with $G_p(s)$ given by

$$G_p(s) = \frac{2(s + 5)(s + 10)}{s(s + 50)(s + 200)}$$

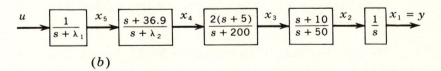

**Fig. 9.4-5**   Example 9.4-3.   (a) Uncompensated plant; (b) compensated plant.

A comparison of the plant and the desired $y(s)/r(s)$ reveals that three changes are necessary to make the two compatible. First, a zero must be removed; second, the pole-zero excess must be increased; and, third, a zero must be moved to the desired location. All three requirements can be met by the addition of a second-order series compensation consisting of two poles and the desired zero, as shown in Fig. 9.4-5$b$.

The added series compensation makes the plant effectively fifth-order with a pole-zero excess of 2 as required. The two extra poles are used to cancel the two unwanted zeros in $G_p(s)$ so that the actual $y(s)/r(s)$ to be implemented is

$$\frac{y(s)}{r(s)} = \frac{(2.17 \times 10^4)(s + 36.9)(s + 5)(s + 10)}{[(s + 37.2)^2 + (37.2)^2](s + 290)(s + 5)(s + 10)}$$

In order to simplify the computational labor, let us select $\lambda_1$ and $\lambda_2$ as 5 and 10, respectively. If this selection, corresponding to two of the poles of $y(s)/r(s)$, is made, these factors also appear in the numerator of $H_{eq}(s)$. Therefore we need only to factor a second-order polynomial to determine the zero of $H_{eq}(s)$. In addition, it is possible to show that these values for $\lambda_1$ and $\lambda_2$ meet the sensitivity criterion. The compensated-plant transfer function is therefore

$$G(s) = \frac{2(s + 5)(s + 10)(s + 36.9)}{s(s + 5)(s + 10)(s + 50)(s + 200)}$$

By completing the routine design steps of finding $y(s)/r(s)$ in terms of $K$ and **k** and equating coefficients, we find that

$$K = 10,850 \qquad \text{and} \qquad \mathbf{k} = \text{col } (1.0, 0.046, -0.047, -0.0025, 0.016)$$

The corresponding open-loop transfer function is given by

$$KG(s)H_{eq}(s)$$
$$= \frac{(9.68 \times 10^{-9})K(s + 5)(s + 10)[(s + 62.7)^2 + (55.6)^2]}{s(s + 5)(s + 10)(s + 50)(s + 200)}$$

The root locus for this system is shown in Fig. 9.4-6. As before, the pole-zero pairs that lie at one point in the open-loop transfer function do not affect the drawing of the root locus. Hence, even though this is a fifth-order problem, drawing the root-locus diagram is relatively simple.

The final example to be considered in this section illustrates the control of a second-order plant that has been modified by the addition of a series compensator before state-variable feedback. In the terminol-

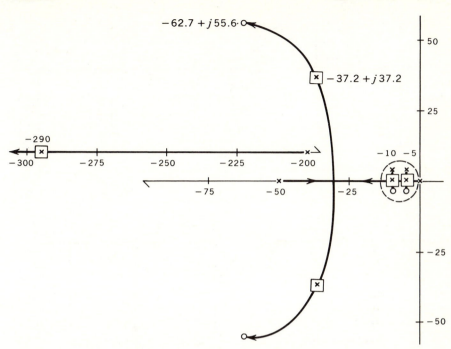

**Fig. 9.4-6**   Root locus for Example 9.4-3.

ogy of the preceding section, a zero has been added.   Of course, the compensating element $G_c(s)$ must contain a pole as well, and Example 9.4-4 illustrates two different methods by which the required pole-zero pair can be realized.

> **Example 9.4-4**   The plant to be controlled is given by the transfer function
>
> $$G_p(s) = \frac{2}{s(s + 5)}$$
>
> and two methods of system realization are indicated in Fig. 9.4-7. In each case the desired closed-loop transfer function is given by
>
> $$\frac{y(s)}{r(s)} = \frac{67.5(s + 4)}{[(s + 3)^2 + 3^2](s + 15)}$$
>
> The problem is to find the feedback coefficients **k** and the gain $K$ for both cases and then compare the two resulting systems.   At the outset observe that in case I the describing state-variable equations are of the form (**Ab**) and the state variables are labeled $x_1$, $x_2$, and

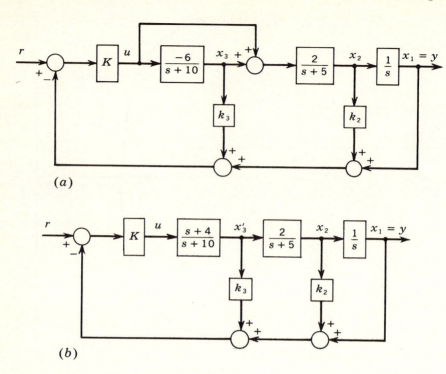

**Fig. 9.4-7**   Two realizations of the same closed-loop system: (a) case I; (b) case II.

$x_3$.   In case II the construction of the pole-zero pair as indicated produces a $\dot{u}$ term in the last state-variable equation.  Hence this last state variable is labeled $x_3'$, to distinguish it from the state variable $x_3$ above.

For case I,

$$KG_c(s)G_p(s) = KG(s) = \frac{2K(s+4)}{s(s+5)(s+10)}$$

$$H_{eq}(s) = \frac{(-3k_3+k_2)s^2 + (1-15k_3+4k_2)s + 4}{s+4}$$

and

$$\frac{y(s)}{r(s)} =$$

$$\frac{2K(s+4)}{s^3 + (15-6Kk_3+2Kk_2)s^2 + (50-30Kk_3+8Kk_2+2K)s + 8K}$$

After equating coefficients in powers of $s$ to like coefficients in the desired $y(s)/r(s)$, it is found that

$$K = 33.75 \quad \text{and} \quad \mathbf{k} = \text{col } (1.0, 0.585, 0.165)$$

and $H_{eq}(s)$ is

$$H_{eq}(s) = \frac{0.090[(s + 4.81)^2 + 4.62^2]}{s + 4}$$

Therefore the open-loop transfer function $KG(s)H_{eq}(s)$ is

$$KG(s)H_{eq}(s) = \frac{0.090K[(s + 4.81)^2 + 4.62^2]}{s(s + 5)(s + 10)}$$

Here note especially that the compensator zero at $s = -4$ does not appear in the open-loop transfer function. The root locus for case I is shown in Fig. 9.4-8, and the result does not appear to be very startling.

Case II, however, results in a rather unusual system. For case II

$$KG(s) = \frac{2K(s + 4)}{s(s + 5)(s + 10)}$$

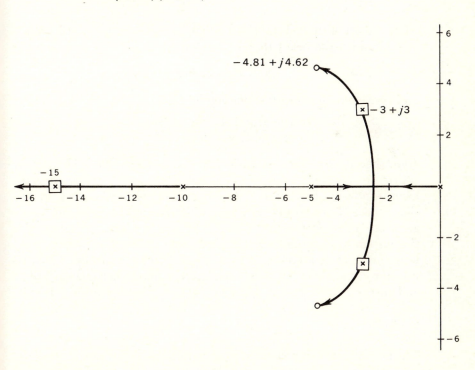

**Fig. 9.4-8**   Root locus for case I of Example 9.4-4.

and $H_{eq}(s)$ is simply

$$H_{eq}(s) = \frac{k_3 s^2 + (5k_3 + 2k_2)s + 2}{2}$$

Here the zero of $G(s)$ does not appear as a pole of $H_{eq}(s)$ since the describing equations are not of the form (**Ab**). The closed-loop transfer function is

$$\frac{y(s)}{r(s)} = \frac{2K(s + 4)}{(1 + Kk_3)s^3 + (15 + 2Kk_2 + 9Kk_3)s^2 + (50 + 8Kk_2 + 20Kk_3 + 2K)s + 8K}$$

The equations that must be solved to find the feedback coefficients are

$$\frac{2K}{1 + Kk_3} = 67.5$$

$$\frac{15 + 2Kk_2 + 9Kk_3}{1 + Kk_3} = 21$$

$$\frac{50 + 8Kk_2 + 20Kk_3 + 2K}{1 + Kk_3} = 108$$

After cross multiplication, these equations become linear in $K$, $Kk_2$, and $Kk_3$ with the result that

$$K = -7.36 \quad \text{and} \quad \mathbf{k} = \text{col } (1.0, 0.585, 0.165)$$

and the open-loop transfer function is

$$KG(s)H_{eq}(s) = \frac{-K(s + 4)(s + 1.1)(s + 11)}{s(s + 5)(s + 10)}$$

Here the negative sign requires that a zero-degree locus be drawn, rather than a 180° locus, as is usual. The resulting zero-degree locus is shown in Fig. 9.4-9. For low gain, the closed-loop configuration of case II is unstable, although of course, at the design value of gain, the input-output behavior of cases I and II are identical. If the system were actually built as indicated in case II, extreme care would have to be exercised when the system was turned on. If any warmup time were required, the system might become unstable and destroy itself before the gain reached the acceptable design value. Hence the system of case I is clearly preferred in this example.

The purpose of this section has been to illustrate some of the features of the state-variable-feedback procedures by the consideration of a

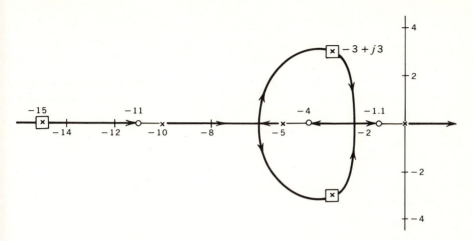

**Fig. 9.4-9**   Zero-degree root locus corresponding to case II of Example 9.4-4.

number of example problems. The reader will find that a digital computer can be used very effectively to assist in the design of high-order
systems. Such a program[1] was used for many of the calculations in this
section.

**Exercises 9.4**   *9.4-1.*   Repeat Example 9.4-1 with $\lambda = 10$; plot the root
locus of the resulting system.

*answer:*

$K = 382.4,$ $\mathbf{k} = \text{col} \ (1, 0.132, -0.095)$

*9.4-2.*   Find the feedback coefficients, $K$, and the series compensation for the plant shown in Fig. 9.4-10 such that the closed-loop

---

[1] J. L. Melsa, An Algorithm for the Analysis and Design of Linear State
Variable Feedback Systems, *Proc. Asilomar Conf., Nov.* 1–3, 1967, *Monterey,
Calif.,* pp. 791–799.
    J. L. Melsa, "Computer Programs for Computational Assistance in the
Study of Linear Control Theory," McGraw-Hill Book Company, New York,
1970.

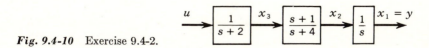

**Fig. 9.4-10**   Exercise 9.4-2.

transfer function is

$$\frac{y(s)}{r(s)} = \frac{5(s+2)}{[(s+1)^2+1](s+5)}$$

Select the pole of the compensator such that the closed-loop system is stable for all $K > 0$.  Show on the root locus of your final design that $\lambda = 1$ is not an acceptable choice.

*answer:*

$K = 5$, $\mathbf{k} = \mathrm{col}\ (1, \frac{1}{2}, -1, (3.5 - \lambda)/(10 - 5\lambda))$

*9.4-3.*  Find the series compensation, controller gain $K$, and $\mathbf{k}$ such that the plant shown in Fig. 9.4-11 has the following closed-loop transfer function:

$$\frac{y(s)}{r(s)} = \frac{4(s+2)}{(s+2)^2 + 2^2}$$

The series compensator should be selected so that the system is stable for all positive values of $K$.  Show the root locus of the resulting system.

*answer:*

If $\lambda = 1$

$K = 1.0 \qquad \mathbf{k} = \mathrm{col}\ (1, -0.5, 3.0)$

*9.4-4.*  Find the series compensation, $K$, and $\mathbf{k}$ for the plant

$$\dot{\mathbf{x}} = \begin{bmatrix} -1 & 1 \\ 0 & -2 \end{bmatrix} \mathbf{x} + \begin{bmatrix} 0 \\ 1 \end{bmatrix} u \qquad y = x_1$$

such that the closed-loop transfer function is

$$\frac{y(s)}{r(s)} = \frac{10(s+\frac{4}{5})}{[(s+1)^2+1](s+4)}$$

Plot the root locus for the system.

*answer:*

If $\lambda = 4$

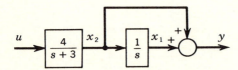

**Fig. 9.4-11**   Exercise 9.4-3.

$$K = 10 \qquad \mathbf{k} = \text{col}\,(-1.5, -1.17, -0.33)$$

*9.4-5.* Find the series compensation, $K$, and $\mathbf{k}$ for the plant

$$\dot{\mathbf{x}} = \begin{bmatrix} 0 & 1 \\ 0 & -4 \end{bmatrix} \mathbf{x} + \begin{bmatrix} 1 \\ 1 \end{bmatrix} u \qquad y = x_1$$

such that the closed-loop transfer function is

$$\frac{y(s)}{r(s)} = \frac{8}{(s+2)^2 + 2^2}$$

Select the compensator-pole location so that the closed-loop system is stable for all positive values of $K$. Plot the root locus for the system.

*answer:*

If $\lambda = 2$

$$K = 8 \qquad \mathbf{k} = \text{col}\,(1, 0, 0.375)$$

## 9.5   *Special procedures*

In the last three sections the general procedures of linear state-variable feedback have been developed and discussed. These concepts may be applied to the design of almost all linear control systems and form the framework on which special procedures may be developed. These special procedures may be used to meet unusual problems or to achieve unusual results. Three of these special procedures are discussed in this section: (1) high-gain design, (2) gain-insensitive design, and (3) feedforward design.

*High-gain design.* The high-gain-design procedure is based on the behavior of the poles of $y(s)/r(s)$ as the forward-path gain tends toward infinity. If the system is represented by Eqs. (**Ab**) and (**c**), $H_{eq}(s)$ has $n - 1$ zeros, and the only asymptote associated with the root locus of $1 + KG(s)H_{eq}(s)$ is at $-180°$. Therefore for large values of $K$ one of the poles of $y(s)/r(s)$ is located far out on the negative real axis, and the remaining closed-loop poles are located close to the zeros of $H_{eq}(s)$.

The location of poles of $y(s)/r(s)$ suggests an approximate way to realize the desired $y(s)/r(s)$. Let us place the $n - 1$ zeros of $H_{eq}(s)$, which are under the control of the designer, at $n - 1$ of the poles of the desired $y(s)/r(s)$. Now, as long as the gain remains high, $n - 1$ of the poles of the actual $y(s)/r(s)$ are close to the $n - 1$ poles of the desired $y(s)/r(s)$. One advantage of this high-gain design is that the exact value of the gain is not important, as long as it is large. Hence the system is relatively gain-insensitive.

Another advantage of the method is that the design procedure is considerably simplified. Since the zeros of $H_{eq}(s)$ are equated with the poles of the desired $y(s)/r(s)$, it is no longer necessary to find $y(s)/r(s)$ in terms of **k** and $K$; only $H_{eq}(s)$ need be formed. In addition, if $y(s)/r(s)$ is given in factored form, as it usually is, the numerator of $H_{eq}(s)$ is also factored, thereby simplifying the plotting of the root locus of $1 + KG(s)H_{eq}(s)$.

One disadvantage of the high-gain design is that the desired $y(s)/r(s)$ is only approximately realized. Since the poles of $y(s)/r(s)$ can be equal to zeros of $H_{eq}(s)$ only for infinite gain, which is a practical impossibility, there is always some error in the realization.

Another problem is that only $n - 1$ poles of $y(s)/r(s)$ may be positioned arbitrarily. The remaining pole is located on the negative real axis and is far out along that axis if the gain is high. Normally this is an acceptable location. In effect, this result simply means that the order of the desired $y(s)/r(s)$ must be reduced by one.

One important disadvantage of the high-gain design is the possibility of saturation. If the gain becomes too large, it is likely that the output of an amplifier or other input device may saturate for large step inputs. When this occurs, the assumption of linearity is violated, and the resulting system performance may not be as good as expected and may even be unacceptable.

**Example 9.5-1**   To illustrate this high-gain-design procedure, let us consider the system shown in Fig. 9.5-1. The desired closed-loop transfer function for this system is

$$\frac{y(s)}{r(s)} = \frac{2\alpha}{[(s + 1)^2 + 1](s + \alpha)}$$

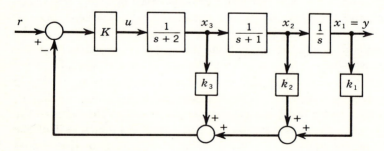

**Fig. 9.5-1**   Example 9.5-1.

where $\alpha \geq 10$. In selecting the desired $y(s)/r(s)$, we have used the knowledge that one of the poles is far out on the negative real axis for high gain. The zeros of $H_{eq}(s)$ are set equal to the set of complex conjugate poles of $y(s)/r(s)$ so that

$$N_h(s) = s^2 + 2s + 2 \qquad\qquad (9.5\text{-}1)$$

On the other hand, $H_{eq}(s)$ may be computed from Fig. 9.5-1 in terms of **k** as

$$H_{eq}(s) = \frac{s^2 + [(k_2 + k_3)/k_3]s + k_1/k_3}{1/k_3} \qquad (9.5\text{-}2)$$

If the two expressions for the numerator of $H_{eq}(s)$ are equated, we find that

$$\frac{k_2 + k_3}{k_3} = 2 \qquad \text{and} \qquad \frac{k_1}{k_3} = 2$$

or

$$k_2 = k_3 \qquad \text{and} \qquad k_1 = 2k_3 \qquad\qquad (9.5\text{-}3)$$

Notice that two of the feedback coefficents are specified in terms of the third. This is always the situation; $n - 1$ coefficients are specified in terms of one of the coefficients. In order to make this solution specific, it is necessary to introduce an additional condition into the problem. This condition is generated by the form selected for $y(s)/r(s)$ which requires zero steady-state position error, so that in this case $k_1 = 1$. The feedback coefficients are therefore

$$\mathbf{k} = \text{col } (1.0, 0.5, 0.5)$$

For this value of **k**, the loop-gain transfer function becomes

$$KG_p(s)H_{eq}(s) = \frac{0.5K(s^2 + 2s + 2)}{s(s + 1)(s + 2)}$$

The root locus for $KG_p(s)H_{eq}(s)$ is shown in Fig. 9.5-2. Gain values are marked along the root locus to indicate the degree of approximation for various values of gain. It is seen that, for $K > 5.0$, the complex conjugate poles are approximated rather closely, and for $K > 20$, the condition that $\alpha$ be greater than 10 is met.

*Gain-insensitive design.* In the preceding development of the high-gain-design procedure, we were able to achieve a degree of gain insensitivity of the closed-loop poles as long as the gain remained high. Here we wish to consider another approach to the sensitivity problem which we shall refer to as the gain-insensitive design. This procedure is partic-

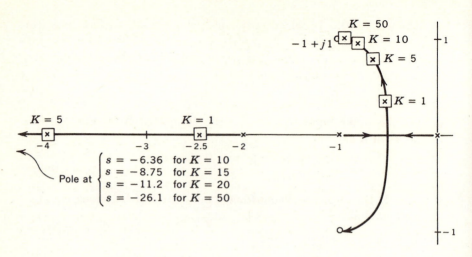

**Fig. 9.5-2**   Root locus for Example 9.5-1.

ularly useful in systems where wide or unknown variations in the forward-path gain $K$ may occur.

The gain-insensitive design is based on a fact that we have observed and commented on several times in the two preceding sections. If an open-loop pole is located at the position of a closed-loop pole, a zero of $H_{eq}(s)$ must also occur at that point, and the closed-loop pole is completely insensitive to gain variations. If all the open-loop poles could be positioned at the locations of the closed-loop poles, it would be possible to achieve complete gain insensitivity of the system. However, since there are only $n - 1$ zeros in $H_{eq}(s)$, only $n - 1$ of the poles of $y(s)/r(s)$ may be made gain-insensitive. Thus the situation is somewhat similar to that encountered in the high-gain design in that only $n - 1$ of the poles of $y(s)/r(s)$ may be positioned arbitrarily. The remaining pole of $y(s)/r(s)$ moves out along the negative real axis toward minus infinity.

Of course, the poles of the plant are not normally located at the poles of the desired $y(s)/r(s)$. Therefore it is first necessary to use series compensation, feedback, or a combination of the two to position $n - 1$ of the open-loop poles at the locations of $n - 1$ of the desired poles of $y(s)/r(s)$. If there is a pole at the origin, this pole is usually not moved, so that zero steady-state position error may be easily achieved. If there is no pole at the origin, any other pole of $G_p(s)$ may be selected. In positioning the open-loop pole especially if feedback is used, it is important that the gain $K$ not be involved; otherwise the gain-insensitive feature is not realized.

***Example 9.5-2***   Let us illustrate the use of the gain-insensitive-design procedure by applying it to the plant shown in Fig. 9.5-3a. Suppose that the desired closed-loop transfer function is

$$\frac{y(s)}{r(s)} = \frac{8}{(s+2)^2 + 2^2}$$

Since the desired $y(s)/r(s)$ is second-order, it is necessary to use series compensation to increase the order of the plant, since only $n-1$ of the poles may be positioned in the gain-insensitive method. In addition, it is necessary to modify the plant so that the open-loop poles lie at $s = -2 \pm j2$. This can most easily be accomplished by means of feedback. In effect, the problem of

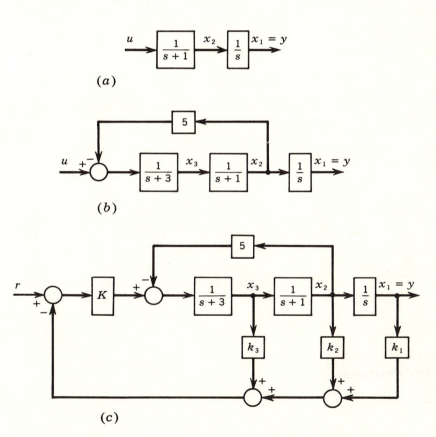

(a)

(b)

(c)

**Fig. 9.5-3**  Example 9.5-2.   (a) Plant to be controlled; (b) plant after modification to move the open-loop poles to the location of the desired closed-loop poles; (c) system configuration.

modifying $G_p(s)$ is just another state-variable-feedback problem. Figure 9.5-3*b* shows one of the infinite number of possible solutions. Observe that the feedback loop has been made without involving *K*.

The forward transfer function has now become

$$KG(s) = \frac{K}{s[(s+2)^2 + 2^2]}$$

The pole at the origin was not moved so that zero steady-state position error could be accomplished simply by setting $k_1 = 1$. The $H_{eq}(s)$ for the system shown in Fig. 9.5-3*c* is

$$H_{eq}(s) = k_1 + (k_2 + k_3)s + k_3 s^2$$

If the zeros of $H_{eq}(s)$ are now positioned at $s = -2 \pm j2$, we find that

$$k_3 = \frac{k_1}{8} \quad \text{and} \quad k_2 = \frac{3k_1}{8}$$

If we let $k_1 = 1$ so that zero steady-state position error is achieved, then **k** becomes

$$\mathbf{k} = \text{col } (1, \tfrac{3}{8}, \tfrac{1}{8})$$

With this value for **k**, the open-loop transfer function is

$$KG(s)H_{eq}(s) = \frac{(K/8)[(s+2)^2 + 2^2]}{s[(s+2)^2 + 2^2]}$$

Remember that the poles of $G(s)$ and the zeros of $H_{eq}(s)$ do not cancel but rather establish branches of the root locus at single points so that the root locus is as shown in Fig. 9.5-4.

The desired $y(s)/r(s)$ has been only approximately realized, since an extra pole exists on the negative real axis. As the gain *K* is increased, however, this pole moves to the left and its effect decreases rapidly. Once the gain becomes larger than, say, 50, the contribution of this pole is negligible and the system performs as though its poles were at $s = -2 \pm j2$ *independent of gain*.

One of the difficulties associated with the gain-insensitive design is the possible necessity of completing feedback loops inside the gain *K*, as was done in Example 9.5-2. This procedure poses a possible problem owing to mismatch of power levels. The output of the transducers used to measure the states may be only a few milliwatts, whereas the required input power of the plant, that of the output of the power amplifier, may

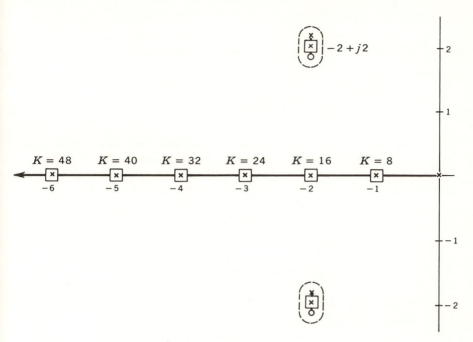

*Fig. 9.5-4*   Root locus for Example 9.5-2.

be of the order of watts or even hundreds of watts. Therefore, if feedback of the states is to be accomplished without involving $K$, some additional source of power amplification may be needed. When such additional power amplification is added, the system may be sensitive to gain of this new amplifier, even though it is insensitive to the original gain. Even in such situations, however, one may find that the gain of the new amplifier can be controlled more readily than that of the original amplifier and therefore the gain-insensitive design is still useful.

Although there are practical limitations on the usefulness of the gain-insensitive-design procedure, there are many cases in which it may be applied with at least partial success. In addition, the method is often useful in reducing the sensitivity of the closed-loop system to other plant parameters.

*Feedforward design.* In all the previous discussion in this and the preceding sections, attention has been directed at the problem of positioning the closed-loop poles, and the closed-loop zeros were manipulated only by the addition of a series compensator. The feedforward-design

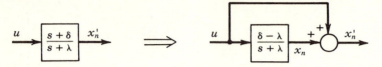

***Fig. 9.5-5***   Customary state-variable representation when the input
block contains a pole-zero pair.

procedure, on the other hand, allows one to move or even create zeros
without affecting the location of the closed-loop poles.

The feedforward process involves transmitting state-variable information to a point closer to the output, as contrasted with feedback, in
which information is transmitted to a point farther from the output,
usually to the input.   The feedforward procedure differs from feedback
in two ways.   Normally only one or two state variables are involved in
a feedforward procedure, and these variables are usually not fed forward
to the output but rather to some other state variable.

The concept of the feedforward-design procedure is one that is not
completely new to us, even though we have used the word "feedforward"
before only in passing.   Throughout the book, whenever the input block
of a given plant has contained a zero, we have assumed that the system
was represented so that the describing equations still had the form (**Ab**).
Specifically, transfer functions of the form $(s + \delta)/(s + \lambda)$ were represented as in Fig. 9.5-5, whether the state variable $x_n$ was a physical variable or not.   In those cases where the input block was not a part of the
plant but an additional $G_c(s)$, it was assumed that the input block was
actually built as in Fig. 9.5-5.   In fact, Example 9.4-4 was devoted to
describing the advantages of building $G_c(s)$ in this manner.

Here we wish to generalize the feedforward idea to include feedforward to an arbitrary place within the plant.   One or more of the state
variables and/or the input is fed forward through feedforward coefficients
and added at one or more points.   By appropriate selection of the feedforward coefficients, the zeros of the closed-loop transfer functions may
be placed as desired.   Because the number of possible feedforward configurations is extremely large even for a simple example, it is difficult to
prescribe any general rules for or properties of the feedforward procedures.
Rather than attempt this formidable task, let us consider a specific but
nontrivial illustration of the procedure.

***Example 9.5-3***   The plant to be controlled is represented in Fig.
9.5-6$a$.   It is assumed that it is possible to create a feedforward loop
from $x_3$ to $x_2$ so that the complete closed-loop system takes the form

shown in Fig. 9.5-6b. The existence of the feedforward loop allows us to create a closed-loop zero at any desired location. The desired closed-loop transfer function is therefore selected to be

$$\frac{y(s)}{r(s)} = \frac{24(s + \frac{4}{3})}{[(s + 2)^2 + 2^2](s + 4)} \tag{9.5-4}$$

Here the zero has been positioned so that the resulting system has zero steady-state error for velocity inputs.

The problem is then to select $\alpha$, $\mathbf{k}$, and $K$ such that the closed-loop transfer function of the system shown in Fig. 9.5-6b is given by Eq. (9.5-4). This problem can be solved by determining $y(s)/r(s)$ for the system shown in Fig. 9.5-6b and simply equating the two forms for $y(s)/r(s)$. The closed-loop transfer function may be found by using either block-diagram manipulations or matrix algebra. Because of the feedforward path, the matrix approach is probably the easier. The state-variable representation of the closed-loop system is

$$\dot{\mathbf{x}} = \begin{bmatrix} 0 & 1 & \alpha \\ 0 & -1 & 2 \\ -Kk_1 & -Kk_2 & (-2 - Kk_3) \end{bmatrix} \mathbf{x} + \begin{bmatrix} 0 \\ 0 \\ K \end{bmatrix} r = \mathbf{A}_k \mathbf{x} + K\mathbf{b}r$$

$$y = [1 \quad 0 \quad 0]\mathbf{x} = \mathbf{c}^T \mathbf{x}$$

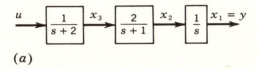

(a)

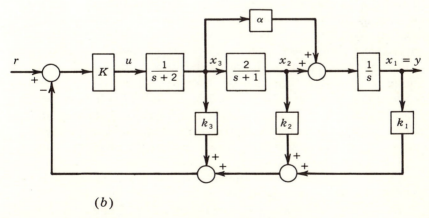

(b)

**Fig. 9.5-6** Example 9.5-3. (a) Plant; (b) complete system.

Making use of the expression

$$\frac{y(s)}{r(s)} = K\mathbf{c}^T(s\mathbf{I} - \mathbf{A}_k)^{-1}\mathbf{b}$$

we find that $y(s)/r(s)$ for this case is given by

$$\frac{y(s)}{r(s)}$$
$$= \frac{\alpha K[s + (\alpha + 2)/\alpha]}{s^3 + (3 + Kk_3)s^2 + (2 + Kk_3 + \alpha Kk_1 + 2Kk_2)s + (2 + \alpha)Kk_1} \tag{9.5-5}$$

The values for $\alpha$, $\mathbf{k}$, and $K$ may now be determined by equating the coefficients in Eqs. (9.5-4) and (9.5-5) and solving the resulting equations. If this is done, one finds that the answer is $\alpha = 6$, $\mathbf{k} = \text{col}\,(1.0, -0.875, 1.25)$, and $K = 4$. It is suggested that the reader verify this solution by the use of block-diagram manipulations.

In the above solution, it should be noted that $\alpha$ may be calculated independently of the closed-loop-pole locations or the values of $\mathbf{k}$ or $K$. On the other hand, even though $\alpha$ does not affect the poles of the open-loop system, the values for $\mathbf{k}$ and $K$ depend on $\alpha$.

Because the feedforward procedure makes it possible to move or create zeros and even to reduce the pole-zero excess, the reader may wonder why feedforward is not used in place of series compensation to achieve all desired zero configurations. A second and closely related question may also arise: How does one decide which feedforward loops to use? Both questions have basically the same answer.

The feedforward procedure cannot be used in all systems because there are often physical limitations that make its use impractical or even impossible. More specifically, it is not always possible to add quantities at any arbitrary point on the block diagram, so that feedforward loops cannot be created at these points. Suppose, for instance, that the plant considered in Example 9.5-3 were a field-controlled-motor positioning system so that $x_1 = y = $ output position. Then $x_2$ would represent output velocity and the feedforward loop shown in Fig. 9.5-6$b$ would require that we add velocities, which would physically present a problem. In general, from the input to the output in an automatic control system, the power level associated with each of the state variables increases. (This is the basic difference between control systems and communication systems, where power levels remain at "signal" levels.) Thus it is very often impossible to feed forward to an arbitrary point within the plant.

Because of these physical limitations on the feedforward procedure, the problem is not which feedforward loop to select but rather whether there are any that may be used. There are only a small number of systems that permit feedforward at all and an even smaller number that allow more than one possibility. If more than one possibility should exist, then all the possible feedforward paths should be investigated and the one that best meets the sensitivity or stability criterion selected.

Before ending this discussion of the feedforward procedure, let us consider the situation in which one may feed forward around a block but only by introducing a dynamical element into the feedforward path. Even in this situation, feedforward may be used to manipulate zeros, although the results are not quite as dramatic. Consider, for example, the simple first-order plant shown in Fig. 9.5-7a. Let us suppose that it is possible to use feedforward on the plant but only by introducing a dynamical element whose transfer function is $G_f(s) = 1/(s + 1)$. The feedforward situation is therefore as shown in Fig. 9.5-7b. By simple block-diagram algebra this system may be reduced to the block diagram shown in Fig. 9.5-7c. Once again we see that the zero may be placed at any position. However, in this case, an additional pole was introduced into the plant so that the pole-zero excess was unaltered.

In this section, three special techniques have been introduced. Although these techniques produce outstanding results for a large number of systems, they were also introduced to illustrate the generality and flexibility of the linear state-variable-feedback method. It is hoped that the results presented here encourage the reader to use his imagination in meeting the challenge of unusual problems that he may encounter.

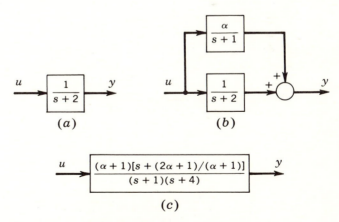

**Fig. 9.5-7**   Feedforward with a dynamical element. (a) Simple plant; (b) feedforward loop added; (c) resulting plant.

**Exercises 9.5**   *9.5-1.*   Use the high-gain-design procedure on the plant

$$\dot{x} = \begin{bmatrix} 0 & 1 & 0 \\ 0 & -2 & 1 \\ 0 & 0 & -4 \end{bmatrix} x + \begin{bmatrix} 0 \\ 0 \\ 2 \end{bmatrix} u \qquad y = x_1$$

to achieve approximately a closed-loop transfer function of

$$\frac{y(s)}{r(s)} = \frac{8}{(s+2)^2 + 2^2}$$

Plot the root locus of the system and determine the value of $K$ necessary to place the real pole at $s = -8$.

*answer:*

$\mathbf{k} = \mathrm{col}\ (1,\tfrac{1}{4},\tfrac{1}{8}),\ K = 19.2$

*9.5-2.*   Repeat Exercise 9.5-1 but use internal feedback as shown in Fig. 9.5-8 to move the open-loop poles at $s = -2$ and $-4$ to $s = -2 \pm j2$ and then apply the gain-insensitive procedure.   What is the minimum value of $K$ such that the real closed-loop pole is to the left of $s = -8$?

*answer:*

$\alpha = -1,\ \beta = 2,\ \mathbf{k} = \mathrm{col}\ (1,\tfrac{1}{4},\tfrac{1}{8}),\ K = 32$

*9.5-3.*   Compare the two designs found in Exercises 9.5-1 and 9.5-2 if the gain in the $x_3(s)/u(s)$ block varies from its nominal value of 2 to 10.   Draw a root locus of the closed-loop poles as a function of this gain for each system.

*9.5-4.*   The closed-loop system

$$\dot{x} = \begin{bmatrix} 0 & 1 \\ 0 & -1 \end{bmatrix} x + \begin{bmatrix} 0 \\ 1 \end{bmatrix} u \qquad y = x_1 \qquad u = 8(r - [1 \quad \tfrac{3}{8}]x)$$

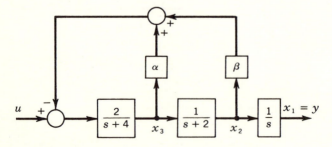

**Fig. 9.5-8**   Exercise 9.5-2.

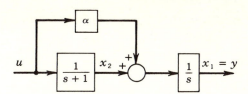

**Fig. 9.5-9**   Exercise 9.5-4.

has been designed to give a satisfactory transient behavior. However, the $K_v$ for this system is only 2. If the feedforward loop shown in Fig. 9.5-9 can be used in this system, find the value of the feedforward gain $\alpha$ such that the velocity-error constant becomes $\infty$. Find the values of $K$ and $\mathbf{k}$ such that the closed-loop poles remain at $s = -2 \pm j2$. Plot the step response of the original and the feedforward systems.

*answers:*

$\alpha = 1$, $K = 4$, $\mathbf{k} = \text{col} (1, -1.4)$

## 9.6   *Conclusions*

In concluding this chapter it is important to emphasize once again the two assumptions made originally in Sec. 9.1 and adhered to throughout the chapter. These assumptions are that the desired closed-loop transfer function is specified and that all the state variables are available for measurement and control. Surprisingly, on the basis of these assumptions, the problem of synthesis becomes a rather trivial one, in fact, not much more than algebra. Even in the case where series compensation must be used along with state-variable feedback, the design procedure is completely mechanical if sensitivity considerations are employed in locating the pole or poles of the necessary compensation network.

In a sense, the ease with which synthesis is accomplished might be something of a disappointment. After struggling through the previous eight chapters of this book, the reader may have been inclined to feel that synthesis was the climax of the control engineer's art. And now it has been demonstrated in this chapter that synthesis can be somewhat anticlimactic and seem almost too simple. Much of this simplicity results from the assumption of the availability of all the state variables. If all the state variables are not available, it may not be possible to realize any desired closed-loop transfer function, and it then becomes necessary to specify the response of the closed-loop system in a manner other than by selecting the closed-loop transfer function.    Chapter 10 deals with this situation.

## 9.7  Problems

*9.7-1.*    Design a linear state-variable-feedback controller for the plant

$$\dot{\mathbf{x}} = \begin{bmatrix} -1 & 1 & 0 \\ 0 & -4 & 2 \\ 0 & 0 & -10 \end{bmatrix} \mathbf{x} + \begin{bmatrix} 0 \\ 0 \\ 1 \end{bmatrix} u \qquad y = x_1$$

such that the closed-loop transfer function is

$$\frac{y(s)}{r(s)} = \frac{180}{[(s+3)^2 + 3^2](s+10)}$$

Plot the root locus of the resulting system.

*9.7-2.*    The closed-loop transfer function of the system shown in Fig. 9.7-1 is

$$\frac{y(s)}{r(s)} = \frac{1.67(s+2)}{s^3 + 4.13s^2 + 5.67s + 3.33}$$

Note that this is not described by Eqs. (**Ab**) and (**c**) since a $\dot{u}$ term appears

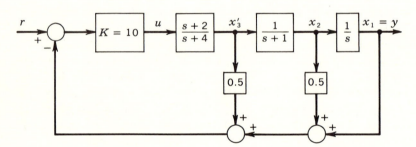

*Fig. 9.7-1*    Problem 9.7-2.

in the $\dot{x}_3$ equation.    If the feedforward scheme is used to represent the pole-zero pair, the plant is described by

$$\dot{\mathbf{x}} = \begin{bmatrix} C & 1 & 0 \\ 0 & -1 & 1 \\ 0 & 0 & -4 \end{bmatrix} \mathbf{x} + \begin{bmatrix} 0 \\ 1 \\ -2 \end{bmatrix} u \qquad y = x_1$$

Show that the closed-loop transfer function given above can be realized by using the same feedback coefficients as shown in Fig. 9.7-1 but with a different value of $K$.   Find the new value of $K$ and plot the root-locus diagram of each system.

**9.7-3.**   Use the gain-insensitive-design technique on the plant

$$\dot{x} = \begin{bmatrix} 0 & 1 & 0 \\ 0 & -2 & 2 \\ 0 & 1 & -4 \end{bmatrix} x + \begin{bmatrix} 0 \\ 0 \\ 1 \end{bmatrix} u \qquad y = [1 \quad 1 \quad 0]x$$

to realize approximately the closed-loop transfer function

$$\frac{y(s)}{r(s)} = \frac{8(s+1)}{(s+2)^2 + (2)^2}$$

Plot the step response of the system for $K = 5, 10, 20, 100$.

**9.7-4.**   Use the technique of Sec. 8.8 to find a closed-loop transfer function that approximates the second-order model

$$\left[ \frac{y(s)}{r(s)} \right]_m = \frac{4(s+2)}{(s+2)^2 + 2^2}$$

if the plant transfer function is

$$G_p(s) = \frac{s+1}{s(s+2)(s+4)}$$

Since the zeros of the model and the plant are not the same, it is necessary to add a series compensation to the plant of the form $G_c(s) = (s+2)/(s+\lambda)$.   If the value of $K$ is selected as 10, find the desired closed-loop transfer function of several different values of $\lambda$ and find the step and frequency responses for each.   Discuss the answers.

**9.7-5.**   Find the series compensation, $K$, and $\mathbf{k}$ for the plant

$$\dot{x} = \begin{bmatrix} -1 & 1 \\ 0 & -4 \end{bmatrix} x + \begin{bmatrix} 0 \\ 1 \end{bmatrix} u \qquad y = x_1$$

such that the closed-loop transfer function is

$$\frac{y(s)}{r(s)} = \frac{10(s+2)}{[(s+1)^2 + 1](s+5)}$$

Select the location of the compensator pole so that the sensitivities of the closed-loop transfer function with respect to $\lambda$ and to $k_3$ are equal.

**9.7-6.**   Use series compensation and the high-gain approach on the plant

$$\dot{x} = \begin{bmatrix} 0 & 1 \\ 0 & -2 \end{bmatrix} x + \begin{bmatrix} 0 \\ 2 \end{bmatrix} u \qquad y = x_1$$

in order to realize approximately the closed-loop transfer function

$$\frac{y(s)}{r(s)} = \frac{2(s+1)}{(s+1)^2 + 1}$$

Consider the use of both the pole-zero pair and feedforward form for the compensation.    Discuss the results.

**9.7-7.**    Find the series compensation, $K$, and **k** for the plant

$$\dot{\mathbf{x}} = \begin{bmatrix} -2 & 1 \\ 0 & -4 \end{bmatrix} \mathbf{x} + \begin{bmatrix} 0 \\ 2 \end{bmatrix} u \qquad y = x_1 + x_2$$

such that the closed-loop transfer function is

$$\frac{y(s)}{r(s)} = \frac{4(s+1)}{[(s+2)^2 + 2^2](s+2)}$$

Select the compensation so that the system is stable for all positive values of $K$.    Plot the root locus of the system.

**9.7-8.**    The unstable plant

$$\dot{\mathbf{x}} = \begin{bmatrix} 0 & 1 & 0 \\ 0 & 4 & 1 \\ 0 & 0 & -1 \end{bmatrix} \mathbf{x} + \begin{bmatrix} 0 \\ 0 \\ 1 \end{bmatrix} u \qquad y = x_1$$

is to be controlled so that the closed-loop transfer function is

$$\frac{y(s)}{r(s)} = \frac{5(s+2)}{[(s+1)^2 + 1](s+5)}$$

Plot the root locus of the system.    Use the Routh-Hurwitz criterion to find the minimum value of $K$ for stable performance.

**9.7-9.**    Find $K$ and **k** for the plant

$$\dot{\mathbf{x}} = \begin{bmatrix} 0 & 1 & 0 & 1 \\ 0 & 0 & 1 & 0 \\ -1 & -2 & -2 & 1 \\ 0 & 0 & -1 & -1 \end{bmatrix} \mathbf{x} + \begin{bmatrix} 0 \\ 0 \\ 0 \\ 1 \end{bmatrix} u \qquad y = x_1$$

such that the closed-loop transfer function is

$$\frac{y(s)}{r(s)} = \frac{8}{(s+2)^2 + 2^2}$$

# ten    *system design with series compensation*

## 10.1  Introduction

The design procedures of the preceding chapter are appealing in the sense that, once $y(s)/r(s)$ has been specified, the problem solution is usually unique, and all one must do to realize the solution is to turn the algebraic crank.   However, it is important to realize that the synthesis methods of Chap. 9 are based directly upon the following two assumptions:

1.  The desired closed-loop transfer function is known in terms of a specific transfer function.
2.  All the state variables are available for measurement and may be used to realize the desired control.

These assumptions are often violated.  It may be too difficult or costly to measure all the state variables, or their measurement may be so noisy that the signal is not usable.  In such cases one may not be able to realize a specific closed-loop transfer function.

In addition, the system specifications are often not given in terms of a specific $y(s)/r(s)$.  Rather, a variety of time- and frequency-domain specifications, discussed in Chap. 8, may be used separately or together to describe the desired closed-loop response.  The use of a variety of time- and frequency-domain specifications leads to a variety of consequences.  At times the given specifications are impossible to realize; that is, one specification conflicts with another.  In such cases the closed-loop system is said to be overspecified.  In other cases the system response is underspecified; the specifications are so loose that a variety of closed-loop systems are capable of realizing the desired response properties.  Here the problem solution is by no means unique, and one may even have difficulty in deciding which of a number of possible solutions is the best one.

Nonuniqueness of the problem solution was alluded to at the beginning of Chap. 9.  There, however, once $y(s)/r(s)$ was chosen, almost all ambiguity ended.  In this chapter, except for the section immediately following this introduction, it is not assumed that $y(s)/r(s)$ is ever given.  Hence the system specification is somewhat more vague, and the problem of nonuniqueness is much more pronounced.  As an extreme example, consider the control system of an experimental aircraft.  The pilot may return from a test flight and say that the plane does not "feel" right.  The responsible engineer must then determine in what way the plane's response does not feel right and transform this information into an engineering performance specification.  The resulting system description may be based upon the engineer's judgment and experience, and clearly the solution is not one unique closed-loop transfer function.  The solution exists only when the plane "feels" right.

In Sec. 10.2 the assumed system specification is still in terms of a unique $y(s)/r(s)$, but one or more of the state variables are not available.  In particular, first one and then several of the state variables are considered not available for measurement and control.  Finally, the output alone is assumed to be the only usable variable with which to effect control.  In these cases the feedback coefficients from the available state variables are no longer constants but are themselves transfer functions.  With only the output available for control, the feedback transfer function may be shifted into the forward loop, which results in a rather complicated series compensator in the general case.

Although it is impossible to realize a specified $y(s)/r(s)$, it is often possible to achieve satisfactory results with less complicated series equalizers.    For example, networks with one pole and one zero are sufficient in some cases.    The use of such simple lead or lag networks is considered in Secs. 10.3 to 10.5, and the more complicated lead-lag network used in series compensation is considered in Sec. 10.6.    In these sections the design philosophy is based upon the frequency response of the open-loop system, rather than selection of the closed-loop transfer function.    The frequency response of the open loop is shaped in such a way as to ensure a satisfactory closed-loop response.

## 10.2   *Inaccessible state variables*

In this section we relax the requirement that all the state variables be available for measurement and control.    Certainly the ability to measure and use all the state variables for control is an idealized situation, and often in practice the idealized textbook situation is not the case.    As noted in the introduction to this chapter, it may often be too difficult or too costly to measure each of the state variables.    In such situations the natural compromise is to measure the state variables that can be measured and use them for control.    It is demonstrated in this section that in this case the feedback of the available state variables is not through constant, frequency-insensitive elements, as in the previous discussion. Rather, feedback is through a transfer function, which is more commonly known as minor-loop feedback.    In the case when all the state variables except the output are unavailable, the resulting minor-loop equalizer may be shifted into the forward path and hence is replaced by a series equalizer. In any case, it is still possible to realize the required closed-loop transfer function; hence throughout this section a specified closed-loop transfer function is still assumed to be the system specification.

Although the describing equations for the plant are still Eqs. (**Ab**) and (**c**), it is convenient to use the block-diagram representation of the system.    This is the case because ultimately $u$ is not $K(r - \mathbf{k}^T\mathbf{x})$ but involves transfer functions.    In the block-diagram representation used throughout this section it is implied that all the state variables are physical variables, since we are interested in using these quantities to effect control.

The design procedure is extremely simple; in fact, the first stages of the design procedure are identical to those of Chap. 9.    Assume all the state variables are available and determine the necessary feedback coefficients to realize the given $y(s)/r(s)$, as in Chap. 9.    The only alteration occurs in the realization of the control input $u(t)$.    Since some of the state variables are not available, they must be created by the feedback structure or controller employed.    A simple example illustrates two means by which the unavailable state variables may be artificially created.

Since the design procedures here are so dependent upon those of Chap. 9, let us consider as an example the system discussed in Sec. 9.2. There the plant had the transfer function

$$G_p(s) = \frac{y(s)}{u(s)} = \frac{2}{s(s+1)(s+3)} \tag{10.2-1}$$

and was described by its physical variables as shown in Fig. 9.2-1a, repeated here for convenience as Fig. 10.2-1a.    The desired $y(s)/r(s)$ was

$$\frac{y(s)}{r(s)} = \frac{20}{[(s+1)^2 + 3](s+5)} \tag{10.2-2}$$

The final system configuration is given as Fig. 10.2-1b.

Let us now assume that the state variable $x_2$ is not available.    If we think in terms of either a field-controlled or an armature-controlled dc motor, this is not an unwarranted assumption.    In this case the output velocity is measured not directly but by means of a tachometer on the output shaft.    Thus the actual realization is as indicated in Fig. 10.2-2a.    In terms of the block diagram of Fig. 10.2-1b, Fig. 10.2-2a is realized by generating $x_2$ from $x_1$ by differentiation.    Figure 10.2-2b is a

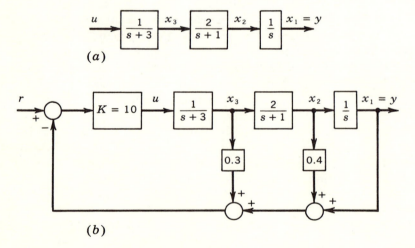

(a)

(b)

**Fig. 10.2-1**    Example problem of Sec. 9.2.    (a) Plant to be controlled; (b) final closed-loop system.

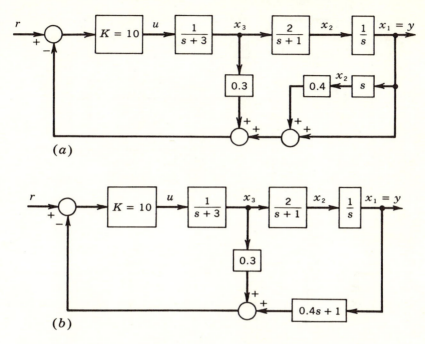

**Fig. 10.2-2** Methods of viewing the generation of the state variable $x_2$. (a) Generation of $x_2$ by differentiation of the output; (b) an equivalent form.

simple redrawing of Fig. 10.2-2a where the generated state variable $x_2$ is no longer used.

If both $x_2$ and $x_3$ had been unavailable, both these variables could be generated from $x_1$, as indicated in Fig. 10.2-3a. Of course, if this figure is redrawn by combining the parallel feedback loops from $x_1$ to the summing junction as shown in Fig. 10.2-3b, the feedback transfer function is nothing more than $H_{eq}(s)$. Note that two differentiations are required to generate the required feedback signal. As mentioned before, the operation of differentiation accentuates noise that may be present. This is the reason why analog computers use integrators rather than differentiators. Thus the realization of the multiple zeros in $H_{eq}(s)$ is not practical. If poles are added to $H_{eq}(s)$ to make it more realistic, the desired closed-loop transfer function cannot be achieved, since the additional poles in $H_{eq}(s)$ increase the order of the system. For these reasons it is usually impractical to attempt to generate missing state variables by the means indicated in the above discussion. This is true

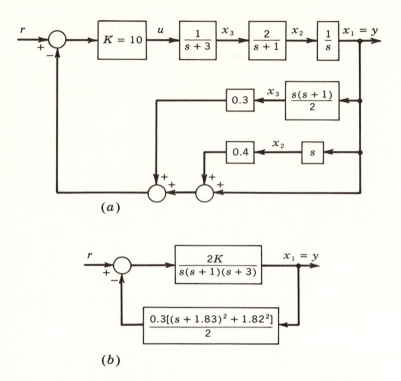

(a)

(b)

**Fig. 10.2-3**   Both state variables $x_2$ and $x_3$ are unavailable.   (a) Both $x_2$ and $x_3$ are generated from the output; (b) $H_{eq}(s)$ form.

not just for the example considered here but in general.   The problem is inherent in generating missing state variables by differentiation.

Let us pursue another attack by returning to the original structure of Fig. 10.2-1b.   As before, let us assume that the state variable $x_2$ is unavailable, only, rather than generate this variable from the output, let us generate it from $x_3$.   This is rather easily done by passing the measured signal at $x_3$ through a block $2/(s+1)$, as indicated in Fig. 10.2-4a.   In this figure the generated state $x_2$ is clearly indicated, and, as before, this is passed through the required gain of 0.4.   The two parallel paths that lead from $x_3$ to the summing junction are added to give the equivalent circuit of Fig. 10.2-4b, where the generated state variable $x_2$ is no longer apparent.   Note that the feedback transfer function of Fig. 10.2-4b is just a simple pole-zero network which is fairly easy to build with passive elements.   Thus the solution pictured in Fig. 10.2-4b is satisfactory.

If both $x_2$ and $x_3$ are unavailable, they may be generated in a manner similar to that used above.   Such a procedure is shown in Fig. 10.2-5a, where both $x_2$ and $x_3$ have been generated from the output of $K$.    Figure 10.2-5b combines the feedback elements into one transfer function, this time with two poles and one zero.    Again this transfer function may be easily realized, and with the use of this minor-loop equalizer, the desired $y(s)/r(s)$ has been achieved.

If this line of reasoning is pursued a bit further, it is apparent that all the state variables could be generated by building a model of the plant and subjecting this model to the output of the controller amplifier $K$. This approach could be used regardless of the order of the system, and none of the actual state variables would have to be fed back.    In reality, of course, the output would have to be fed back, unless the model were a *perfect* representation of the plant, and this is never possible.    Any change in the physical plant due to variations in temperature or aging, for example, would not be reflected by changes in the model.    Hence the system would, in effect, be running open loop, even though all the model variables

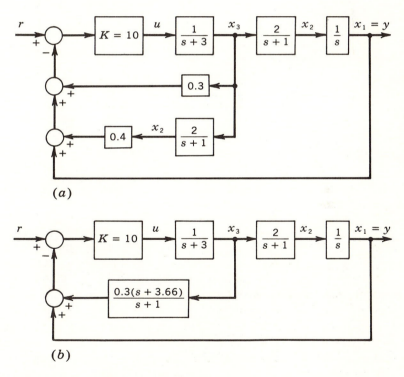

(a)

(b)

**Fig. 10.2-4**   Utilization of minor-loop feedback with unavailable state variables.    (a) State $x_2$ is generated from $x_3$; (b) equivalent transfer-function feedback.

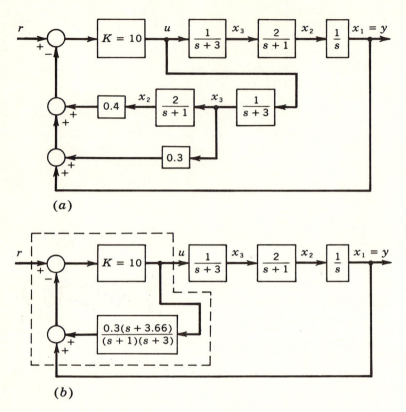

*(a)*

*(b)*

**Fig. 10.2-5**   Utilization of minor-loop feedback when two state variables
are unavailable.   (a) Both $x_2$ and $x_3$ are generated from the
output of $K$; (b) equivalent minor-loop-feedback form.

were fed back.   For this reason, it is always assumed that the output
state variable or variables are available, although the remaining variables
may be generated from a system model, much as in Fig. 10.2-5a or b.

Before we proceed, it is important to stress the identical nature of
the closed-loop transfer functions of Figs. 10.2-1b to 10.2-5b.   If viewed
only from the input-output relations, these five systems appear to be
identical; that is, their response to any input is the same from one system
to the next.   If none of the system parameters in any of these systems
changes, then, from an input-output point of view, these systems remain
identical.   If one or more system parameters should change, however,
the systems are not identical.   Perhaps this is most easily seen from Fig.
10.2-4a.   There the variable $x_2$ is generated from $x_3$ by passing the signal
at $x_3$ through the transfer function $2/(s + 1)$.   Suppose that the plant

pole at $s = -1$ was actually at some other place, say at $s = -a$. Then the value of $x_2$ that is generated is in error, since the pole is at $-a$ rather than at $-1$. Hence the control signal $u$ is not the correct one, and the resulting performance is different from that desired.

On the other hand, if the control signal is fed back from the actual $x_2$ state location, as in Fig. 10.2-1b, for instance, any change in the pole location at $s = -1$ is indeed reflected in the measured value of $x_2$. The resulting performance is more closely related to the desired one, although, of course, still not ideal.

Thus far the unavailability of one or more state variables has resulted in the utilization of minor-loop feedback. The use of minor-loop feedback had the effect of keeping the control signal $u$ the same for each of the systems of Figs. 10.2-1b to 10.2-5b. Series equalization can also be considered, and the rest of this section is devoted to that approach. The use of series compensation no longer preserves the form of the input to the plant.

As we did earlier in this section, let us begin by considering the control of the plant of Fig. 10.2-1a, which is ideally controlled as in Fig. 10.2-1b. As before, we assume initially that the state variable $x_2$ is inaccessible. However, in order to indicate the method of series compensation to be used, let us redraw Fig. 10.2-4a as Fig. 10.2-6a. The dotted portion of this figure may be reduced by the usual block-diagram methods to yield

$$\frac{x_3(s)}{u'(s)} = \frac{10(s + 1)}{s^2 + 4s + 11} \tag{10.2-3}$$

This must be the transfer function included within the dotted area if the specified closed-loop response is to be realized. Of course the plant pole at $s = -3$ cannot be altered; this is always assumed to be a fixed property of the plant. However, if a series compensator equal to

$$KG_c(s) = \frac{10(s + 1)(s + 3)}{s^2 + 4s + 11} \tag{10.2-4}$$

is inserted prior to the plant, the zero of this compensator cancels the pole at $s = -3$, and the transfer function included within the dotted region of Fig. 10.2-6b is that required in Eq. (10.2-3). It is immaterial whether or not the controller gain $K$ is included with the series compensator. Notice that the reference input as well as the feedback signal from both $x_3$ and $x_1$ now pass through the series equalizer in order to create the correct control input $u(t)$. This is the outstanding difference between the series-compensation form and the five previous realizations of the same closed-loop transfer function.

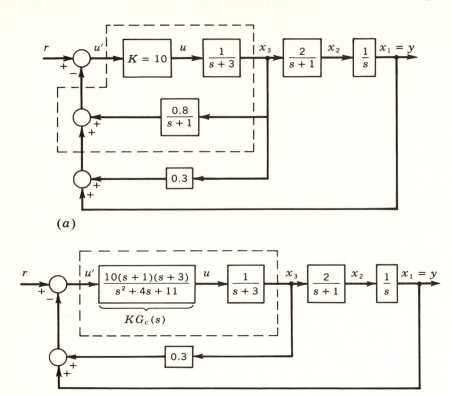

**Fig. 10.2-6** Use of series compensation when the state variable $x_2$ is unavailable. (a) Redrawing of Fig. 10.2-4a; (b) an equivalent system configuration.

Note that the addition of the series compensator has caused the system to become fifth-order. However, since two zeros have also been introduced, the effective order of the system from input to output remains three. Nevertheless, there are two states associated with the controller, and one should not forget this fact. Consider, for example, the two frequency-insensitive systems of Fig. 10.2-7. The transfer function from

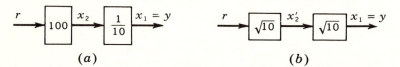

**Fig. 10.2-7** Two examples of realizing the same transfer function in different ways: (a) system $a$; (b) system $b$.

input to output of each is identical.    If the two blocks in each system act in a linear fashion over the range of inputs expected, the two systems perform equally well, and it is not important that in system *a* the state variable $x_2$ is more than 30 times that of system *b*.    On the other hand, it is likely that the high-gain unit in the first block of system *a* will cause saturation before the lower-gain unit of system *b* will do so; thus if a large range of inputs is expected, the behavior of the two might be substantially different.    It is important to examine the behavior of all the state variables, not just the output variable.

Had both the state variables $x_2$ and $x_3$ been inaccessible, the feedback path from $x_3$ in Fig. 10.2-6*b* would not be present, and it would become necessary to lump this quantity into the dotted portion of that figure. If this is done, the resulting series compensator is

$$KG_c(s) = \frac{10(s+1)(s+3)}{s^2 + 7s + 14}$$

Utilization of this series compensation is indicated in Fig. 10.2-8.

The series compensator of Fig. 10.2-8 could have been obtained by the use of the $G_{eq}(s)$ realization of the ideal control system of Fig. 10.2-1*b*. Through simple block-diagram manipulations, $G_{eq}(s)$ is found from that figure to be

$$G_{eq}(s) = \frac{20}{s(s^2 + 7s + 14)}$$

The plant transfer function given in Eq. (10.2-3) is known, and with

$$G_{eq}(s) = KG_c(s)G_p(s)$$

the required compensator $KG_c(s)$ is easily found.    Again the gain has not been separated from the compensator itself, since in all likelihood they are built as one unit.

It is interesting to examine Fig. 10.2-8 more closely.    It is seen that the zeros of the series compensator correspond to the two poles of the plant; that is, these zeros cancel the plant poles.    The plant poles are then replaced by the poles that are required to give the desired closed-

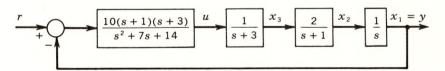

**Fig. 10.2-8**    Use of series compensation when no variables other than the output are available.

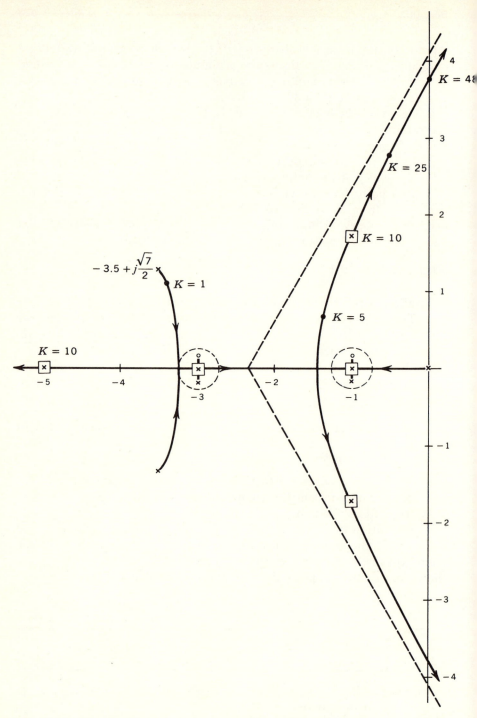

**Fig. 10.2-9** Root locus of the series-compensation system of Fig. 10.2-8.

*524*

loop response.    Had the given plant been seventh-order, six of the plant poles would have to be canceled by six zeros in the compensator to ensure that the specified closed-loop response was realized.    In the $n$th-order case, an equalizer requiring $n - 1$ poles and $n - 1$ zeros is required.

In addition to the hardware problem associated with the realization of this $(n - 1)$st-order series compensation, there is another shortcoming of this approach.    In order to illustrate this difficulty graphically, let us plot the root locus for the poles of $y(s)/r(s)$ versus the gain $K$ for the series-compensation system of Fig. 10.2-8 and any of the systems of Figs. 10.2-1$b$ to 10.2-5$b$.    All the latter systems have identical root-locus plots. The two root-locus plots are shown in Figs. 10.2-9 and 10.2-10.

We note that these two plots are strikingly different even though they both yield the correct closed-loop poles when $K$ has its design value of 10.    However, in the series-compensation form, as $K$ changes from its design value the poles vary widely and the system becomes unstable if $K > 48$.    In the minor-loop form, on the other hand, the poles of

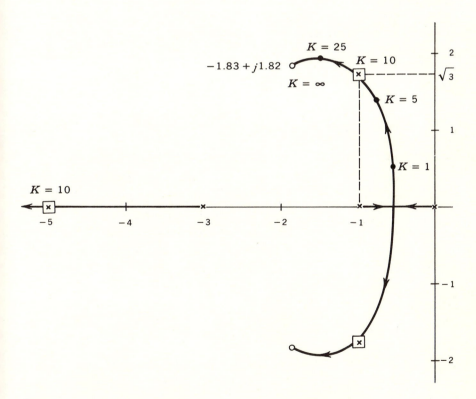

***Fig. 10.2-10***    Root locus of the systems of Figs. 10.2-1$b$ to 10.2-5$b$.

$y(s)/r(s)$ remain relatively unchanged for wide variations in $K$ from the design value, and the system is stable for all values of $K$.

The behavior noted here for this specific example is typical of the series-compensation approach and is one of its serious shortcomings. In fact, one may show that the minor-loop approach is, in general, superior from a sensitivity standpoint for all parameter variations, not only $K$. There are, however, situations in which the series-compensation approach provides an acceptable solution.

This last statement brings up a point that is the basis for all the succeeding material in this chapter. By specifying less than the complete closed-loop transfer function, is it possible to use a less complicated series-compensation network? A simple compensation network could not cancel all the unwanted poles of the plant and, indeed, may not cancel any of them. But the outstanding feature of the simple compensation network is its simplicity. In the seventh-order system alluded to above, for example, a clever designer may use a network with only two poles and two zeros and be able to produce a closed-loop system that is entirely satisfactory. It may be satisfactory because the desired seventh-order response is not unique; a number of different seventh-order responses may be satisfactory, as long as the dominant system characteristics remain similar.

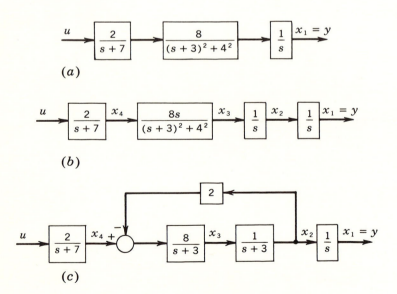

**Fig. 10.2-11** Fourth-order plant with complex conjugate poles. (*a*) Original plant; (*b*) phase-variable method of representation; (*c*) alternative method of representation.

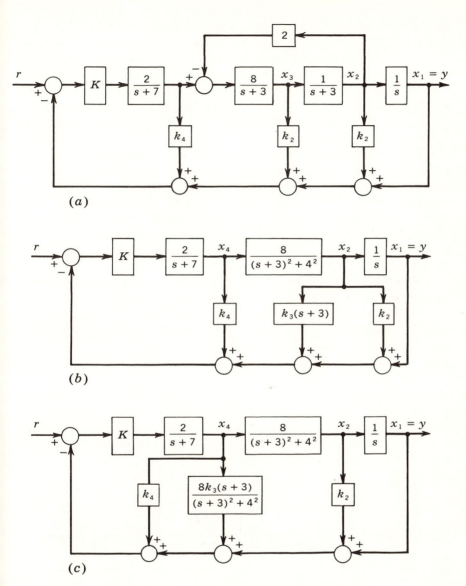

**Fig. 10.2-12**  State-variable feedback in which an inaccessible state variable is associated with a pair of complex conjugate poles.

Before proceeding to the use of simple compensation networks, let
us consider one further example dealing with the existence of an inaccessible state variable associated with complex conjugate poles. The problem considered here is pictured in Fig. 10.2-11. The unalterable plant of

Fig. 10.2-11*a* is fourth-order, and the complex conjugate poles are inherent, so that it is difficult to define state variables at least in terms of this block diagram. Figure 10.2-11*b* and *c* indicate two methods by which the missing state variable could be defined. In Fig. 10.2-11*b* the phase-variable approach is considered as a means of describing the complex conjugate roots of the given plant, and Fig. 10.2-11*c* uses a rather fictitious approach that assumes that the complex conjugate roots stem from two real roots with feedback around them. The two real roots are assumed to have the same real part as the real part of the complex conjugate roots being considered although this is not necessary. By adjusting the feedback element, the correct imaginary part can be realized.

The procedure to be used is similar to that already described in this section. We proceed as though the state variables are all available and determine the required values of **k** and $K$ to meet the specified $y(s)/r(s)$. Block-diagram manipulations are then used to remove the feedback from the unavailable state.

For the example under consideration, assume that we have defined the state variables as in Fig. 10.2-11*c*. Control of this physical plant is illustrated in Fig. 10.2-12*a*, where feedback is indicated from all the state variables, including the inaccessible state variable $x_3$. Figure 10.2-12*b* and *c* indicate the two-step process by which the equivalent required minor-loop compensator is determined. In Fig. 10.2-12*b* the feedback path from the state variable $x_3$ is moved to the $x_2$ location, and the artificial representation of the pair of complex conjugate roots is replaced by the original representation. In Fig. 10.2-12*c* the unrealizable transfer function $k_3(s + 3)$ is moved back around the complex conjugate poles, as indicated. The combination of the two parallel branches that now originate from the location of the state variable $x_4$ yields the final realization of Fig. 10.2-13. The minor-loop equalizer of Fig. 10.2-13 has com-

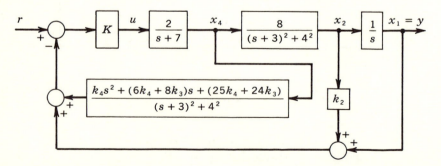

*Fig. 10.2-13*    Final form of the compensated closed-loop system of Fig. 10.2-11.

plex conjugate poles itself; these may be realized with passive components or with active circuits, such as an operational amplifier.   Often the latter method is superior, especially where weight considerations are important.

This section has considered a simple procedure for generating inaccessible state variables, based on block-diagram manipulations.   There are many other techniques[1] for accomplishing this task but they are beyond the scope of the treatment intended here.

**Exercises 10.2**   *10.2-1.*   Draw block diagrams for the system

$$\dot{\mathbf{x}} = \begin{bmatrix} 0 & 1 & 0 \\ 0 & -2 & 1 \\ 0 & 0 & -4 \end{bmatrix} \mathbf{x} + \begin{bmatrix} 0 \\ 0 \\ 1 \end{bmatrix} u$$

$$y = x_1 \qquad u = 5(r - [1 \quad 0.5 \quad 0.5]\mathbf{x})$$

if (a) $x_2$ is inaccessible; (b) $x_3$ is inaccessible; (c) $x_2$ and $x_3$ are inaccessible.   Use the minor-loop technique to generate the inaccessible states.

*10.2-2.*   Find the equivalent series compensation for the system

$$\dot{\mathbf{x}} = \begin{bmatrix} -1 & 1 & 0 \\ -1 & 0 & 2 \\ 0 & 0 & -2 \end{bmatrix} \mathbf{x} + \begin{bmatrix} 0 \\ 0 \\ 1 \end{bmatrix} u$$

$$y = x_1 \qquad u = 10(r - [1 \quad 1 \quad 1]\mathbf{x})$$

if $x_2$ and $x_3$ are inaccessible.

*answer:*

$$G_c(s) = \frac{s^3 + 13s^2 + 14s + 24}{s^3 + 13s^2 + 34s + 44}$$

*10.2-3.*   Draw block diagrams to illustrate how the minor-loop technique may be used to generate the inaccessible state for each of the systems given below if $x_2$ is inaccessible.

$$(a) \quad \dot{\mathbf{x}} = \begin{bmatrix} 0 & 1 & 0 \\ 0 & 0 & 2 \\ -1 & 0 & -2 \end{bmatrix} \mathbf{x} + \begin{bmatrix} 0 \\ 0 \\ 1 \end{bmatrix} u$$

$$y = x_1 \qquad u = 10(r - [1 \quad 0.5 \quad 0.4]\mathbf{x})$$

$$(b) \quad \dot{\mathbf{x}} = \begin{bmatrix} 0 & 1 & 0 \\ 0 & -2 & 1 \\ 0 & 0 & -4 \end{bmatrix} \mathbf{x} + \begin{bmatrix} 0 \\ 1 \\ 1 \end{bmatrix} u$$

$$y = x_1 \qquad u = 10(r - [1 \quad 1 \quad 0.5]\mathbf{x})$$

[1] See, for example, A. P. Sage, "Optimum Systems Control," chaps. 8–11, Prentice-Hall, Inc., Englewood Cliffs, N.J., 1968.

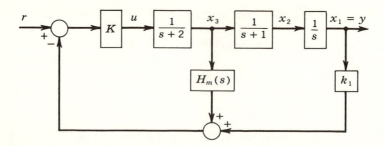

**Fig. 10.2-14**    Exercise 10.2-4.

*10.2-4.*    Find the transfer function of the minor loop $H_m(s)$ and the values of $k_1$ and $K$ for the system shown in Fig. 10.2-14 such that the closed-loop transfer function of the system is

$$\frac{y(s)}{r(s)} = \frac{10}{[(s+1)^2 + 1](s+5)}$$

*answer:*

$$H_m(s) = \frac{(0.4)(s + \frac{11}{4})}{s+1} \qquad K = 10 \qquad k_1 = 1$$

*10.2-5.*    Find the values of $K$, $k_1$, $k_3$, and $k_4$ for the system shown in Fig. 10.2-15 such that the closed-loop transfer function is

$$\frac{y(s)}{r(s)} = \frac{10}{[(s+1)^2 + 1](s+5)}$$

Note that Fig. 10.2-15 is another method of solving the problem of inaccessible states.

HINT: The answer of Prob. 10.2-4 may be helpful.

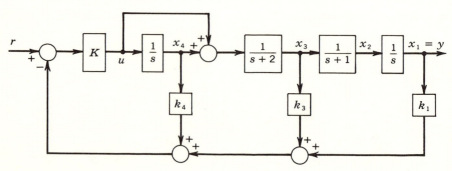

**Fig. 10.2-15**    Exercise 10.2-5.

*answer:*

$K = 10$    $\mathbf{k} = \text{col } (1,0,-0.3,0.8)$

## 10.3   *Series compensation*

In this and the following three sections, we shall restrict our attention to
the unity-ratio-feedback system with series compensation, shown in
Fig. 10.3-1.   In other words, we assume that either the output $y$ is the
only variable that is available for measurement and control or equivalently that the output is the only variable that we wish to use.   The
latter situation is probably the more practical one since it is as rare, in
a practical sense, to have only the output available as it is to have all
the state variables available.   Neither of these extremes is common in
practical control problems, and some middle position where some of but
not all the state variables are available for measurement is perhaps the
most common situation.

The restriction to the configuration of Fig. 10.3-1 is made not
because it is advocated that only $y$ be fed back, although this configuration has been effective in a great number of practical systems, but rather
because it leads to considerable simplification of the treatment.   This
series-compensation approach also requires a new method of specifying
system behavior which is simpler, although less precise, than an exact
specification of $y(s)/r(s)$.   In addition, historically the series compensator
has played a prominent role in system design; hence in order to analyze
many existing systems, a knowledge of series compensation may be
helpful.   The "best" solution in any practical problem is a combination of
the series-compensation and state-variable-feedback concepts.   Because
of the almost limitless number of variations possible in combining series
compensation and state-variable feedback and because each problem must,
in general, be treated individually, this subject is not discussed in detail
here but is left for the problems at the end of this chapter.

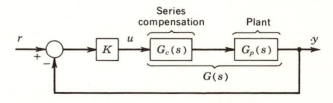

*Fig. 10.3-1*   Series-compensation configuration.

The basic vehicle for our development of series compensation is the straight-line Bode diagram of the magnitude of the frequency response of the open-loop $KG_p(s)$ and the compensated open-loop $KG(s) = KG_c(s)G_p(s)$ since $H_{eq}(s) = 1$. In order to work with the open-loop transfer function, it is necessary to translate the specifications for the *closed-loop performance* into specifications on the *open-loop transfer function*. This is an important difference in approach. We shall discuss this problem in detail later in this section. The approach here should be contrasted with that of Chap. 9 where the specifications were translated into a desired closed-loop transfer function $y(s)/r(s)$.

The design procedure could also be based on the Nyquist diagram or root-locus plot for the plant. However, it is felt that the Bode-diagram approach leads to simpler and more systematic procedures that are therefore easier to understand at first exposure. We shall, however, use the root locus to explain the concepts and consequences of series compensation as clearly as possible.

For simplicity, we assume that the plant is minimum-phase; this allows us to work only with the magnitude plot of $KG_p(j\omega)$ or $KG(j\omega)$. The extension to the nonminimum-phase plant is relatively simple once the basic concepts have been mastered. The reader is reminded, however, that pole-zero cancellation should never be used in the right half of the $s$ plane. On the other hand, feedback may be used to move a pole from the right- to the left-half plane.

*Specifications.* As mentioned previously, our design procedure is no longer to be based on realizing a desired closed-loop transfer function that is specified by the desired closed-loop behavior. In the series-compensation approach, the characteristics of the desired closed-loop behavior are translated into some general properties of the desired open-loop transfer function. The open-loop transfer function is then realized by the addition of a series compensator.

Of course, a series compensation could be designed, as discussed in Sec. 10.2, by specifying a desired $y(s)/r(s)$ and then using $G_{eq}(s)$ reduction to find $KG_c(s)$. The disadvantage of this approach is that the resulting $KG_c(s)$ is usually of high order, normally $(n - 1)$st. As we shall see in the sequel, a $KG_c(s)$ of first- or second-order is often acceptable to achieve closed-loop performance that is almost identical to the more complex $KG_c(s)$. In addition, in the case of high-order systems, it is often difficult to specify uniquely a desired $y(s)/r(s)$ from the given closed-loop specifications.

In general, the specification of the desired open-loop transfer function $KG(s)$ takes the form of one or more of the following items:

1. Low-frequency asymptote
2. Middle-frequency gain or crossover frequency $\omega_c$
3. High-frequency attenuation
4. Phase margin or $M$ peak

The relationship of each of these four items to closed-loop behavior is discussed below.

The specification of the low-frequency asymptote of the open-loop transfer function implies the specification of a system type, that is, the number of free integrations, and an associated error constant.[1] This specification then describes the steady-state behavior of the system for polynomial-in-$t$-type inputs. For example, if we desire the closed-loop system to have zero steady-state error for step inputs and an error of 0.002 for a unit ramp input, we would specify that the plant have a pole at the origin and sufficient gain to ensure a velocity-error constant of $1/0.002 = 500$.

The specification of a middle-frequency gain or crossover frequency is used to establish the useful frequency range of the closed-loop system. Suppose, for example, that from experimental data or user prediction it is known that the input signal contains no significant frequency components above 1 rad/sec. In order to reproduce the input signal accurately, let us require that the (steady-state) error for sinusoidal inputs up to 1 rad/sec be less than or equal to 2 percent or 0.02 of the input amplitude.

The method of translating this requirement into a middle-frequency-gain specification can be illustrated by considering the error transfer function

$$\frac{e(s)}{r(s)} = \frac{r(s) - y(s)}{r(s)} = 1 - \frac{y(s)}{r(s)} \tag{10.3-1}$$

In the case of the unity-feedback series-compensated system of Fig. 10.3-1, this equation becomes

$$\frac{e(s)}{r(s)} = 1 - \frac{KG(s)}{1 + KG(s)} = \frac{1}{1 + KG(s)} \tag{10.3-2}$$

For frequencies within the signal range, $|KG(j\omega)|$ should be large so that

$$\left| \frac{e(j\omega)}{r(j\omega)} \right| \sim \frac{1}{|KG(j\omega)|} \qquad \text{for } |KG(j\omega)| \gg 1 \tag{10.3-3}$$

Now, in order to ensure that $|e(j\omega)/r(j\omega)| \leq 0.02$ for $\omega \leq 1$ rad/sec, we must have $|KG(j\omega)| \geq 1/0.02 = 50$ for $\omega \leq 1$ rad/sec. This specifies a

---

[1] See Chap. 8 for a discussion of system type and error constants.

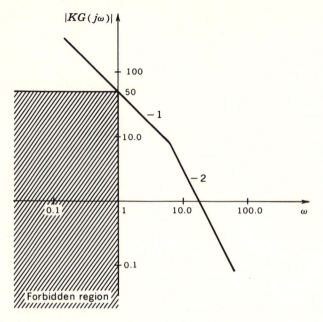

**Fig. 10.3-2**   Middle-frequency-gain specification.

region through which the magnitude diagram of $KG(j\omega)$ cannot pass, as shown in Fig. 10.3-2.

Another method of specifying the frequency range of interest is to specify the crossover frequency $\omega_c$.   Since, at $\omega_c$, $|KG(j\omega_c)| = 1$,

$$\left|\frac{y(j\omega_c)}{r(j\omega_c)}\right| = \frac{1}{|1 + KG(j\omega_c)|} \sim \frac{1}{\sqrt{2}} \qquad (10.3\text{-}4)$$

and $\omega_c$ is a fair approximation to the bandwidth of the closed-loop system. As shown in Chap. 8, the specification of $\omega_c$ provides a method for translating a speed-of-response requirement into a specification on $KG(s)$.

High-frequency input noise may be reduced at the output by specifying a desired high-frequency attenuation.   To illustrate this fact, let us consider the system shown in Fig. 10.3-3 where the input consists of

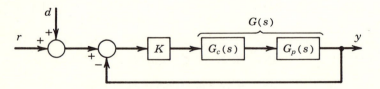

**Fig. 10.3-3**   System with a disturbance added.

the signal $r$ plus a noise or disturbance $d$. Let us consider the transfer function from $d(s)$ to $y(s)$, which is obviously the same as that from $r(s)$ to $y(s)$, or

$$\frac{y(s)}{d(s)} = \frac{KG(s)}{1 + KG(s)} \tag{10.3-5}$$

Let us suppose now that we wish to have the disturbance inputs at frequencies greater than 1,000 rad/sec reduced at the output to no more than 1 percent or 0.01 of their value at the input. In other words, we desire that

$$\left| \frac{y(j\omega)}{d(j\omega)} \right| \leq 0.01 \qquad \text{for } \omega \geq 1{,}000 \text{ rad/sec} \tag{10.3-6}$$

In order to meet this requirement, it is obvious that $|KG(j\omega)| \ll 1$ for $\omega \geq 1{,}000$ rad/sec so that we may make the approximation

$$\left| \frac{y(j\omega)}{d(j\omega)} \right| \sim |KG(j\omega)| \tag{10.3-7}$$

since $|1 + KG(j\omega)| \sim 1$ if $|KG(j\omega)| \ll 1$. The high-frequency–noise-rejection requirement has therefore been translated into the high-frequency-attenuation requirement that

$$|KG(j\omega)| \leq 0.01 \qquad \text{for } \omega \geq 1{,}000 \text{ rad/sec} \tag{10.3-8}$$

Once again this requirement generates a region through which the magnitude plot cannot pass. If the low-frequency-asymptote, middle-frequency-gain, and high-frequency-attenuation requirements discussed above are placed on the same Bode plot, the picture takes the form of Fig. 10.3-4. We see that the general shape of the compensated-plant transfer function has become fairly well established by these three specifications.

All the closed-loop behavior that we have discussed so far has been related to the steady-state performance of the system. No mention has been made of the transient behavior. In general, the character of the transient response may be determined by a specification of the phase margin of the compensated system. When phase margin is used in this manner, it is common to assume the system can be reasonably well represented by a second-order system. In this case, the relationship, discussed in Chap. 6, between phase margin and overshoot, for example, can be used to translate the desired behavior to a phase-margin specification. Of course, not all systems can be adequately represented by a second-order model. In any case, one should check the resulting design by the use of root locus or simulation to ensure that the design specifications have

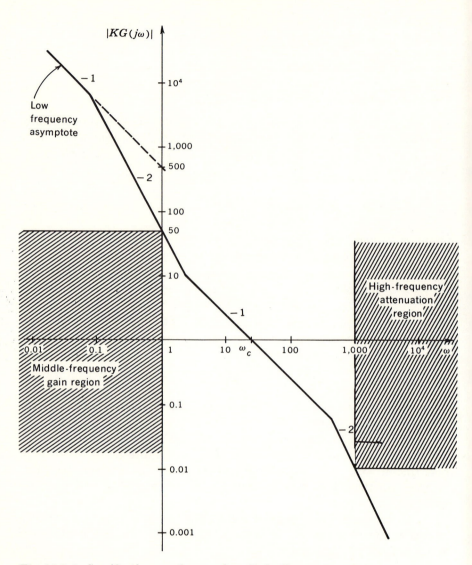

*Fig. 10.3-4* Specifications on the open-loop Bode diagram.

been met. It should be noted that this is the first occasion in our present discussion of specifications where any assumption has been made concerning the order of the system.

We could also use $M$ peak to establish the desired character of the transient response. However, since the use of the $M$-peak criterion requires that $|KG(j\omega)|$ and arg $KG(j\omega)$ be checked at a number of frequencies near crossover, rather than just one, as in the case of phase margin, it is normally more difficult to deal with $M$ peak. In addition, the translation of transient specifications to a desired $M$ peak is still approximate although more reliable than the phase-margin approach, and it is still necessary to check the final design. Design procedures based on $M$ peak are usually accomplished on the Nyquist diagram with the aid of $M$ circles.[1] An alternative and also frequently used approach is to judge transient response on the basis of the damping ratio $\zeta$ of the dominant set of complex conjugate closed-loop poles. When this approach is adopted, the root-locus method is normally employed.[2] Because of the simplicity of the Bode-diagram approach, phase margin is used here as a measure of the desirability of the resulting transient response. Regardless of what approach is used, it is always necessary to check the final design.

One additional comment should be made with regard to checking the results of any series-compensation design. Even though the design may be based solely on input-output properties, one should always check the behavior of all the state variables to ensure that the performance is within design or physical limitations of the devices that comprise the system. The reader will recall that this problem was discussed in Secs. 4.4 and 8.7.

*Types of compensation.* The following four types of series compensation are considered in the next three sections:

1. Gain adjustment (usually attenuation)
2. Lag equalizer
3. Lead equalizer
4. Lead-lag equalizer

The characteristics of these four types of series compensation are discussed briefly below.

The gain-adjustment compensation is included only for the sake of initiating our discussion. As we shall see in the next section, a simple gain adjustment is usually not an acceptable solution.

The transfer function of the lag equalizer (or compensator) consists

[1] H. Chestnut and R. W. Mayer, "Servomechanisms and Regulating System Design," John Wiley & Sons, Inc., New York, 1951.

[2] C. J. Savant, "Control System Design," 2d ed., McGraw-Hill Book Company, New York, 1964.

of a simple pole-zero pair and is given by

$$G_c(s) = \frac{1 + s/\alpha\omega_1}{1 + s/\omega_1} \qquad \alpha > 1 \tag{10.3-9}$$

With $\alpha > 1$ as required, the pole-zero plot takes on the form indicated in Fig. 10.3-5a, with the pole always nearer the origin of the $s$ plane than the zero. The magnitude and phase of the frequency response of the transfer function of Eq. (10.3-9) are given in Fig. 10.3-5b and c. Note that the magnitude response has an initial gain of unity and a slope of zero. Between the break frequency of the pole at $s = -\omega_1$ and that of the zero at $s = -\alpha\omega_1$ ($\alpha > 1$), the slope is $-1$. At high frequencies the gain is $1/\alpha$, and the slope is again equal to zero. The phase characteristic has its maximum negative value at the geometric mean of the pole and zero break frequencies, or at $\sqrt{\alpha}\,\omega_1$, and the phase shift approaches zero at both high and low frequencies.

A lag equalizer may be realized with the network of Fig. 10.3-5d,

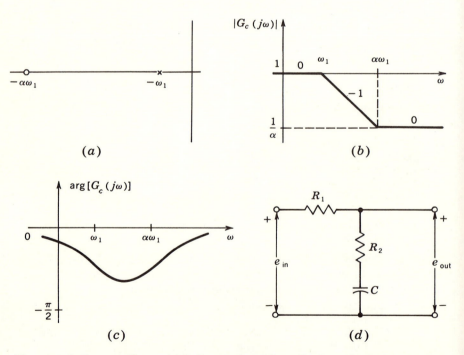

**Fig. 10.3-5**  Lag equalizer.  (*a*) Pole-zero plot; (*b*) amplitude of frequency response; (*c*) phase plot; (*d*) network realization.

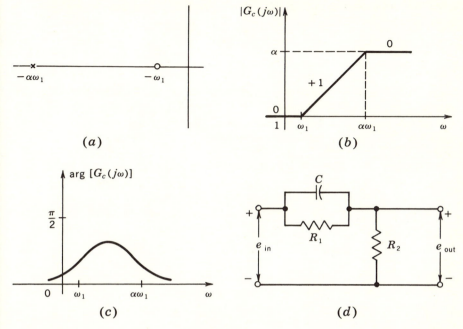

**Fig. 10.3-6**  Lead equalizer.  (a) Pole-zero plot; (b) magnitude plot; (c) phase plot; (d) network realization.

and it is easily shown that

$$\omega_1 = \frac{1}{(R_1 + R_2)C} \qquad \text{and} \qquad \alpha\omega_1 = \frac{1}{R_2 C}$$

The use of lag equalizers is discussed in the next section, along with a brief discussion of simple gain adjustment.

The transfer function of a lead equalizer is given as

$$G_c(s) = \frac{1 + s/\omega_1}{1 + s/\alpha\omega_1} \qquad \alpha > 1 \tag{10.3-10}$$

and the pole and zero locations are just reversed from those of the lag equalizer, as indicated in Fig. 10.3-6a.  The frequency-response curves are shown in Fig. 10.3-6b and c.  It is seen that the required amplitude at high frequencies is $\alpha$, so that it is impossible to realize the transfer function of Eq. (10.3-10) with purely passive elements.  As was the case with the lag equalizer, the maximum phase shift occurs at $\sqrt{\alpha}\,\omega_1$, but this time, of course, the phase angle is leading.

The electrical network of Fig. 10.3-6d is a means of realizing a pole-zero pair as required in Fig. 10.3-6a.   The transfer function of this network is

$$\frac{e_{\text{out}}(s)}{e_{\text{in}}(s)} = \frac{R_2}{R_1 + R_2} \frac{1 + R_1 C s}{1 + [R_2/(R_1 + R_2)]R_1 C s} = \frac{1}{\alpha} \frac{1 + s/\omega_1}{1 + s/\alpha\omega_1} \qquad (10.3\text{-}11)$$

where

$$\omega_1 = \frac{1}{R_1 C} \qquad \alpha = \frac{R_1 + R_2}{R_2}$$

The transfer function of this network [Eq. (10.3-11)] is not identical to the required lead-network transfer function [Eq. (10.3-10)].   The difference is the attenuation factor $1/\alpha$.   At low frequencies the gain of this network is $1/\alpha$, and at high frequencies the gain is unity.   If this network is used to realize the required pole-zero pair, it must be accompanied by an amplifier of gain $\alpha$, and thus the realization is not completely passive.   Often this is not important, as the electrical network is built as an integral part of the free amplifier $K$ that precedes the plant.   When such is the case, the additional gain required to compensate for the attenuation in the network is made up by simply increasing $K$.   For purposes of discussion, however, we need not be concerned with exactly how the desired lead-equalizer transfer function of Eq. (10.3-10) is realized.   Thus, in the extensive discussion of lead equalizers in Sec. 10.5, the gain $K$ is the usual free gain that precedes the plant, and the lead-compensator transfer function is assumed to be that of Eq. (10.3-10) rather than that of Eq. (10.3-11).   If the network of Fig. 10.3-6d is used to realize the required pole-zero pair, then an additional gain of $\alpha$ is needed.

The lag-lead or composite equalizer is exactly what the name implies, a combination of a lead and a lag equalizer.   The transfer function is

$$G_c(s) = \frac{1 + s/\beta\omega_1}{1 + s/\omega_1} \frac{1 + s/\omega_2}{1 + s/\alpha\omega_2} \qquad \alpha, \beta > 1 \qquad (10.3\text{-}12)$$

and the pole-zero plot is indicated in Fig. 10.3-7a, with the accompanying amplitude and phase diagrams given in Fig. 10.3-7b and c.   The realization indicated in Fig. 10.3-7d is the obvious one, with an isolation amplifier included between the lag and the lead networks to prevent loading of the first network by the second.   If this isolation amplifier is given a gain of $\alpha$, the attenuation in the lead network is eliminated, and the transfer function is as desired in Eq. (10.3-12).

If $\beta\omega_1 \leq \omega_2$ and $\beta \geq \alpha$, the case pictured in the amplitude diagram of Fig. 10.3-7b, it is possible to approximate the transfer function of the composite equalizer whose transfer function is Eq. (10.3-12) with the electrical

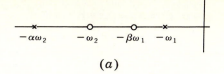

$(a)$

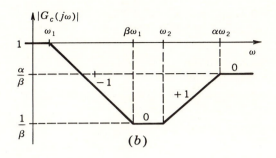

$(b)$

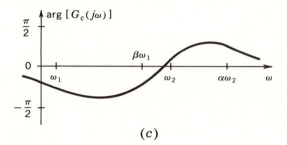

$(c)$

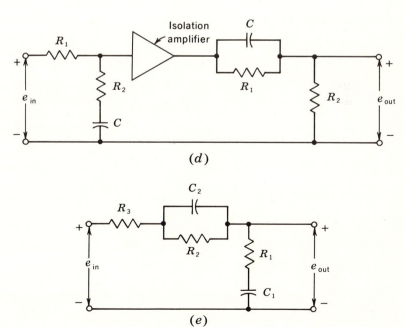

$(d)$

$(e)$

**Fig. 10.3-7** Lead-lag equalizer. $(a)$ Pole-zero plot; $(b)$ magnitude plot; $(c)$ phase plot; $(d)$ network realization; $(e)$ approximate realization.

network of Fig. 10.3-7e.  It is easily seen that this network approximates both the lag and the lead networks at low and high frequencies, respectively.  At low frequencies the capacitors look like large impedances, so that $C_2$ in parallel with $R_2$ may be considered open, and the usual lag network results.  At high frequencies $C_1$ appears as a short, and a lead network results.  In the case $\alpha$ equals $\beta$, $R_3$ is zero, and if $\alpha$ is greater than $\beta$, $R_3$ is again zero but now additional gain is needed to realize the required response at high frequencies.

The actual transfer function of the network of Fig. 10.3-7e is

$$\frac{e_{\text{out}}(s)}{e_{\text{in}}(s)} =$$

$$\frac{(1 + R_1 C_1 s)(1 + R_2 C_2 s)}{(R_1 C_1 R_2 C_2 + R_3 C_1 R_2 C_2)s^2 + (R_1 C_1 + R_3 C_1 + R_2 C_1 + R_2 C_2)s + 1}$$

so that

$$\beta \omega_1 = \frac{1}{R_1 C_1} \qquad \omega_2 = R_2 C_2$$

but the pole locations are not immediately evident.  The design of closed-loop systems using lag-lead equalizers will be considered in Sec. 10.6.

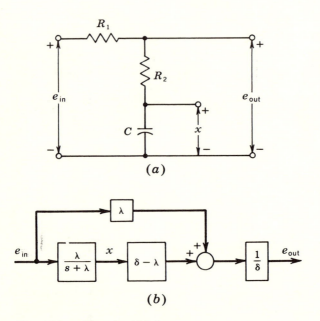

(a)

(b)

*Fig. 10.3-8*  Exercise 10.3-1.

The division of series compensation into these four forms is rather arbitrary since in any design often one or more forms may be used. This division was made solely to assist the reader in understanding the basic concepts.

**Exercise 10.3**   *10.3-1.*   Show that the lag network in Fig. 10.3-8a with the internal variable $x$ labeled can be represented by the block diagram of Fig. 10.3-8$b$ where $1/\lambda = C(R_1 + R_2)$ and $1/\delta = CR_2$. Note that this is a method of obtaining the feedforward form of the pole-zero pair used in Chap. 9. Show that the lead network of Fig. 10.3-6 can also be represented by a feedforward form if the proper internal state is defined.

## 10.4   *Lag compensation*

The basic concept underlying both the gain-adjustment and lag-equalizer design is that the desired degree of stability (that is, phase margin) may often be achieved by a reduction of gain. In order to initiate our discussion of lag compensation, let us consider the use of a simple gain adjustment, usually attenuation. Although pure attenuation is seldom an adequate solution to a design problem, it serves to illustrate the basic approach of lag compensation. In addition, our discussion serves to point out the inadequacy of pure attenuation and the need for the lag equalizer.

For our gain-adjustment example design, we assume that the only specification is a desired phase margin of 45°. Let us consider the system shown in Fig. 10.4-1$a$ where the transfer function of the uncompensated plant is

$$G_p(s) = \frac{100}{s(1 + s/10)^2} \tag{10.4-1}$$

The problem is to select the value of $K = 1/\alpha$ for the attenuator such that the compensated open-loop transfer function $KG_p(s)$ has a phase margin of 45°. Note here $G_c(s) = H_{eq}(s) = 1$. In order to meet the phase-margin specification, we need to find the frequency for which the phase shift is $-135°$ and then select $K$ such that crossover occurs at that point. A simple application of the Bode approximation to the magnitude plot shown in Fig. 10.4-1$b$ for the uncompensated plant reveals that the phase shift is $-180°$ at $\omega = 10$. Therefore the frequency for which the phase shift is $-135°$ must be below the break point at 10 rad/sec and we

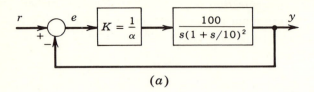

(a)

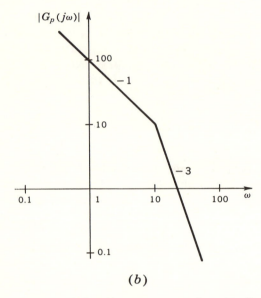

(b)

**Fig. 10.4-1**  Gain-adjustment example. (a) System
           block diagram; (b) magnitude plot of the
           uncompensated plant.

may use the arctangent approximation to write

$$\arg G_p(j\omega) \sim -\frac{\pi}{2} - 2\frac{\omega}{10} \qquad \text{for } \omega < 10$$

The value of $\omega$ for which $\arg G_p(j\omega) = -135°$ is therefore obtained by
solving the equation

$$-\frac{3\pi}{4} = -\frac{\pi}{2} - 2\frac{\omega}{10}$$

from which we find that

$$\omega = \tfrac{5}{4}\pi = 3.925$$

The magnitude of $G_p(j\omega)$ at this frequency is

$$|G_p(j3.925)| \sim \frac{100}{3.925} = 25.5$$

In order for crossover to occur at $\omega = 3.925$ so that the phase margin is 45°, it is necessary that the gain be attenuated by a factor of 25.5 to 1.[1] In other words, $K$ should be selected as $1/25.5$ so that $KG_p(s)$ becomes

$$KG_p(s) = \frac{1}{25.5}\frac{100}{s(1 + s/10)^2} = \frac{3.925}{s(1 + s/10)^2} \qquad (10.4\text{-}2)$$

The resulting magnitude diagram for the compensated-plant transfer function $KG_p(s)$ is shown in Fig. 10.4-2 along with the original uncompensated-plant transfer function $G_p(s)$.

An examination of the compensated- and uncompensated-plant

[1] Since the arctangent and straight-line approximations have been used in obtaining the answer, the actual attenuation needed is somewhat less than 25.5 to 1 but the change is relatively small.

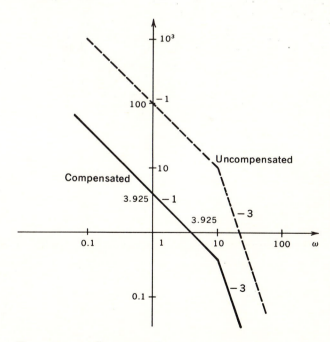

**Fig.** *10.4-2* Compensated- and uncompensated-plant (open-loop) magnitude plots.

transfer functions reveals that, although the desired phase margin has been achieved, the low-frequency gain has also been reduced by a factor of 25.5.   This means, for example, that the velocity-error constant has been decreased by 25.5.   In other words, the positional error for a ramp input was originally 1 percent, but it would now be 25.5 percent, a significant change.   Because of its effect on low-frequency accuracy, decreasing the gain is generally not an acceptable solution.

The reader will recall from our discussion of the Bode approximation in Sec. 5.4 that the phase angle at any frequency is dependent, in a practical sense, only on the slope of the magnitude plot for one decade above and below the frequency of interest.   This means that the phase shift at crossover, and hence the phase margin, is dependent only on the magnitude plot one decade above and below the crossover frequency.

Let us consider the magnitude plots shown in Fig. 10.4-3.   In addi-

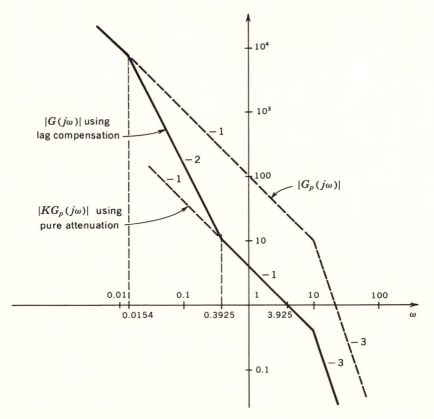

*Fig. 10.4-3*   Magnitude plots for the plant.

tion to the Bode diagrams for the uncompensated plant $G_p(s)$ and the plant with the pure-attenuation type of compensation, a third Bode diagram labeled as lag compensation is shown. This last diagram has the same low-frequency asymptote as the original plant and hence the same velocity-error constant. Thus $K$ is 1 in this case. The lag-compensation plot is the same as the pure-attenuation plot in the decade around crossover and hence should have approximately the same phase margin as the pure-attenuation result.

To show that the phase margin is approximately the same, let us determine the phase margin of the lag-compensated system. Using the arctangent approximation,[1] we can write the phase shift for the lag-compensated plant as

$$\arg G(j\omega) = -\frac{\pi}{2} - \left(\frac{\pi}{2} - \frac{0.0154}{\omega}\right) + \left(\frac{\pi}{2} - \frac{0.3925}{\omega}\right) - 2\frac{\omega}{10}$$

$$= -\frac{\pi}{2} - \frac{0.3771}{\omega} - \frac{\omega}{5} \qquad \text{for } 0.3925 < \omega < 10 \qquad (10.4\text{-}3)$$

At crossover, therefore, the phase shift is

$$\arg G(j3.925) = -\frac{\pi}{2} - \frac{0.3771}{3.925} - \frac{3.925}{5} = 2.451 \text{ rad}$$

and the phase margin is 0.69 rad or 39.5°. This result compares very favorably with the desired phase margin of 45°. In many practical problems this result might well be of sufficient accuracy.

The transfer function of the lag compensator may be obtained analytically by dividing $G(s)$ by $G_p(s)$:

$$G_c(s) = \frac{G(s)}{G_p(s)} = \frac{\dfrac{100(1 + s/0.3925)}{s(1 + s/0.0154)(1 + s/10)^2}}{\dfrac{100}{s(1 + s/10)^2}}$$

$$= \frac{1 + s/0.3925}{1 + s/0.0154} \qquad (10.4\text{-}4)$$

The magnitude plot for this transfer function is shown in Fig. 10.4-4. This magnitude plot could also have been obtained by subtracting, in the usual geometric sense, the plot for $G_p(s)$ from $G(s)$ since subtraction on the log-log plot corresponds to division.

Let us summarize what can be accomplished with this lag-compensation form. By retaining the same low-frequency asymptote as the original

---

[1] It would be just as easy to use the exact arctangent expression here, but in order to compare this result with the previous one, the arctangent approach has been retained.

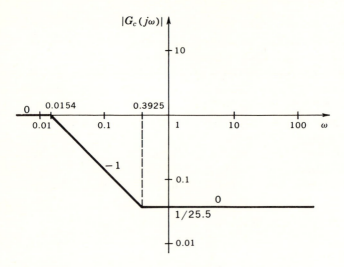

**Fig. 10.4-4** Magnitude plot for the lag compensation.

uncompensated plant, the low-frequency accuracy—and in particular the velocity-error constant—of the system is maintained at its original high value of 100. At the same time, the use of attenuation before crossover permits the phase margin to be raised to approximately the desired value. These statements form the basic philosophy of lag compensation even though we have proceeded so far in a rather casual fashion.

To establish the details of the lag-compensation design more clearly, let us consider the use of lag compensation on the following uncompensated-plant transfer function:

$$G_p(s) = \frac{5}{s(1 + s/10)(1 + s/50)} \tag{10.4-5}$$

After compensation, the compensated open-loop transfer function $G(s)$ must satisfy the following specifications:

1. The velocity-error constant $K_v$ must be equal to 50.
2. The phase margin must be equal to 45°.

As the first step in our design procedure, let us sketch the magnitude and phase plots of the uncompensated plant, as shown in Fig. 10.4-5. This figure allows us to determine not only if compensation is necessary but also if it is possible (provided that it is needed) and some preliminary details of the design. From the solid line in Fig. 10.4-5a we see that the

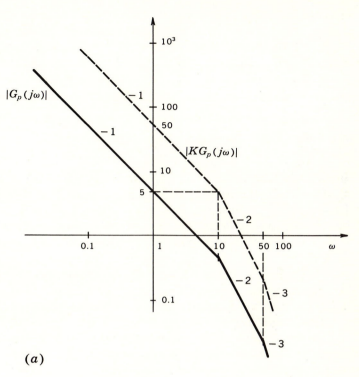

*(a)*

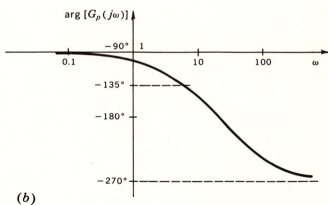

*(b)*

**Fig. 10.4-5** Magnitude and phase diagrams for the plant.

slope at crossover is $-1$, and from the arctan approximation the phase margin is found to be $58.2°$. This is a more than adequate phase margin; however, the velocity-error constant for the given uncompensated plant has yet to be calculated. Since this is a type 1 system, that is, it has one integration in the plant, the velocity-error constant is found by taking the limit of $sG_p(s)$ as $s$ goes to zero; the velocity-error constant is

$$K_v = \lim_{s \to 0} sG_p(s) = \lim_{s \to 0} \frac{5}{(1 + s/10)(1 + s/50)} = 5$$

In order to increase the velocity-error constant to 50, $K$ must be equal to 10. If this simple gain adjustment is made, the dashed line in Fig. 10.4-5$a$ results. Now not only is the phase margin less than $45°$, but, in fact, the gain-compensated closed-loop system is nearly unstable since the phase shift at crossover is very nearly $-180°$. However, we observe from the phase-shift diagram of Fig. 10.4-5$b$, which applies to both amplitude diagrams of Fig. 10.4-5$a$, that for frequencies less than about 7 rad/sec the phase lag is less than $135°$. Thus, in addition to a gain adjustment, we may use lag compensation to reduce the crossover frequency to about 5 or 6 rad/sec.

What we have accomplished thus far is to establish a low-frequency asymptote that ensures the proper velocity-error constant. Now let us assume that the given problem is actually represented by the dashed line in the Bode amplitude diagram of Fig. 10.4-5$a$ and attack the problem of determining the necessary lag compensator, much as was done in the previous example. That is, as a second step of the design, it is often helpful to carry out a pure-attenuation design on the gain-compensated system. It may seem strange to increase the gain first and then perform a pure-attenuation design. This is done here because we are attempting to simplify the procedure by first finding the required gain $K$ and then finding the necessary lag compensator $G_c(s)$ rather than finding the combination $KG_c(s)$ in one step.

Since the frequency for which the arg $KG_p(j\omega) = -135°$ is less than 10 rad/sec, the arctangent expression for arg $KG_p(j\omega)$ becomes

$$\arg KG_p(j\omega) \sim -\frac{\pi}{2} - \frac{\omega}{10} - \frac{\omega}{50} \qquad (10.4\text{-}6)$$

so that the frequency for which arg $KG_p(j\omega) = -135°$ is given by the equation

$$-\frac{\pi}{2} - \frac{\omega}{10} - \frac{\omega}{50} = -\frac{3\pi}{4} \qquad (10.4\text{-}7)$$

or

$$\omega = \frac{25\pi}{12} = 6.55 \text{ rad/sec}$$

At this frequency the magnitude of $KG_p(j\omega)$ as given by the straight-line approximation is

$$|KG_p(j6.55)| \sim \frac{50}{6.55} = 7.63 \tag{10.4-8}$$

Therefore, if pure attenuation were to be used, the gain would be decreased by 7.63. This gain reduction establishes a lower limit on the value of $\alpha$ which we may choose for the lag compensator, which has the form

$$G_c(s) = \frac{1 + s/\alpha\omega_1}{1 + s/\omega_1} \tag{10.4-9}$$

In addition, the crossover frequency, 6.55 rad/sec, for the pure-attenuation case establishes an upper limit on the crossover frequency for the lag-compensated system. Because of the additional phase lag introduced by the series compensation, the value for $\alpha$ must be greater than 7.63, and the crossover frequency consequently is less than 6.55 rad/sec.

With this information concerning the pure-attenuation design for the gain-compensated system, it is possible to sketch the general shape of the magnitude plot of the transfer function

$$KG(s) = \frac{50(1 + s/\alpha\omega_1)}{s(1 + s/10)(1 + s/50)(1 + s/\omega_1)}$$

for the lag-compensation design as shown in Fig. 10.4-6. For comparison, the magnitude plot of the gain-compensated plant is also shown. The problem is to select the values of $\omega_1$ and $\alpha$ such that the desired phase-margin specification is met. The low-frequency-asymptote requirement is satisfied by the previous addition of gain. Since the crossover frequency is unknown until the attenuation factor $\alpha$ is selected,[1] we shall treat the problem as if there are three unknowns: $\omega_1$, $\alpha$, and $\omega_c$. The procedure we shall follow is to write three equations that may be solved to obtain the three unknowns.

As the first equation, we may simply write, using the straight-line approximation, the condition for crossover:

$$|KG(j\omega_c)| = 1 \sim \frac{50(\omega_c/\alpha\omega_1)}{\omega_c(\omega_c/\omega_1)} = \frac{50}{\alpha\omega_c} \tag{10.4-10}$$

or

$$\omega_c \sim 50\alpha$$

[1] The crossover frequency depends only on $\alpha$ since we assume that $\alpha\omega_1 < \omega_c$.

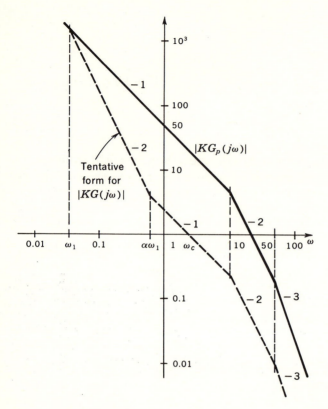

**Fig. 10.4-6**  Tentative sketch of the lag-compensation design.

Note that we have assumed that $\omega_c > \alpha\omega_1$, as shown in Fig. 10.4-6.   This assumption is necessary to ensure that the phase-margin requirement can be met by having a $-1$ slope in the range of crossover.

The phase-margin specification may be used to form another equation.   Here again we use the arctangent approximation to avoid transcendental equations.   The phase shift of the compensated plant at crossover is given by

$$\arg KG(j\omega_c) \sim -\frac{\pi}{2} - \left(\frac{\pi}{2} - \frac{\omega_1}{\omega_c}\right) + \left(\frac{\pi}{2} - \frac{\alpha\omega_1}{\omega_c}\right) - \frac{\omega_c}{10} - \frac{\omega_c}{50}$$

$$\sim -\frac{\pi}{2} - \frac{(\alpha - 1)\omega_1}{\omega_c} - \frac{3\omega_c}{25} \tag{10.4-11}$$

In order to have a phase margin of 45°, it is necessary that

$$-\frac{\pi}{2} - \frac{(\alpha - 1)\omega_1}{\omega_c} - \frac{3\omega_c}{25} = -\frac{3\pi}{4}$$

or

$$\frac{(\alpha - 1)\omega_1}{\omega_c} + \frac{3\omega_c}{25} - \frac{\pi}{4} = 0 \qquad\qquad (10.4\text{-}12)$$

The complete solution of the problem requires that a third equation be written so that we may solve for $\omega_1$, $\alpha$, and $\omega_c$. Three possible methods of determining this third equation are considered.

1.  Select the $\omega_c/\alpha\omega_1$ ratio.
2.  Select the attenuation factor $\alpha$.
3.  Minimize middle-frequency attenuation.

Each of these three methods is discussed below.

*Specified $\omega_c/\alpha\omega_1$ ratio.* The concept of selecting the ratio of $\omega_c$ to $\alpha\omega_1$ is a direct outgrowth of the initial discussion of lag compensation in this section. As an introduction to lag compensation, we placed the attenuation one decade away from crossover in order that the effect of the lag compensation on the phase shift at crossover would be small. Although in the present development we are assured of satisfying the phase-margin specification as long as Eqs. (10.4-10) and (10.4-12) are satisfied, we may nevertheless still complete the specification of the solution by selecting the frequency ratio of the last break point of the lag compensation $\alpha\omega_1$ to the crossover frequency $\omega_c$. Although any frequency ratio may be chosen, engineering practice has indicated that a ratio of one decade, that is,

$$\omega_c = 10\alpha\omega_1 \qquad\qquad (10.4\text{-}13)$$

generally leads to a satisfactory design.

The use of Eq. (10.4-13) along with Eqs. (10.4-10) and (10.4-12) generates a complete solution to the lag-compensation design. In order to obtain the solution, let us substitute Eq. (10.4-13) for $\omega_c$ into Eqs. (10.4-10) and (10.4-12). After a few simple algebraic reductions, the result is the following two equations in the two unknowns $\alpha$ and $\omega_1$:

$$\alpha^2\omega_1 = 5 \qquad\qquad (10.4\text{-}14)$$

$$\frac{\alpha - 1}{10} + \frac{30}{25}\alpha^2\omega_1 - \frac{\pi}{4}\alpha = 0 \qquad\qquad (10.4\text{-}15)$$

We may now substitute Eq. (10.4-14) into Eq. (10.4-15) to obtain the following equation for $\alpha$:

$$\frac{\alpha - 1}{10} + 6 - \frac{\pi}{4}\alpha = 0$$

This equation is easily solved for $\alpha$ as

$$\alpha = 8.62$$

and then

$$\alpha\omega_1 = \frac{5}{\alpha} = 0.58 \text{ rad/sec}$$

$$\omega_c = 10\alpha\omega_1 = 5.80 \text{ rad/sec}$$

and

$$\omega_1 = \frac{\alpha\omega_1}{\alpha} = 0.067 \text{ rad/sec}$$

The desired lag compensation is therefore

$$G_c(s) = \frac{1 + s/0.58}{1 + s/0.067}$$

*Specified* $\alpha$.    The design in terms of a specified value for $\alpha$ is probably the simplest of the three methods.    The approach is to select a value for $\alpha$ based on some practical engineering criterion such as desired high-frequency attenuation and then to find the associated values for $\omega_1$ and $\omega_c$.    Once $\alpha$ has been selected, $\omega_c$ may be obtained directly from Eq. (10.4-10) as

$$\omega_c = \frac{50}{\alpha} \tag{10.4-16}$$

and $\omega_1$ may then be found by solving Eq. (10.4-12):

$$\omega_1 = \frac{\omega_c}{\alpha - 1}\left(\frac{\pi}{4} - \frac{3\omega_c}{25}\right) \tag{10.4-17}$$

The value of $\alpha$ for the pure-attenuation design, namely, $\alpha = 7.63$, provides a lower bound on the values that may be selected for $\alpha$.    If this value is used, $\omega_c$ is equal to 6.55 rad/sec, $\omega_1$ is equal to zero, and the pure-attenuation design results.

In order to examine the consequence of the selection of $\alpha$ on the final design, let us complete the design for a number of different values of $\alpha$.    We have just obtained the solution for $\alpha = 8.62$ in the previous development, and so let us consider the design for $\alpha = 10, 15,$ and $20$.

For $\alpha = 10$, $\omega_c$ as given by Eq. (10.4-16) is

$$\omega_c = \frac{50}{\alpha} = 5 \text{ rad/sec}$$

and $\omega_1$ is

$$\omega_1 = \frac{\omega_c}{\alpha - 1}\left(\frac{\pi}{4} - \frac{3\omega_c}{25}\right) = \frac{5}{9}\left(\frac{\pi}{4} - \frac{3}{5}\right) = 0.103 \text{ rad/sec}$$

The other values of $\alpha$ are treated in a similar fashion, and the results are summarized in Table 10.4-1 and Fig. 10.4-7 for comparison.

*Table 10.4-1*  *Lag-compensation Designs for Various Values of $\alpha$*

| $\alpha$ | $\omega_c$ | $\omega_1$ | $\alpha\omega_1$ |
|---|---|---|---|
| 8.62 | 5.80 | 0.067 | 0.58 |
| 10.0 | 5.0 | 0.103 | 1.03 |
| 15.0 | 3.33 | 0.092 | 1.37 |
| 20.0 | 2.50 | 0.0595 | 1.19 |

An examination of Fig. 10.4-7 reveals a very interesting phenomenon. As the value of $\alpha$ is increased from its minimum of 7.63, the $-2$ slope below crossover initially moves to the right. However, as $\alpha$ is further increased to values of 15 or 20, we see that the low-frequency $-2$ slope begins to move back to the left. This behavior is due to the increase in the phase lag caused by the $-2$ slope as $\alpha$ is increased, and consequently the length of the $-2$ slope increases. The minimum amount of middle-frequency attenuation and hence the best-frequency-response characteristic occur when the $-2$ slope is moved as far as possible to the right. This condition is equivalent to selecting $\alpha$ to maximize the value of $\omega_1$. This is the approach taken by the third method discussed below.

*Minimum middle-frequency attenuation.* In order to move the $-2$ slope associated with the lag compensation as far to the right as possible and thereby minimize the middle-frequency attenuation caused by the lag compensation, one needs only to maximize the value of $\omega_1$. The truth of this statement is easily verified by a consideration of Fig. 10.4-7. In other words, the minimization of middle-frequency attenuation is equivalent to maximization of $\omega_1$. Our problem is therefore to select $\alpha$ such that $\omega_1$ is maximized. This may be accomplished by writing $\omega_1$ as a func-

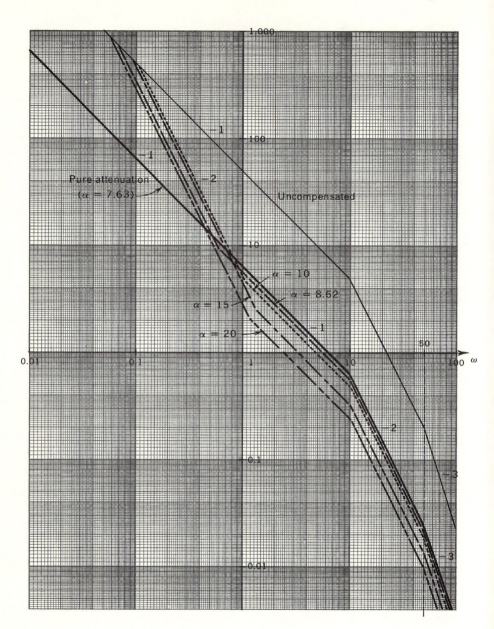

*Fig. 10.4-7*   Magnitude plots for various values of $\alpha$.

tion of $\alpha$, setting $d\omega_1(\alpha)/d\alpha$ equal to zero, and then solving for the value of $\alpha$ that satisfies this equation. The desired expression for $\omega_1(\alpha)$ may be obtained by substituting Eq. (10.4-16) into Eq. (10.4-17) as

$$\omega_1(\alpha) = \frac{50/\alpha}{\alpha - 1}\left[\frac{\pi}{4} - \frac{3(50/\alpha)}{25}\right]$$

or

$$\omega_1(\alpha) = \frac{25\pi\alpha - 600}{2\alpha^2(\alpha - 1)} \tag{10.4-18}$$

The derivative of $\omega_1(\alpha)$ with respect to $\alpha$ is therefore

$$\frac{d\omega_1(\alpha)}{d\alpha} = \frac{2\alpha^2(\alpha - 1)25\pi - (25\pi\alpha - 600)(6\alpha^2 - 4\alpha)}{4\alpha^4(\alpha - 1)^2}$$
$$= \frac{-100\pi\alpha^2 + (3{,}600 + 50\pi)\alpha - 2{,}400}{4\alpha^3(\alpha - 1)^2}$$

The desired value of $\alpha$ that minimizes the middle-frequency attenuation is a solution of the equation

$$-100\pi\alpha^2 + (3{,}600 + 50\pi)\alpha - 2{,}400 = 0 \tag{10.4-19}$$

A simple use of the quadratic formula reveals that

$$\alpha = 0.206 \qquad \text{and} \qquad \alpha = 11.7$$

are solutions of this equation. Since $\alpha = 0.206$ is less than the minimum value of 7.63, this solution is meaningless, and the solution that we are seeking is $\alpha = 11.7$. The values for $\omega_c$ and $\omega_1$ may now be obtained by the use of Eqs. (10.4-16) and (10.4-18):

$$\omega_c = 4.27 \qquad \text{and} \qquad \omega_1 = 0.108$$

The resulting lag compensation is therefore

$$G_c(s) = \frac{1 + s/1.27}{1 + s/0.108} \tag{10.4-20}$$

The magnitude plots for the compensated- and uncompensated-plant transfer functions are shown in Fig. 10.4-8 along with the magnitude plot for the lag compensator.

It is interesting to note that the value of $\alpha$ obtained above is not very different from the value of 8.62 obtained by the much simpler first method of selecting $\alpha\omega_1$ as 0.1 of $\omega_c$. Since both designs are approximate in the sense of achieving desired closed-loop behavior, the simpler method might well be justified, particularly as a first attempt at the design.

On the other hand, if one had some reason for desiring a given value

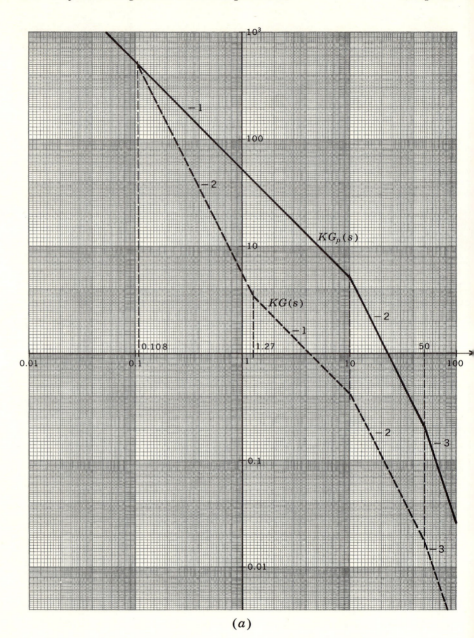

$(a)$

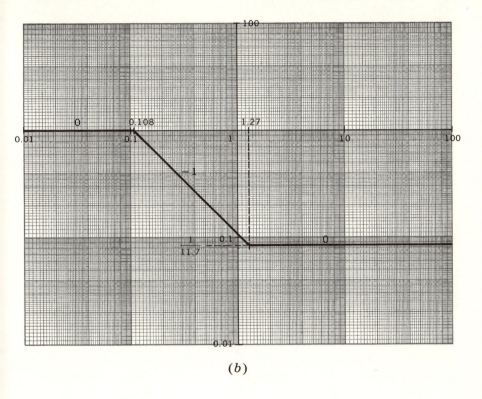

(b)

**Fig. 10.4-8** Magnitude plots. (a) Compensated and uncompensated open-loop transfer function, $K = 10$; (b) lag compensation.

## Table 10.4-2    Summary of Lag-compensation Design

| Step No. | Procedure | Applicable Equation Nos. |
|---|---|---|
| 1 | Sketch $\|G_p(j\omega)\|$ and arg $G_p(j\omega)$ and check if compensation is necessary and possible. | |
| 2 | Select the value of $K$ to meet the low-frequency-asymptote requirement. | |
| 3 | Carry out pure-attenuation design to establish bounds on the lag compensation.<br>(a) Use the arctangent approximation to determine the necessary $\omega_c$.<br>(b) Compute the value of $\alpha$ necessary to cause crossover at $\omega_c$. | (10.4-7)<br><br>(10.4-8) |
| 4 | Sketch the approximate shape of $\|KG(j\omega)\|$ | |
| 5 | Write the expression for $\|KG(j\omega_c)\| = 1$ | (10.4-10) |
| 6 | Use the arctangent approximation and set arg $KG(j\omega_c)$ $= -\pi + \phi_m$. | (10.4-12) |
| 7a | Method 1: *Specified $\omega_c/\alpha\omega_1$ ratio.*<br>(a) Select the ratio of $\alpha\omega_1$ to $\omega_c$.<br>(b) Solve for $\alpha$, $\omega_1$, and $\omega_c$. | (10.4-13) |
| 7b | Method 2: *Specified $\alpha$.*<br>(a) Select $\alpha$.<br>(b) Solve the equations in steps 4 and 5 for $\omega_1$ and $\omega_c$. | |
| 7c | Method 3: *Minimum middle-frequency attenuation.*<br>(a) Maximize $\omega_1$ with respect to $\alpha$.<br>(b) Solve for $\omega_1$ and $\omega_c$. | (10.4-19) |
| 8 | Write the expression for the required lag compensation $$G_c(s) = \frac{1 + s/\alpha\omega_1}{1 + s/\omega_1}$$ | |
| 9 | Check the final design to ensure that the desired closed-loop behavior has been achieved. | |

of attenuation, the second method might well be the best to follow. In particular, if a certain amount of high-frequency attenuation were needed to eliminate an undesired noise signal, one might need a larger value for $\alpha$ than might be obtained with either of the other methods.

The complete lag-compensation procedure is summarized in Table 10.4-2 for easy reference. It is *not* intended that this table be followed in a step-by-step fashion but rather that it should be used to focus attention on the basic items of the lag compensation. Indeed, as the reader gains familiarity with compensation procedures, he may wish to modify or eliminate some of the steps listed. This table and the entire discussion of lag compensation have been oriented about a specific example. It should be emphasized that the procedures developed are general and may be applied with only minor alterations to a wide range of plants.

The reader will note that the last step in Table 10.4-2 suggests that an evaluation of the compensated closed-loop system be carried out to ensure that the desired behavior has been obtained. This is an important step since, as has been pointed out numerous times previously, the design in terms of the open-loop transfer function is approximate and the resulting design should always be checked. This is particularly true if phase margin has been used as a specification, because of the possible lack of reliability in any phase-margin specification.

A convenient tool for an initial evaluation of the compensated closed-loop system is the root locus. Figures 10.4-9a and b present the root-locus diagrams associated with the original gain-compensated plant $KG_p(s)$ and the compensated plant $KG(s) = KG_c(s)G_p(s)$. An examination of these two diagrams reveals a number of interesting facts.

At the required design gain of 10, the compensated system has closed-loop poles at $s = -1.89$, $-3.61 \pm j3.87$, and $-50.99$ and a zero at $s = -1.27$. The system behavior, therefore, is fairly well dominated by the complex conjugate pair since the effect of the pole at $s = -1.89$ is minimized by the zero at $s = -1.27$. The set of complex poles has a damping ratio $\zeta$ of 0.73 and an undamped natural frequency $\omega_n$ of 5.3 rad/sec, which indicate a relatively well-behaved system. Hence we can conclude that our design is probably acceptable. If the damping ratio were too large, we could reduce the phase margin and repeat the design.

On the root locus associated with the uncompensated system, we observe that at a gain of 10 the closed loop is very close to being unstable, the damping ratio is almost zero. This confirms the need for compensation that we had determined previously from the Bode-diagram information.

It is also interesting to note that the closed-loop poles of the compensated system correspond quite closely to those of the uncompensated

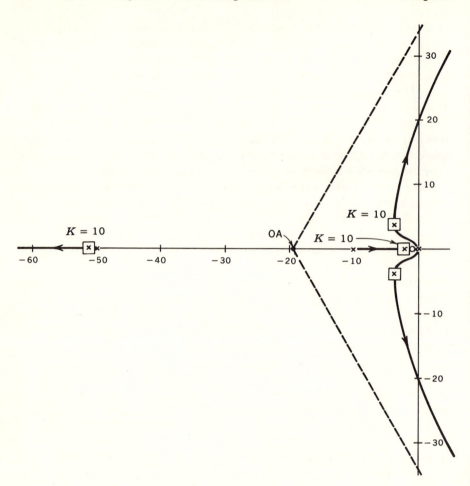

***Fig. 10.4-9a*** Root-locus diagrams for the lag-compensation example—compensated.

system with the gain reduced by a factor of 11.7. Of course, this is not surprising since the basic stabilizing vehicle of lag compensation is attenuation.

The synthesis problem has been solved, and in our discussion of the results on the root-locus diagram, we have begun to analyze the resulting closed-loop system. For example, in addition to knowing that the system is stable because the phase margin is 45°, we also know the locations of the closed-loop poles. Before we may conclude that the solution is actually satisfactory, we make a complete analysis of the resulting closed-loop system. For this example, the final system includes three closed-loop poles and one zero, and so the charts of Chap. 8 may be used to determine percent overshoot and rise time, for example. Before it is possible to conclude that the system is satisfactory, it is necessary to ensure that none of the state variables saturates over the given range of expected inputs. Thus we must return to the complete analysis techniques discussed in Chap. 4. We could also obtain the closed-loop frequency response by using the methods of Chap. 5.

In this section we have considered the first of three types of compensation: the lag compensator. The basic philosophy of lag compensation is to reduce the gain rapidly before crossover in order to achieve the

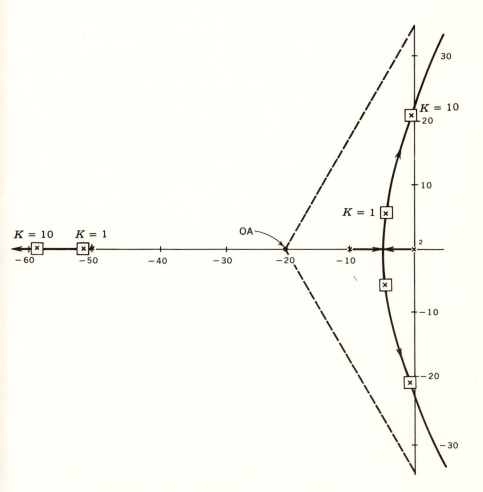

***Fig. 10.4-9b***   Root-locus diagrams for the lag-compensation example—uncompensated.

desired phase margin, while still meeting the low-frequency-asymptote requirement. The lag compensator has the advantage of being a simple passive device and also of providing high-frequency attenuation to help eliminate noise. The basic disadvantage of the lag compensator is the fact that it does not work on all plants. The uncompensated plant must possess sufficiently small phase lag at an acceptably high frequency before lag compensation may be contemplated. In addition, because lag compensation makes use of attenuation, it must necessarily lower the crossover frequency and the middle-frequency gain, leading to a more sluggish closed-loop response.

**Exercises 10.4**   *10.4-1.*   A 20:1 lag network is to be designed for a unity-feedback system whose open-loop transfer function is

$$G_p(s) = \frac{10}{s(s+1)}$$

The resulting phase margin should be 45°.

*answer:*

$$G_c(s) = \frac{1}{20} \frac{s + 0.142}{s + 0.0071}$$

*10.4-2.*   Design a lag compensation for the plant

$$G_p(s) = \frac{20}{s(1 + s/10)^2}$$

such that the compensated system has $K_v = 20$ and $\phi_m = 35°$. Select $\alpha$ for minimum middle-frequency attenuation.

*answer:*

$$K = 1 \qquad G_c(s) = \frac{1 + s/2.52}{1 + s/0.787}$$

*10.4-3.*   Repeat Exercise 10.4-2 but use the criterion that $\omega_c/\alpha\omega_1 = 10$ rather than minimum middle-frequency attenuation.

*answer:*

$$K = 1 \qquad G_c(s) = \frac{1 + s/0.517}{1 + s/0.133}$$

*10.4-4.*   Design a 5:1 lag network to compensate the plant

$$G_p(s) = \frac{5}{s(1 + s/20)}$$

such that the following requirements are met:

(a)   Errors due to sinusoidal signals for $\omega \leq 2$ should be less than 10 percent.

(b)   $\phi_m$ is as large as possible.

(c)   $K_v = 50$.

What is the resulting phase margin?

*answers:*

$$K = 10 \qquad G_c(s) = \frac{1 + s/4}{1 + s/0.8} \qquad \phi_m = 43°$$

**10.4-5.**   Find the value of $K$ for the closed-loop system

$$\dot{x} = \begin{bmatrix} 0 & 1 & 0 \\ 0 & 0 & -1,990 \\ 0 & 0 & -2,000 \end{bmatrix} x + \begin{bmatrix} 0 \\ 200 \\ 200 \end{bmatrix} u \qquad y = x_1 \qquad u = K(r - x_1)$$

such that the system has maximum-phase margin. What is the phase margin, and what is the crossover frequency?

*answers:*

$$K = 10^3 \qquad \phi_m = 78.5° \qquad \omega_c = 100 \text{ rad/sec}$$

**10.4-6.**   Use pure-attenuation compensation on the plant

$$G_p(s) = \frac{20}{s(1 + s/10)^2}$$

such that the phase margin is 45°.

*answer:*

$$K = 1/5.1$$

## 10.5   *Lead compensation*

The use of lag compensation is not always possible. In order to be able to apply lag compensation, there must be a frequency or preferably a range of frequencies for which the phase shift of the uncompensated plant is sufficient to provide the desired phase margin. For example, if a phase margin of 45° is desired, there must be at least one frequency for which the phase lag is no more than 135°. Of course, many uncompensated plants do not satisfy such a requirement. If $G_p(s)$ is $1/s^2$, for instance, the plant can never be made stable with lag compensation. In the case of systems for which lag compensation is not applicable, the lead compensator may prove to be a satisfactory solution.

The basic concept of lead compensation is to use series compensation to contribute phase lead *at crossover* in order to meet the phase-margin specification. This philosophy is to be contrasted with that of lag compensation where one utilizes gain reduction to improve the phase margin. In the lag design we attempt to minimize the phase lag attendant with the gain reduction. Here, by contrast, we attempt to minimize the gain increase associated with achieving the desired phase lead. In other words, in the lead design the phase lead added is the item of interest, whereas in the lag compensation the gain attenuation is of primary interest.

To illustrate the lead-compensation technique let us consider the following uncompensated, open-loop plant:

$$G_p(s) = \frac{10}{s(1 + s/10)} \tag{10.5-1}$$

The design specifications for this problem are as follows:

1. The velocity-error constant $K_v$ must be 100.
2. Phase margin should be approximately 45°.
3. Sinusoidal inputs of up to 1 rad/sec should be reproduced with $\leq 2$ percent error.

The amplitude diagram for the given plant is indicated by the solid line in Fig. 10.5-1a, and from the phase-angle diagram it is seen that the given plant has a phase margin of 45°, as required. However, the velocity-error constant of the given plant is only 10, and hence $K$ must be equal to 10 in order to ensure the desired low-frequency asymptote of $KG_p(s)$. The gain-compensated magnitude plot is indicated by the dashed line in the amplitude diagram of Fig. 10.5-1a. Now the phase margin is not adequate, and our first thought is the application of a lag compensator. However, the specification of a middle-frequency-gain requirement necessitates that $|KG(j\omega)| \geq 50$ for $\omega \leq 1$ rad/sec. This requirement is indicated by the shaded area in Fig. 10.5-1a, and the use of lag compensation would require that we enter this forbidden region. Thus, in this problem, lag compensation cannot be used, not because the phase lag is too large at all frequencies, as in a type 2 system, but rather because the phase lag is too large in the frequency range that is necessary to meet the middle-frequency specification.

Although lag compensation may not be used as a solution to the current problem, the situation is almost ideal for the application of lead compensation.    Lead compensation may be successfully applied here because no high-frequency-attenuation specification is given and because the phase lag of the uncompensated plant increases slowly after crossover. The importance of this last statement will be discussed in detail later.

Before beginning the development of the lead-compensation procedure, let us examine in more detail the lead compensator that was discussed briefly at the end of Sec. 10.3.    The transfer function of the

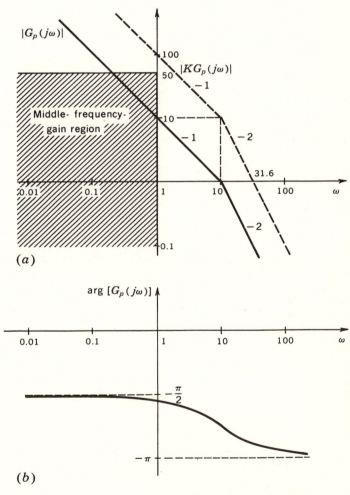

*(a)*

*(b)*

**Fig. 10.5-1**    Uncompensated plant.    (a) Magnitude plot; (b) phase plot.

lead compensator is

$$G_c(s) = \frac{1 + s/\omega_1}{1 + s/\alpha\omega_1} \qquad \alpha > 1$$

and the frequency-response plots take the form shown in Fig. 10.5-2. Of particular interest in the present development is the phase characteristic. The exact expression for the phase shift of $G_c(j\omega)$ is given by

$$\arg G_c(j\omega) = \arctan \frac{\omega}{\omega_1} - \arctan \frac{\omega}{\alpha\omega_1} \qquad (10.5\text{-}2)$$

In order to determine the frequency for which the maximum of the phase plot exists, we may set $d[\arg G_c(j\omega)]/d\omega$ equal to zero and solve for $\omega$:

$$\frac{d[\arg G_c(j\omega)]}{d\omega} = \frac{1/\omega_1}{1 + (\omega/\omega_1)^2} - \frac{1/\alpha\omega_1}{1 + (\omega/\alpha\omega_1)^2}$$

Therefore the frequency at which the maximum occurs is the solution of the equation

$$\frac{1/\omega_1}{1 + (\omega/\omega_1)^2} - \frac{1/\alpha\omega_1}{1 + (\omega/\alpha\omega_1)^2} = 0$$

for which it is easy to show that

$$\omega_{\max} = \sqrt{\alpha}\,\omega_1 \qquad (10.5\text{-}3)$$

In other words, the maximum phase lead of the compensation occurs at the geometric mean of the two break frequencies of the compensator. The actual value of the maximum phase lead may be determined

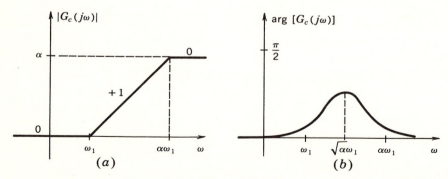

**Fig. 10.5-2**   Frequency-response plots for lead compensation.   (a) Magnitude plot; (b) phase plot.

by substituting $\omega_{max}$ from Eq. (10.5-3) into Eq. (10.5-2) as

$$\phi_{max} = \arg G_c(j\omega_{max}) = \arctan\sqrt{\alpha} - \arctan\frac{1}{\sqrt{\alpha}} \qquad (10.5\text{-}4)$$

From this result it is possible to show (see Exercise 10.5-1) that

$$\sin\phi_{max} = \frac{\alpha - 1}{\alpha + 1} \qquad (10.5\text{-}5)$$

or

$$\phi_{max} = \arcsin\frac{\alpha - 1}{\alpha + 1}$$

Let us return now to the problem of designing a lead compensator for the uncompensated plant of Eq. (10.5-1). As mentioned previously, the basic concept of lead compensation is the addition of phase lead at crossover in order to achieve the phase-margin specification. To obtain an estimate of the amount of phase lead needed, let us compute the phase shift at crossover of the gain-compensated plant. The crossover frequency may be obtained by setting $|KG_p(j\omega_c)| = 1$ and solving for $\omega_c$. Using the straight-line approximation and assuming that $10 < \omega_c$, we have

$$|KG_p(j\omega_c)| \sim \frac{100}{\omega_c(\omega_c/10)} = \frac{1{,}000}{\omega_c{}^2} = 1$$

or

$$\omega_c \sim 10\sqrt{10} = 31.6 \qquad (10.5\text{-}6)$$

The phase shift at crossover is therefore

$$\arg G_p(j\omega_c) \sim -\frac{\pi}{2} - \arctan\frac{10\sqrt{10}}{10} \sim -162.5°$$

In order to meet the phase-margin requirement of 45°, 162.5° − 135° or 27.5° of phase lead must be added. In actuality, however, more than 27.5° of phase lead must be added. When the lead compensator is added, the crossover frequency of the compensated plant increases, causing additional phase lag to be contributed by the pole at $s = -10$. To compensate for this added phase lag, we can increase slightly the amount of phase lead required, to 30°, for example. An alternative method of compensating for the increase in phase lag caused by the increase in crossover frequency is to compute the $\alpha$ associated with the phase lead needed at crossover of the gain-compensated plant and then to increase $\alpha$ slightly.

The larger the amount of phase lead becomes, the greater is the amount of high-frequency gain that is needed, as indicated in Sec. 10.3.

In order to minimize the gain that must be added, it is necessary to make optimal use of the phase lead created by the compensation. This may be done by placing the lead compensation so that it contributes its maximum lead at the crossover frequency of the compensated plant.[1] Stated in another manner, we wish the crossover frequency of the compensated plant to be at the geometric mean of the break frequencies of the lead compensation, since the geometric mean is the frequency at which the phase lead is maximum.

If we assume that the amount of phase lead needed is the maximum phase lead of the compensation, the necessary value of $\alpha$ may be obtained by solving Eq. (10.5-5) for $\alpha$:

$$\alpha = \frac{1 + \sin \phi_{max}}{1 - \sin \phi_{max}} \tag{10.5-7}$$

For the problem at hand, therefore, $\alpha$ becomes

$$\alpha = \frac{1 + \sin 30°}{1 - \sin 30°} = \frac{1.5}{0.5} = 3.0 \tag{10.5-8}$$

So far we know that the compensation has the form

$$G_c(s) = \frac{1 + s/\omega_1}{1 + s/3\omega_1} \tag{10.5-9}$$

and the compensated frequency response takes the form shown in Fig. 10.5-3. The problem now is to select the value of $\omega_1$. The general shape of the compensated plot shown in Fig. 10.5-3 is selected because it is known that $\omega_c$ must lie between $\omega_1$ and $\alpha\omega_1 = 3\omega_1$. The exact value for $\omega_1$ can be obtained by making use of the fact that $\omega_c$ is the geometric mean of $\omega_1$ and $3\omega_1$, or

$$\omega_c = \sqrt{\alpha}\,\omega_1 = \sqrt{3}\,\omega_1 \tag{10.5-10}$$

and that, at the crossover frequency $\omega_c$, $|KG(j\omega_c)| = 1$, or

$$|KG(j\omega_c)| \sim \frac{100(\omega_c/\omega_1)}{\omega_c(\omega_c/10)} = 1 \tag{10.5-11}$$

so that

$$\omega_1\omega_c \sim 1{,}000 \tag{10.5-12}$$

---

[1] This procedure is not exactly correct, and one should actually set up the problem of minimizing the value of $\alpha$ directly. The latter procedure is much more complicated, and the difference in $\alpha$ when calculated in both ways is usually negligible, particularly if the phase lag of the uncompensated system increases only slowly after crossover.

The simultaneous solution of Eqs. (10.5-10) and (10.5-12) yields

$\omega_c = 41.6$ rad/sec      and      $\omega_1 = 24$ rad/sec

The required lead compensation is therefore[1]

$$G_c(s) = \frac{1 + s/24}{1 + s/72} \qquad\qquad (10.5\text{-}13)$$

[1] If this transfer function is to be realized by the network of Fig. 10.3-6*d*, an additional gain of 3 is needed so that $K$ becomes 30.

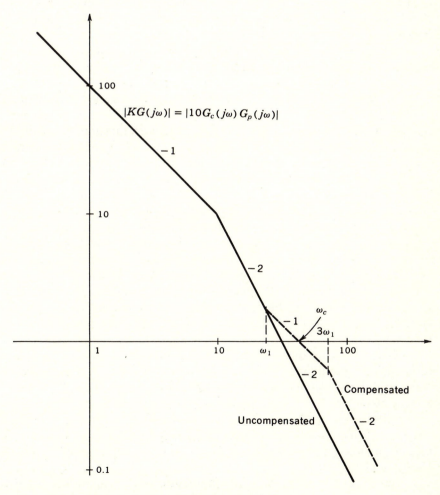

*Fig. 10.5-3*   Magnitude plots for lead compensation.

Since we made a rather arbitrary increase from 27.5 to 30° in the phase lead needed, the actual phase margin obtained should be checked. The phase shift at crossover for the compensated plant is

$$\arg KG(j\omega_c) = -\frac{\pi}{2} - \arctan\frac{41.6}{10} + \arctan\frac{41.6}{24} - \arctan\frac{41.6}{72}$$
$$= -136.5°$$

Hence the phase margin is only 43.5° rather than the desired 45°. The reason for this discrepancy lies in the fact that the phase shift of the uncompensated plant at the compensated crossover frequency increased by 1.5° more than the 2.5° allowed by increasing the phase lead from 27.5 to 30°. Note that we have used the straight-line approximations to establish the condition that $|KG(j\omega)| = 1$.

If the phase-margin requirement is very important, one should increase the value of the phase lead to be added to 33 or 35°, recalculate the value for $\alpha$, and so forth.

Since the phase margin achieved in the above design is already quite close to the desired 45°, rather than to rework the problem to realize more closely the 45° specification, a better approach would be actually to check the closed-loop system to see if it behaves as desired. The phase-margin specification is, at best, approximate, and repeated attempts at exactly satisfying it are hence somewhat foolish. In order to check the closed-loop behavior, the root-locus diagram and/or simulation often prove to be useful tools. In low-order systems, such as the present example, the use of the root-locus diagram is usually adequate because the time response of the output may be easily and accurately approximated from the root locus. In the case of high-order systems, simulation on either an analog or digital computer is the only recourse since the pole-zero diagram is so complex. Even in such systems, the root locus is often of assistance in analyzing the nature of the response and possible corrective techniques. Eventually the response of all state variables must be examined, not only the output.

The root-locus diagrams for the lead-compensated and the gain-compensated plants are shown in Figs. 10.5-4a and b. A comparison of these two diagrams once again reveals a number of interesting facts concerning the use of lead compensation. The gain-compensated system has a damping ratio of 0.158 and an undamped natural frequency of 31.6 rad/sec. Hence the need for additional compensation other than simple gain adjustment is obvious since with $\zeta = 0.158$ the system would be highly oscillatory.

At the design gain, $K = 10$, the compensated closed-loop system has poles at $s = -23.7 \pm j39.0$ and $-34.6$ and a zero at $s = -24.0$.

If one considers only the set of complex conjugate poles, the value of $\zeta$ is 0.52 and $\omega_n = 45.5$ rad/sec. It would be reasonable to expect that the closed-loop system would be well behaved and probably acceptable. If the system were too oscillatory, the value of the specified phase margin could be increased and the design repeated.

Approximately the same value of damping, $\zeta = 0.52$, can be achieved in the gain-compensated system by simply using the uncompensated plant

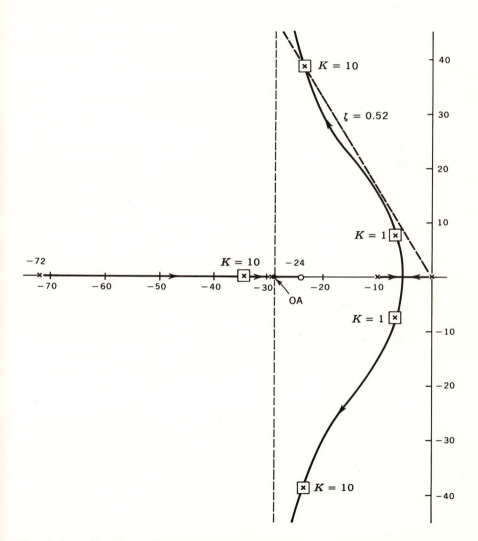

**Fig. 10.5-4a**   Root-locus diagrams for lead-design example—lead-compensated.

with $K = 1$.  Of course, the low-frequency-asymptote specification is violated, although this difficulty could be rectified by employing lag compensation rather than just pure attenuation.  If lag compensation were employed, the middle-frequency gain would be reduced more than is allowable, however.  With $K = 1$, the natural frequency $\omega_n$ of the gain-compensated plant is reduced to 1.0 rad/sec vs. 45.5 rad/sec for the lead-compensated system.

A comparison of Figs. 10.5-4$a$ and $b$ reveals that compensation has been achieved by modification of the asymptote location from $s = -5$

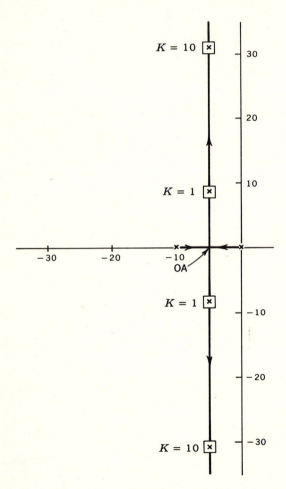

**Fig. 10.5-4b**    Root-locus diagrams for lead-design
example—gain-compensated.

to $-29$.   This fact should be contrasted with the lag-compensation case where the root locus remained relatively unchanged, and attenuation was the basic compensation tool.

One other item should be noted concerning the root locus of the lead-compensated system shown in Fig. 10.5-4a.   For gain values from $K = 1$ to $K = 10$, the damping ratio of the complex conjugate poles remains almost unchanged although, naturally, the undamped frequency decreases as the gain is decreased.   Although this feature is not necessarily inherent in the lead-compensation procedure, it is a desirable facet of the current design and indicates that lead compensation may be an acceptable solution for this system.

Since our development has centered about a specific example, in order to indicate the basic procedure of lead compensation, Table 10.5-1 summarizes the basic steps involved in the use of lead compensation.

*Table 10.5-1*   ***Summary of Lead-compensation Design***

| Step No. | Procedure | Applicable Equation Nos. |
|---|---|---|
| 1 | Sketch $|G_p(j\omega)|$ and arg $G_p(j\omega)$ and check if compensation is necessary and possible. | |
| 2 | Apply what gain compensation is required by the low-frequency-asymptote specification. | |
| 3 | Determine crossover frequency and phase angle at crossover for $KG_p(j\omega)$. | (10.5-6) |
| 4 | Estimate the amount of phase lead needed to achieve the phase-margin specification. | |
| 5 | Compute the value of the gain $\alpha$ for the lead compensation. | (10.5-7) |
| 6 | Place the compensation so that maximum phase lead is contributed at crossover. | (10.5-10) (10.5-11) |
| 7 | Check the phase-margin specification and repeat the design if necessary. | |
| 8 | Check the final design to ensure that the desired closed-loop behavior has been achieved. | |

The reader is once again cautioned not to think of this table as an exact design procedure to be followed step by step but rather as an aid to understanding the general design procedure.

The reader may wonder why we do not exactly satisfy the phase-margin specification here as we did in the preceding section. By using the arctangent approximations in the equation

$$\arg KG(j\omega_c) = -\pi + \phi_m \tag{10.5-14}$$

it is possible to form a third equation involving $\alpha$, $\omega_1$, and $\omega_c$. If this equation is used along with Eqs. (10.5-10) and (10.5-11) to obtain the value of $\alpha$ rather than determining $\alpha$ by estimating the amount of phase lead needed, the phase-margin specification would be met. This approach has two problems, however. First, the computational labor involved in this approach is far greater than that of the method presented above. Second, the accuracy associated with the arctangent approximation is not much greater than the accuracy with which the phase-margin specification can be met by using the approximate approach with the exact arctangent expressions. In addition, since phase margin is an approximate specification for all but simple second-order systems, the need to realize exactly a given phase-margin specification is questionable.

Two basic disadvantages are inherent in the lead-compensation procedure. First, the lead compensator increases the high-frequency gain and hence accentuates any noise that may be present. In addition, because of the gain required, the compensator can no longer be composed of solely passive elements.

The second disadvantage is that there are some plants for which lead compensation may not be used. In particular, if the phase shift of the uncompensated plant increases rapidly at and above the crossover frequency, then lead compensation may be impractical if not impossible. To illustrate this problem, let us consider the following plant:

$$G_p(s) = \frac{10}{s^2(1 + s/10)^2}$$

The magnitude and phase plots for this plant are shown in Fig. 10.5-5. Lag compensation cannot be used because there is no point where the phase lag is less than 180°. The phase shift at crossover for the uncompensated plant is

$$\arg G_p(j\omega_c) = -\pi - 2 \arctan 0.316 = -215°$$

Therefore the minimum value that $\alpha$ could have if the desired phase margin is 45° would be

$$\alpha = \frac{1 + \sin 80°}{1 - \sin 80°} = 132$$

Clearly the value is already prohibitive. However, if this value of $\alpha$ were to be used, the crossover frequency would increase so much that the phase lag at crossover would be greater than 225°. Since the maximum amount of lead that may be obtained from a first-order lead compensator is 90°, compensation by this means would clearly be impossible. At present, it would appear there is no method of compensating this plant, other than to return to the use of state-variable feedback. However, we shall see in the next section that by combining the basic lead- and lag-compensation techniques it is possible to achieve acceptable performance even for systems of this nature.

The basic advantage of lead compensation is that it allows one to achieve the desired closed-loop transient behavior without sacrificing the middle-frequency gain. In fact, the design increases the crossover frequency of the open-loop transfer function and in general improves the speed-of-response characteristics, by increasing the bandwidth of the closed-loop system.

A simple comparison of the advantages and disadvantages of lead and lag compensation indicates that the advantages of one tend to be the disadvantages of the other. The lead-lag-compensation technique discussed in the next section makes use of this fact to offset the disadvantages

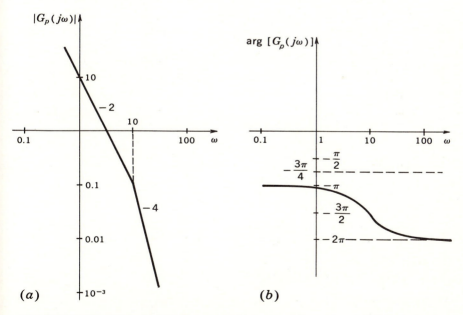

**Fig. 10.5-5** Plant for which lead compensation cannot be used. (*a*) Magnitude plot; (*b*) phase plot.

of one method with the advantages of the other.  By doing so, not only are a greater number of systems capable of being handled but in addition the results are better.

**Exercises 10.5**    *10.5-1.*    Show that

$$\sin \phi_{\max} = \frac{\alpha - 1}{\alpha + 1}$$

for the lead compensator by making use of Eq. (10.5-4).

*10.5-2.*    Repeat the lead-compensation design of this section for the plant of Eq. (10.5-1) but make use of Eq. (10.5-14) and the arctangent approximations to achieve "exactly" the 45° phase-margin specification.

*10.5-3.*    Use a lead network to compensate a unity-feedback system whose open-loop uncompensated transfer function is

$$G_p(s) = \frac{100}{s(s + 1)}$$

The velocity-error constant should be 100, and the phase margin should be approximately 45°.  Draw a root locus for the final design.

*answers:*

$$K = 6 \qquad G_c(s) = \frac{s + 6.5}{s + 39}$$

*10.5-4.*    Design a 10:1 lead network to compensate the plant

$$G_p(s) = \frac{100}{s(s + 1)}$$

so that the compensated system satisfies the following specifications:

(*a*)    The crossover frequency should be maximized.
(*b*)    $\phi_m \sim 45°$.

*answers:*

$$K = 10 \qquad G_c(s) = \frac{s + 4}{s + 40}$$

## 10.6    *Lead-lag compensation*

In the two previous sections lead- and lag-compensation techniques have been developed as separate approaches.  Here, by combining the two approaches to generate a composite lead-lag-compensation method, we show how it is possible to achieve better results than when either of the methods is used separately.

Lag compensation is based on the concept that the phase lag is usually monotonically increasing with frequency and that by reducing the crossover frequency the phase margin may be increased. The basic stabilizing tool is therefore attenuation. Stated in another manner, the stability of most systems can be improved by a reduction of the middle-frequency gain. Although lag compensation is simple and passive, it has the disadvantage that the closed-loop system may become sluggish due to the reduction in bandwidth caused by decreasing the crossover frequency of the open-loop transfer function. In addition, lag compensation cannot be applied if there is not an acceptable range of frequency for which the phase lag is sufficiently small to satisfy the phase-margin specification. Because of the attenuation associated with lag design, the high-frequency attenuation is increased, thereby improving the noise rejection of the system.

On the other hand, the basic concept of lead compensation is to add phase lead near crossover in order to increase the phase margin. Because the lead compensation requires gain in order to generate phase lead, the advantages of a passive compensation are lost. In addition, the attendant increase in the crossover frequency may cause the approach to fail if the inherent phase lag of the given plant increases rapidly near crossover. The increase in the middle-frequency gain, however, improves the speed-of-response characteristics of the system although the high-frequency attenuation is also reduced, leading to greater noise transmission through the system.

A simple comparison of the features of lead and lag compensation reveals that the two methods are almost exactly complementary. Lead compensation increases crossover frequency whereas lag compensation decreases it; lead compensation decreases high-frequency attenuation whereas lag increases it; and so forth. The two approaches are therefore ideally suited to be used together.

In the composite lead-lag compensation, attenuation is used below crossover, as in lag compensation, in order that phase lead can be added at crossover, as in lead compensation. Although this procedure requires that the middle-frequency gain be reduced, the reduction is much less than in pure lag design because of the added phase lead. The reduction of the crossover frequency, on the other hand, decreases the amount of phase lead needed as opposed to the pure lead design where the increase in crossover frequency requires additional lead to be used.

To illustrate the characteristics of the composite compensation more clearly, let us consider the general lead-lag compensation discussed briefly at the end of Sec. 10.3.   The transfer function of composite compensation is

$$G_c(s) = \frac{(1 + s/\beta\omega_1)(1 + s/\omega_2)}{(1 + s/\omega_1)(1 + s/\alpha\omega_2)} \tag{10.6-1}$$

where $\alpha$, $\beta > 1$ and $\beta\omega_1 \leq \omega_2$.   The frequency-response plots of the compensation are given in Fig. 10.6-1.   Although the compensation is passive, $|G_c(j\omega)| < 1$ for all $\omega$ if $\alpha < \beta$, phase lead is generated by the $+1$ slope between $\omega_2$ and $\alpha\omega_2$.   If the compensation is placed so that the crossover

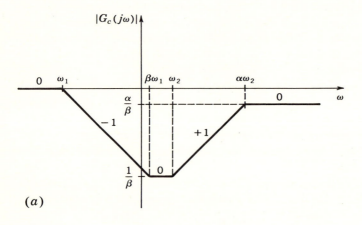

(a)

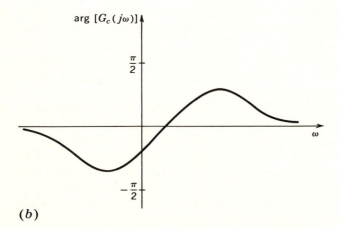

(b)

**Fig. 10.6-1**   Frequency response of lead-lag compensation.   (a) Magnitude; (b) phase.

of the compensated open-loop transfer function $KG(s)$ occurs near $\alpha\omega_2$, phase lead is added and the phase margin is increased.  Since the compensation has a gain of less than 1 at all frequencies, the reduction in crossover frequency that occurs also increases the phase margin if the plant has the usual monotonically increasing phase-lag characteristic.

The design procedure that we shall use for lead-lag compensation is almost completely trial and error in form.  Although a purely analytic procedure is possible, the computational labor far exceeds that of a trial-and-error approach, since usually two or three trial designs are sufficient to reach a final design.  In addition, we shall rely heavily on graphical methods based on the straight-line magnitude plots of $KG_p(s)$ and $KG(s)$.

Because of the large number of possible forms that $G_p(s)$ and hence $G(s)$ may take, the steps in the design procedure must remain rather general and nonspecific.  The basic concept of the design is to achieve the maximum possible phase margin consistent with the simple second-order lead-lag compensation of Eq. (10.6-1) and the specifications.  If the phase margin obtained is too large, adjustments may be made to strengthen the specifications and/or simplify the design.  If the required phase margin is not obtained, then either the specifications must be adjusted or a more complex compensation must be used.

The general design procedure is outlined in Table 10.6-1.  In using this procedure, one must remember that the slopes of $|KG_p(j\omega)|$ and $|KG(j\omega)|$ can never differ by more than 1 at any given frequency since $|G_c(j\omega)|$ never has a slope of magnitude greater than 1.  Since the initial and final slopes of $|G_c(j\omega)|$ are zero, the initial and final slopes of $|KG_p(j\omega)|$ and $|KG(j\omega)|$ must be identical.

To illustrate the design procedure for lead-lag compensation, let us consider once again the plant discussed in Sec. 10.5 in terms of lead compensation, namely,

$$G_p(s) = \frac{10}{s(1 + s/10)}$$

The specifications were as follows:

1.  The velocity-error constant $K_v$ must be 100.
2.  Phase margin must be approximately 45°.
3.  Sinusoidal inputs of up to 1 rad/sec should be reproduced with $\leq 2$ percent error.

To this set of specifications, let us add the high-frequency attenuation:

4.  Sinusoidal inputs of greater than 100 rad/sec should be attenuated at the output to 5 percent of their value at the input.

## Table 10.6-1    Outline of Lead-lag-compensation Design

| Step No. | Procedure |
|----------|-----------|
| 1 | Sketch the straight-line magnitude plot of $G_p(j\omega)$ and transfer the low-, middle-, and high-frequency specifications onto the plot. |
| 2 | Make the necessary gain adjustment to satisfy the low-frequency-asymptote specification and sketch $|KG_p(j\omega)|$. |
| 3 | Sketch a plot of $|KG(j\omega)|$ in the low- and middle-frequency ranges, using as much middle-frequency attenuation as possible consistent with the middle-frequency-gain specifications. |
| 4 | Sketch a plot of $|KG(j\omega)|$ in the high-frequency range, using as much high-frequency gain as possible consistent with the required high-frequency attenuation. |
| 5 | Complete the sketch of $|KG(j\omega)|$ by connecting the middle- and high-frequency plots in such a fashion as to maximize the phase margin. |
| 6 | Check the design and make any desired adjustments. |

Specification 4 requires that the $|KG(j\omega)| \leq 0.05$ for $\omega \geq 100$ rad/sec. Figure 10.6-2 illustrates the magnitude plot of $|G_p(j\omega)|$ with a solid line and the desired specifications. Again adjustment is made immediately to satisfy the $K_v$ requirement, and, as before, a gain of 10 is required. The plot of $KG_p(s)$ is indicated by the dotted line in Fig. 10.6-2; this amplitude diagram is used in the remainder of the discussion. The high-frequency-attenuation specification rules out the use of lead compensation since no high-frequency gain may be used. We have seen in Sec. 10.5 that lag also cannot be used. However, this problem can be solved quite easily with the use of lead-lag compensation.

The third step of the design procedure of Table 10.6-1 is shown in Fig. 10.6-3a. For $\omega \leq 10$ rad/sec, the slope of $|KG(j\omega)|$ cannot be more negative than $-2$. Hence the maximum middle-frequency attenuation is achieved by moving the $-2$ slope until the middle-frequency-gain specification is met. Of course, $KG(s)$ must merge with $KG_p(s)$ in the low-frequency range in order to have $K_v = 100$. From this plot we can read $\omega_1$ for $G_c(s)$ as 0.5.

In Fig. 10.6-3b the tentative plot of $|KG(j\omega)|$ in the high-frequency range has been made. This final $-2$ slope has been moved as far to the right as possible consistent with the high-frequency-attenuation requirement.

The completion of the fourth step of the design procedure of Table 10.6-1 requires that the two partial plots of $|KG(j\omega)|$ shown in Fig. 10.6-3 be joined in such a manner as to maximize $\phi_m$. In terms of the present problem, the two plots may be joined by the use of a segment of $-1$ slope, as shown in Fig. 10.6-4. The object is to place the $-1$-slope segment so that the phase margin is maximized. One way of accomplishing this is to select various values for the crossover frequency and to evaluate the phase margin associated with each configuration. The value of $\omega_c$ for which the phase margin is maximized is the desired design.

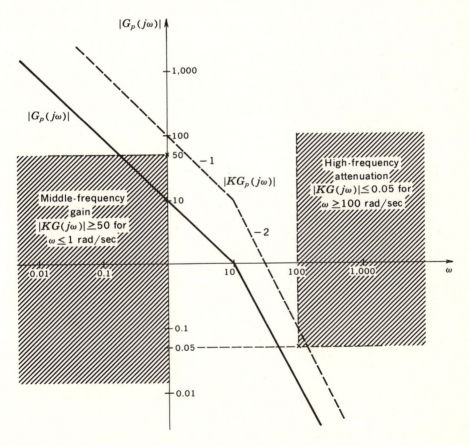

*Fig. 10.6-2*   Magnitude plot of the plant and the specifications.

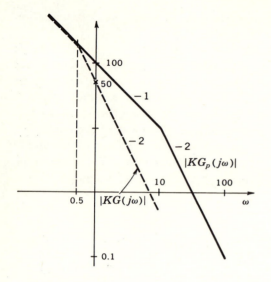

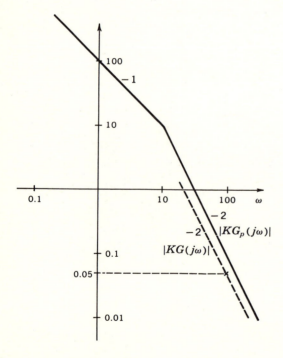

**Fig. 10.6-3** Tentative form for $|KG(j\omega)|$. (a) Low and middle frequency; (b) high frequency.

For example, if $\omega_c = 10$ rad/sec, then $KG(s)$ becomes

$$KG(s) = \frac{100(1 + s/5)}{s(1 + s/0.5)(1 + s/50)}$$

The exact phase shift at crossover is $\phi_c = -125°$ so that the phase margin is $\phi_m = 55°$. This phase margin is already more than adequate to meet the phase-margin specification and, in that sense, the design is complete. However, let us examine some other selections for $\omega_c$ and see if a larger

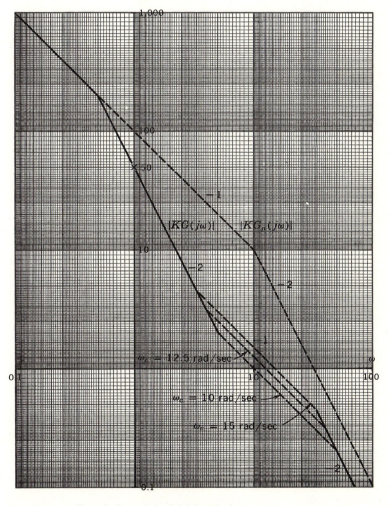

***Fig. 10.6-4***    Completion of the lead-lag design.

value for $\phi_m$ may be obtained.    For $\omega_c = 12.5$ rad/sec, the phase margin is 59.5°, and for $\omega_c = 15$ rad/sec, $\phi_m = 56.2°$.

Hence let us select the design with $\omega_c = 12.5$ rad/sec, for which $KG(s)$ is

$$KG(s) = \frac{100(1 + s/4)}{s(1 + s/0.5)(1 + s/40)}$$

The required series compensation is therefore

$$G_c(s) = \frac{KG(s)}{KG_p(s)} = \frac{(1 + s/4)(1 + s/10)}{(1 + s/0.5)(1 + s/40)}$$

The magnitude plot for $G_c(j\omega)$ is shown in Fig. 10.6-5 where we see that $\omega_1 = 0.5$, $\beta = 8$, $\omega_2 = 10$, and $\alpha = 4$.    Since the magnitude of $G_c(j\omega)$ is never greater than 1, $G_c(s)$ may be realized by means of a passive network. By the use of lead-lag compensation, we have been able to achieve a larger phase margin than that obtained by pure lead compensation, without the addition of any high-frequency gain with its inherent realization and noise-rejection problems.    Of course, we have paid for this improvement with reduced middle-frequency gain and crossover frequency, although the middle-frequency-gain specification has been met.

Since the phase margin is more than 10° greater than required, the design could be adjusted to improve the middle-frequency gain or the high-frequency gain.    If, for example, the middle-frequency-gain spec-

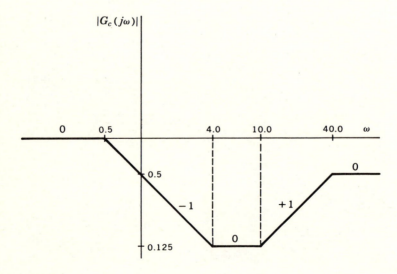

**Fig. 10.6-5**   Magnitude plot of $G_c(j\omega)$.

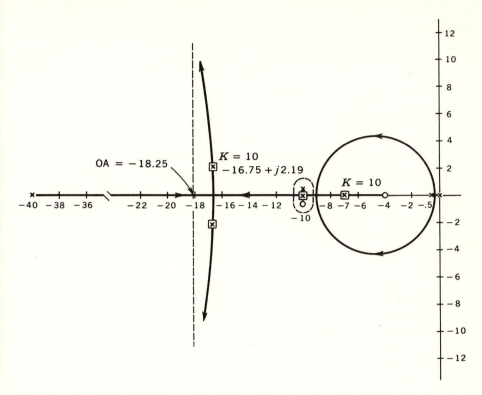

*Fig. 10.6-6*   Root locus for the lead-lag design.

ification were increased so that $|KG(j\omega)| \geq 100$ for $\omega \leq 1$ rad/sec, the problem could be reworked to see if the phase margin could still be met.

Let us assume that the additional phase margin is acceptable and consider the analysis of the closed-loop system by the use of a root-locus diagram for the compensated system, as shown in Fig. 10.6-6. The root locus for the gain-compensated system is given in Sec. 10.5 as Fig. 10.5-4b. From Fig. 10.6-6, we see the effect of the additional phase margin; here the damping ratio has become approximately 1 and the system may be too sluggish. If this were a practical application, one would probably recompute the design with either larger middle-frequency gain, greater high-frequency attenuation, or larger $K_v$.

Since the lead-lag network degenerates to the simple lead and lag compensators as special cases, the composite compensation has all the advantages of the separate forms. Because of the flexibility of the lead-lag compensation, it tends to be the most practical of the series-compensa-

tion methods.  The only major disadvantage of the method is the need
for the more complicated second-order compensation.

**Exercises 10.6**   *10.6-1.*   Design a series compensation for a unity-
feedback system with an open-loop transfer function of

$$G_p(s) = \frac{10}{s(s + 1)}$$

The compensated system must meet the following specifications:

(a)   Errors due to sinusoidal inputs for frequencies less than 0.1
rad/sec must be less than 1 percent.
(b)   Inputs of frequencies $\omega \geq 100$ rad/sec must be attenuated 10
to 1 at the output.
(c)   $K_v = 100$.
(d)   $\phi_m$ as large as possible.

What is the phase margin of the resulting system?

*answer:*

$K = 100, \ \phi_m = 86°$

$$G_c(s) = \frac{(s + 0.18)(s + 1)}{(s + 0.01)(s + 180)}$$

*10.6-2.*   Find a series compensation for the plant

$$G_p(s) = \frac{500}{s(1 + s/10)}$$

such that:

(a)   The error for frequencies less than 2 rad/sec are less than 1
percent.
(b)   The error due to noise inputs is less than 1 percent for
$\omega \geq 1,000$ rad/sec.
(c)   $\phi_m$ is as large as possible.
(d)   $K_v = 1,000$.

*answer:*

$$K = 2 \qquad G_c(s) = \frac{(s + 8.6)(s + 10)}{(s + 0.41)(s + 210)}$$

*10.6-3.*   It is proposed to compensate the plant

$$G_p(s) = \frac{100}{s(1 + s/10)}$$

with the series compensation

$$G_c(s) = \frac{(1 + s/5)(1 + s/10)}{(1 + s)(1 + s/50)}$$

Evaluate this design by the following procedure:

(a) Plot the root locus of the compensated and uncompensated systems.

(b) From the root locus, estimate the locations of the closed-loop poles of the two systems.

(c) Evaluate the effect of the series compensation.

*answer:*

(b) Uncompensated closed-loop poles are at $s = -5 \pm j31.23$. Compensated closed-loop poles are at $s = -6.61, -22.2 \pm j16.2$.

*10.6-4.* For each of the three open-loop transfer functions given below

(1)    $G_p(s) = \dfrac{1{,}000}{s^2(1 + s/10)}$

(2)    $G_p(s) = \dfrac{100}{s^2}$

(3)    $G_p(s) = \dfrac{100(1 - s)}{s(1 + s)(1 + s/10)}$

determine which of the following types of compensation can be used to stabilize each system:

(a) Pure gain adjustment

(b) Lag

(c) Lead

(d) Lead-lag

*answers:*

(1) $d$; (2) $c, d$; (3) $a, b, d$

## 10.7   *Conclusions*

This chapter has been involved with the problem of applying the state-variable-feedback method of Chap. 9 in the important practical case where all the state variables are not available for direct measurement. The early portions of this chapter dealt with direct approaches to the problem of generating the missing state variables, and the latter portions treated series-compensation methods for the case when only the output is available.

The problem of control-system synthesis is multifaceted, involving complex interrelations of cost, weight, size, and, in many cases least importantly, performance. In the treatment of synthesis in Chaps. 9 and 10, we have considered only the aspects of design related to practical performance capability. Even with this rather limited outlook, we have not been able to cover the entire subject of the synthesis of linear constant-coefficient systems. We have discussed the methods that we believe offer the greatest promise for future success and/or have been used most successfully in the past.

There are also many aspects of modern control theory that are beyond the scope of this modest treatment,[1] such as optimization theory, nonlinear and time-varying systems, sampled-data systems, estimation theory, and adaptive systems. Ideally, this book has served to whet the reader's interest in the study of these advanced topics and at the same time prepared him to undertake such a study.

## 10.8   Problems

*10.8-1.*   Use both lead- and lag-compensation methods to compensate the open-loop transfer function

$$G_p(s) = \frac{5}{s(s + 10)}$$

The system should have $K_v = 50$ and $\phi_m = 45°$ in both cases. Draw the root locus for both systems as well as for the uncompensated system. Make use of the design charts in Secs. 8.4 and 8.5 to compare the behavior of the three systems.

*10.8-2.*   Find a lead-lag compensation for the system associated with the open-loop transfer function

$$G_p(s) = \frac{10}{s^2(1 + s/10)^2}$$

so that the system has the following:

(a)   $\phi_m = 45°$
(b)   $K_a = 10$
(c)   No high-frequency gain

---

[1] See, for example, Donald G. Schultz and James L. Melsa, "State Functions and Linear Control Systems," McGraw-Hill Book Company, New York, 1967.

It was shown in Sec. 10.5 that this system cannot be compensated with either pure lead or pure lag compensation.

**10.8-3.** Show that lag compensation cannot be used on the plant

$$\dot{x} = \begin{bmatrix} 0 & 1 & 0 \\ 0 & 2 & 1 \\ 0 & 0 & -10 \end{bmatrix} x + \begin{bmatrix} 0 \\ 0 \\ 1 \end{bmatrix} u \qquad y = x_1$$

Use linear state-variable feedback to control the plant so that the closed-loop transfer function is

$$\frac{y(s)}{r(s)} = \frac{80}{[(s+2)^2 + 2^2](s+10)}$$

Can a series compensator be used to obtain the same closed-loop transfer function?

**10.8-4.** Find a series compensation for the plant

$$G_p(s) = \frac{5(1-s)}{s(s+10)}$$

such that

(a) $\phi_m = 45°$
(b) $K_v = 50$
(c) No high-frequency gain is added.

**10.8-5.** Find a closed-loop-system configuration for the plant

$$\dot{x} = \begin{bmatrix} -1 & 1 & 0 \\ 0 & -2 & 1 \\ 0 & 0 & -10 \end{bmatrix} x + \begin{bmatrix} 0 \\ 0 \\ 1 \end{bmatrix} u \qquad y = x_1$$

such that the closed-loop transfer function is

$$\frac{y(s)}{r(s)} = \frac{500}{[(s+5)^2 + 5^2](s+10)}$$

Plot the root locus for the system. What is the equivalent series compensator that realizes the desired closed-loop transfer function? Draw the Bode diagrams for $KG(s)$ and $KG_p(s)$ and compare to lead-lag design.

**10.8-6.** Repeat Prob. 10.8-5 but assume that $x_2$ is inaccessible.

**10.8-7.** A satellite in orbit can be accurately approximated as a purely inertial system having a transfer function $G_p(s) = 1/Js^2$. For this problem let $J = 0.01$. Find a closed-loop-system configuration such that the step response of the closed-loop system has 10 percent overshoot and a rise time of approximately 1 sec.

# appendix

## *A* *the Laplace transform— a summary*

This appendix is designed to serve as a brief review of the properties of the Laplace transform method that are needed in the developments of this book. As such, it does not contain a complete treatment of all phases of the Laplace transform nor does it serve as an introduction to that subject. The reader who is not familiar with Laplace transforms is directed to one of the many available textbooks for an introductory treatment.[1]

---

[1] See, for example, R. V. Churchill, "Operational Mathematics," McGraw-Hill Book Company, New York, 1958; D. L. Holl, C. G. Maple, and B. Vinograde, "Introduction to the Laplace Transform," Appleton-Century-Crofts, Inc., New York, 1959; or S. C. Gupta, "Transform and State Variable Methods in Linear Systems," John Wiley & Sons, Inc., New York, 1966.

*Definitions.*    The Laplace transform of a time function $f(t)$, written as $f(s) = \mathcal{L}[f(t)]$, is defined by the following integral operation:

$$f(s) = \mathcal{L}[f(t)] = \int_0^\infty f(t)e^{-st}\, dt \tag{A-1}$$

In dealing with the Laplace transform definition of Eq. (A-1), we assume that $f(t)$ is zero for $t < 0$.  In order for $\mathcal{L}[f(t)]$ to exist, it is necessary and sufficient that $f(t)$ be *sectionally continuous* and of *exponential order*.

A function $f(t)$ is sectionally continuous if it possesses at most a finite number of finite discontinuities in any finite interval.  A function $f(t)$ is of exponential order if $|f(t)| \leq Me^{-\alpha t}$ for some numbers $M$ and $\alpha$ and all values of $t$ larger than some value $b$.  In effect, this last statement simply means that $f(t)e^{-\alpha t}$ approaches zero as $t$ approaches infinity; all functions of the form $f(t) = t^n e^{\lambda t} \sin \omega t$ are of exponential order.

The Laplace transform possesses a unique inverse transform $f(t) = \mathcal{L}^{-1}[f(s)]$ given by the expression

$$f(t) = \mathcal{L}^{-1}[f(s)] = \frac{1}{2\pi j} \int_{\sigma - j\infty}^{\sigma + j\infty} f(s)e^{st}\, ds \tag{A-2}$$

Here $\sigma$ is chosen to be greater than the real part of any of the poles of $f(s)$, thus ensuring the existence of an inverse.  In practice, rather than use Eq. (A-2), one normally obtains a partial-fraction expansion[1] of $f(s)$ and makes use of a table of Laplace transforms such as the one given in Appendix B.

*Transform properties.*    Some of the more common properties of the Laplace transform are given below.  The proofs for these have been kept simple; the reader is again directed to the available textbooks for a more complete delineation of the properties of the Laplace transform and rigorous proofs of these properties.

1.  Superposition.    For any constants $a_1$ and $a_2$ and any two functions $f_1(t)$ and $f_2(t)$ that possess Laplace transforms,

$$\begin{aligned}
\mathcal{L}[a_1 f_1(t) + a_2 f_2(t)] &= a_1 \mathcal{L}[f_1(t)] + a_2 \mathcal{L}[f_2(t)] \\
&= a_1 f_1(s) + a_2 f_2(s)
\end{aligned} \tag{A-3}$$

---

[1] See Sec. 4.2 for a detailed discussion of partial-fraction expansion.

Proof.   By the use of the definition of Eq. (A-1), $\mathcal{L}[a_1 f_1(t) + a_2 f_2(t)]$ is given by

$$\mathcal{L}[a_1 f_1(t) + a_2 f_2(t)] = \int_0^\infty [a_1 f_1(t) + a_2 f_2(t)]e^{-st}\, dt$$

$$= a_1 \int_0^\infty f_1(t)e^{-st}\, dt + a_2 \int_0^\infty f_2(t)e^{-st}\, dt$$

The last two integrals may be recognized as the Laplace transforms of $f_1(t)$ and $f_2(t)$, respectively, so that we have established that

$$\mathcal{L}[a_1 f_1(t) + a_2 f_2(t)] = a_1 f_1(s) + a_2 f_2(s)$$

In simple terms, this first property is nothing more than a statement that the Laplace transform is a linear operator.

2.   Derivatives.   The Laplace transform of the time derivative of a function $f(t)$ is given by $s$ times the Laplace transform of $f(t)$ minus the initial value of $f(t)$; that is,

$$\mathcal{L}\left[\frac{df(t)}{dt}\right] = \mathcal{L}[f'(t)] = sf(s) - f(t)\,\Big|_{t=0} \qquad \text{(A-4)}$$

where $f(s) = \mathcal{L}[f(t)]$.
Proof.   By definition, $\mathcal{L}[f'(t)]$ is given by

$$\mathcal{L}[f'(t)] = \int_0^\infty f'(t)e^{-st}\, dt \qquad \text{(A-5)}$$

After applying the integration-by-parts formula

$$\int_0^\infty u\, dv = uv\,\Big|_0^\infty - \int_0^\infty v\, du$$

to Eq. (A-5) with $u = e^{-st}$ and $dv = f'(t)\, dt$, we obtain

$$\mathcal{L}[f'(t)] = f(t)e^{-st}\,\Big|_{t=0}^{t=\infty} + s\int_0^\infty f(t)e^{-st}\, dt$$

$$= -f(t)\,\Big|_{t=0} + sf(s)$$

In the last step we have made use of the fact that $f(t)$ is of exponential order and have set $\lim_{t \to \infty} f(t)e^{-st} = 0$.

By continuing the above procedure $k$ times, one can easily establish the general rule of derivatives as

$$\mathcal{L}\left[\frac{d^k f(t)}{dt^k}\right] = s^k f(s) - s^{k-1} f(t)\,\Big|_{t=0} - s^{k-2}\frac{df(t)}{dt}\,\Big|_{t=0} - \frac{d^{k-1} f(t)}{dt}\,\Big|_{t=0}$$

$$\text{(A-6)}$$

3.  Integration.   The Laplace transform of the integral of a function $f(t)$ is given by $1/s$ times the Laplace transform of $f(t)$ if $f(t)\big|_{t=0}$ is assumed to be finite; that is,

$$\mathcal{L}\left[\int_0^t f(\tau)\,d\tau\right] = \frac{1}{s} f(s) \tag{A-7}$$

Proof.   By the use of the definition, $\mathcal{L}\left[\int_0^t f(\tau)\,d\tau\right]$ becomes

$$\mathcal{L}\left[\int_0^t f(\tau)\,d\tau\right] = \int_0^\infty e^{-st}\left[\int_0^t f(\tau)\,d\tau\right] dt$$

By applying integration by parts to this expression with $u = \int_0^t f(\tau)\,d\tau$ and $dv = e^{-st}\,dt$, we obtain

$$\mathcal{L}\left[\int_0^t f(\tau)\,d\tau\right] = -\frac{1}{s} e^{-st} \int_0^t f(\tau)\,d\tau \Big|_0^\infty + \frac{1}{s}\int_0^\infty f(t)e^{-st}\,dt$$

$$= \frac{1}{s} f(s)$$

4.  Real translation.   The Laplace transform of a time function delayed by time $T$ is given by $e^{-sT}$ times the Laplace transform of $f(t)$; that is,

$$\mathcal{L}[f(t-T)\mu(t-T)] = e^{-sT}f(s) \tag{A-8}$$

where $\mu(t)$ is the unit step function.
Proof.   By definition, $\mathcal{L}[f(t-T)\mu(t-T)]$ is given by

$$\mathcal{L}[f(t-T)\mu(t-T)] = \int_0^\infty f(t-T)\mu(t-T)e^{-st}\,dt$$

$$= \int_T^\infty f(t-T)e^{-st}\,dt \tag{A-9}$$

Now let $\tau = t - T$ so that Eq. (A-9) becomes

$$\mathcal{L}[f(t-T)\mu(t-T)] = \int_0^\infty f(\tau)e^{-s(\tau+T)}\,d\tau$$

$$= e^{-sT}\int_0^\infty f(\tau)e^{-s\tau}\,d\tau = e^{-sT}f(s)$$

5.  Complex translation.   The Laplace transform of $e^{-\alpha t}$ times $f(t)$ is equal to the Laplace transform of $f(t)$ with $s$ replaced by $s + \alpha$; that is,

$$\mathcal{L}[e^{-\alpha t}f(t)] = f(s + \alpha) \tag{A-10}$$

Proof. By definition, $\mathcal{L}[e^{-\alpha t}f(t)]$ is

$$\mathcal{L}[e^{-\alpha t}f(t)] = \int_0^\infty e^{-\alpha t}f(t)e^{-st}\,dt$$

$$= \int_0^\infty f(t)e^{-(s+\alpha)t}\,dt = f(s+\alpha)$$

6. Initial value. If the initial value of $f(t)$ is finite, then $f(t)\big|_{t=0}$ is equal to the limit of $sf(s)$ as $s$ approaches infinity; that is,

$$f(t)\,\big|_{t=0} = \lim_{s\to\infty} sf(s) \tag{A-11}$$

Note that this is an *equality*, not a transform pair.

Proof. Let us consider the Laplace transform of $f'(t)$ given by

$$\mathcal{L}[f'(t)] = sf(s) - f(t)\,\big|_{t=0} = \int_0^\infty f'(t)e^{-st}\,dt \tag{A-12}$$

Taking the limit as $s$ approaches $\infty$ on both sides of Eq. (A-12) we obtain

$$\lim_{s\to\infty} sf(s) - f(t)\,\big|_{t=0} = \lim_{s\to\infty} \int_0^\infty f'(t)e^{-st}\,dt$$

Now, assuming that the limit and integration can be interchanged, we obtain

$$\lim_{s\to\infty} sf(s) - f(t)\,\big|_{t=0} = \int_0^\infty f'(t)\,\lim_{s\to\infty} e^{-st}\,dt = 0$$

7. Final value. If both limits exist, the limit of $f(t)$ as $t$ approaches $\infty$ is equal to the limit of $sf(s)$ as $s$ approaches zero; that is,

$$\lim_{t\to\infty} f(t) = \lim_{s\to 0} sf(s) \tag{A-13}$$

Proof. Let us begin with Eq. (A-12) and in this case take the limit as $s$ approaches zero on both sides so that we obtain

$$\lim_{s\to 0} sf(s) - f(t)\,\big|_{t=0} = \lim_{s\to 0} \int_0^\infty f'(t)e^{-st}\,dt$$

Once again interchanging the limit and integration operations, we obtain

$$\lim_{s\to 0} sf(s) - f(t)\,\big|_{t=0} = \int_0^\infty f'(t)\,\lim_{s\to 0} e^{-st}\,dt$$

$$= \int_0^\infty f'(t)\,dt = f(t)\,\Big|_0^\infty$$

$$= \lim_{t\to\infty} f(t) - f(t)\,\big|_{t=0}$$

Therefore we have established that

$$\lim_{s \to 0} sf(s) = \lim_{t \to \infty} f(t)$$

One must be very careful in applying the above results to ensure that both limits exist. For example, the Laplace transform of $f(t) = e^t$ is $f(s) = 1/(s - 1)$ and

$$\lim_{s \to 0} sf(s) = \lim_{s \to 0} \frac{s}{s - 1} = 0$$

Since $\lim_{t \to \infty} f(t) = \infty$, the result above is meaningless. In terms of the stability development of Chap. 6, we know that both limits exist if $f(s)$ has all its poles in the interior of the left-half $s$ plane.

8. Convolution. The inverse Laplace transform of the product of two transformed functions is given by

$$\mathcal{L}^{-1}[f_1(s)f_2(s)] = \int_0^\infty f_1(\tau)f_2(t - \tau) \, d\tau$$
$$= \int_0^\infty f_1(t - \tau)f_2(\tau) \, d\tau \qquad (A\text{-}14)$$

Proof. The proof of this result is presented in detail in Sec. 4.5 and is not repeated here.

# *appendix*

# *B* *Laplace transform table*[1]

| | $F(s)$ | $f(t)$     $0 \le t$ |
|------|------|------|
| 1. | $1$ | $\delta(t)$     unit impulse at $t = 0$ |
| 2. | $\dfrac{1}{s}$ | $1$ or $\mu(t)$     unit step at $t = 0$ |
| 3. | $\dfrac{1}{s^2}$ | $t\mu(t)$     ramp function |
| 4. | $\dfrac{1}{s^n}$ | $\dfrac{1}{(n-1)!}\, t^{n-1}$     $n$ is a positive integer |

[1] This table is reproduced with permission from J. J. D'Azzo and C. H. Houpis, "Feedback Control System Analysis and Synthesis," 2d ed., McGraw-Hill Book Company, New York, 1966.

| $F(s)$ | $f(t)$        $0 \leq t$ |
|---|---|
| 5. $\dfrac{1}{s} e^{-as}$ | $\mu(t - a)$        unit step starting at $t = a$ |
| 6. $\dfrac{1}{s}(1 - e^{-as})$ | $\mu(t) - \mu(t - a)$        rectangular pulse |
| 7. $\dfrac{1}{s + a}$ | $e^{-at}$        exponential decay |
| 8. $\dfrac{1}{(s + a)^n}$ | $\dfrac{1}{(n - 1)!} t^{n-1} e^{-at}$        $n$ is a positive integer |
| 9. $\dfrac{1}{s(s + a)}$ | $\dfrac{1}{a}(1 - e^{-at})$ |
| 10. $\dfrac{1}{s(s + a)(s + b)}$ | $\dfrac{1}{ab}\left(1 - \dfrac{b}{b - a} e^{-at} + \dfrac{a}{b - a} e^{-bt}\right)$ |
| 11. $\dfrac{s + \alpha}{s(s + a)(s + b)}$ | $\dfrac{1}{ab}\left[\alpha - \dfrac{b(\alpha - a)}{b - a} e^{-at} + \dfrac{a(\alpha - b)}{b - a} e^{-bt}\right]$ |
| 12. $\dfrac{1}{(s + a)(s + b)}$ | $\dfrac{1}{b - a}(e^{-at} - e^{-bt})$ |
| 13. $\dfrac{s}{(s + a)(s + b)}$ | $\dfrac{1}{a - b}(ae^{-at} - be^{-bt})$ |
| 14. $\dfrac{s + \alpha}{(s + a)(s + b)}$ | $\dfrac{1}{b - a}[(\alpha - a)e^{-at} - (\alpha - b)e^{-bt}]$ |
| 15. $\dfrac{1}{(s + a)(s + b)(s + c)}$ | $\dfrac{e^{-at}}{(b - a)(c - a)} + \dfrac{e^{-bt}}{(c - b)(a - b)} + \dfrac{e^{-ct}}{(a - c)(b - c)}$ |
| 16. $\dfrac{s + \alpha}{(s + a)(s + b)(s + c)}$ | $\dfrac{(\alpha - a)e^{-at}}{(b - a)(c - a)} + \dfrac{(\alpha - b)e^{-bt}}{(c - b)(a - b)} + \dfrac{(\alpha - c)e^{-ct}}{(a - c)(b - c)}$ |
| 17. $\dfrac{\omega}{s^2 + \omega^2}$ | $\sin \omega t$ |
| 18. $\dfrac{s}{s^2 + \omega^2}$ | $\cos \omega t$ |
| 19. $\dfrac{s + \alpha}{s^2 + \omega^2}$ | $\dfrac{\sqrt{\alpha^2 + \omega^2}}{\omega} \sin(\omega t + \phi)$        $\phi = \arctan\dfrac{\omega}{\alpha}$ |

| | $F(s)$ | $f(t) \qquad 0 \le t$ |
|---|---|---|
| 20. | $\dfrac{s \sin \theta + \omega \cos \theta}{s^2 + \omega^2}$ | $\sin (\omega t + \theta)$ |
| 21. | $\dfrac{1}{s(s^2 + \omega^2)}$ | $\dfrac{1}{\omega^2} (1 - \cos \omega t)$ |
| 22. | $\dfrac{s + \alpha}{s(s^2 + \omega^2)}$ | $\dfrac{\alpha}{\omega^2} - \dfrac{\sqrt{\alpha^2 + \omega^2}}{\omega^2} \cos (\omega t + \phi) \qquad \phi = \arctan \dfrac{\omega}{\alpha}$ |
| 23. | $\dfrac{1}{(s + a)(s^2 + \omega^2)}$ | $\dfrac{e^{-at}}{a^2 + \omega^2} + \dfrac{1}{\omega \sqrt{a^2 + \omega^2}} \sin (\omega t - \phi) \qquad \phi = \arctan \dfrac{\omega}{a}$ |
| 24. | $\dfrac{1}{(s + a)^2 + b^2}$ | $\dfrac{1}{b} e^{-at} \sin bt$ |
| 24a. | $\dfrac{1}{s^2 + 2\zeta\omega_n s + \omega_n{}^2}$ | $\dfrac{1}{\omega_n \sqrt{1 - \zeta^2}} e^{-\zeta\omega_n t} \sin \omega_n \sqrt{1 - \zeta^2}\, t$ |
| 25. | $\dfrac{s + a}{(s + a)^2 + b^2}$ | $e^{-at} \cos bt$ |
| 26. | $\dfrac{s + \alpha}{(s + a)^2 + b^2}$ | $\dfrac{\sqrt{(\alpha - a)^2 + b^2}}{b} e^{-at} \sin (bt + \phi)$ <br><br> $\phi = \arctan \dfrac{b}{\alpha - a}$ |
| 27. | $\dfrac{1}{s[(s + a)^2 + b^2]}$ | $\dfrac{1}{a^2 + b^2} + \dfrac{1}{b \sqrt{a^2 + b^2}} e^{-at} \sin (bt - \phi)$ <br><br> $\phi = \arctan \dfrac{b}{-a}$ |
| 27a. | $\dfrac{1}{s(s^2 + 2\zeta\omega_n s + \omega_n{}^2)}$ | $\dfrac{1}{\omega_n{}^2} - \dfrac{1}{\omega_n{}^2 \sqrt{1 - \zeta^2}} e^{-\zeta\omega_n t} \sin (\omega_n \sqrt{1 - \zeta^2}\, t + \phi)$ <br><br> $\phi = \arccos \zeta$ |
| 28. | $\dfrac{s + \alpha}{s[(s + a)^2 + b^2]}$ | $\dfrac{\alpha}{a^2 + b^2} + \dfrac{1}{b} \sqrt{\dfrac{(\alpha - a)^2 + b^2}{a^2 + b^2}}\, e^{-at} \sin (bt + \phi)$ <br><br> $\phi = \arctan \dfrac{b}{\alpha - a} - \arctan \dfrac{b}{-a}$ |

| | $F(s)$ | $f(t) \qquad 0 \leq t$ |
|---|---|---|
| 29. | $\dfrac{1}{(s+c)[(s+a)^2+b^2]}$ | $\dfrac{e^{-ct}}{(c-a)^2+b^2} + \dfrac{e^{-at}\sin(bt-\phi)}{b\sqrt{(c-a)^2+b^2}}$ <br><br> $\phi = \arctan\dfrac{b}{c-a}$ |
| 30. | $\dfrac{1}{s(s+c)[(s+a)^2+b^2]}$ | $\dfrac{1}{c(a^2+b^2)} - \dfrac{e^{-ct}}{c[(c-a)^2+b^2]}$ <br><br> $+ \dfrac{e^{-at}\sin(bt-\phi)}{b\sqrt{a^2+b^2}\sqrt{(c-a)^2+b^2}}$ <br><br> $\phi = \arctan\dfrac{b}{-a} + \arctan\dfrac{b}{c-a}$ |
| 31. | $\dfrac{s+\alpha}{s(s+c)[(s+a)^2+b^2]}$ | $\dfrac{\alpha}{c(a^2+b^2)} - \dfrac{(c-\alpha)e^{-ct}}{c[(c-a)^2+b^2]}$ <br><br> $+ \dfrac{\sqrt{(\alpha-a)^2+b^2}}{b\sqrt{a^2+b^2}\sqrt{(c-a)^2+b^2}}e^{-at}\sin(bt+\phi)$ <br><br> $\phi = \arctan\dfrac{b}{\alpha-a} - \arctan\dfrac{b}{-a} - \arctan\dfrac{b}{c-a}$ |
| 32. | $\dfrac{1}{s^2(s+a)}$ | $\dfrac{1}{a^2}(at-1+e^{-at})$ |
| 33. | $\dfrac{1}{s(s+a)^2}$ | $\dfrac{1}{a^2}(1-e^{-at}-ate^{-at})$ |
| 34. | $\dfrac{s+\alpha}{s(s+a)^2}$ | $\dfrac{1}{a^2}[\alpha-\alpha e^{-at}+a(a-\alpha)te^{-at}]$ |
| 35. | $\dfrac{s^2+\alpha_1 s+\alpha_0}{s(s+a)(s+b)}$ | $\dfrac{\alpha_0}{ab} + \dfrac{a^2-\alpha_1 a+\alpha_0}{a(a-b)}e^{-at} - \dfrac{b^2-\alpha_1 b+\alpha_0}{b(a-b)}e^{-bt}$ |
| 36. | $\dfrac{s^2+\alpha_1 s+\alpha_0}{s[(s+a)^2+b^2]}$ | $\dfrac{\alpha_0}{c^2} + \dfrac{1}{bc}[(a^2-b^2-\alpha_1 a+\alpha_0)^2$ <br><br> $\qquad\qquad + b^2(\alpha_1-2a)^2]^{\frac{1}{2}}e^{-at}\sin(bt+\phi)$ <br><br> $\phi = \arctan\dfrac{b(\alpha_1-2a)}{a^2-b^2-\alpha_1 a+\alpha_0} - \arctan\dfrac{b}{-a}$ <br><br> $c^2 = a^2+b^2$ |

| | $F(s)$ | $F(t) \qquad 0 \leq t$ |
|---|---|---|
| 37. | $\dfrac{1}{(s^2 + \omega^2)[(s + a)^2 + b^2]}$ | $\dfrac{(1/\omega) \sin{(\omega t + \phi_1)} + (1/b)e^{-at} \sin{(bt + \phi_2)}}{[4a^2\omega^2 + (a^2 + b^2 - \omega^2)^2]^{\frac{1}{2}}}$ <br><br> $\phi_1 = \arctan{\dfrac{-2a\omega}{a^2 + b^2 - \omega^2}}$ <br><br> $\phi_2 = \arctan{\dfrac{2ab}{a^2 - b^2 + \omega^2}}$ |
| 38. | $\dfrac{s + \alpha}{(s^2 + \omega^2)[(s + a)^2 + b^2]}$ | $\dfrac{1}{\omega}\left(\dfrac{\alpha^2 + \omega^2}{c}\right)^{\frac{1}{2}} \sin{(\omega t + \phi_1)}$ <br><br> $+ \dfrac{1}{b}\left[\dfrac{(\alpha - a)^2 + b^2}{c}\right]^{\frac{1}{2}} e^{-at} \sin{(bt + \phi_2)}$ <br><br> $c = (2a\omega)^2 + (a^2 + b^2 - \omega^2)^2$ <br><br> $\phi_1 = \arctan{\dfrac{\omega}{\alpha}} - \arctan{\dfrac{2a\omega}{a^2 + b^2 - \omega^2}}$ <br><br> $\phi_2 = \arctan{\dfrac{b}{\alpha - a}} + \arctan{\dfrac{2ab}{a^2 - b^2 + \omega^2}}$ |
| 39. | $\dfrac{s + \alpha}{s^2[(s + a)^2 + b^2]}$ | $\dfrac{1}{c}\left(\alpha t + 1 - \dfrac{2\alpha a}{c}\right)$ <br><br> $+ \dfrac{[b^2 + (\alpha - a)^2]^{\frac{1}{2}}}{bc} e^{-at} \sin{(bt + \phi)}$ <br><br> $c = a^2 + b^2$ <br><br> $\phi = 2 \arctan{\dfrac{b}{a}} + \arctan{\dfrac{b}{\alpha - a}}$ |
| 40. | $\dfrac{s^2 + \alpha_1 s + \alpha_0}{s^2(s + a)(s + b)}$ | $\dfrac{\alpha_1 + \alpha_0 t}{ab} - \dfrac{\alpha_0(a + b)}{(ab)^2} - \dfrac{1}{a - b}\left(1 - \dfrac{\alpha_1}{a} + \dfrac{\alpha_0}{a^2}\right)e^{-at}$ <br><br> $- \dfrac{1}{a - b}\left(1 - \dfrac{\alpha_1}{b} + \dfrac{\alpha_0}{b^2}\right)e^{-bt}$ |

# *C* *polynomial factoring*

The topic of polynomial factoring is mentioned several times in this book. Polynomial factoring is sometimes necessary to find the roots of the characteristic equation and thus the location of the closed-loop poles of a system. Polynomial factoring is almost always necessary to find the zeros of $H_{eq}(s)$, since the numerator of $H_{eq}(s)$ is a polynomial of order $n - 1$.

Section 7.6 deals with root-locus methods for polynomial factoring. Since the root locus is a graphical approach, the resulting factors are approximate. Here analytic methods are discussed for determining a better estimate of the roots of a polynomial. The answers obtained from the root locus may well serve as a starting place from which better estimates of the actual root locations may be obtained.

The magnitude of the problem should not be underestimated.   The factoring of high-order polynomials is a difficult job, particularly if a number of pairs of complex conjugate roots are involved.   No doubt the wisest approach is to assign the problem to a computer.[1]   If this is not possible, the following methods may be used.   The difficulty increases rapidly with the order of the system.

*The second-order case.*   The most general second-order polynomial is

$$f(s) = As^2 + Bs + C$$

and the roots of the polynomial are given by the familiar quadratic formula as

$$s = \frac{-B \pm \sqrt{B^2 - 4AC}}{2A} \tag{C-1}$$

In the discussion of the higher-order cases to follow, it will always be assumed that the coefficient of the highest power in $s$ is simply unity. If it is not unity, all coefficients are divided by the coefficient of the highest power in $s$ in order to force this situation.   Hence, for the second-order case, $A$ would be 1.   Assume that $A$ is 1, so that $f(s)$ is

$$f(s) = s^2 + Bs + C$$

Assume the roots as given by Eq. (C-1) are $s = -\gamma_1, -\gamma_2$ so that

$$\begin{aligned} f(s) &= (s + \gamma_1)(s + \gamma_2) = s^2 + (\gamma_1 + \gamma_2)s + \gamma_1\gamma_2 \\ &= s^2 + Bs + C \end{aligned} \tag{C-2}$$

Note that in Eq. (C-2) the coefficient of $s$ is just the negative of the sum of the roots and that the magnitude of the constant term is the product of the magnitude of the roots.   If $n$ factors of the form $s + \gamma_i$ are multiplied, the coefficient of $s^{n-1}$ is the negative of the sum of the roots, and the magnitude of the constant term is the product of the magnitudes of the roots.   Use of this fact will be made in discussing the third- and higher-order cases below.

*The third-order case.*   In the third-order case the most general polynomial is

$$f(s) = s^3 + as^2 + bs + c \tag{C-3}$$

[1] See, for example, J. L. Melsa, "Computer Programs for Computational Assistance in the Study of Linear Control Theory," McGraw-Hill Book Company, New York, 1970.

with the three roots $s = -\gamma_1$, $-\gamma_2$, and $-\gamma_3$. This polynomial has at least one real root, and the procedure for factoring Eq. (C-3) is to find the real root and reduce the polynomial to second order. The resulting second-order polynomial is factored by the use of the quadratic formula Eq. (C-1).

A simple and straightforward way of finding the real root is to plot $f(s)$ versus $s$ for real values of $s$. When $s = 0$, $f(s)$ is $c$, and this establishes one point on the plot. If all three roots lie in the left-half $s$ plane, as determined by the Routh-Hurwitz criterion, the real root lies between $0$ and $-a$. This is true since the negative sum of the roots is $a$. Hence, in the search for the real root of a stable polynomial, $f(s)$ need be evaluated only for the restricted range of $s$ from $0$ to $-a$.

> ***Example C-1***   Determine the roots of the third-order polynomial
>
> $$f(s) = s^3 + 9s^2 + 24s + 70$$
>
> Since from the Routh criterion $(9)(24) - 70$ is greater than zero, the roots are all in the left-half $s$ plane, and the negative real root must lie between $0$ and $-9$. For $s = 0$, $f(s) = 70$, and at $s = -9$, $f(s) = -146$. On the basis of a straight-line approximation between these two points, the real root appears to be closer to zero than to $-9$. At $s = -4$, $f(s) = 54$, and the straight-line approximation

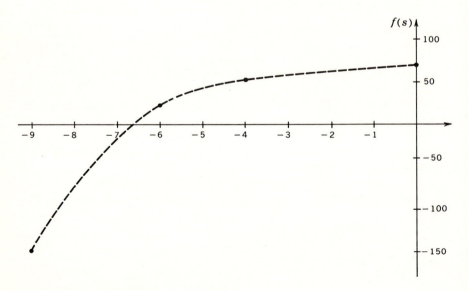

**Fig. C-1**   A plot of $f(s)$ of Example C-1.

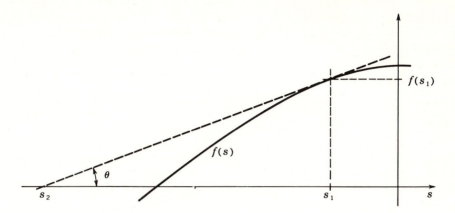

**Fig. C-2**   The basis for Newton's method.

between the two extreme end points was not a good approximation. A second rough guess, based on the values of $f(s)$ at $s = -9$ and $s = -4$, is selected as $s = -6$, and the resulting $f(s)$ is 34.

A plot of the points tested thus far is shown in Fig. C-1; these points are connected by a smooth curve. It appears that the real root is near $-7$, and this is the correct value. Long division of the given $f(s)$ by $s + 7$ yields

$$f(s) = (s + 7)(s^2 + 25s + 10)$$

and application of the quadratic formula shows that the roots of the quadratic are located at $s = -1 \pm j3$. Thus $f(s)$ is

$$f(s) = (s + 7)(s + 1 + j3)(s + 1 - j3)$$
$$= (s + 7)[(s + 1)^2 + 3^2]$$

In the above example a number of rather rough guesses were made before we actually converged on the root located at $s = -7$. This guessing may be systematized by the use of *Newton's method*. The basis for the method is shown in Fig. C-2. The method is based on the slope of $f(s)$ at the trial point. Let us call the first trial point $s_1$ and attempt to find a good choice for the second trial point, $s_2$. From the figure

$$\tan \theta = \left.\frac{df(s)}{ds}\right|_{s = s_1} = f'(s_1) = \frac{f(s_1)}{s_1 - s_2}$$

This equation may be solved for $s_2$ in terms of the value of the function and its derivative at $s_1$, or

$$s_2 = s_1 - \frac{f(s_1)}{f'(s_1)} \tag{C-4}$$

Successive guesses are made in the same way, where the $(i+1)$th approximation of the real root is given as

$$s_{i+1} = s_i - \frac{f(s_i)}{f'(s_i)} \tag{C-5}$$

*Higher-order polynomials.* Polynomials of fourth-order or higher may be either even or odd. If the polynomial is odd, the real root is first removed, much as in the third-order example just discussed. The resulting polynomial is then of even-order.

The discussion here centers on means of factoring even-order polynomials. Because it is possible that an even-order polynomial may have only complex conjugate roots, the approach needed is one that removes quadratic factors from the polynomial. For example, a sixth-order polynomial would first be reduced to the product of a second- and a fourth-order polynomial. The fourth-order polynomial would be reduced to two second-order ones. The quadratic formula is then used on each of the second-order factors to determine the actual root locations. Not until this final step is taken is one aware whether the roots actually are all complex or not.

A procedure for removing quadratic factors from high-order polynomials is known as *Lin's method*. Assume that the even-order polynomial to be factored is

$$f(s) = s^6 + as^5 + bs^4 + cs^3 + ds^2 + es + f$$

Lin's method attempts to remove the quadratic factor by a purely mechanical approach using long division. In one form of Lin's method, the first trial divisor is obtained directly from the last three terms of the polynomial being factored. This trial divisor is $s^2 + es/d + f/d$. A typical long division is indicated below:

$$
\begin{array}{r}
s^4 + \cdots \\
s^2 + \dfrac{e}{d}s + \dfrac{f}{d} \overline{\smash{\big)}\ s^6 + as^5 + bs^4 + cs^3 + ds^2 + es + f} \\
s^6 + \cdots \\
\hline
\cdots\cdots\cdots\cdots \\
d_1 s^2 + e_1 s + f \\
d_1 s^2 + e_2 s + f_1 \\
\hline
\text{Remainder}
\end{array}
$$

If the remainder is not small, a new trial divisor is chosen to be $s^2 + (e_1/d_1)s + f/d_1$. The long-division procedure is repeated with this new trial divisor and another remainder determined. The trial divisor for the

$i$th trial is $s^2 + (e_{i-1}/d_{i-1})s + f/d_{i-1}$. The procedure is repeated until the remainder becomes small or, equivalently, until there is little change from one trial divisor to the next. The final trial divisor is then one quadratic factor of the given polynomial.

A procedure such as Lin's method is satisfactory from an engineering point of view only if the method converges rapidly. There is no mathematical means of ensuring that Lin's method actually does converge, although experience has shown that the approach is satisfactory if the roots are widely separated on the $s$ plane.

A means of ensuring more rapid convergence is to start the entire long-division process with a trial divisor that is a "good guess." A way of obtaining a first-trial divisor that is a good guess is to make use of the root-locus methods of Sec. 7.6. The root locus may be drawn incompletely and with very poor accuracy, as indicated in the example to follow. Once the first division has been made with this improved trial divisor, the succeeding steps in the procedure are exactly as above.

***Example C-2***    The problem is to factor the polynomial

$$f(s) = s^5 + 16s^4 + 127s^3 + 572s^2 + 1{,}376s + 1{,}156$$

The roots of the polynomial are actually

$$\begin{aligned} f(s) &= (s + 2)(s^2 + 8s + 17)(s^2 + 6s + 34) \\ &= (s + 2)[(s + 4)^2 + 1^2][(s + 3)^2 + 5^2] \end{aligned}$$

We have given the answer simply for use in the discussion of the problem solution. The real root at $s = -2$ is first found by plotting $f(s)$ for $s$ in the range from 0 to $-16$, as the Routh-Hurwitz test indicates that all the poles are in the left-half $s$ plane. The resulting $f(s)$ is then

$$f(s) = (s + 2)(s^4 + 14s^3 + 99s^2 + 374s + 578)$$

and we now have a fourth-order (even) polynomial to factor. Instead of using the last three terms of this polynomial as the first trial divisor, we consider factoring the equation

$$s^4 + 14s^3 + 99s^2 + 374s + 578 = 0 \tag{C-6}$$

by the root-locus methods of Sec. 7.6. Equation (C-6) may be rewritten as

$$s^2(s^2 + 14s + 99) = -374(s + 578/374)$$

or

$$\frac{\alpha(s + 1.55)}{s^2[(s + 7)^2 + 7.07^2]} = -1 = 1\underline{/180°} \tag{C-7}$$

with $\alpha = 374$. The pole-zero plot corresponding to Eq. (C-7) is shown in Fig. C-3, and we are ready to draw the root locus. A guess of the root-locus shape is made in Fig. C-4, and since our interest is only in finding an approximate value for *one* pair of poles, let us ignore completely the branches of the root locus that originate at the complex conjugate poles. A gain determination along the negative real axis indicates a point of relatively minimum gain near $s = -4$, where $K$ is approximately 370. This is fairly near our required value of 374 from Eq. (C-7), and on this basis it appears that two poles exist at $s = -4$, so that a guess for one quadratic term is $(s + 4)^2 = s^2 + 8s + 16$. This is near the actual answer, although the indicated gain values are only approximate.

Suppose we had been further off in our rough root-locus dia-

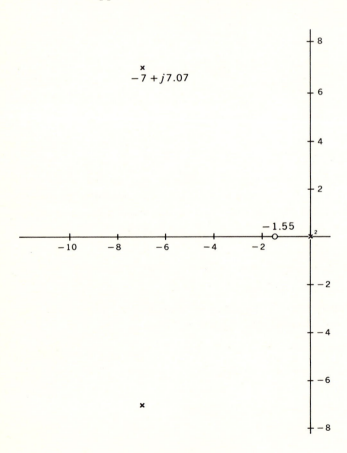

*Fig. C-3*  The pole-zero plot for Eq. (C-7).

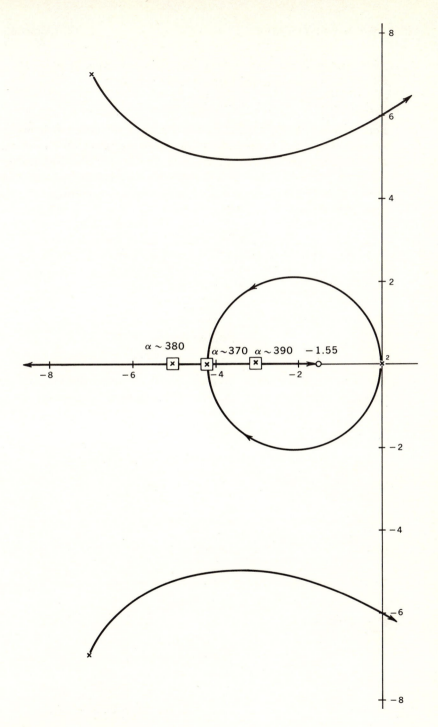

***Fig. C-4***   An approximate root locus.

gram and had actually determined pole locations at $s = -3$ and $s = -5$. Our quadratic term would then be

$$(s + 3)(s + 5) = s^2 + 8s + 15$$

For purposes of illustration, let us assume that our root locus is in greater error than it actually is and use this latter quadratic term as our "good guess" for a first-trial divisor. Note that we have assumed two real poles, although we know from the given answer to this problem that the actual roots are complex conjugates. The first long division is indicated below:

$$
\begin{array}{r}
s^2 + 6s^2 + 36 \\
s^2 + 8s + 15 \overline{\smash{\big)}\ s^4 + 14s^3 + 99s^2 + 374s + 578} \\
s^4 + \phantom{0}8s^3 + 15s^2 \\
\hline
6s^3 + 84s^2 + 374s \\
6s^3 + 48s^2 + \phantom{0}90s \\
\hline
36s^2 + 284s + 578 \}\text{\scriptsize New trial-divisor term} \\
36s^2 + 288s + 540 \\
\hline
-4s + \phantom{0}38 \}\text{\scriptsize Remainder}
\end{array}
$$

The new trial divisor is divided by 36 to make the leading term unity, so that the new trial divisor is actually

$$s^2 + \frac{284s}{36} + \frac{578}{36} = s^2 + 7.90s + 16$$

The trial divisor that results from this long division is

$$s^2 + \frac{276.4}{34.9}s + \frac{578}{34.9} = s^2 + 7.91s + 16.5$$

and the next long division yields the trial divisor

$$s^2 + \frac{274}{34.4}s + \frac{578}{34.4} = s^2 + 7.96s + 16.8$$

Use of this trial divisor in long division is indicated below:

$$
\begin{array}{r}
s^2 + 6.04s + 34.2 \\
s^2 + 7.96s + 16.8 \overline{\smash{\big)}\ s^4 + \phantom{00}14s^3 + \phantom{00}99s^2 + 374s + 578} \\
s^4 + \phantom{0}7.96s^3 + 16.8s^2 \\
\hline
6.04s^3 + 82.2s^2 + 374s \\
6.04s^3 + 48.0s^2 + 101s \\
\hline
34.2s^2 + 273s + 578 \}\text{\scriptsize New trial-divisor term} \\
34.2s^2 + 272s + 575 \\
\hline
s + \phantom{00}3 \}\text{\scriptsize Remainder}
\end{array}
$$

The next trial divisor would be

$$s^2 + \frac{273}{34.2} s + \frac{578}{34.2} = s^2 + 7.98s + 16.9$$

Note that the remainder is now small and that the new trial divisor differs only slightly from the previous one.   Thus we assume the last trial divisor is the actual quadratic factor.   Its roots are located at

$$s = -3.99 \pm j0.99$$

which is very near $s = -4 \pm j1$.

# *index*